86 Springer Series in Solid-State Sciences

Edited by Peter Fulde

Springer Series in Solid-State Sciences

Editors: M. Cardona P. Fulde K. von Klitzing H.-J. Queisser

Managing Editor: H. K. V. Lotsch

Volumes 1–89 are listed at the end of the book

90 **Earlier and Recent Aspects of Superconductivity**
Editors: J. G. Bednorz and K. A. Müller

91 **Electronic Properties of Conjugated Polymers III**
Basic Models and Applications
Editors: H. Kuzmany, M. Mehring, and S. Roth

92 **Physics and Engineering Applications of Magnetism**
Editors: Y. Ishikawa and N. Miura

93 **Quasicrystals**
Editors: T. Fujiwara and T. Ogawa

94 **Electronic Conduction in Oxides**
By N. Tsuda, K. Nasu, A. Yanase, and K. Siratori

E. Fick G. Sauermann

The Quantum Statistics of Dynamic Processes

Translated by W. D. Brewer

With 54 Figures

Springer-Verlag
Berlin Heidelberg New York London
Paris Tokyo Hong Kong Barcelona

Professor Dr. Eugen Fick
Professor Dr. Günter Sauermann
Institut für Festkörperphysik, Technische Hochschule Darmstadt, Hochschulstrasse 2
D-6100 Darmstadt, Fed. Rep. of Germany

Translator:
Professor William D. Brewer, Ph. D.
Freie Universität Berlin, Fachbereich Physik, Arnimallee 14
D-1000 Berlin 33

Series Editors:
Professor Dr., Dres. h. c. Manuel Cardona
Professor Dr., Dr. h. c. Peter Fulde
Professor Dr., Dr. h. c. Klaus von Klitzing
Professor Dr. Hans-Joachim Queisser
Max-Planck-Institut für Festkörperforschung, Heisenbergstrasse 1
D-7000 Stuttgart 80, Fed. Rep. of Germany

Managing Editor:
Dr. Helmut K. V. Lotsch
Springer-Verlag, Tiergartenstrasse 17
D-6900 Heidelberg, Fed. Rep. of Germany

Enlarged and revised translation of the original German edition:
E. Fick, G. Sauermann: *Quantenstatistik dynamischer Prozesse*
Band I: Generelle Aspekte

Band IIa: Antwort- und Relaxationstheorie

ISBN-13 : 978-3-642-83717-3 e-ISBN-13 : 978-3-642-83715-9
DOI : 10.1007 / 978-3-642-83715-9

Library of Congress Cataloging-in-Publication Data. Fick, Eugen. [Quantenstatistik dynamischer Prozesse. English] The quantum statistics of dynamic processes / E. Fick, G. Sauermann ; translated by W. D. Brewer. p. cm.–(Springer series in solid-state sciences ; 86) Translation of: Quantenstatistik dynamischer Prozesse. Includes bibliographical references. 1. Quantum statistics. I. Sauermann, Günter. II. Title. III. Series. QC173.97.F5313 1990 530.1'33–dc20 89-21689

Typesetting: Macmillan India Ltd., India

2154/3150-543210 – Printed on acid-free paper

Preface

The methods of statistical physics have become increasingly important in recent years for the treatment of a variety of diverse physical problems. Of principal interest is the microscopic description of the dynamics of dissipative systems. Although a unified theoretical description has at present not yet been achieved, we have assumed the task of writing a textbook which summarizes those of the most important methods which are self-contained and complete in themselves. We cannot, of course, claim to have treated the field exhaustively. A microscopic description of physical phenomena must necessarily be based upon quantum theory, and we have therefore carried out the treatment of dynamic processes strictly within a quantum-theoretical framework. For this reason alone it was necessary to omit a number of extremely important theories which have up to now been formulated only in terms of classical statistics.

The goal of this book is, on the one hand, to give an introduction to the general principles of the quantum statistics of dynamical processes, and, on the other, to provide readers who are interested in the treatment of particular phenomena with methods for solving specific problems.

The theory is for the most part formulated within the calculational framework of Liouville space, which, together with projector formalism, has become an expedient mathematical tool in statistical physics.

In Part 1, we treat the general aspects of quantum statistics and, in particular, the various possibilities for transforming the exact statistical operator for dynamics to operators suitable for particular physical problems. The physical arguments are emphasized more than is a detailed analysis of the validity of some of the mathematical suppositions. The generalized canonical statistical operator is introduced on an information-theoretical basis; here, we have been at pains to use a description which avoids the possibility of a subjective interpretation. Its physical meaning as well as the possibility of a multiplicity of entropies in dynamic processes are discussed.

Parts 2 and 3 treat response and relaxation theory for small perturbations of equilibrium; this is particularly important for the description of numerous experiments. In Part 2, we begin with a general formulation of quantum-statistical theory of the response of a system to time-dependent external fields. The Kubo formalism is presented using the Mori product in Liouville space, which is especially appropriate for the description of all linear theories.

Part 3 concerns itself with the central theme of the book. We treat the Mori theory in particular detail, giving special emphasis to the discussion of approximations which form the basis for the quantum-mechanical expressions for damping constants (relaxation times or linewidths) of linear, dissipative systems. Calculations are described using selected physical examples. A further point is the formulation of a linear response theory of internal parameters which arise from the generalized canonical statistical operator; this theory represents the link to irreversible thermodynamics. The emphasis on the Mori theory seems reasonable to us, since it is intuitively clear and since the residual forces which it contains lead to the formulation of the second fluctuation-dissipation theorem.

In Part 4, various projector techniques for rearranging the von Neumann equation are treated from a unified point of view. First, the Nakajima–Zwanzig theory is described and treated as a linear projection onto the relevant part of the statistical operator. Then we develop the structure of a general non-linear projection theory. The latter contains the Nakajima–Zwanzig theory and the Robertson theory as special cases. The relationships, differences, and advantages of the various techniques are illustrated.

The exercises and hints for solving them are intended to help the reader become more familiar with the methods described and with their application.

The present English edition is the translation of a two-volume work in German, *Quantenstatistik dynamischer Prozesse, Vols. I and IIa.* In combining these two volumes into a single book, we have rewritten the chapter on the relevant statistical operator and have added appendices on the relationship between two projector techniques and on symmetry relations.

We thank Prof. W.D. Brewer for the careful translation and for useful suggestions.

Finally, we wish to thank the publishers for their considerate cooperation and for the format of the book.

Darmstadt, May 1990

Eugen Fick
Günter Sauermann

Contents

Part 1 General Aspects

Part 2 Response to Time-Dependent External Fields

Part 3 Equations of Motion for Observables in the Case of Small Deviations from Equilibrium

Part 4 Equations of Motion of the Relevant Parts of the Statistical Operator

Appendix

Part 1

General Aspects

1. The Concept of Statistical Physics

The dynamics of a physical system are described by differential equations for the time variation of the quantities related to the observables. In classical mechanics, these equations are Newton's laws of motion, or, in a more elegant mathematical form, Hamilton's equations for the position and momentum coordinates of the particles. One may also mention Maxwell's equations from classical field theory, which describe the time variation of electromagnetic fields. In a quantum theoretical treatment of a problem, the time variation of the observables is given by the Heisenberg equations of motion of the corresponding operators (or, which is mathematically equivalent, by the Schrödinger equation of motion of the state vector).

In addition to the particular structure of the differential equations, the determination of the time dependence requires knowledge of the initial conditions. In a *pure state*, all of the initial conditions which are required in principle have been or may be specified. Taking the example of the classical mechanics of an N-particle system, this means that at time $t=0$, all $6N$ position and momentum coordinates are known; in the case of quantum mechanics, a measurement at $t=0$ of a complete set of commuting observables is required to determine the initial state vector.

However, in practice, a pure state can seldom be produced experimentally, neither in the classical nor in the quantum-theoretical sense. In particular, when the observable in question yields continuous values in a measurement, it is impossible to attain an arbitrarily precise determination of a certain value experimentally. Even in the classical motion of a single point mass in a force field, the imprecision inherent in the measurement of the initial position and momentum has the consequence that the actual path of the particle may deviate considerably from that calculated using the slightly incorrect initial conditions (Fig. 1.1). The situation in quantum theory is similar. In this case, one could in principle determine the state vector exactly by measurement of a discrete, nondegenerate eigenvalue. In practice, however, a continuous or quasi-continuous eigenvalue spectrum is nearly always present (e.g. when one considers the interactions within a many-body system). An even slightly imprecise determination of the initial state vector again causes errors in the calculated state vector at later times.

If one considers a macroscopic system with many degrees of freedom, e.g. a gas with $N\approx 10^{23}$ particles, it is completely impossible to determine the pure

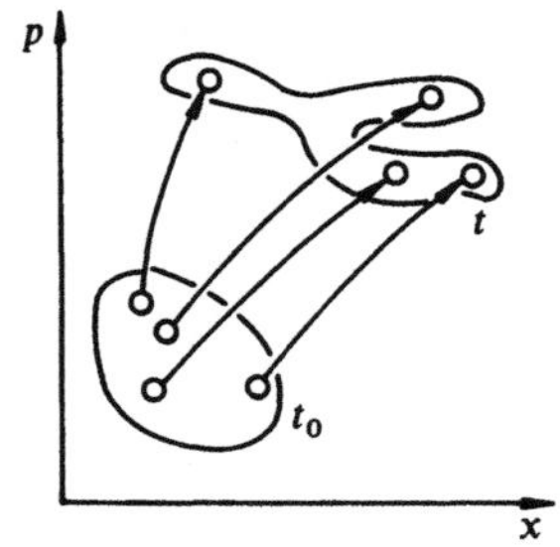

Fig. 1.1. Classical dynamics ($t_0 \to t$) of phase-space points

initial state of the system, i.e. all the classical position and momentum coordinates, or the quantum-theoretical state vector, experimentally. To describe the state of the system, one instead has only the values of a few macroscopic observables, e.g. the total energy.

When it is not possible to specify the pure state of a system, one speaks of a *mixed state* or a *mixture.* The goal of *statistical physics* is to describe the properties of such mixtures theoretically. To this end, one introduces probabilities for the pure states. The equations of motion then determine the expectation values of the observables at a later time.

The question may be raised as to the significance of the calculated expectation values for the result of a measurement. One cannot assume that the expectation values of the coordinates of individual molecules in a macroscopic system have a physical meaning, since their values generally exhibit considerable fluctuations. However, predictions of such observables are completely unimportant for a macroscopic system; only the expectation values of *macroscopic observables,* whose relative fluctuation for a large number of particles is vanishingly small, are of interest (Chap. 4). The probability of observing a particular value of a macroscopic observable is thus sharply peaked around the mean, so that the calculated expectation value gives a good prediction of the result of an experimental determination. *For macroscopic observables, the theoretically calculated expectation values may be set equal to the experimental results.*

Because statistical physics is based on mixtures, i.e. on a lack of knowledge of the pure state, it contains new concepts unknown in the physics of pure states, such as the *entropy*. The latter obtains a microscopic definition, in contrast to its phenomenological definition in thermodynamics, and thus acquires a direct, clear meaning.

Quantum theory provides the appropriate tools for describing matter at an atomic level, and we shall therefore base our treatment of statistical physics on quantum theory.

2. Summary of Quantum Theory

In order to facilitate the study of quantum statistics for the reader, we start by summarizing the important results and methods of quantum mechanics in this chapter, and at the same time we define the notation which will be used throughout the book. We assume, however, that the reader is already familiar with quantum theory.

2.1 Observables as Operators. Commutation Relations

The physical *observables* of a quantum-theoretical system Σ are described by Hermitian operators[1] $\mathscr{F} = \mathscr{F}^\dagger$ in a unitary space $\mathfrak{U}$ (cf. Sect. 2.2). Important examples of such observables are the energy, given by the Hamiltonian $\mathscr{H}$, as well as the position, momentum, orbital angular momentum, and spin operators

$$\mathscr{x}_{iv}, \mathscr{p}_{iv}, \mathscr{l}_{iv}, \mathscr{s}_{iv} \quad (i=x, y, z;\ v=1, \ldots, N)$$

of the N particles of the system.

These operators obey *commutation relations.* The various components of the position operator commute with one another, as do those of the momentum, while

$$[\mathscr{p}_{iv}, \mathscr{x}_{j\mu}] = \frac{\hbar}{\mathrm{i}} \delta_{ij} \delta_{v\mu} \mathscr{I} \tag{2.1.1}$$

is found for position and momentum[2].

For the angular momentum operators $\mathscr{j}_{iv}$, we have the commutation relations

$$[\mathscr{j}_{xv}, \mathscr{j}_{y\mu}] = -\frac{\hbar}{\mathrm{i}} \delta_{v\mu} \mathscr{j}_{z\mu}, \ldots, \tag{2.1.2}$$

where ... means that two additional relations are produced by cyclic permutation of the indices x, y, and z.

---

[1] We write operators in $\mathfrak{U}$ using script letters.

[2] $[\mathscr{F}, \mathscr{G}] = \mathscr{F}\mathscr{G} - \mathscr{G}\mathscr{F}$.

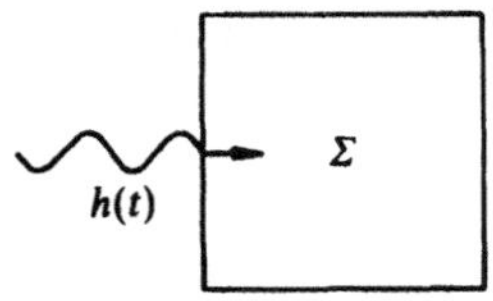

Fig. 2.1. A system Σ driven by an external, time-dependent c-number field $h(t)$

The commutator of the Hamiltonian $\mathscr{H}$ with another observable $\mathscr{F}$ yields the observable "time variation of the observable $\mathscr{F}$" [3]

$$\boxed{\mathring{\mathscr{F}} = \frac{\mathrm{i}}{\hbar}[\mathscr{H}, \mathscr{F}] + \left(\frac{\partial \mathscr{F}}{\partial t}\right)_{\mathrm{ex}}} \quad . \tag{2.1.3}$$

Here, $(\partial\mathscr{F}/\partial t)_{\mathrm{ex}}$ means the time derivative of $\mathscr{F}$, insofar as the observable $\mathscr{F} = \mathscr{F}(\mathscr{x}, \mathscr{p}, \mathscr{s}, t)$ is *explicitly* time dependent. For example, the Hamiltonian $\mathscr{H}(\mathscr{x}, \mathscr{p}, \mathscr{s}, h(t))$ has such an explicit time dependence when the physical system Σ considered is subject to external forces or fields $h(t)$ (Fig. 2.1), which are themselves not observables of the system considered and are therefore not operators, but rather c-numbers. In Part 2, we will consider the response of systems to the action of such external fields in more detail.

2.2 The Unitary Space $\mathfrak{U}$ of States. Expectation Values

The operators $\mathscr{F}$ act on the vectors $|u\rangle, |v\rangle, \ldots$ of a unitary space $\mathfrak{U}$ (the Hilbert space) and generate a linear *correspondence* onto the vectors. A complex scalar product $\langle u|v\rangle$ is defined in the space. The usual linear properties of a vector space are valid.

The eigenvectors $|u_F\rangle$ and eigenvalues F of an operator $\mathscr{F}$ obey the eigenvalue equation

$$\mathscr{F}|u_F\rangle = F|u_F\rangle \tag{2.2.1}$$

or, in the case that F is degenerate,

$$\mathscr{F}|u_F^\mu\rangle = F|u_F^\mu\rangle \quad (\mu = 1, \ldots, d_F = \text{degeneracy of } F) \ . \tag{2.2.2}$$

They are specified by the commutation relations of the operators. We require the eigenvectors to be normalized to one or at least to be normalizable to a Dirac

[3] $\mathring{\mathscr{F}}$ is thus not a derivative of $\mathscr{F}$, but rather represents another observable. For example, if $\mathscr{x}$ is the "position" observable, then $\mathring{\mathscr{x}}$ is the operator of the observable "time variation of the position observable", i.e. the operator of the observable "velocity".

function. The eigenvectors $|v_k\rangle$ of a complete, commuting set of observables (e.g. x_{iv}, σ_{zv})[4] span the unitary space $\mathfrak{U}$. They form a basis set, i.e. the $|v_k\rangle$ fulfill not only the orthonormalization relations

$$\langle v_k | v_{k'} \rangle = \delta(k, k') \tag{2.2.3}$$

but also the completeness relation

$$\sum\!\!\!\!\!\!\int_k |v_k\rangle \langle v_k| \, dk = 1 \ , \tag{2.2.4}$$

where $\delta(k, k')$ stands for the Kronecker delta symbol $\delta_{kk'}$ or for the Dirac delta function $\delta(k-k')$ and $\sum\!\!\!\!\!\!\int$ means a sum or an integral over k, depending on whether k represents a discrete or a continuous eigenvalue.

For a system of fermions (e.g. electrons), or of bosons (e.g. photons), the corresponding unitary space contains only vectors which are antisymmetric or symmetric with respect to particle permutation, respectively.

The *pure state* of a quantum-theoretical system is described by a *state vector* $|\Phi\rangle$ which is normalized to one. The state vector is specified by the measurement of a non-degenerate observable $\mathcal{G}$ with a discrete eigenvalue G. Immediately after such a measurement, $|\Phi\rangle$ points in the direction of the eigenvector $|u_G\rangle$,

$$|\Phi\rangle = |u_G\rangle \ . \tag{2.2.5}$$

As we mentioned in Chap. 1, however, such a precise specification of $|\Phi\rangle$ is practically never possible. The formulation of a quantum theory for mixed states is therefore necessary, and we shall describe it in Sect. 2.3.

If the state vector $|\Phi\rangle$ is known at a time t, one can calculate the *quantum-theoretical expectation value* $\langle \mathcal{F} \rangle$ – i.e. the mean value of the measurements of the observable $\mathcal{F}$ in the state of $|\Phi\rangle$ – from the scalar product

$$\langle \mathcal{F} \rangle = \langle \Phi | \mathcal{F} \Phi \rangle \ . \tag{2.2.6}$$

Instead of the state vector $|\Phi\rangle$, one can also employ the *projection operator*

$$\mathcal{P}_{|\Phi\rangle} = |\Phi\rangle\langle\Phi| \ . \tag{2.2.7}$$

With the help of the projection operator, the expectation value may be written as a trace

$$\boxed{\langle \mathcal{F} \rangle = \mathrm{Tr}(\mathcal{P}_{|\Phi\rangle} \mathcal{F})} \ . \tag{2.2.8}$$

The operator $\mathcal{P}_{|\Phi\rangle}$ is thus called the *statistical operator* (*von Neumann*) *of the pure state.*

---

[4] E.g. in a spin system for which the orbital motion of the particles is not of interest, the eigenvectors of the spin operators σ_{zv} form a complete basis set.

The *trace* of an operator $\mathscr{F}$,

$$\mathrm{Tr}\,\mathscr{F} = \sum\!\!\!\!\!\!\int_k \langle v_k | \mathscr{F} v_k \rangle \, dk \ ,^5 \tag{2.2.9}$$

is independent of the basis set chosen. As may be readily seen from (2.2.9), even noncommuting operators can undergo *cyclic permutation*

$$\mathrm{Tr}(\mathscr{F}\mathscr{G}) = \mathrm{Tr}(\mathscr{G}\mathscr{F}) \ , \qquad \mathrm{Tr}(\mathscr{F}\mathscr{G}\mathscr{L}) = \mathrm{Tr}(\mathscr{L}\mathscr{F}\mathscr{G}) \ldots , \tag{2.2.10}$$

within a trace without changing its value; from this property follows the useful relation

$$\mathrm{Tr}(\mathscr{F}[\mathscr{G}, \mathscr{L}]) = \mathrm{Tr}([\mathscr{F}, \mathscr{G}]\mathscr{L}) \ . \tag{2.2.11}$$

The statistical operator $\mathscr{P}_{|\Phi\rangle}$ also satisfies relation (2.1.3) with the Hamiltonian. Because of the continuing existence of a correspondence between the state vector and the physical system Σ, however, we also have $\mathring{\mathscr{P}}_{|\Phi\rangle} = \mathscr{O}$, so that the relation

$$\boxed{\mathring{\mathscr{P}}_{|\Phi\rangle} = \frac{\mathrm{i}}{\hbar}[\mathscr{H}, \mathscr{P}_{|\Phi\rangle}] + \left(\frac{\partial \mathscr{P}_{|\Phi\rangle}}{\partial t}\right)_{\mathrm{ex}} = \mathscr{O}} \tag{2.2.12}$$

is fulfilled. Equations (2.1.3) and (2.2.12) form the *fundamentals of quantum theoretical dynamics of pure states.* The expectation values $\langle \mathscr{F} \rangle(t)$ are determined at some time $t > t_0$ according to (2.2.8), so long as $\mathscr{P}_{|\Phi\rangle}(t_0)$ [or $|\Phi\rangle(t_0)$ via (2.2.5)] is given as an initial condition. Due to the invariance of the expectation values (2.2.8) towards time-dependent unitary transformations which are carried out simultaneously in $\mathscr{P}_{|\Phi\rangle}$ and in $\mathscr{F}$, various *formulations* of the dynamics may be introduced (cf. Sect. 3.2) without changing the physical content of the theory.

2.3 The Statistical Operator of a Mixed State

The quantum-theoretical dynamics of a pure state are based upon the measurement of a discrete, nondegenerate eigenvalue which determines the state vector $|\Phi\rangle$ of the system considered at a particular initial time (2.2.5). Such a complete quantum-theoretical specification of a system, however, is practically impossible to carry out, as pointed out in Chap. 1. One therefore must develop theoretical descriptions which take into account the *missing information about the state vector* of the system. This means that over and above the quantum-theoretical

---

[5] When we refer to Tr $\mathscr{F}$ in the following, we assume its (unique) existence. For some operators $\mathscr{F}$, this assumption in an infinite-dimensional space $\mathfrak{U}$ would require a specific, thorough mathematical investigation.

probability laws (2.2.8), which cannot be circumvented by more detailed measurements, we must apply further statistical considerations which take into account the *indeterminateness* of the state vector. This *quantum-theoretical mixture* is already required for the description of the state of a microscopic object, when the degeneracy of the measured eigenvalues does not permit the specification of the state vector $|\Phi\rangle$. However, it attains critical importance for macroscopic systems where no complete measurement is possible.

When it is impossible to determine the statistical operator $\mathscr{P}_{|\Phi\rangle}$ of the pure state of the system Σ, one considers a set of statistical operators[6]

$$\mathscr{P}_{|\Phi^\alpha\rangle} \text{ with } weights\ p^\alpha \ , \tag{2.3.1}$$

which describe the probability that the system Σ possesses a $\mathscr{P}_{|\Phi_\alpha\rangle}$. The following relation must thus hold:

$$\sum_\alpha p^\alpha = 1 \quad \text{and} \quad 0 \leqq p^\alpha \leqq 1 \ . \tag{2.3.2}$$

Taking the weighted mean of the $\mathscr{P}_{|\Phi_\alpha\rangle}$ with the weights p^α leads to the *statistical operator of the mixed state or the mixture* (*density operator*)

$$\rho = \sum_\alpha p^\alpha \mathscr{P}_{|\Phi^\alpha\rangle} \ , \tag{2.3.3}$$

from which the *expectation value of the observables* $\mathscr{F}$ of Σ is calculated according to

$$\boxed{\langle \mathscr{F} \rangle = \mathrm{Tr}(\rho \mathscr{F})} \ . \tag{2.3.4}$$

In (2.3.3), an average is taken over the statistical operators $\mathscr{P}_{|\Phi^\alpha\rangle}$ and *not* over the states $|\Phi_\alpha\rangle$; therefore, (2.3.4) represents an *averaging of the particular quantum-theoretical expectation values* $\langle \mathscr{F} \rangle^\alpha = \mathrm{Tr}(\mathscr{P}_{|\Phi^\alpha\rangle} \mathscr{F})$ *with weights* p^α.

From (2.3.2) and (2.3.3) it follows immediately that ρ is a Hermitian operator in $\mathfrak{U}$ with the property

$$\mathrm{Tr}\,\rho = 1 \ . \tag{2.3.5}$$

The eigenvalue equation for ρ,

$$\rho |v_r^\gamma\rangle = r |v_r^\gamma\rangle \ , \tag{2.3.6}$$

leads to eigenvalues r with

$$\sum_{r,\gamma} r = 1 \quad \text{and} \quad 0 \leqq r \leqq 1 \ . \tag{2.3.7}$$

[6] The states $|\Phi^\alpha\rangle$ need not be orthogonal.

In quantum theory, the probability w_F for the occurrence of a measured value (eigenvalue) F of an observable $\mathscr{F}$ is given by the expectation value of the projection operator on the eigenspace u_F:

$$\mathscr{P}_F = \sum\!\!\!\!\!\!\int_{\mu} |u_F^\mu\rangle\langle u_F^\mu|\, d\mu \ . \tag{2.3.8}$$

Thus, if we replace $\mathscr{F}$ by $\mathscr{P}_F$ in (2.3.4), we obtain

$$w_F = \mathrm{Tr}(\rho\mathscr{P}_F) = \sum\!\!\!\!\!\!\int_{\mu} \langle u_F^\mu|\rho u_F^\mu\rangle\, d\mu \ , \tag{2.3.9}$$

i.e. the *diagonal elements* of ρ in the $\mathscr{F}$ representation yield the *probability* for the occurrence of the measured value F.

In quantum-theoretical dynamics, the system Σ is described by a Hamiltonian $\mathscr{H}$. Since therefore no measurements are carried out on the system, the information about its state does not change, i.e. the weights p^α are time-independent,

$$\frac{d}{dt} p^\alpha = 0 \ . \tag{2.3.10}$$

On the other hand, in a mixture, each $\mathscr{P}_{|\Phi_\alpha\rangle}$ is a possible pure state, for which (2.2.12) would hold. Thus we obtain for the statistical operator ρ of a mixture the following fundamental relation:

$$\boxed{\mathring{\rho} = \frac{i}{\hbar}[\mathscr{H}, \rho] + \left(\frac{\partial \rho}{\partial t}\right)_{\mathrm{ex}} = \mathcal{O}} \ .^{7} \tag{2.3.11}$$

The statement $\mathring{\rho} = \mathcal{O}$ then again denotes the permanent correspondence of ρ to the system Σ.

To conclude, we summarize the results of this section: the quantum theory of a mixture, i.e. *quantum statistics* in the sense of this book, is distinguished from the quantum theory of a pure state only through the fact that the statistical operator ρ replaces $\mathscr{P}_{|\Phi\rangle}$ [(2.3.4) and (2.3.11)]. All other quantum-theoretical relations, in particular also (2.1.3), remain valid, independently of whether the system is in a pure or a mixed state. The existence of a Hamiltonian operator $\mathscr{H}$ in both cases means that the system has no effects on its surroundings, i.e. it is not a *subsystem* of a larger system[8]. In the language of thermodynamics, this means that Σ is adiabatically isolated. (The treatment of subsystems is introduced in Sect. 3.3.)

---

[7] This relation is the quantum-theoretical analogy to the Liouville equation for the phase-space density in classical statistical mechanics.

[8] The effects of given time-dependent external fields on Σ via an explicitly time-dependent Hamiltonian are naturally permitted.

The *quantum-theoretical dynamics* of a mixture is based on (2.1.3) and (2.3.11). From $\mathscr{F}(t)$ and $\rho(t)$ in a particular representation, all the statistical results may be calculated as expectation values

$$\langle \mathscr{F} \rangle (t) = \mathrm{Tr}(\rho(t)\mathscr{F}(t)) \tag{2.3.12}$$

at a time t. A detailed discussion of the formulations of the dynamics will be given in Sect. 3.2.

For the *construction of the statistical operator* ρ, e.g. at the beginning of a dynamical process, one can make use of the results of the quantum theory of measurements. These are useful for microscopic objects, but for macroscopic objects another approach is expedient, namely using information theory (Chaps. 5 and 6).

Exercise 2.1 Show that $\mathrm{Tr}\,\rho^2 \leqq 1$ holds.

3. Quantum Theory in Liouville Space

In the formulation of quantum theory which we have just given, the vectors $|v\rangle$ in the unitary space $\mathfrak{U}$ hardly made an appearance. They were only required for the explicit calculation of expectation values as traces in $\mathfrak{U}$. Corresponding to the physics of the problem, the operators in $\mathfrak{U}$ played a more important role in the theory. It is therefore expedient to use a mathematical device, the Liouville space $\mathfrak{L}$, in which the importance of the operators is appropriately taken into account. This procedure, which is described in the following sections, may at first appear to be overly formal and unnecessary. However, its usefulness will later become apparent, particularly for quantum statistical problems.

3.1 The Liouville Space $\mathfrak{L}$ (Without Scalar Product)

3.1.1 The Elements in Liouville Space

The Liouville space $\mathfrak{L}$ is a linear vector space over the field of complex numbers, with mathematical properties which are familiar e.g. from quantum theory. However, in the latter, the elements $|v\rangle$ of the unitary space $\mathfrak{U}$ consist of the quantum-theoretically permitted state vectors; in contrast, the elements $|\mathscr{F})$ of the Liouville space $\mathfrak{L}$ are the linear operators $\mathscr{F}$ from $\mathfrak{U}$,

$$\boxed{\text{Linear operators } \mathscr{F} \text{ in } \mathfrak{U} \leftrightarrow \text{elements } |\mathscr{F}) \text{ of } \mathfrak{L}} \ . \tag{3.1.1}$$

The linear operators $\mathscr{F}$ in $\mathfrak{U}$ fulfill the axioms of a linear vector space such as commutativity and associativity of the sum and the distributive property for multiplication with complex numbers. The powers $\mathscr{F}^n$ of the operators $\mathscr{F}$ correspond to separate elements $|\mathscr{F}^n)$ in $\mathfrak{L}$.

Examples: Position $x \leftrightarrow |x)$, momentum $p \leftrightarrow |p)$, and the Hamiltonian

$$\mathscr{H} = \frac{p^2}{2m} + \mathscr{V}(x) \leftrightarrow |\mathscr{H}) = \frac{1}{2m}|p^2) + |\mathscr{V}(x)) \ .$$

When we wish to emphasize the operators $\mathscr{F}$ as elements of $\mathfrak{L}$, we write $|\mathscr{F})$, but for the most part, we shall leave off the symbol $|\ )$. A scalar product in $\mathfrak{L}$ will not be introduced in this initial treatment (cf. Chap. 8).

If the unitary space $\mathfrak{U}$ of the states is n-dimensional, the corresponding Liouville space $\mathfrak{L}$ is n^2-dimensional,

$$\dim \mathfrak{L} = (\dim \mathfrak{U})^2 \ . \tag{3.1.2}$$

An arbitrary operator $\mathscr{F}$ may thus be written in a basis $|v_k\rangle$ of $\mathfrak{U}$

$$\mathscr{F} = \sum_{kk'} |v_k\rangle\langle v_k|\mathscr{F} v_{k'}\rangle\langle v_{k'}| = \sum_{kk'} F_{kk'}|v_k\rangle\langle v_{k'}| \ ,$$

i.e. in the Liouville space

$$|\mathscr{F}) = \sum_{kk'}^{nn} F_{kk'}\big|\,|v_k\rangle\langle v_{k'}|\big) \ . \tag{3.1.3}$$

The n^2 dyads $|v_k\rangle\langle v_{k'}|$ in $\mathfrak{U}$ are n^2 *basis set elements* $\big|\,|v_k\rangle\langle v_{k'}|\big)$ in $\mathfrak{L}$. Instead of the basis consisting of these dyads, one can naturally also span the Liouville space with other sets of n^2 linearly independent elements $|\mathscr{V}_\alpha)$,

$$|\mathscr{F}) = \sum_{\alpha=1}^{n^2} c_\alpha|\mathscr{V}_\alpha) \ . \tag{3.1.4}$$

Equations (3.1.3) and (3.1.4) are also valid when the original unitary space $\mathfrak{U}$ is a Hilbert space of infinite dimensions. This is, however, not an appropriate place to take up the mathematical problems associated with these infinities.

3.1.2 Operators in $\mathfrak{L}$ (Superoperators)

If an element $|\mathscr{F})$ of the Liouville space $\mathfrak{L}$ is mapped onto another element $|\mathscr{G})$, we describe the correspondence by means of a superoperator S in $\mathfrak{L}$ (Fig. 3.1),

$$\boxed{|\mathscr{F}) \xrightarrow{\mathsf{S}} |\mathscr{G}) = \mathsf{S}|\mathscr{F}) = |\mathsf{S}\mathscr{F})} \ .^{1} \tag{3.1.5}$$

This correspondence is required to have the mathematical properties of *linear operators*; their product need not be commutative. These *operators in* $\mathfrak{L}$ are also called *superoperators*, in order to distinguish them from operators in $\mathfrak{U}$ (which themselves play the role of elements in $\mathfrak{L}$). In the language of the unitary space $\mathfrak{U}$, (3.1.5) means that the superoperator S effects a linear mapping of the operator $\mathscr{F}$ in $\mathfrak{U}$ onto an operator $\mathscr{G}$ in $\mathfrak{U}$.

[1] For the *operators* in $\mathfrak{L}$ (superoperators), we use Grotesque type, e.g. S, A, L, M, U, P, Q, 1, ... We thus avoid confusion with the *operators* in $\mathfrak{U}$, which are printed as script symbols. This also allows us to leave off the prefix "super" when no misunderstandings will result.

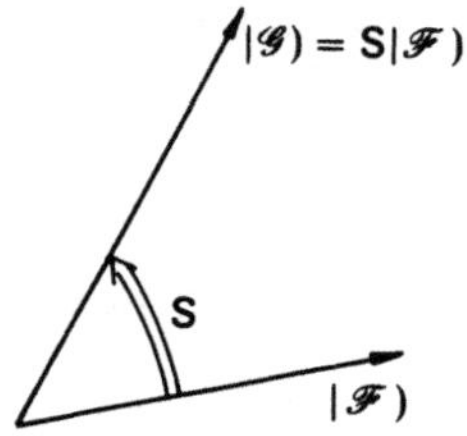

Fig. 3.1. The mapping of elements in the Liouville space by operators (superoperators)

The inverse operator to S is denoted by S^{-1} (in case it exists):

$$\mathsf{SS}^{-1} = \mathsf{S}^{-1}\mathsf{S} = 1 \ .^{2} \tag{3.1.6}$$

In the following, we introduce two types of superoperators and discuss their properties, which are suggested by quantum theory.

1) The transformation of observables $\mathscr{F}$ by linear operators $\mathscr{A}$ in $\mathfrak{U}$ according to

$$\mathscr{F} \to \mathscr{A}\mathscr{F}\mathscr{A}^{-1} \tag{3.1.7}$$

may be represented by an operator A in $\mathfrak{L}$,

$$\boxed{\mathsf{A}\mathscr{F} = \mathscr{A}\mathscr{F}\mathscr{A}^{-1}} \ . \tag{3.1.8}$$

It is easy to convince oneself that A, like $\mathscr{A}$, is linear. The inverse of A, A^{-1}, has the following effect:

$$\mathsf{A}^{-1}\mathscr{F} = \mathscr{A}^{-1}\mathscr{F}\mathscr{A} \ . \tag{3.1.9}$$

The possibility of cyclic permutation (2.2.10) within the trace in $\mathfrak{U}$

$$\mathrm{Tr}\,(\mathscr{G}\mathscr{A}\mathscr{F}\mathscr{A}^{-1}) = \mathrm{Tr}\,(\mathscr{A}^{-1}\mathscr{G}\mathscr{A}\mathscr{F}) \tag{3.1.10}$$

leads to the useful rule

$$\boxed{\mathrm{Tr}\,(\mathscr{G}\cdot\mathsf{A}\mathscr{F}) = \mathrm{Tr}\,(\mathsf{A}^{-1}\mathscr{G}\cdot\mathscr{F})} \tag{3.1.11}$$

for superoperators of the type shown in (3.1.8). The dot within the parentheses indicates the point where the action of the superoperator ends.

[2] Adjoint, Hermitian, and unitary superoperators can only be defined after we introduce a scalar product in $\mathfrak{L}$ (cf. Chap. 8).

2) The most important superoperator in quantum statistics is the socalled *Liouville operator* L. It mediates the mapping of $\mathcal{F}$ onto the commutation relation $(1/\hbar)[\mathcal{H}, \mathcal{F}]$,

$$\boxed{\mathcal{F} \xrightarrow{\mathsf{L}} \mathsf{L}\mathcal{F} = \frac{1}{\hbar}[\mathcal{H}, \mathcal{F}]}\ .^3 \tag{3.1.12}$$

The higher powers of L refer to nested commutation relations with $\mathcal{H}$, e.g.

$$\mathsf{L}^2\mathcal{F} = \frac{1}{\hbar^2}[\mathcal{H}, [\mathcal{H}, \mathcal{F}]]\ , \tag{3.1.13}$$

whereby many quantum-theoretical formulas become considerably clearer. From (3.1.12), it follows that (cf. Exercise 3.2)

$$\boxed{e^{c\mathsf{L}}\mathcal{F} = e^{c\mathcal{H}/\hbar}\mathcal{F}\, e^{-c\mathcal{H}/\hbar}}\ , \tag{3.1.14}$$

i.e. $e^{c\mathsf{L}}$ is a superoperator of the type given in (3.1.8).

Cyclic permutations within the trace in $\mathfrak{U}$ yield the following important rule for every superoperator of type (3.1.12) [or more generally, of type (3.1.12a)]:

$$\boxed{\mathrm{Tr}(\mathcal{G}\cdot \mathsf{L}\mathcal{F}) = -\mathrm{Tr}(\mathsf{L}\mathcal{G}\cdot\mathcal{F})} \tag{3.1.15}$$

and, generalizing to application on a function $\varphi(\mathsf{L}) = \sum c_n \mathsf{L}^n$ (c_n complex),

$$\boxed{\mathrm{Tr}(\mathcal{G}\cdot \varphi(\mathsf{L})\mathcal{F}) = \mathrm{Tr}(\varphi(-\mathsf{L})\mathcal{G}\cdot\mathcal{F})}\ , \tag{3.1.16}$$

thus, for example,

$$\boxed{\mathrm{Tr}(\mathcal{G}\cdot e^{c\mathsf{L}}\mathcal{F}) = \mathrm{Tr}(e^{-c\mathsf{L}}\mathcal{G}\cdot\mathcal{F})} \tag{3.1.17}$$

corresponding to (3.1.11).

Finally, we mention the *eigenvalue problem of the Liouville operator* L: if the Hamiltonian $\mathcal{H}$ obeys the eigenvalue equation

$$\mathcal{H}|u_k\rangle = E_k|u_k\rangle\ , \tag{3.1.18}$$

---

[3] Such a commutation relation need not necessarily contain $\mathcal{H}$; thus, more generally,

$$\mathcal{F} \xrightarrow{\mathsf{M}} \mathsf{M}\mathcal{F} = \frac{1}{\hbar}[\mathcal{M}, \mathcal{F}]. \tag{3.1.12a}$$

then it follows from (3.1.12) with $\mathscr{F} = |u_k\rangle\langle u_{k'}|$ on writing out the commutation relation and taking the Hermitian property of $\mathscr{H}$ into account, that L in $\mathfrak{L}$ obeys the eigenvalue equation

$$\mathsf{L}\big|\,|u_k\rangle\langle u_{k'}|\big) = \frac{E_k - E_{k'}}{\hbar}\big|\,|u_k\rangle\langle u_{k'}|\big) \ . \tag{3.1.19}$$

The eigenelements of L are thus the dyads $|u_k\rangle\langle u_{k'}|$ of the energy eigenvectors, and the corresponding eigenvalues are the *energy differences* divided by $\hbar$.

Exercise 3.1 Write the statistical operator ρ using the spin operators and their expectation values for $s = 1/2$.

Exercise 3.2 Prove (3.1.14).

3.2 The Formulations of Quantum-Theoretical Dynamics

3.2.1 The Fundamental Equations of Time Evolution

All of quantum-theoretical dynamics is contained in the fundamental relations (2.1.3) and (2.3.11) for the observables $\mathscr{F}$ and the statistical operator ρ. If we replace the commutation relations with $\mathscr{H}$ in them by the Liouville operator L as in (3.1.12) (whereby a possible explicit time dependence in $\mathscr{H}$ is transferred to L), they become more compact:

$$\mathring{\mathscr{F}} = \mathrm{i}\mathsf{L}\mathscr{F} + \left(\frac{\partial \mathscr{F}}{\partial t}\right)_{\text{ex}} , \tag{3.2.1}$$

$$\mathring{\rho} = \mathrm{i}\mathsf{L}\rho + \left(\frac{\partial \rho}{\partial t}\right)_{\text{ex}} = \mathcal{O} \ . \tag{3.2.2}$$

When the initial conditions are given, the expectation values at time t

$$\langle \mathscr{F} \rangle(t) = \mathrm{Tr}\,(\rho(t)\mathscr{F}(t)) \tag{3.2.3}$$

are completely determined by these equations. $\mathscr{F}$ and ρ may be subjected together to the same unitary transformation in $\mathfrak{A}$, leaving the Hermitian property of $\mathscr{F}$ and ρ as well as the expectation value (3.2.3), as an invariant trace, unchanged. This is true even for *time dependent* transformations. The resulting equations of motion for the *total time variation* of the operators $\mathscr{F}(t)$ and $\rho(t)$ are different in the different possible *formulations* of quantum-theoretical dynamics. The time variations of the expectation values (3.2.3), which are the only physically significant quantities, however remain unchanged.

3.2.2 Dynamics in the Schrödinger Formulation

The Schrödinger formulation is defined by the requirement that the total time variation d/dt of all operators be given by their explicit time variations $(\partial/\partial t)_{\mathrm{ex}}$. Equations (3.2.1) and (3.2.2) lead then to the equations of motion in the Schrödinger formulation,

$$\frac{d\mathscr{F}^S(t)}{dt} = \left(\frac{\partial \mathscr{F}^S}{\partial t}\right)_{\mathrm{ex}} , \tag{3.2.4}$$

$$\frac{d\rho^S(t)}{dt} = -\mathrm{i}\mathsf{L}^S(t)\rho^S(t) \qquad \text{(von Neumann equation)} . \tag{3.2.5}$$

Equation (3.2.4) shows that explicitly time-independent operators are constant with respect to time in the Schrödinger formulation. The Hamiltonian $\mathscr{H}^S(t)$, and thus also the Liouville operator $\mathsf{L}^S(t)$, are therefore dependent on the time t only through a (possible) explicit time dependence.

Differential equation (3.2.5) for $\rho^S(t)$ may be formally integrated by a superoperator $\mathsf{U}(t, t_0)$:

$$\rho^S(t) = \mathsf{U}(t, t_0)\rho(t_0) \ .[4] \tag{3.2.6}$$

By substitution in (3.2.5), one can show that $\mathsf{U}(t, t_0)$ is determined by the differential equation

$$\frac{d}{dt}\mathsf{U}(t, t_0) = -\mathrm{i}\mathsf{L}^S(t)\mathsf{U}(t, t_0) \quad \text{with} \quad \mathsf{U}(t_0, t_0) = 1 \ . \tag{3.2.7}$$

In the special case that L is explicitly time-independent, (3.2.7) has the solution

$$\mathsf{U}(t, t_0) = \mathrm{e}^{-\mathrm{i}\mathsf{L}\cdot(t-t_0)} \ . \tag{3.2.8}$$

If $\mathsf{L}^S(t)$ is explicitly time-dependent, the integration of (3.2.7) in this simple manner is not possible, because $\mathsf{L}^S(t)$ does not, in general, commute at all times. The integration must then be carried out with the aid of the time-ordering operator.

---

[4] Leaving off the superoperator notation, this equation in $\mathfrak{U}$ reverts to the well-known relation

$$\rho^S(t) = \mathscr{U}(t, t_0)\rho(t_0)\mathscr{U}(t, t_0)^{-1} , \tag{3.2.6a}$$

where $\mathscr{U}(t, t_0)$ is a unitary operator in $\mathfrak{U}$. $\mathsf{U}(t, t_0)$ is thus of type (3.1.8). The unitarity of $\mathscr{U}(t, t_0)$ in $\mathfrak{U}$ need not mean that $\mathsf{U}(t, t_0)$ in $\mathfrak{L}$ is unitary; the latter property has not yet even been defined.

The time dependence of the physical *expectation values* is then given by the general result from (3.2.3),

$$\boxed{\langle \mathscr{F} \rangle(t) = \mathrm{Tr}\,(\mathsf{U}(t, t_0)\rho(t_0)\cdot \mathscr{F}^S(t))} \; . \tag{3.2.9}$$

3.2.3 Dynamics in the Heisenberg Formulation

In the Heisenberg formulation, one identifies the overall time variation d/dt of all operators with °, so that (3.2.1) and (3.2.2) become the equations of motion

$$\frac{d\mathscr{F}^H(t)}{dt} = \mathrm{i}\mathsf{L}^H(t)\mathscr{F}^H(t) + \left(\frac{\partial \mathscr{F}^H}{\partial t}\right)_{\mathrm{ex}} \quad \text{(Heisenberg equation)} \; , \tag{3.2.10}$$

$$\frac{d\rho^H(t)}{dt} = 0 \; . \tag{3.2.11}$$

The dynamics are thus reflected only in the time variation (3.2.10) of the observables $\mathscr{F}^H(t)$.[5] The statistical operator, on the other hand, remains constant in time and always stays the same as the initial statistical operator at the beginning of the dynamic process, $\rho(t_0)$,

$$\rho^H(t) = \rho^H(t_0) \; . \tag{3.2.12}$$

The Heisenberg equation, (3.2.10), may be formally integrated using the superoperator $\mathsf{U}^{-1}(t, t_0)$ [which is the inverse of $\mathsf{U}(t, t_0)$, Eqs. (3.2.7) and (3.2.8)]

$$\mathscr{F}^H(x^H(t), p^H(t), t) = \mathsf{U}^{-1}(t, t_0)\mathscr{F}(x(t_0), p(t_0), t) \; , \tag{3.2.13}$$

if one uses the (possible) explicit time-dependence of the observables $\mathscr{F}$ on the right side to calculate them at the time t. By differentiation of (3.2.13), one may show that it is obeyed by the Heisenberg equation (3.2.10). Here, one uses the fact that $\mathsf{U}^{-1}(t, t_0)$ obeys the differential equation

$$\frac{d}{dt}\mathsf{U}^{-1}(t, t_0) = \mathrm{i}\mathsf{U}^{-1}(t, t_0)\mathsf{L}^S(t) = \mathrm{i}\mathsf{L}^H(t)\mathsf{U}^{-1}(t, t_0) \tag{3.2.14}$$

due to $d/dt\, \mathsf{U}^{-1} = -\mathsf{U}^{-1}\dot{\mathsf{U}}\mathsf{U}^{-1}$, where $\mathsf{L}^H(t)$ is defined via the commutator with $\mathscr{H}^H(t)$ and the overall time dependence of $\mathscr{H}^H(t)$ is determined by (3.2.13). In the

[5] The Heisenberg equation, (3.2.10), plays an analogous role to the Hamilton equations in classical mechanics, which employ calculations with Poisson brackets. In quantum mechanics, however, one cannot dispense with the operator character of the observables, and thus an essential difference from classical physics remains. This is true also for macroscopic systems (cf. Chap. 4) when they are described microscopically.

case of explicitly time-dependent Hamiltonians, the time function $\mathscr{H}^S(t)$ is not identical to that of $\mathscr{H}^H(t)$ (cf. Exercise 3.3) [and therefore, $\mathsf{L}^H(t)$ is also not identical to $\mathsf{L}^S(t)$]; it is then expedient to use the first part of (3.2.14) or (3.2.7) to determine U^{-1}. For explicitly time-independent Hamiltonians, we find immediately from (3.2.8)

$$\mathsf{U}^{-1}(t, t_0) = \mathrm{e}^{+\mathrm{i}\mathsf{L}\cdot(t-t_0)} . \tag{3.2.15}$$

The *equivalence* of the Schrödinger and the Heisenberg formulations for the expectation values follows immediately from (3.2.9) when one observes relation (3.1.11),

$$\langle \mathscr{F} \rangle(t) = \mathrm{Tr}\,(\mathsf{U}(t, t_0)\rho(t_0)\cdot \mathscr{F}^S(t)) = \mathrm{Tr}\,(\rho(t_0)\cdot \mathsf{U}^{-1}(t, t_0)\mathscr{F}^S(t)) , \tag{3.2.16}$$

since $\mathsf{U}^{-1}(t, t_0)\mathscr{F}^S(t)$ is just the right-hand side of (3.2.13).

We shall not employ the *interaction formulation* in explicit form in this book. It will, however, be used implicitly in applying operator identities (Sect. 3.4).

3.2.4 The Ehrenfest Theorem and Its Consequences

For the time variation of the expectation value

$$\langle \mathscr{F} \rangle(t) = \mathrm{Tr}\,(\rho(t)\mathscr{F}(t)) \tag{3.2.17}$$

of an arbitrary observable $\mathscr{F}$, it follows directly from (3.2.10 and 11)[6] that

$$\boxed{\frac{d}{dt}\langle \mathscr{F} \rangle = \langle \mathring{\mathscr{F}} \rangle = \mathrm{i}\langle \mathsf{L}\mathscr{F} \rangle + \left\langle \left(\frac{\partial \mathscr{F}}{\partial t}\right)_{\mathrm{ex}} \right\rangle} . \tag{3.2.18}$$

This Ehrenfest Theorem thus states that the time variation of the expectation value of an observable $\mathscr{F}$ is equal to the expectation value of the observable $\mathring{\mathscr{F}}$. According to (3.1.15), we may also write for the former

$$\frac{d}{dt}\langle \mathscr{F} \rangle = -\mathrm{i}\,\mathrm{Tr}\,(\mathsf{L}\rho\cdot \mathscr{F}) + \left\langle \left(\frac{\partial \mathscr{F}}{\partial t}\right)_{\mathrm{ex}} \right\rangle . \tag{3.2.19}$$

One may immediately recognize two important results of the Ehrenfest theorem:

1) If a statistical operator ρ_0 has the property of commuting with $\mathscr{H}$ for all t,

$$\mathsf{L}\rho_0 = \mathscr{O} , \tag{3.2.20}$$

[6] This line of reasoning in the Heisenberg formulation for the expectation-value equation (3.2.18) naturally leads to a result which is independent of the formulation. It may thus also be applied e.g. to the Schrödinger formulation, in which (3.2.19) arises.

it is termed *stationary*. The expectation value of *any* explicitly time-*in*dependent observable $\mathscr{F}$ is then constant in time, following (3.2.19),

$$\mathrm{Tr}(\rho_0 \mathscr{F}) = \mathrm{const} \ . \tag{3.2.21}$$

An important example of a stationary ρ_0 is the canonical statistical operator, which will be introduced in Sect. 5.3.1.

2) When an observable $\mathscr{F}_c$ has the property that

$$\mathring{\mathscr{F}}_c = 0 \tag{3.2.22}$$

then (3.2.18) yields, for arbitrary statistical operators,

$$\mathrm{Tr}(\rho \mathscr{F}_c) = \mathrm{const} \ . \tag{3.2.23}$$

One then refers to $\mathscr{F}_c$ as a constant of the motion.

Note the difference in the conditions for validity of (3.2.21) and (3.2.23)!

Exercise 3.3 Give the operator $\mathscr{H}^H(t)$ of a harmonic oscillator which is subject to the influence of an explicitly time-dependent force $F(t)\mathit{1}$.

3.3 Subsystems

3.3.1 Combined Systems

Let two systems Σ^1 and Σ^2 be initially completely independent of one another. They are thus described in two different unitary spaces[7], $\mathfrak{U}^1$ and $\mathfrak{U}^2$. If we imagine that we have combined these two independent systems Σ^1 and Σ^2 to a total system Σ (Fig. 3.2a), the unitary space of Σ is the direct product space

$$\mathfrak{U} = \mathfrak{U}^1 \times \mathfrak{U}^2 \ . \tag{3.3.1}$$

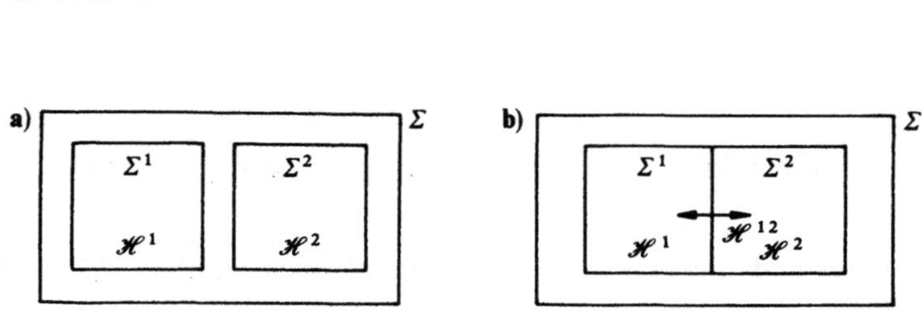

Fig. 3.2a, b. A combined system Σ. (**a**) Σ^1 and Σ^2 independent; (**b**) Σ^1 and Σ^2 in energetic contact via $\mathscr{H}^{12}$

---

[7] $\mathfrak{U}^1$ belongs to a fixed number of particles and to the volume given by Σ^1; the same is true of $\mathfrak{U}^2$.

All the operators $\mathscr{F}^1$ of system Σ^1 commute with all the operators $\mathscr{M}^2$ of Σ^2,

$$[\mathscr{F}^1, \mathscr{M}^2] = \mathcal{O} \quad . \tag{3.3.2}$$

In the product space, (3.3.1), however, it is also possible to take into account a mutual energy contact between Σ^1 and Σ^2, since any arbitrary linear operator $\mathscr{X}$ in $\mathfrak{U}$ is a linear combination of products of $\mathscr{V}^1_\alpha$ (from $\mathfrak{U}^1$) with $\mathscr{W}^2_\beta$ (from $\mathfrak{U}^2$) (cf. (3.1.4)[8])

$$\mathscr{X} = \sum_{\alpha\beta} c_{\alpha\beta} \mathscr{V}^1_\alpha \mathscr{W}^2_\beta \; . \tag{3.3.3}$$

The Hamiltonian of Σ with the internal interaction $\mathscr{H}^{12}$ (Fig. 3.2b) between Σ^1 and Σ^2,

$$\mathscr{H} = \mathscr{H}^1 + \mathscr{H}^2 + \mathscr{H}^{12} \; , \tag{3.3.4}$$

has the form (3.3.3), because the observables from Σ^1 and Σ^2 enter $\mathscr{H}^{12}$ and thus also $\mathscr{H}$.[9]

We can thus treat the *energy exchange* (*heat conduction*) between two reservoirs Σ^1 and Σ^2. The two subsystems do not, however, have to be spatially adjacent. One can, for example, consider the spin subsystem as Σ^1 and the lattice as Σ^2 within a solid, with the spin-lattice interaction $\mathscr{H}^{12}$, dependent on spin and lattice observables, coupling them. The energy exchange between Σ^1 and Σ^2 which is then possible is manifested as the spin-lattice relaxation (cf. Chap. 16).

3.3.2 The Product Liouville Space

We attribute a product Liouville space according to Sect. 3.1,

$$\mathfrak{L} = \mathfrak{L}^1 \times \mathfrak{L}^2 \tag{3.3.5}$$

to the unitary space containing the states of the combined system Σ

$$\mathfrak{U} = \mathfrak{U}^1 \times \mathfrak{U}^2 \; .$$

It has the operators (3.3.3) as elements:

$$|\mathscr{X}) = \sum_{\alpha\beta} c_{\alpha\beta} |\mathscr{V}^1_\alpha \mathscr{W}^2_\beta) \; . \tag{3.3.6}$$

Owing to the commutation relations (3.3.2), the order in $|\mathscr{V}^1_\alpha \mathscr{W}^2_\beta)$ is unimportant, as required by the mathematical properties of the elements of a

[8] An exchange of particles between Σ^1 and Σ^2 is not permitted, due to the choice of $\mathfrak{U}^1$ and $\mathfrak{U}^2$, i.e. there is a barrier between Σ^1 and Σ^2 which allows energy to pass, but not particles.

[9] The extension of these considerations to many systems Σ^1, $\Sigma^2, \ldots$ with an interaction energy $\mathscr{H}^{12\cdots}$ is evident.

product space. In the following, we leave off the symbol $|\ )$ for the most part, as we have already done above.

The mapping of $\mathscr{X}$ onto the commutator with $\mathscr{H}$ (3.3.4) is performed by the Liouville operator

$$\mathsf{L}=\mathsf{L}^1+\mathsf{L}^2+\mathsf{L}^{12} . \tag{3.3.7}$$

3.3.3 Expectation Values in a Subsystem. The Reduced Statistical Operator

In case Σ^1 and Σ^2 are independent of one another, the statistical operator ρ of the combined system Σ consists of a product $\rho^1\rho^2$. However, if an interaction (3.3.4) exists between Σ^1 and Σ^2, ρ in general does not have such a simple form, but is given by the more general form (3.3.3) or (3.3.6),

$$\rho=\sum_{\alpha\beta} r_{\alpha\beta}\mathscr{V}^1_\alpha\mathscr{W}^2_\beta . \tag{3.3.8}$$

If we are particularly interested in the expectation value $\langle\mathscr{F}^1\rangle$ of an observable $\mathscr{F}^1$ in the subsystem Σ^1,

$$\langle\mathscr{F}^1\rangle=\mathrm{Tr}(\rho\mathscr{F}^1) , \tag{3.3.9}$$

we may rewrite (3.3.9) utilizing the decomposition of the trace in the product space $\mathfrak{U}=\mathfrak{U}^1\times\mathfrak{U}^2$,

$$\mathrm{Tr}=\mathrm{Tr}_1\mathrm{Tr}_2=\mathrm{Tr}_2\mathrm{Tr}_1 \tag{3.3.10}$$

in the form

$$\boxed{\langle\mathscr{F}^1\rangle=\mathrm{Tr}_1\,\rho^1\mathscr{F}^1} , \tag{3.3.11}$$

where ρ^1 is the reduced ρ:

$$\boxed{\rho^1=\mathrm{Tr}_2\,\rho} .^{10} \tag{3.3.12}$$

This *reduced statistical operator* ρ^1 *is an operator in* $\mathfrak{U}^1$, *which permits the calculation of all expectation values* $\langle\mathscr{F}^1\rangle$. Like ρ, ρ^1 is Hermitian. From $\mathrm{Tr}\,\rho=1$, it follows that

$$\mathrm{Tr}_1\,\rho^1=1 . \tag{3.3.14}$$

---

[10] Written out in full according to (3.3.8), this reads

$$\rho^1=\sum_{\alpha\beta} r_{\alpha\beta}\mathscr{V}^1_\alpha\mathrm{Tr}_2\,\mathscr{W}^2_\beta . \tag{3.3.13}$$

In a similar manner, we may also define the statistical operator ρ^2 (no connection to the square of ρ!)

$$\rho^2 = \mathrm{Tr}_1\, \rho \ , \tag{3.3.15}$$

which is reduced in $\mathfrak{U}^1$ and can be used to calculate the expectation values $\langle \mathscr{M}^2 \rangle$ of the subsystem Σ^2. We note, however, that in general

$$\rho \neq \rho^1 \rho^2 \tag{3.3.16}$$

holds.

3.3.4 The Time Variation of the Reduced Statistical Operator

In the global system Σ, the dynamics of ρ (in the Schrödinger formulation) are described by the von Neumann equation

$$\frac{d\rho}{dt} = -\mathrm{i}\mathsf{L}\rho = -\mathrm{i}(\mathsf{L}^1 + \mathsf{L}^2 + \mathsf{L}^{12})\rho \ .^{11} \tag{3.3.17}$$

Due to $\mathrm{Tr}_2\, \mathsf{L}^2 \rho = 0$ [cf. (3.3.18)], it follows from this that the *time variation of the reduced statistical operator* is given by

$$\boxed{\frac{d\rho^1}{dt} = -\mathrm{i}\mathsf{L}^1 \rho^1 - \mathrm{i}\mathrm{Tr}_2(\mathsf{L}^{12}\rho)} \ . \tag{3.3.19}$$

This equation, however, is not a closed differential equation for $\rho^1(t)$, owing to the interaction term L^{12}. In order to determine $\rho^1(t)$, one instead requires the integration of the full Eq. (3.3.17). The reason for this is that L^{12} describes a *dynamic coupling between the observables of* Σ^1 *and of* Σ^2. We shall, however, later (Sect. 17.2.4) meet an integro-differential equation for $\rho^1(t)$.

Equation (3.3.19) manifests a general difficulty: every real physical system (here Σ^1) is not fully adiabatically isolated from its surroundings (here Σ^2), so that a von Neumann equation, strictly speaking, does not exist for such a system. Only when the interaction L^{12} is negligibly small can we describe $\rho^1(t)$ by a von Neumann equation during a limited time interval in which Σ^1 is practically isolated. In order to consider this point in more detail, we integrate (3.3.19),

---

[11] Using the form (3.3.8), the right-hand side of (3.3.17) means that L^1 only operates as a commutator on the $\mathscr{V}^1_\alpha$ and L^2 only on the $\mathscr{W}^2_\beta$, but L^{12} on the products $\mathscr{V}^1_\alpha \mathscr{W}^2_\beta$,

$$(\mathsf{L}^1 + \mathsf{L}^2 + \mathsf{L}^{12})\rho = \frac{1}{\hbar} \sum_{\alpha\beta} r_{\alpha\beta} \{[\mathscr{H}^1, \mathscr{V}^1_\alpha]\mathscr{W}^2_\beta + \mathscr{V}^1_\alpha[\mathscr{H}^2, \mathscr{W}^2_\beta] + [\mathscr{H}^{12}, \mathscr{V}^1_\alpha \mathscr{W}^2_\beta]\} \ . \tag{3.3.18}$$

treating $-\mathrm{i}\,\mathrm{Tr}_2(\mathsf{L}^{12}\rho)$ formally as an inhomogeneous term[12],

$$\rho^1(t)=\mathsf{U}^1(t,t_0)\,\rho^1(t_0)-\mathrm{i}g\int_{t_0}^{t}\mathsf{U}^1(t,t')\mathrm{Tr}_2\mathsf{L}^{12}\rho(t')\,dt' \ . \tag{3.3.20}$$

Here, $\mathsf{U}^1(t,t_0)$ describes the dynamics of the isolated system Σ^1 according to L^1

$$\frac{d}{dt}\mathsf{U}^1(t,t_0)=-\mathrm{i}\mathsf{L}^1(t)\mathsf{U}^1(t,t_0) \quad \text{with} \quad \mathsf{U}^1(t_0,t_0)=1 \ , \tag{3.3.21}$$

and g refers to a coupling constant factored out of L^{12}. For the expectation value $\langle\mathscr{F}^1\rangle(t)$ of an observable $\mathscr{F}^1$ in Σ^1, (3.3.11) and (3.3.20) yield

$$\langle\mathscr{F}^1\rangle(t)=\mathrm{Tr}_1(\mathscr{F}^1\cdot\mathsf{U}^1(t,t_0)\rho^1(t_0))-g\int_{t_0}^{t}f(t,t')\,dt' \tag{3.3.22}$$

with

$$f(t,t')=\mathrm{i}\,\mathrm{Tr}(\mathscr{F}^1\cdot\mathsf{U}^1(t,t')\mathsf{L}^{12}\rho(t')) \ . \tag{3.3.23}$$

If we assume that the function $f(t,t')$ does not diverge in the interval $t_0\leqq t'\leqq t$, and thus $|f(t,t')|$ has an upper bound c, then the second term in (3.3.22) is smaller than $gc\cdot(t-t_0)$. For time intervals $t-t_0$ which are much shorter than the *isolation time* τ_{is}, which is a function of the coupling g, $\langle\mathscr{F}^1\rangle(t)$ is determined only by the first term in (3.3.22), i.e. by the dynamics of the isolated system Σ^1. Conversely, this means that the formal results $\langle\mathscr{F}\rangle(t)=\mathrm{Tr}(\mathscr{F}\rho(t))$ for a system treated as isolated are physically valid only for times $t-t_0<\tau_{\mathrm{is}}$. For long times $t-t_0$, even a very weak coupling to the surroundings affects the dynamics; in other words, adiabatic isolation cannot be maintained for arbitrarily long times. In this case, the system must be enlarged to include the coupling to the surroundings.

3.3.5 Transfer of Work and Heat into a Subsystem

We consider a system Σ^1, for which the Hamiltonian $\mathscr{H}^1$ is explicitly time-dependent due to given external c-number fields $h_\alpha(t)$ (Schrödinger formulation),

$$\mathscr{H}^1(t)=\mathscr{H}^1[h_\alpha(t)] \ , \tag{3.3.24}$$

and which is dynamically coupled to the observables of another system Σ^2 by an interaction operator $\mathscr{H}^{12}$ (Fig. 3.3).

The *energy* of Σ^1 at the time t is, according to (3.3.11), given by

$$\langle\mathscr{H}^1\rangle(t)=\mathrm{Tr}_1(\rho^1(t)\,\mathscr{H}^1(t)) \tag{3.3.25}$$

[12] One can best convince oneself of the validity of (3.3.20) by differentiating it with respect to t.

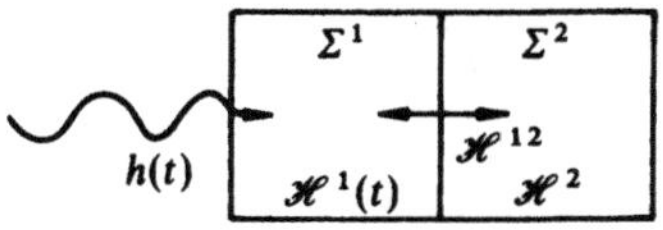

Fig. 3.3. Dynamic transfer of heat and work to Σ^1

or the energy change relative to an initial time t_0 by

$$\langle\mathscr{H}^1\rangle(t)-\langle\mathscr{H}^1\rangle(t_0)=\mathrm{Tr}_1(\rho^1(t)\mathscr{H}^1(t))-\mathrm{Tr}_1(\rho^1(t_0)\mathscr{H}^1(t_0)) \ . \tag{3.3.26}$$

Following Ehrenfest's theorem (3.2.18), applied to $\mathscr{H}^1$, the differential change in $\langle\mathscr{H}^1\rangle(t)$ obeys the relation

$$\frac{d}{dt}\langle\mathscr{H}^1\rangle=\frac{\mathrm{i}}{\hbar}\langle[\mathscr{H}^{12},\mathscr{H}^1]\rangle+\left\langle\left(\frac{\partial\mathscr{H}^1}{\partial t}\right)_{\mathrm{ex}}\right\rangle , \tag{3.3.27}$$

where we replace $\mathscr{H}$ in the commutator by $\mathscr{H}^{12}$, since, according to (3.3.2), $\mathscr{H}^1+\mathscr{H}^2$ commutes with $\mathscr{H}^1$.

The time dependence of $\mathscr{H}^1$ due to the explicit time dependence (3.3.24), $\sum_\alpha\partial\mathscr{H}/\partial h_\alpha\,\dot{h}_\alpha$, implies that the fields $h_\alpha(t)$ do mechanical work on Σ^1. Thus, the *work done* upon the system Σ^1 in the time interval from t_0 to t, denoted by A^1, is given by

$$\boxed{\begin{aligned}A^1(t,t_0)&=\int_{t_0}^{t}\left\langle\left(\frac{\partial\mathscr{H}^1}{\partial t'}\right)_{\mathrm{ex}}\right\rangle dt'\\&=\int_{t_0}^{t}\mathrm{Tr}_1\left(\rho^1(t')\left(\frac{\partial\mathscr{H}^1}{\partial t'}\right)_{\mathrm{ex}}\right)dt'\end{aligned}} \ . \tag{3.3.28}$$

The energy change in Σ^1 which is expressed by the first term of (3.3.27) is due to the energetic contact $\mathscr{H}^{12}$ between Σ^1 and another system Σ^2 (the heat bath), and in the language of thermodynamics it represents the *heat input* Q^1 to Σ^1,

$$\boxed{\begin{aligned}Q^1(t,t_0)&=\frac{\mathrm{i}}{\hbar}\int_{t_0}^{t}\langle[\mathscr{H}^{12},\mathscr{H}^1]\rangle\,dt'\\&=\int_{t_0}^{t}\mathrm{Tr}_1(\dot{\rho}^1(t')\mathscr{H}^1(t'))\,dt'\end{aligned}} \ , \tag{3.3.29}$$

where we have used (3.3.19) in the second line.

Of course, the work A^1 also depends on the interaction energy $\mathscr{H}^{12}$ (via $\rho^1(t')$). Conversely, the heat Q^1 also depends upon the explicit time dependence $\mathscr{H}^1(t')$.

The relation given by Ehrenfest's theorem (3.3.27),

$$\boxed{\langle\mathscr{H}^1\rangle(t)-\langle\mathscr{H}^1\rangle(t_0)=Q^1(t,t_0)+A^1(t,t_0)} \tag{3.3.30}$$

is the most *general formulation of the First Law for dynamic processes* (without particle exchange). Here it is not required that the expectation values $\langle\mathscr{H}^1\rangle$ be state variables in the sense of thermostatics (cf. Sects. 5.3.1 and 6.3.1).

Exercise 3.4 Is the reduced $\rho^1=\mathrm{Tr}_2\,\mathscr{P}_{|\Phi\rangle}$ of a pure state over a subsystem Σ^2 also a pure state?

3.4 Useful Operator Identities

3.4.1 Operator Identities for Time Evolution

When a Liouville operator $\mathsf{L}(t)$ is given, (3.2.7) defines the time-evolution operator $\mathsf{U}(t,t_0)$ in $\mathfrak{L}$:

$$\frac{d}{dt}\mathsf{U}(t,t_0)=-\mathrm{i}\mathsf{L}(t)\mathsf{U}(t,t_0)\quad\text{with}\quad \mathsf{U}(t_0,t_0)=\mathsf{1}\ . \tag{3.4.1}$$

It will prove to be expedient in later parts of this book to decompose $\mathsf{L}(t)$ into the sum of two operators,

$$\mathsf{L}(t)=\mathsf{L}_1(t)+\mathsf{L}_2(t)\ , \tag{3.4.2}$$

whereby we have not yet specified the particular nature of $\mathsf{L}_1(t)$ and $\mathsf{L}_2(t)$. With $\mathsf{L}_1(t)$, the differential equation

$$\frac{d}{dt}\mathsf{U}_1(t,t_0)=-\mathrm{i}\mathsf{L}_1(t)\mathsf{U}_1(t,t_0)\quad\text{with}\quad \mathsf{U}_1(t_0,t_0)=\mathsf{1} \tag{3.4.3}$$

defines another operator $\mathsf{U}_1(t,t_0)$ in $\mathfrak{L}$.

The exact operator identities

$$\mathsf{U}(t,t_0)=\mathsf{U}_1(t,t_0)-\mathrm{i}\int_{t_0}^{t}\mathsf{U}(t,t')\mathsf{L}_2(t')\mathsf{U}_1(t',t_0)\,dt'\ , \tag{3.4.4a}$$

$$\mathsf{U}(t,t_0)=\mathsf{U}_1(t,t_0)-\mathrm{i}\int_{t_0}^{t}\mathsf{U}_1(t,t')\mathsf{L}_2(t')\mathsf{U}(t',t_0)\,dt'\ . \tag{3.4.4b}$$

then hold between $\mathsf{U}(t,t_0)$ and $\mathsf{U}_1(t,t_0)$.

Proof of (3.4.4a): We define the operator

$$\mathsf{X}(t, t_0) \equiv \mathsf{U}_1(t, t_0) - \mathrm{i} \int_{t_0}^{t} \mathsf{U}(t, t')\mathsf{L}_2(t')\mathsf{U}_1(t', t_0)\, dt' \tag{3.4.5}$$

and show that it obeys the differential equation (3.4.1). Differentiating w.r.t. t, we obtain

$$\dot{\mathsf{X}}(t, t_0) = \dot{\mathsf{U}}_1(t, t_0) - \mathrm{i}\mathsf{U}(t, t)\mathsf{L}_2(t)\mathsf{U}_1(t, t_0) - \mathrm{i}\int_{t_0}^{t} \dot{\mathsf{U}}(t, t')\mathsf{L}_2(t')\mathsf{U}_1(t', t_0)\, dt' \ .$$

Using (3.4.1) and (3.4.3), this equation becomes

$$\dot{\mathsf{X}}(t, t_0) = -\mathrm{i}(\mathsf{L}_1 + \mathsf{L}_2)\mathsf{U}_1 - \mathsf{L}(t) \int_{t_0}^{t} \mathsf{U}(t, t')\mathsf{L}_2(t')\mathsf{U}_1(t', t_0)\, dt' \ .$$

Applying (3.4.2) and (3.4.5), we derive from this

$$\dot{\mathsf{X}} = -\mathrm{i}\mathsf{L}\mathsf{X} \ ,$$

i.e. X obeys the same first-order differential equation as U. Furthermore, X has the same initial conditions as U, according to (3.4.5),

$$\mathsf{X}(t_0, t_0) = \mathsf{U}(t_0, t_0) = 1 \ ,$$

so that $\mathsf{X}(t, t_0)$ is identical to $\mathsf{U}(t, t_0)$; q.e.d.

Proof of (3.4.4b): Differentiation of (3.4.4b) leads to

$$\dot{\mathsf{U}}(t, t_0) = \dot{\mathsf{U}}_1(t, t_0) - \mathrm{i}\mathsf{L}_2(t)\mathsf{U}(t, t_0) - \mathsf{L}_1(t) \int_{t_0}^{t} \mathsf{U}_1(t, t')\mathsf{L}_2(t')\mathsf{U}(t', t_0)\, dt' \ .$$

If we eliminate the integral by using (3.4.4b), we obtain, with (3.4.3),

$$\dot{\mathsf{U}}(t, t_0) = -\mathrm{i}\mathsf{L}(t)\mathsf{U}(t, t_0) \ ,$$

i.e. (3.4.4b) obeys the differential equation (3.4.1).

For later reference, we collect some special cases of (3.4.4a and b):

If L_1 is time *in*dependent, (3.4.3) can be integrated to

$$\mathsf{U}_1(t, t_0) = \mathrm{e}^{-\mathrm{i}\mathsf{L}_1 \cdot (t - t_0)}$$

and (3.4.4a and b) take on the forms

$$\mathsf{U}(t, t_0) = \mathrm{e}^{-\mathrm{i}\mathsf{L}_1 \cdot (t - t_0)} - \mathrm{i} \int_{t_0}^{t} \mathsf{U}(t, t')\mathsf{L}_2(t')\mathrm{e}^{-\mathrm{i}\mathsf{L}_1 \cdot (t' - t_0)}\, dt' \ , \tag{3.4.6a}$$

$$\mathsf{U}(t, t_0) = \mathrm{e}^{-\mathrm{i}\mathsf{L}_1 \cdot (t - t_0)} - \mathrm{i} \int_{t_0}^{t} \mathrm{e}^{-\mathrm{i}\mathsf{L}_1 \cdot (t - t')}\mathsf{L}_2(t')\mathsf{U}(t', t_0)\, dt' \ . \tag{3.4.6b}$$

If *both* L_1 and L_2 are time *in*dependent, the operator identities can be further specialized to

$$e^{-i\mathsf{L}\cdot(t-t_0)}=e^{-i\mathsf{L}_1\cdot(t-t_0)}-i\int_{t_0}^{t} e^{-i\mathsf{L}\cdot(t-t')}\mathsf{L}_2\, e^{-i\mathsf{L}_1\cdot(t'-t_0)}\,dt' \;, \tag{3.4.7a}$$

$$e^{-i\mathsf{L}\cdot(t-t_0)}=e^{-i\mathsf{L}_1\cdot(t-t_0)}-i\int_{t_0}^{t} e^{-i\mathsf{L}_1\cdot(t-t')}\mathsf{L}_2\, e^{-i\mathsf{L}\cdot(t'-t_0)}\,dt' \;. \tag{3.4.7b}$$

In the case of explicitly time-independent operators, the Heisenberg equation of motion can be integrated using $\exp[+i\mathsf{L}\cdot(t-t_0)]$. For a time-independent decomposition $\mathsf{L}=\mathsf{L}_1+\mathsf{L}_2$, we thus need only replace $-i$ in (3.4.7a and b) by i:

$$e^{+i\mathsf{L}\cdot(t-t_0)}=e^{+i\mathsf{L}_1\cdot(t-t_0)}+i\int_{t_0}^{t} e^{+i\mathsf{L}\cdot(t-t')}\mathsf{L}_2\, e^{+i\mathsf{L}_1\cdot(t'-t_0)}\,dt' \;, \tag{3.4.8a}$$

$$e^{+i\mathsf{L}\cdot(t-t_0)}=e^{+i\mathsf{L}_1\cdot(t-t_0)}+i\int_{t_0}^{t} e^{+i\mathsf{L}_1\cdot(t-t')}\mathsf{L}_2\, e^{+i\mathsf{L}\cdot(t'-t_0)}\,dt' \;. \tag{3.4.8b}$$

3.4.2 Differentiation of Exponential Operators

In connection with generalized canonical statistical operators in $\mathfrak{U}$ (cf. Sect. 5.3.1), it will prove necessary to differentiate exponential operators $\exp[-\mathscr{X}(a)]$ in $\mathfrak{U}$ w.r.t. parameters a.

If we read the operator identity (3.4.7a) in $\mathfrak{U}$ with the formal substitutions $t_0\to 0$, $i\mathsf{L}t\to\mathscr{X}$, and $t'\to\mu t$, we obtain the identity

$$\boxed{e^{-\mathscr{X}}=e^{-\mathscr{X}_1}-e^{-\mathscr{X}}\int_0^1 e^{\mu\mathscr{X}}\mathscr{X}_2\, e^{-\mu\mathscr{X}_1}\,d\mu} \quad (\text{with} \quad \mathscr{X}=\mathscr{X}_1+\mathscr{X}_2)\;. \tag{3.4.9}$$

If $\mathscr{X}$ depends on a parameter a, $\mathscr{X}=\mathscr{X}(a)$, we obtain from (3.4.9) with the substitutions

$$\mathscr{X}\to\mathscr{X}(a)+\frac{\partial\mathscr{X}}{\partial a}\Delta a, \quad \mathscr{X}_1\to\mathscr{X}(a), \quad \mathscr{X}_2\to\frac{\partial\mathscr{X}}{\partial a}\Delta a$$

and using the definition of the partial derivative, the following relation:

$$\boxed{\frac{\partial}{\partial a}e^{-\mathscr{X}(a)}=-e^{-\mathscr{X}(a)}\int_0^1 e^{\mu\mathscr{X}(a)}\frac{\partial\mathscr{X}}{\partial a}e^{-\mu\mathscr{X}(a)}\,d\mu}\;.\text{[13]} \tag{3.4.10}$$

[13] Because $\mathscr{X}$ does not need to be commutative for different parameters a, the differentiation of $e^{-\mathscr{X}(a)}$ is described by (3.4.10), which is more complicated than for the differentiation of a c-number exponential function. In the special case that $\mathscr{X}$ commutes with $\partial\mathscr{X}/\partial a$, the right side simplifies to the c-number result, $-\exp[-\mathscr{X}(a)]\,(\partial\mathscr{X}(a)/\partial a)$.

If we set $\mathscr{X}(a)$ equal to $\beta\mathscr{H}(a)$ with $\mu\beta=\lambda$, we find, in particular,

$$\boxed{\frac{\partial}{\partial a}\mathrm{e}^{-\beta\mathscr{H}(a)}=-\mathrm{e}^{-\beta\mathscr{H}(a)}\int\limits_{0}^{\beta}\mathrm{e}^{\lambda\mathscr{H}(a)}\frac{\partial\mathscr{H}}{\partial a}\mathrm{e}^{-\lambda\mathscr{H}(a)}\,d\lambda} \quad , \tag{3.4.11}$$

which is an operator relation that will prove useful in linear response theory (see Sect. 2.2).

Exercise 3.5 Show that the evolution operator $\mathsf{U}(t, t_0)$ defined by (3.4.1) also obeys the differential equation

$$\frac{\partial}{\partial t_0}\mathsf{U}(t,t_0)=+\mathsf{U}(t,t_0)\,\mathrm{i}\mathsf{L}(t_0) \quad .$$

4. Systems of Many Particles

If a system having a macroscopic volume consists of many particles, only macroscopic observables, which combine globally the properties of a large number of particles, are measurable. Although the state of such a macroscopic system[1] is not precisely determined by this type of measurement, quantum statistics nevertheless permits the microscopic calculation of the time dependence of the experimentally measured values of the macroscopic observables. In order to understand this fact, one requires a knowledge of some of the properties of macroscopic observables, which we shall introduce in this chapter. A completely satisfying quantum theory of macroscopic observables does not at present exist, and thus the following discussion is necessarily qualitative.

4.1 The Mean Square Deviations of Macroscopic Observables

According to quantum theory, only statistical predictions may be made concerning the results of the measurement of observables. The eigenvalues of operators are in general found to give the measured values of the corresponding observables with a certain probability only; the measured values exhibit deviations from the mean values. The uncertainty relations give inequalities between the deviations for different observables which, in principle, cannot be overcome.

In this section, we wish to show that in macroscopic systems, treated quantum statistically, i.e. retaining all the statistical results of quantum mechanics, the deviations of macroscopic observables are practically negligible. In the following, this will be demonstrated using the example of spatial densities.

The considerations of Sect. 4.1 are taken to apply at any particular time in the course of a quantum-statistical dynamical process.

4.1.1 Microscopic Densities and Their Correlation Functions

Along with integral observables of a system in a volume V, when considering processes which are spatially inhomogeneous (e.g. heat conduction), one needs

[1] By "macroscopic system", we mean primarily solids, liquids, gases, or plasmas. Although atomic nuclei are not macroscopic bodies, their internal collective motions (in nuclei with large numbers of nucleons) can be treated approximately using statistical-thermodynamic methods as described in this book.

to take into account their densities $\mathscr{f}(\mathfrak{r})$ at various points $\mathfrak{r}$ (c-number parameter) in space. The extensive quantities are found by calculating volume integrals

$$\int_V \mathscr{f}(\mathfrak{r})\, d^3x \quad (t = \text{fixed}) \; . \tag{4.1.1}$$

These *microscopic densities* $\mathscr{f}(\mathfrak{r})$ are Hermitian operators in $\mathfrak{U}$, which are parametrized by the location of the observation, $\mathfrak{r}$. They are thus functions of the operators for the microscopic particles contained in the system.

We mention the following simple examples[2, 3, 4]

$$\begin{aligned}
&\text{particle density} && \mathscr{n}(\mathfrak{r}) = \sum_{\nu=1}^{N} \delta(\mathfrak{r} - \vec{x}_\nu) \quad \text{or} \\
& && \sum_m \psi_m^\dagger(\mathfrak{r})\, \psi_m(\mathfrak{r}) \; ; \\
&\text{current density} && \vec{\mathscr{j}}(\mathfrak{r}) = \sum_{\nu=1}^{N} \frac{\vec{p}_\nu}{2m} \delta(\mathfrak{r} - \vec{x}_\nu) + \text{h.c.} \quad \text{or} \\
& && \sum_m \psi_m^\dagger(\mathfrak{r}) \frac{\hbar}{2im} \nabla \psi_m(\mathfrak{r}) + \text{h.c.} \; ; \\
&\text{spin density} && \vec{\sigma}(\mathfrak{r}) = \sum_{\nu=1}^{N} \vec{\sigma}_\nu\, \delta(\mathfrak{r} - \vec{x}_\nu) \quad \text{or} \\
& && \sum_{mm'} \psi_m^\dagger(\mathfrak{r}) \frac{\hbar}{2} \sigma_{mm'} \psi_{m'}(\mathfrak{r}) \; .
\end{aligned} \tag{4.1.2}$$

The densities are not necessarily sums of single-particle operators; e.g. the energy density for particles in an external potential $V(\mathfrak{r})$ with an interaction

---

[2] $\psi_m(\mathfrak{r})$ and $\psi_m^\dagger(\mathfrak{r})$ are annihilation and creation operators for particles at the position $\mathfrak{r}$ with spin $m\hbar$ in the unitary space $\mathfrak{U}^\pm$ with variable particle number; they obey the well-known commutation or anticommutation relations, depending on whether the particles are fermions or bosons. The right-hand expressions in (4.1.2) are therefore somewhat more general than those in the middle, which are valid for a fixed number of particles. In the language of the field representation (e.g. the Schrödinger matter field), they can be obtained from the local conservation laws of the field. The field-theoretical approach is particularly important for relativistic problems, for example for the radiation field, in which the particle picture is insufficient.

[3] h.c. = Hermitian conjugate.

[4] Another important example is the operator of the stress tensor or of the pressure at the position $\mathfrak{r}$.

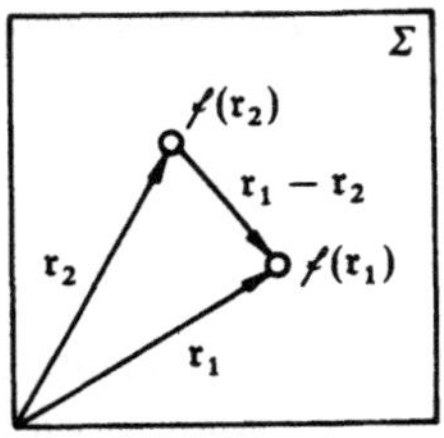

Fig. 4.1. The definition of a correlation function

$V(\mathrm{r},\mathrm{r}')$ between the particles is

$$\mathscr{h}(\mathrm{r})=\psi^{\dagger}(\mathrm{r})\left(-\frac{\hbar^2}{2m}\Delta+V(\mathrm{r})+\frac{1}{2}\int\psi^{\dagger}(\mathrm{r}')\,V(\mathrm{r},\mathrm{r}')\psi(\mathrm{r}')\,d^3x'\right)\psi(\mathrm{r})\ .^5 \quad (4.1.3)$$

The integral over this density as in (4.1.1) gives the total energy $\mathscr{H}$.

The operator for the *fluctuations* of an observable is the deviation of the operator from its expectation value. It is

$$\delta\not{f}(\mathrm{r})=\not{f}(\mathrm{r})-\langle\not{f}(\mathrm{r})\rangle\ ; \quad (4.1.4)$$

here, the operator for the fluctuations of the density $\not{f}$ at the position r. If we consider two observation points r_1 and r_2 with the densities $\not{f}(\mathrm{r}_1)$ and $\not{f}(\mathrm{r}_2)$ (Fig. 4.1), then the operator of the correlation between $\not{f}(\mathrm{r}_1)$ and $\not{f}(\mathrm{r}_2)$ is the product of the fluctuations,

$$\delta\not{f}(\mathrm{r}_1)\,\delta\not{f}(\mathrm{r}_2)\ .^6 \quad (4.1.5)$$

Its expectation value is the *correlation function* of the densities between the two points

$$\langle\delta\not{f}(\mathrm{r}_1)\,\delta\not{f}(\mathrm{r}_2)\rangle=\langle\not{f}(\mathrm{r}_1)\not{f}(\mathrm{r}_2)\rangle-\langle\not{f}(\mathrm{r}_1)\rangle\langle\not{f}(\mathrm{r}_2)\rangle\ . \quad (4.1.6)$$

If the points r_1 and r_2 in a macroscopic system are *far enough from one another*, then for *short-range forces* between the particles, the probability $w_{f_1 f_2}$ for observing the value f_1 for the density $\not{f}$ at point r_1 *and* the value f_2 for the density $\not{f}$ at the point r_2 will, to a good approximation, be equal to the probability of the independent results f_1 and f_2 [7]

$$w_{f_1 f_2}\approx w_{f_1}\,w_{f_2}\ .$$

---

[5] Spin indices, which would be important for spin-dependent forces, have been suppressed here.

[6] If $\not{f}(\mathrm{r}_1)$ and $\not{f}(\mathrm{r}_2)$ do not commute, the product must be symmetrized

$$\tfrac{1}{2}\delta\not{f}(\mathrm{r}_1)\delta\not{f}(\mathrm{r}_2)+\mathrm{h.c.} \quad (4.1.5\mathrm{a})$$

Equations (4.1.10) and following are however independent of this symmetrization.

[7] w_{f_1} is the probability for observing f_1 independent of f_2 and w_{f_2} vice versa.

In the language of quantum theory, it follows from this (when the densities commute at sufficiently large distances, cf. Exercise 4.1)

$$\mathrm{Tr}(\rho\mathscr{P}_{f_1}\mathscr{P}_{f_2}) \approx \mathrm{Tr}(\rho\mathscr{P}_{f_1})\mathrm{Tr}(\rho\mathscr{P}_{f_2})\ ^{8} \qquad \text{and}$$

$$\langle f(\mathbf{r}_1)f(\mathbf{r}_2)\rangle = \sum_{f_1 f_2} f_1 f_2 w_{f_1 f_2} \approx \langle f(\mathbf{r}_1)\rangle\langle f(\mathbf{r}_2)\rangle\ .$$

The limiting distance $|\mathbf{r}_1 - \mathbf{r}_2|$, above which this factorization is roughly valid, i.e. the correlation function (4.1.6) nearly vanishes, is called the *correlation length* ξ of $\delta f(\mathbf{r})$,

$$\boxed{\langle\delta f(\mathbf{r}_1)\delta f(\mathbf{r}_2)\rangle \approx 0 \quad \text{for} \quad |\mathbf{r}_1 - \mathbf{r}_2| > \xi}\ . \tag{4.1.7}$$

The value of ξ depends on the observable f and on the state of the system, i.e. on the statistical operator. It is the order of magnitude of several times the average particle spacing[9].

4.1.2 Macroscopic Densities and Their Fluctuations

The microscopic density $f(\mathbf{r})$ in a mathematical point is not physically measurable. Rather, only its mean value over a given volume element $\Delta V(\mathbf{r})$ (= c-number), which includes the neighborhood of $\mathbf{r}$, will be accessible to measurement (Fig. 4.2),

$$\boxed{\mathscr{F}(\mathbf{r}) = \frac{1}{\Delta V(\mathbf{r})}\int_{\Delta V(\mathbf{r})} f(\mathbf{r}')d^3x'}\ . \tag{4.1.8}$$

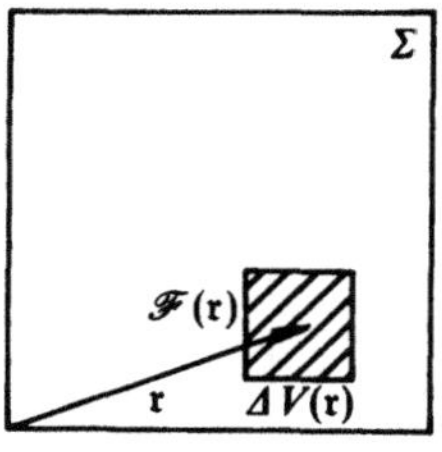

Fig. 4.2. The definition of macroscopic density, $\mathscr{F}(\mathbf{r})$

[8] $\mathscr{P}_{f_1}$ is the projector on the eigenspace u_{f_1} belonging to the eigenvalue f_1 of $f(\mathbf{r}_1)$; $\mathscr{P}_{f_2}$ correspondingly for f_2.

[9] In the case of phase transitions, ξ can become very large, but remains finite in systems of finite particle number.

In macroscopic systems, ΔV should, on the one hand, be so small in macroscopic units that macroscopically detectable inhomogeneities are not averaged out; but on the other hand, so large in microscopic units that the mean value is obtained over several orders of magnitude of the average particle spacing. $\mathscr{F}(\mathfrak{r})$ is thus the operator of a *macroscopic density*. Examples for the expectation values of such densities are the magnetization $\mathfrak{M}(\mathfrak{r})$ and the polarization $\mathfrak{P}(\mathfrak{r})$, which occur in Maxwell's theory in the presence of matter.

If one substitutes the Fourier transform of $f(\mathfrak{r})$,

$$f(\mathfrak{r}) = \int_{-\infty}^{-\infty} f(\mathfrak{k})\,e^{-i\mathfrak{k}\mathfrak{r}} d^3k$$

into $\mathscr{F}(\mathfrak{r})$, then (4.1.8) yields[10]

$$\mathscr{F}(\mathfrak{r}) \approx \int_{|k_i| \lesssim 2\pi/\Delta x_i} f(\mathfrak{k})\,e^{-i\mathfrak{k}\mathfrak{r}} d^3k, \quad (\Delta x_1\,\Delta x_2\,\Delta x_3 = \Delta V)\ , \tag{4.1.9}$$

i.e. only the Fourier components $f(\mathfrak{k})$ with *small propagation vectors* $\mathfrak{k}$, or with large wavelengths, contribute to $\mathscr{F}(\mathfrak{r})$.

We shall now proceed to investigate the quality of the expectation value of a macroscopic density $\langle \mathscr{F}(\mathfrak{r}) \rangle$ for predicting the results of measurements (Fig. 4.3). To this end, we calculate the mean square deviation from $\mathscr{F}(\mathfrak{r})$,

$$\langle (\mathscr{F}(\mathfrak{r}) - \langle \mathscr{F}(\mathfrak{r}) \rangle)^2 \rangle = \langle \mathscr{F}^2(\mathfrak{r}) \rangle - \langle \mathscr{F}(\mathfrak{r}) \rangle^2\ ,$$

for which we obtain from (4.1.4) and (4.1.8) the result

$$\frac{1}{(\Delta V(\mathfrak{r}))^2} \iint_{\Delta V(\mathfrak{r})} \langle \delta f(\mathfrak{r}_1)\,\delta f(\mathfrak{r}_2) \rangle\, d^3x_1\, d^3x_2\ ; \tag{4.1.10}$$

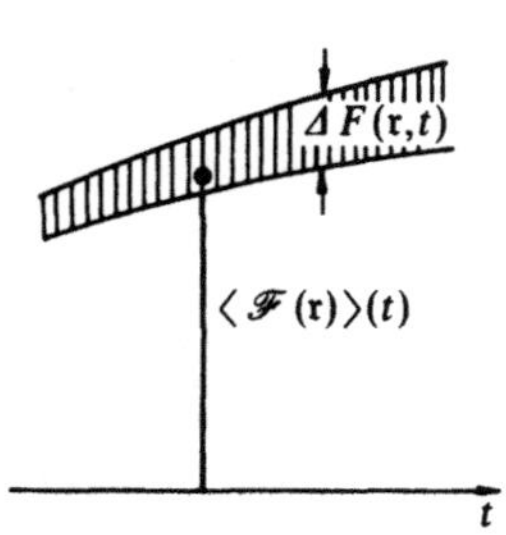

Fig. 4.3. Expectation value and mean square deviation of a macroscopic density $\mathscr{F}(\mathfrak{r})$

---

[10] The function $\sin(k_i \Delta x_i/2)\cdot(k_i \Delta x_i/2)^{-1}$ has been approximated by 1 in the interval $|k_i|\Delta x_i/2 < \pi$ and set equal to zero outside this interval. Other averaging procedures (4.1.8) lead to equivalent results.

this is the correlation function of the microscopic density $\mathscr{f}(\mathbf{r})$, (4.1.6), averaged over the volume $\Delta V(\mathbf{r})$. According to (4.1.7), this is negligibly small for $|\mathbf{r}_1 - \mathbf{r}_2| > \xi$, so that the double integral extends only over the region $|\mathbf{r}_1 - \mathbf{r}_2| < \xi$ if we choose ΔV *so large* that

$$\boxed{\Delta V \gg \xi^3} \tag{4.1.11}$$

is valid. The double integral is therefore *not* proportional to $(\Delta V)^2$, but rather to $\xi^3 \Delta V$; thus

$$\boxed{\langle \mathscr{F}^2(\mathbf{r}) \rangle - \langle \mathscr{F}(\mathbf{r}) \rangle^2 \sim \frac{\xi^3}{\Delta V(\mathbf{r})} = \varrho(\mathbf{r}) \frac{\xi^3}{\Delta N(\mathbf{r})}} \;.^{11} \tag{4.1.12}$$

When ΔV fulfills condition (4.1.11), the mean square deviation of $\mathscr{F}(\mathbf{r})$ will be very small[12]. We conclude: *for macroscopic observables, the quantum-statistically calculated expectation values are practically identical to the experimentally measured results*:

$$\boxed{\langle \mathscr{F}(\mathbf{r}) \rangle \approx \text{measured value}} \;. \tag{4.1.13}$$

When the mean square deviation is negligible, the probability distribution $w_{F(\mathbf{r})} = \mathrm{Tr}(\rho \mathscr{P}_{F(\mathbf{r})})$ for the eigenvalues $F(\mathbf{r})$ in the neighborhood of the expectation value $\langle \mathscr{F}(\mathbf{r}) \rangle$ will exhibit a sharp maximum. This results from the interaction of the properties of the observables (number of states in an interval [cf. (4.2.21)] and those of the statistical operator.

For the integral quantities $\mathscr{F}^{\text{tot}}$ of the overall system in the volume V

$$\mathscr{F}^{\text{tot}} = \int_V \mathscr{f}(\mathbf{r}) dV = \sum_{\Delta V(\mathbf{r})} \mathscr{F}(\mathbf{r}) \Delta V(\mathbf{r}) \;, \tag{4.1.14}$$

the same chain of reasoning which led to (4.1.12) yields the result

$$\langle \mathscr{F}^{\text{tot}\,2} \rangle - \langle \mathscr{F}^{\text{tot}} \rangle^2 \sim \xi^3 V \tag{4.1.15}$$

(since in the definition of $\mathscr{F}^{\text{tot}}$, in contrast to that of the density $\mathscr{F}(\mathbf{r})$, there is no division by V).

---

[11] $\varrho(\mathbf{r}) = [\Delta N(\mathbf{r})/\Delta V(\mathbf{r})]$ is the macroscopic particle density at the point $\mathbf{r}$, where $\Delta N(\mathbf{r})$ is the number of particles in $\Delta V(\mathbf{r})$ ($\Delta N(\mathbf{r})$ is not to be confused with the uncertainty in the number of particles).

[12] This does not eliminate the possibility of also measuring the correlation function $(1/2)\langle \delta \mathscr{f}(\mathbf{r}_1) \delta \mathscr{f}(\mathbf{r}_2) + \text{h.c.} \rangle$ by means of a suitable experimental apparatus.

The *relative mean square deviation* of the integral quantities $\mathscr{F}^{\text{tot}}$,

$$\frac{\langle \mathscr{F}^{\text{tot}\,2}\rangle - \langle \mathscr{F}^{\text{tot}}\rangle^2}{\langle \mathscr{F}^{\text{tot}}\rangle^2} \sim \frac{\xi^3}{V} = \varrho\frac{\xi^3}{N} \ll 1 \ , \tag{4.1.16}$$

which is important for the accuracy of a measurement, becomes vanishingly small for large volumes V or numbers of particles N, because $\langle \mathscr{F}^{\text{tot}}\rangle \propto V$ and, with (4.1.11), $V \gg \xi^3$ holds. Thus, the expectation values $\langle \mathscr{F}^{\text{tot}}\rangle$ show practically perfect agreement with the measured values, which fact is already contained in (4.1.13) as a special case.

Exercise 4.1 Give the commutation relations for the components at different points in space for

a) the spin density $\vec{\mathfrak{s}}(\mathbf{r})$, and
b) the momentum current density $\vec{j}_p(\mathbf{r})$.

What is found for the corresponding macroscopic density $\vec{\mathscr{J}}_p(\mathbf{r})$?

4.2 General Properties of the Time Evolution of Expectation Values

We consider a system Σ which is *isolated* from its environment with a time-independent Hamiltonian $\mathscr{H}$. The system has a finite volume V and contains N particles. The finite volume means that the energy eigenvalues E_n

$$\mathscr{H}\,|u_n\rangle = E_n\,|u_n\rangle \ , \tag{4.2.1}$$

(n enumerates the states $|u_n\rangle$ and thus includes the degeneracy indices of the energy eigenvalues) are *discrete*. The time variation of the expectation value of an (explicitly time-independent) observable $\mathscr{F}$ in Σ is given by

$$\begin{aligned}\langle \mathscr{F}\rangle(t) &= \mathrm{Tr}(\mathscr{F}\rho(t)) = \mathrm{Tr}(\mathscr{F}\mathrm{e}^{-\mathrm{i}\mathrm{L}\cdot(t-t_0)}\rho(t_0)) \\ &= \sum_{mn}\langle u_m|\mathscr{F}u_n\rangle\langle u_n|\rho(t_0)u_m\rangle\,\mathrm{e}^{(-\mathrm{i}/\hbar)(E_n-E_m)(t-t_0)} \ .\end{aligned} \tag{4.2.2}$$

The function $\langle\mathscr{F}\rangle(t)$ naturally depends upon the particular observable $\mathscr{F}$ (selection rules!) and the statistical operator $\rho(t_0)$. (When $\mathscr{F}$ or $\rho(t_0)$ commute with $\mathscr{H}$, $\langle\mathscr{F}\rangle(t)$ has no time dependence at all.)

In this section, we shall demonstrate some general aspects of the time variation of $\langle\mathscr{F}\rangle(t)$ which follow from the structure of the energy eigenvalue spectrum E_n. We shall see that for macroscopic systems, there are certain time scales on which $\langle\mathscr{F}\rangle(t)$ behaves quite differently, but only one time scale is physically relevant.

If the energy differences divided by $\hbar$,

$$\omega_{nm} \equiv \frac{1}{\hbar}(E_n - E_m) \tag{4.2.3}$$

are integer multiples of a quantity ω_0 (e.g. as in the case of independent spins or independent oscillators),

$$\omega_{nm} = N_{nm}\omega_0 \quad (N_{nm} = \text{integer}) \ , \tag{4.2.4}$$

then $\langle \mathscr{F} \rangle(t)$ is *periodic* with a *recurrence time*

$$\tau_w = \frac{2\pi}{\omega_0} \ , \tag{4.2.5}$$

so that we find

$$\langle \mathscr{F} \rangle(t + \tau_w) = \langle \mathscr{F} \rangle(t) \ . \tag{4.2.6}$$

Equation (4.2.2) is the corresponding Fourier series.

In general, however, assumption (4.2.4) is *not* fulfilled. The Hamiltonian contains interactions between the constituents of Σ, e.g. between the particles, spins, or oscillators. The energy differences ω_{nm} are *real* numbers in some particular units. Then, under certain mathematical preconditions [e.g. the uniform convergence of the partial sums], (4.2.2) represents an *almost periodic function*[13].

In order to approximate the given numbers ω_{nm}, we first introduce an energy E_{max} which is of the order of the largest energies which occur in (4.2.2), as determined by the properties of $\rho(t_0)$[14]. With E_{max}, we specify an upper cutoff to the energy spectrum in (4.2.2). A lower bound is in any case given by the smallest eigenvalue E_0. Within the interval

$$\delta E = E_{max} - E_0 \tag{4.2.8}$$

we wish to approximate the (finite number of) energy differences ω_{nm} as well as possible as integer multiples of a fundamental frequency ω_0 (which is chosen to be sufficiently small and may be smaller than the smallest energy difference; cf.

---

[13] The *almost periodicity* of $\langle \mathscr{F} \rangle(t)$ implies the existence of *translation times* $\tau(\varepsilon)$, so that for a given value of ε we find [cf. Fig. 4.5 (p. 41)]

$$|\langle \mathscr{F} \rangle(t + \tau(\varepsilon)) - \langle \mathscr{F} \rangle(t)| < \varepsilon \ . \tag{4.2.7}$$

There is furthermore an interval $T(\varepsilon)$ with the property that each interval $T(\varepsilon)$ contains at least one τ. These statements are an extension of the properties of periodic functions. [Bohr, H.: *Almost Periodic Functions* (New York, 1951)].

[14] E_{max} can naturally also be determined by the matrix elements of $\mathscr{F}$.

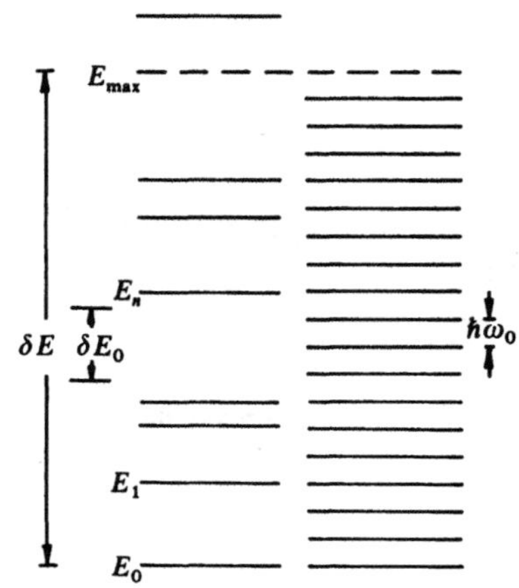

Fig. 4.4. The energy differences $E_n - E_m$ in the interval δE, approximated by integer multiples of $\hbar\omega_0$. Mean energy spacing δE_0

Fig. 4.4),

$$\omega_{nm} \approx N_{nm}\omega_0 \left(\begin{array}{l} N_{nm} = \text{integer} \\ E_{n,}\ E_m \text{ within } \delta E \end{array} \right) . \tag{4.2.9}$$

Due to the cutoff at E_{max}, (4.2.2) becomes a finite sum with a large number of terms, which are proportional to $\exp[-iN_{nm}\omega_0(t-t_0)]$. In this approximation, $\langle \mathscr{F} \rangle(t)$ becomes a periodic function of period

$$\tau_w = \frac{2\pi}{\omega_0} . \tag{4.2.10}$$

In the sense of (4.2.7), τ_w is a translation time. The deviation ε (4.2.7) from an exact periodicity is produced by the errors which result from the cutoff at energy E_{max} and from the rational approximation (4.2.9).

The error in the approximation (4.2.9) can, for example, be represented by a theorem from the theory of irrational numbers[15]: M real numbers α_μ $(\mu = 1, \ldots, M)$ may be simultaneously approximated by quotients p_μ/q $(p_\mu, q = \text{integers})$, whereby

$$|\alpha_\mu q - p_\mu| < \frac{1}{r} \qquad (1 \leqq |q| \leqq r^M) \tag{4.2.11}$$

[15] O. Perron: *Irrationalzahlen* (New York, 1948).

can be fulfilled for an arbitrarily chosen integer r. Applied to the $\omega_{nm} = \alpha_\mu \Omega$ (Ω = unit),

$$\left|\omega_{nm}\frac{1}{\omega_0} - N_{nm}\right| < \frac{1}{r} , \tag{4.2.12}$$

this leads to a bound for the error in (4.2.9),

$$\delta\omega_{nm} \equiv \omega_{nm} - N_{nm}\omega_0 , \tag{4.2.13}$$

which is given by

$$|\delta\omega_{nm}| < \frac{\omega_0}{r} . \tag{4.2.14}$$

The quantities r and $E_{\max}$ yield the value of ε in (4.2.7). To see this, we consider the series (4.2.2)

$$\langle \mathscr{F} \rangle(t) = \sum_{mn}^{E_{\max}} F_{mn}\rho_{nm}(t_0)\mathrm{e}^{-\mathrm{i}\omega_{nm}(t-t_0)} + R(t) .$$

When this series converges absolutely, then $|R(t)| < \varepsilon'$. By substituting (4.2.13) and (4.2.14), we obtain the result

$$\begin{aligned}\left|\langle \mathscr{F} \rangle\left(t + \frac{2\pi}{\omega_0}\right) - \langle \mathscr{F} \rangle(t)\right| &< \sum_{mn}^{E_{\max}} \left| F_{mn}\rho_{nm}(t_0) 2\pi \frac{\delta\omega_{nm}}{\omega_0}\right| + 2\varepsilon' \\ &< \frac{2\pi}{r} \sum_{mn}^{E_{\max}} |F_{mn}\rho_{nm}(t_0)| + 2\varepsilon' = \varepsilon .\end{aligned}$$

We shall now estimate qualitatively the value of a physically reasonable ω_0 from the properties of the energy spectrum E_n. If z is the number of energy eigenvalues within δE, then the *mean spacing*, δE_0, *of neighboring energies* in this interval is defined by

$$\delta E_0 = \frac{\delta E}{z} . \tag{4.2.15}$$

The number $g_{\delta E}$ of energy levels in δE is the sum over all the degeneracies d_n,

$$g_{\delta E} = \sum_{n=1}^{z} d_n , \tag{4.2.16}$$

so that $g_{\delta E}$ lies within the bounds

$$z \leqq g_{\delta E} \leqq d_{\max} \cdot z . \tag{4.2.17}$$

The lower bound is reached (or approximately reached) when the energies are not (or only slightly) degenerate. The upper bound is an estimate from the largest degeneracy in the interval δE. Approximation (4.2.9) is only sufficiently good, i.e. ε is only small enough, when

$$\hbar\omega_0 \leqq \delta E_0 \tag{4.2.18}$$

holds (cf. Fig. 4.4)[16]. From (4.2.15) and (4.2.18), we obtain then the estimate for ω_0,

$$\omega_0 \leqq \frac{\delta E}{\hbar} \frac{d_{\max}}{g_{\delta E}} \tag{4.2.19}$$

and from it, the time τ_w after which the function $\langle \mathscr{F} \rangle(t)$ approximately repeats itself; it is called the *recurrence time* for short:

$$\tau_w = \frac{2\pi}{\omega_0} \geqq \frac{2\pi\hbar}{\delta E} \frac{g_{\delta E}}{d_{\max}} \ . \tag{4.2.20}$$

Our considerations up to now applied to an arbitrary, isolated physical system. We now turn to *macroscopic* systems in particular, with large numbers N of particles and with interactions. As one knows e.g. from the partition function for combined systems, the number of energy levels $g_{\delta E}$ increases exponentially with N,

$$g_{\delta E} \approx \left[f\left(\frac{E_{\max}}{N} \right) \right]^N \tag{4.2.21}$$

[cf. Footnote 32 (p. 69)]. In case the energy degeneracy $d_{\max}$ does not increase exponentially with N, it follows from (4.2.20) that

$$\tau_w \gtrapprox c^N \ , \tag{4.2.22}$$

where c depends to a good approximation only on the ratio $E_{\max}/N$. The recurrence time τ_w is thus enormous for macroscopic systems. (Estimates show that it can even be longer than the age of the universe.) *For all realistic systems, one can therefore completely forget the recurrence time* τ_w[17].

The phenomenon of recurrence is due to the fact that all the individual terms in the series (4.2.2) return to near their initial phases after the time τ_w. Long *before* this time, there is a time interval in which the summation over $\exp(-i\omega_{nm}t)$ causes a *destructive interference* in (4.2.2), which manifests itself only in *small fluctuations* of $\langle \mathscr{F} \rangle(t)$ (cf. Fig. 4.5).

The *physically relevant time interval* of a system which is not driven externally extends from t_0 into the region of destructive interference. One can naturally make no general, qualitative statements about the behavior of $\langle \mathscr{F} \rangle(t)$

[16] In most cases, the considerations which led to (4.2.9) even imply a $\ll$ sign in (4.2.19).

[17] This is all the more true since we have assumed that the system considered is completely isolated from its surroundings. The interaction with the environment, which is practically never negligible, gives rise to a time τ_{is} (see Sect. 3.3.4), after which our present considerations become invalid.

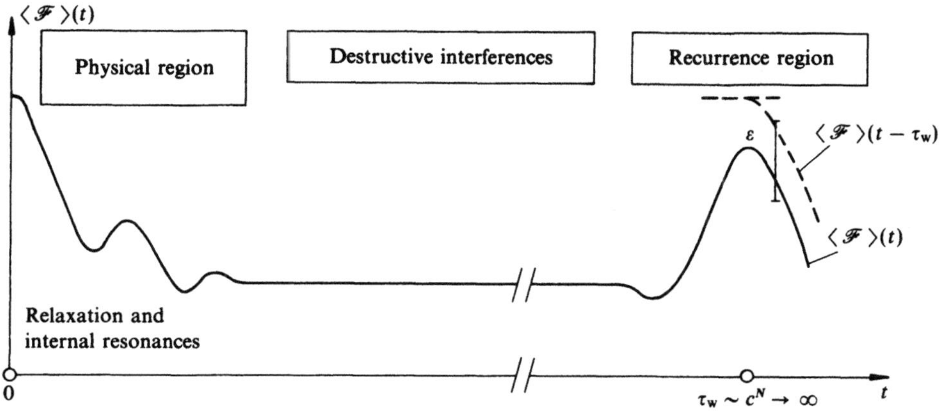

Fig. 4.5. Qualitative time behavior of the expectation value $\langle \mathcal{F} \rangle(t)$ of a macroscopic observable in an adiabatically isolated, non-driven system. Rapid, small fluctuations lie within the thickness of the curve. The translation time τ_w of the almost periodic function $\langle \mathcal{F} \rangle(t)$ becomes very long when the number of particles N is large, so that the recurrence interval no longer has a physical meaning. The observable relaxation and internal resonance phenomena are found in an interval which is "short" compared to τ_w, the physical interval

in the physical time interval. The detailed time dependence (4.2.2) depends essentially on the observable $\mathcal{F}$ under consideration and on the initial state $\rho(t_0)$. In this interval, the product of the time and the difference between neighboring energy levels obeys $(t-t_0)\delta E_0/\hbar \ll 1$, so that in a theoretical treatment, it is permissible to approximate the series (4.2.2) by an integral over energies.

Mathematically, this may be carried out in an elegant fashion by taking the limit of the number of particles $N \to \infty$ with fixed density N/V, since then the energy spectrum becomes, in the limit, continuous[18].

This limit

$$\lim_{\substack{N\to\infty \\ N/V = \text{const}}} \langle \mathcal{F} \rangle(t) \tag{4.2.23}$$

is called the *thermodynamic limit*. It means that a macroscopic system with finite volume is described in terms of an infinitely extended system having the same density, N/V. As the relation (4.2.23) indicates, the expectation value (trace) is to be evaluated for fixed N and V *before* taking the limit.

The behavior of the time dependence of (4.2.2) which we have discussed generally here can be made intuitively clear by a simple *mathematical example*,

[18] When the thermodynamic limit is taken, the region of destructive interference is extended to $t \to \infty$, so that there is no longer a recurrence.

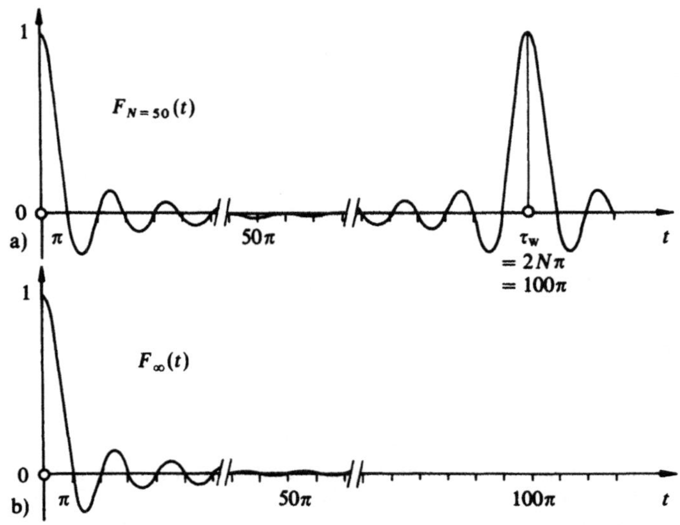

Fig. 4.6a, b. The mathematical example $F_N(t)$ (4.2.24). (**a**) $N = 50$, $\tau_w = 100\pi$; (**b**) $N \to \infty$ (4.2.25)

that of a finite cosine-series

$$F_N(t) = \frac{1}{N}\sum_{n=1}^{N} \cos\frac{nt}{N} = \frac{1}{2N}(\cos t - 1) + \frac{1}{2N}\cot\left(\frac{t}{2N}\right)\cdot \sin t \ . \tag{4.2.24}$$

Along with the recurrence time $\tau_w = 2\pi N$, for large N it illustrates the region of destructive interference mentioned above (Fig. 4.6a).

The limit $N \to \infty$,

$$F_\infty(t) = \int_0^1 \cos xt \, dx = \frac{\sin t}{t} \tag{4.2.25}$$

corresponds to the thermodynamic limit (Fig. 4.6b).

Exercise 4.2 A system containing N spins localized at the sites $\mathfrak{r}_i$ with $s = 1/2$ is described by the Hamiltonian

$$\mathscr{H} = \sum_{i \neq j}^{N} J(\mathfrak{r}_i - \mathfrak{r}_j)\, \sigma_{zi}\sigma_{zj} \ .$$

Calculate the expectation value

$$\left\langle \sum_{i=1}^{N} \sigma_{xi} \right\rangle(t) \ ,$$

when at time $t = 0$ the statistical operator is given approximately [for high temperatures, cf. (6.3.10)] by

$$\rho(0) = \frac{1}{\operatorname{Tr} 1} - \beta \frac{1}{\operatorname{Tr} 1}\left(\mathscr{H} - H_x \gamma \sum_{i=1}^{N} s_{xi} \right) .$$

What is found for a site-independent model $J = J_0$? How must one choose $J_0(N)$, in order that $\langle \mathscr{H} \rangle \sim N$? With this, determine the thermodynamic limit of

$$\frac{1}{N}\left\langle \sum_{i=1}^{N} s_{xi} \right\rangle (t) .$$

5. Information-Theoretical Construction of the Statistical Operator

5.1 The Uncertainty Measure of the Statistical Operator

5.1.1 Definition of the Uncertainty Measure $\eta[\rho]$

In the description of the state of a physical system Σ by means of the statistical operator ρ, the greater or lesser degree of *indeterminateness* of the state vector $|\Phi\rangle$ of Σ is taken into account (compare Sect. 2.3, and, concerning subsystems, Sect. 3.3). This indeterminateness may exist for microscopic systems, but is always particularly significant for macroscopic systems. It is important to have a reasonable measure of the degree to which a statistical operator ρ of a mixture is "distant" from a pure state $\mathscr{P}_{|\Phi\rangle}$. This measure $\eta[\rho]$, which we shall call the *uncertainty measure* of ρ,[1] summarizes the indeterminateness of $\mathscr{P}_{|\Phi\rangle}$ as expressed in ρ. The uncertainty measure has, as we shall show in the following Sect. 5.1.3, a close connection with information theory, so that one may also denote the indeterminateness as 'missing information' and thus the uncertainty measure $\eta[\rho]$ also as the *measure of the missing information about* $\mathscr{P}_{|\Phi\rangle}$. The information-theoretical justification of the expression for $\eta[\rho]$ is plausible, but not compelling. There have been other attempts to derive $\eta[\rho]$ on a formally axiomatic basis, which we shall not treat here, however.

We have therefore chosen an heuristic approach at this point. The uncertainty measure is supposed to be a positive *number* η, which is a *functional* $\eta[\rho]$ of the statistical operator ρ. It increases when the indeterminateness of the statistical operator ρ becomes "larger" relative to a pure state $\mathscr{P}_{|\Phi\rangle}$. One represents η as the trace of a function $f(\rho)$, since it is then invariant with respect to unitary transformations in $\mathfrak{U}$: $\eta = \mathrm{Tr}(f(\rho))$. Following information theory (cf. Sect. 5.1.3), one chooses this function to be $f(\rho) = -k\rho \ln \rho$. The *uncertainty measure* $\eta[\rho]$, i.e. *the measure of the indeterminateness of* $\mathscr{P}_{|\Phi\rangle}$ *due to* ρ, *is defined by*

$$\boxed{\eta[\rho] = -k\mathrm{Tr}(\rho \ln \rho)}\ .^{2,3} \qquad (5.1.1)$$

[1] The word "uncertainty measure" has here nothing to do with the concept of "uncertainty" in the sense of the mean square deviation of measured values of observables.

[2] Seen historically, the expression (5.1.1) is the result of numerous physical ideas (e.g. those of L.E. Boltzmann and M. Planck), which have found their general formulation in the information-theoretical interpretation of C. Shannon and E.T. Jaynes.

[3] E. Ruch and A. Mead have suggested an extension of the uncertainty concept. Their *mixing character* of ρ is based on a different concept of ordering.

The parameter k is a constant which initially may be chosen at will; thermodynamics shows it to be identical to the Boltzmann constant.

5.1.2 Properties of the Uncertainty Measure $\eta[\rho]$

In order to convince the reader that (5.1.1) represents a reasonable definition of the uncertainty measure, we will list some of its properties, showing that they are suitable for a measure of the indeterminateness of $\mathscr{P}_{|\Phi\rangle}$ and that they are analogous to the properties of Shannon's uncertainty measure in classical information theory (see Sect. 5.1.3)[4].

1) In the eigen-representation of ρ (2.3.6) and (2.3.7),

$$\rho|v_m\rangle = r_m|v_m\rangle, \;^{5} \quad \text{with} \quad 0 \leqq r_m \leqq 1, \quad \sum_m r_m = 1 \tag{5.1.2}$$

equation (5.1.1) takes the form

$$\eta[\rho] = -k\sum_m r_m \ln r_m \geqq 0 \; . \tag{5.1.3}$$

2) If the system is in a *pure* state, $\rho = \mathscr{P}_{|\Phi\rangle}$, one eigenvalue is equal to one (that which belongs to the eigenvector $|\Phi\rangle$), and the others are all zero. It then follows from $\lim_{r\to 0} r \ln r = 0$ that

$$\eta[\mathscr{P}_{|\Phi\rangle}] = 0 \; . \tag{5.1.4}$$

The uncertainty measure is zero, since there is no indeterminateness in $\mathscr{P}_{|\Phi\rangle}$.

3) On the other hand, if $\rho = 1/\mathrm{Tr}\, 1 = 1/n$ in a space $\mathfrak{U}_n$ of dimensionality n, then $\mathscr{P}_{|\Phi\rangle}$ is *completely indetermined*. In order to simplify the mathematics, we limit ourselves to the case of a unitary space $\mathfrak{U}_n$ of finite dimensionality n, so that $\mathrm{Tr}\, 1 = n$. The corresponding uncertainty measure is then, according to (5.1.1)

$$\eta\left[\frac{1}{\mathrm{Tr}\, 1}\right] = k \ln n \; . \tag{5.1.5}$$

4) *Every other statistical operator ρ in $\mathfrak{U}_n$ has a smaller uncertainty measure,*

$$\eta[\rho] \leqq \eta\left[\frac{1}{\mathrm{Tr}\, 1}\right] = k \ln n \; . \tag{5.1.6}$$

---

[4] The trace of the *square* of ρ would also be a measure of the deviation from a pure state, since $\mathrm{Tr}\, \rho^2 \leqq 1$, where the equals sign applies only to the case of a pure state (cf. Exercise 2.1). However, $\mathrm{Tr}\, \rho^2$ does not possess the other properties of an information-theoretical uncertainty measure as listed above.

[5] The index m denumerates the eigen*states* $|r_m\rangle$ of ρ and thus also expresses the degeneracy of the eigenvalues r_m.

Using (5.1.3) and

$$\sum_m r_m = 1 \ , \quad \sum_m 1 = n \ ,$$

we have

$$\begin{aligned}\eta[\rho] &= k\sum_m r_m \ln \frac{1}{r_m} = k\sum_m r_m \ln \frac{n}{r_m n} \\ &= k\sum_m r_m \ln n + k\sum_m r_m \ln \frac{1}{r_m n} \\ &\leqq k \ln n + k\sum_m r_m\left(\frac{1}{r_m n} - 1\right) = k \ln n .\end{aligned}$$

Here, we have made use of the inequality

$$\ln x \leqq x - 1 \tag{5.1.7}$$

whose validity is readily apparent on geometrical grounds.

5) If it is certain that the state vector is contained in the eigenspace $\mathfrak{u}_F$ of the eigenvalue F of an observable $\mathscr{F}$ – e.g. due to a measurement of $\mathscr{F}$ with the result F – then [cf. Sect. 5.3.4] ρ_F is proportional to the projector $\mathscr{P}_F$ [compare (2.3.8)] onto the eigenspace $\mathfrak{u}_F$,

$$\rho_F = \frac{\mathscr{P}_F}{\mathrm{Tr}\,\mathscr{P}_F} = \frac{\mathscr{P}_F}{d_F} \ , \tag{5.1.8}$$

where d_F is the dimensionality of the eigenspace $\mathfrak{u}_F$. It follows from (5.1.1) for η

$$\eta\left[\rho_F = \frac{\mathscr{P}_F}{d_F}\right] = k \ln d_F \ ^{6} \tag{5.1.9}$$

i.e. *the uncertainty measure is then proportional to the logarithm of the degree of degeneracy* d_F *of the eigenvalue* F.

6) If we consider two *independent* physical systems Σ^1 and Σ^2 with unitary spaces $\mathfrak{U}^1$ and $\mathfrak{U}^2$ and statistical operators ρ^1 and ρ^2, the combined system will be described in the product space $\mathfrak{U} = \mathfrak{U}^1 \times \mathfrak{U}^2$ [cf. Sect. 3.3]. The statistical operator for the combined system is then

$$\rho = \rho^1 \rho^2 \ . \tag{5.1.10}$$

[6] L.E. Boltzmann and M. Planck used the name "thermodynamic probability" for d_F.

Due to the fact that the trace in the product space is equal to the product of the traces in the subspaces, it follows for the uncertainty measure of $\rho^1\rho^2$ that

$$\eta[\rho^1\rho^2]=\mathrm{Tr}\{\rho^1\rho^2\ln\rho^1+\rho^1\rho^2\ln\rho^2\}$$
$$=\mathrm{Tr}_1(\rho^1\ln\rho^1)\cdot\mathrm{Tr}_2\rho^2+\mathrm{Tr}_1\rho^1\cdot\mathrm{Tr}_2(\rho^2\ln\rho^2)\ ,$$

thus

$$\eta[\rho^1\rho^2]=\eta[\rho^1]+\eta[\rho^2]\ . \tag{5.1.11}$$

That is, the uncertainty measure of a product of independent statistical operators is given by the sum of the uncertainty measures of the individual factors.

7) For the *time variation of the uncertainty measure* $\eta[\rho]$ *during a quantum-theoretical dynamical process in an isolated system*, we have

$$\boxed{\frac{d}{dt}\eta[\rho]=0}\ . \tag{5.1.12}$$

This may be seen most simply using the Heisenberg formulation (3.2.11), but it may also of course be derived in the Schrödinger formulation from the von Neumann equation. Equation (5.1.12) thus leads to the result

$$\eta(t)=\eta(t_0)\ , \tag{5.1.13}$$

i.e. the *dynamics*, without a measurement, *leave the uncertainty measure unchanged.* The indeterminateness of the state vector remains constant during a dynamic process. This may be immediately understood by referring to (2.3.10), because the weights p^α of the pure states $\mathscr{P}_{|\Phi^\alpha\rangle}$ are constant in time.

5.1.3 The Relationship Between Information Theory and the Uncertainty Measure $\eta[\rho]$

In the preceding sections, we have presented plausibility arguments to support expression (5.1.1) as giving the uncertainty measure $\eta[\rho]$ for the indeterminateness of a state vector $|\Phi\rangle$ in a mixture. We offer the following considerations for the reader who wishes to better understand these statements in the light of information theory[7].

If a measurement F has

$$\left.\begin{array}{ll}\text{the possible results} & F_1,\ldots,F_\alpha,\ldots,F_n\\ \text{with the probabilities} & w_1,\ldots,w_\alpha,\ldots,w_n\\ & \left(0\leqq w_\alpha\leqq 1,\ \sum\limits_{\alpha=1}^{n} w_\alpha=1\right)\ ,\end{array}\right\} \tag{5.1.14}$$

[7] The following sections of this book may be read without studying this section.

then, in *information theory*, one ascribes to the *totality* of these probabilities the *information-theoretical uncertainty measure* (Shannon)[8]. (The applications of information theory go far beyond those in physics, although it grew out of statistical physics [cf. Footnote 2 (p. 44)])

$$\boxed{\eta_F[w_1, \ldots, w_n] = -k \sum_{\alpha=1}^{n} w_\alpha \ln w_\alpha} \, . \tag{5.1.15}$$

It is a *global measure for the indeterminateness of the results of the measurement F*, which one can also interpret as a *measure of the missing information.*[9] Although (5.1.3) and (5.1.15) appear to be formally similar, they in fact express *different* statements: Equation (5.1.15) gives a measure for the indeterminateness of the result of an *experiment*, while (5.1.3) is a measure for the indeterminateness of a *state vector* $|\Phi\rangle$ in a mixture. The w_α cannot be identified with the p^α in (2.3.3), since in the latter, a mixture of arbitrary *non-orthogonal* pure states $\mathscr{P}_{|\Phi^\alpha\rangle}$ with weights p^α is implied. The following considerations are meant to show that it is *nevertheless possible*, on the basis of Shannon's formula (5.1.15), to obtain the indeterminateness $\eta[\rho]$ of a state vector in a mixture (5.1.3).

Shannon's uncertainty measure η_F may be immediately applied to the indeterminacy of the result of the *measurement of a quantum-theoretical observable* $\mathscr{F}$. If the system is described by a statistical operator ρ before the measurement, then the probabilities w_F for observing the values F are given by (2.3.9)

$$w_F = \mathrm{Tr}(\rho \mathscr{P}_F) \, , \tag{5.1.16}$$

where $\mathscr{P}_F$ is the projector onto the eigenspace $\mathfrak{u}_F$ of the measured value F. Shannon's uncertainty measure (5.1.15) for measuring the observable $\mathscr{F}$ is then

$$\eta_{\mathscr{F}} = -k \sum_F w_F \ln w_F = -k \sum_F \mathrm{Tr}(\rho \mathscr{P}_F) \ln \mathrm{Tr}(\rho \mathscr{P}_F) \, . \tag{5.1.17}$$

This quantity depends both on the state ρ of the quantum-theoretical system and on the observable to be measured, $\mathscr{F}$.

We now consider, given a particular state ρ of the system, a *complete measurement* in the quantum-theoretical sense, which corresponds to the measurement of a *nondegenerate observable* $\mathscr{M}$.[10] We require of $\mathscr{M}$ in particular that

---

[8] This quantity is also called the "information-theoretical entropy" for the result of a measurement F. However, to avoid confusion, we shall not use the concept 'entropy' in this most general sense.

[9] This is not the place to expand upon this information-theoretical interpretation. For such details, and for a justification of Shannon's expression (5.1.15), we refer the reader to the literature on information theory.

[10] The fact that this can practically not be carried out in systems with many degrees of freedom is another question. Here, we are only concerned with the possibility in principle.

it commutes with ρ (at some instant),

$$\boxed{[\mathcal{M},\rho]=0}\ .^{11} \tag{5.1.18}$$

Thus, $\mathcal{M}$ and ρ have the same eigenvectors $|v_m\rangle$, so that the quantum-theoretical probability for measuring a value M of the observable $\mathcal{M}$,

$$w_M=\langle v_m|\rho v_m\rangle=r_m\ , \tag{5.1.19}$$

is an eigenvalue r_m of the statistical operator. The uncertainty measure for the result of the measurement of $\mathcal{M}$ is therefore

$$\eta_{\mathcal{M}}=-k\sum_M w_M\ln w_M=-\sum_m r_m \ln r_m\ . \tag{5.1.20}$$

It may be shown (cf. Exercise 5.2) that Shannon's uncertainty measure $\eta_{\mathcal{F}}$ for the measurement of an *arbitrary nondegenerate observable* $\mathcal{F}$ can only be larger than or equal to the uncertainty measure for the measurement of a nondegenerate observable $\mathcal{M}$ *which commutes with* ρ,

$$\boxed{\eta_{\mathcal{F}}\geqq\eta_{\mathcal{M}}}\ . \tag{5.1.21}$$

This so-called *Klein's inequality* is a typical result of the statistical properties of *quantum theory* and has nothing to do with the indeterminateness of the state vector[12]. In contrast to $\eta_{\mathcal{F}}$, $\eta_{\mathcal{M}}$ may be written as the trace of *one* operator according to (5.1.2) and (5.1.20), namely

$$\eta_{\mathcal{M}}=-k\mathrm{Tr}(\rho\ln\rho)\ . \tag{5.1.22}$$

For a given ρ, relations (5.1.21) and (5.1.22) tell us that the result of a measurement of $\mathcal{M}$ is less indeterminate, and that the corresponding uncertainty measure $\eta_{\mathcal{M}}$ *depends only on the statistical operator* ρ. A measurement of $\mathcal{M}$ thus yields the best quantum-theoretical information about the state of the system. It is therefore reasonable to use Shannon's uncertainty measure $\eta_{\mathcal{M}}$ for the measurement of $\mathcal{M}$ as an uncertainty measure for the statistical operator ρ of the system,

$$\eta[\rho]\equiv\eta_{\mathcal{M}}=-k\mathrm{Tr}(\rho\ln\rho)\ , \tag{5.1.23}$$

---

[11] We do not require that ρ possess a nondegenerate spectrum.

[12] For a pure state $\rho=\mathcal{P}_{|\Phi\rangle}$, Klein's inequality can be specialized to

$$\eta_{\mathcal{F}}\geqq\eta_{\mathcal{M}}=0\ , \tag{5.1.21a}$$

in which the *quantum-theoretical* probability results manifest themselves in a particularly significant form.

as is required by Jaynes's formula (5.1.1). *We have thus related the uncertainty measure $\eta[\rho]$ for the indeterminateness of the state vector to Shannon's information-theoretical uncertainty measure $\eta_{\mathcal{M}}$ for the measurement of an observable $\mathcal{M}$.* In other words, we have shown how the spectral representation (5.1.3) is related to the information-theoretical uncertainty measure (5.1.15) for the measurement of an observable.

Exercise 5.1 For an arbitrary statistical operator ρ in a space $\mathfrak{U}_3$, discuss the uncertainty measure $\eta[\rho]$ as a function of the eigenvalues r_1, r_2, and r_3.

Exercise 5.2 Prove Klein's inequality (5.1.21).

Exercise 5.3 Assume that it is known that at a particular place on June 1, the probability of rain is 0.4 and that of no precipitation is 0.6. At the same place on January 1, the probability of rain is 0.65, that of snow is 0.15, and that of no precipitation is thus 0.2.

On which of the two days is the weather prediction more indeterminate, if the prediction is taken to include the information [Jaglom and Jaglom]:

a) whether it will rain, snow, or be without precipitation on the day in question?
b) whether there will be any precipitation at all on the day in question?

5.2 The Generalized Canonical Statistical Operator $\mathcal{R}$

In the preceding Sects. 5.1.1 and 5.1.2, we started with a given statistical operator ρ and determined its uncertainty measure $\eta[\rho]$. We now wish to *reverse* this procedure and use the physical properties of a system which are present at a particular time, with the aid of the uncertainty measure, to *construct* a statistical operator which describes these properties.

5.2.1 Observation Levels

In order to characterize the state of a system, we require the expectation values $\langle \mathcal{G}_\nu \rangle$ of a set of observables $\mathcal{G}_\nu$ ($\nu = 1, \ldots, n$), which are supposed to be known from a measurement or a calculation. Using such a set $\{\mathcal{G}\}$, we define an "*observation level*". It also includes all the linear combinations of these observables (compare Sect. 5.2.3). We do not assume that the operators of $\{\mathcal{G}\}$ are mutually commuting or even that some of them commute with each other. An observation level consists in general of only a small portion of all the possible observables of the system, since usually only a few properties of the system are physically accessible and also interesting. This is especially true of macroscopic systems. We refer to Sect. 4.1, in which it was shown that the expectation values of macroscopic observables which are calculated in quantum statistics are practically identical with their measured values.

The question as to whether a particular observation level, i.e. the chosen set of observables $\mathscr{G}_\nu$, is sufficient to provide a physically satisfying description of a given system or whether the observation level must be extended to accomplish this will be discussed in Sect. 5.2.5. We shall introduce some examples of observation levels in Sect. 5.3.

5.2.2 Determination of the Statistical Operator by Maximization of the Uncertainty Measure. Entropy with Respect to an Observation Level

If we consider a particular observation level $\{\mathscr{G}\}=\mathscr{G}_\nu$ $(\nu=1, \ldots, n)$ of a system Σ, the statistical operator ρ of the system is naturally not uniquely determinable from the few expectation values $\langle\mathscr{G}_\nu\rangle=G_\nu$. Instead, there is a *multiplicity of possible statistical operators*, which all yield these numbers G_ν as expectation values,

$$\mathrm{Tr}\,\rho=1 \ , \tag{5.2.1}$$

$$\mathrm{Tr}(\rho\mathscr{G}_\nu)=G_\nu \ .\text{[13]} \tag{5.2.2}$$

Each of these (possible) statistical operators ρ can, however, have a different uncertainty measure $\eta[\rho]$. We wish to use *only* the values G_ν of the chosen observation level for determining the statistical operator of the system Σ, so we must try to select a particular statistical operator $\rho=\mathscr{R}$ from this set of possible operators (Fig. 5.1) in an unbiased manner. The operator $\mathscr{R}$ must be that statistical operator which has the largest uncertainty measure,

$$\eta[\mathscr{R}]=\text{maximum}, \tag{5.2.3}$$

with the constraints

$$\mathrm{Tr}\,\mathscr{R}=1 \ , \tag{5.2.4}$$

$$\mathrm{Tr}(\mathscr{R}\mathscr{G}_\nu)=G_\nu \ .\text{[14]} \tag{5.2.5}$$

For every other statistical operator ρ meeting the condition (5.2.2), $\eta[\rho]<\eta[\mathscr{R}]$ holds. By selecting such operators, we would claim to possess more information about the state vector, *without sufficient justification*, than is actually contained in the values G_ν.

[13] If the given numbers G_ν are the mean values of actual physical measurements, they cannot, *a priori*, violate the quantum-theoretical uncertainty relations. They are expectation values $\mathrm{Tr}(\rho\mathscr{G}_\nu)$ of the observables $\mathscr{G}_\nu$, taken with a statistical operator ρ.

[14] Let the operators $\mathscr{G}_\nu$ be linearly *in*dependent. In case they already include the unit operator 1 as a linear combination, the constraint (5.2.4) is unnecessary [cf. for example Sect. 5.3.4].

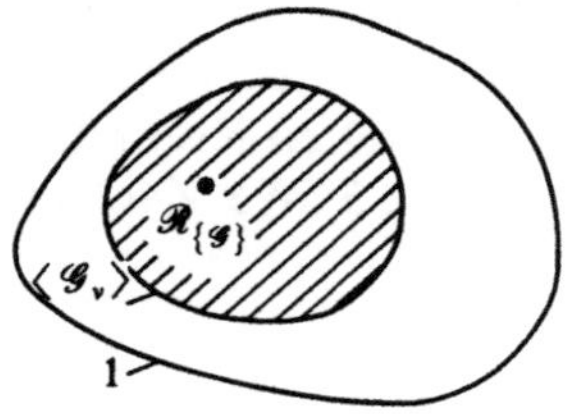

Fig. 5.1. The construction of the generalized canonical statistical operator $\mathscr{R}_{\{\mathscr{G}\}}$. The set of all ρ ($\mathrm{Tr}\,\rho = 1$). The shaded area represents the set of all ρ for a particular $\langle\mathscr{G}_\nu\rangle = \mathrm{Tr}(\rho\mathscr{G}_\nu)$; η is a maximum for $\mathscr{R}_{\{\mathscr{G}\}}$

This is *Jaynes's principle of the maximum uncertainty measure* within an observation level of observables $\{\mathscr{G}\}$. It leads to the unbiased determination of a statistical operator $\mathscr{R}$, taking into account the existing expectation values $G_\nu = \langle\mathscr{G}_\nu\rangle$ of the observables $\mathscr{G}_\nu$ in the chosen observation level $\{\mathscr{G}\}$.

The mathematical procedure required to carry out this program can make use of the method of Lagrange multipliers, as employed in the differential calculus of c-number functions; the constraints are included in the determination of extrema with the aid of Lagrange multipliers λ_ν. On applying this method to operators, however, one must take account of the fact that variations $\delta\rho$ which do not commute with ρ may be encountered. Furthermore, when the $\mathscr{R}$ which represents an extremum has been identified in this way, it is still necessary to demonstrate that the extremum is, in fact, a maximum.

To abbreviate this procedure, we give the expression for the statistical operator $\mathscr{R}$,[15] and show that, among all possible statistical operators which fulfill the constraints (5.2.1) and (5.2.2), it in fact has the *maximum* uncertainty measure η.

This statistical operator, as a function of the observables $\mathscr{G}_\nu$ of the *chosen observation level*, is given by

$$\boxed{\mathscr{R}_{\{\mathscr{G}\}} = \frac{\exp\left(-\sum_\nu \lambda_\nu \mathscr{G}_\nu\right)}{\mathrm{Tr}\exp\left(-\sum_\nu \lambda_\nu \mathscr{G}_\nu\right)}} \,. \tag{5.2.6}$$

It is called the *generalized canonical statistical operator*. The trace which occurs in the denominator,

[15] Here we assume that an operator $\mathscr{R}_{\{\mathscr{G}\}}$ as defined by (5.2.6) *exists* for the chosen set of expectation values G_ν.

$$\boxed{Z_{\{\mathscr{G}\}}(\lambda_1, \ldots, \lambda_n) = \operatorname{Tr}\exp\left(-\sum_\nu \lambda_\nu \mathscr{G}_\nu\right)}\ ^{16} \tag{5.2.7}$$

is referred to as the *generalized partition function.* It guarantees that $\operatorname{Tr}\mathscr{R} = 1$ is obeyed.

The n equations (5.2.5) can be employed to express the n parameters λ_n in terms of the given expectation values $G_\nu = \langle \mathscr{G}_\nu \rangle$ of the observation level $\{\mathscr{G}\}$[17],

$$\lambda_\nu = \lambda_\nu(\langle \mathscr{G}_1 \rangle, \ldots, \langle \mathscr{G}_n \rangle) . \tag{5.2.9}$$

We now *demonstrate* that the uncertainty measure derived from the generalized canonical statistical operator $\mathscr{R}_{\{\mathscr{G}\}}$,

$$\eta[\mathscr{R}_{\{\mathscr{G}\}}] = -k \operatorname{Tr}(\mathscr{R}_{\{\mathscr{G}\}} \ln \mathscr{R}_{\{\mathscr{G}\}}) \tag{5.2.10}$$

is in fact *larger than the uncertainty measure $\eta[\rho]$ of any other statistical operator ρ which fulfills the constraints (5.2.1) and (5.2.2) for the observation level $\{\mathscr{G}\}$.*

The *proof* is carried out as follows: from (5.2.6) we have

$$\ln \mathscr{R} = -\sum_\nu \lambda_\nu \mathscr{G}_\nu - \ln Z \cdot \mathit{1} . \tag{5.2.11}$$

From this we obtain

$$\operatorname{Tr}(\mathscr{R} \ln \mathscr{R}) = -\sum_\nu \lambda_\nu \operatorname{Tr}(\mathscr{R}\mathscr{G}_\nu) - \ln Z \quad \text{and}$$

$$\operatorname{Tr}(\rho \ln \mathscr{R}) = -\sum_\nu \lambda_\nu \operatorname{Tr}(\rho \mathscr{G}_\nu) - \ln Z ,$$

so that, with the given expectation values G_ν and (5.2.2 and 5), we obtain the equality

$$\operatorname{Tr}(\mathscr{R} \ln \mathscr{R}) = \operatorname{Tr}(\rho \ln \mathscr{R}) . \tag{5.2.12}$$

---

[16] The expectation values $\langle \mathscr{G}_\nu \rangle$, i.e. (5.2.5), may be represented as a derivative of Z,

$$\langle \mathscr{G}_\nu \rangle = -\frac{1}{Z}\frac{\partial Z}{\partial \lambda_\nu} = -\frac{\partial}{\partial \lambda_\nu} \ln Z . \tag{5.2.8}$$

The justification of this statement for noncommuting operators $\mathscr{G}_\nu$ makes use of the identity (3.4.10) and the invariance of the trace towards permutation of its arguments.

[17] We shall not consider here the problem of whether or not the set of equations has a unique solution.

The eigenvalue equation of an arbitrary ρ is given by

$$\rho|v_m\rangle = r_m|v_m\rangle \quad ^{18} \quad \left(0 \leqq r_m \leqq 1, \sum_m r_m = 1\right) \tag{5.2.13}$$

and that of $\mathcal{R}$ by

$$\mathcal{R}|w_n\rangle = R_n|w_n\rangle \quad ^{18} \quad \left(0 \leqq R_n \leqq 1, \sum_n R_n = 1\right) . \tag{5.2.14}$$

Thus, in the basis $|v_m\rangle$, we obtain for the difference

$$\mathrm{Tr}(\rho \ln \rho) - \mathrm{Tr}(\rho \ln \mathcal{R}) = \sum_m r_m(\ln r_m - \langle v_m | \ln \mathcal{R} | v_m\rangle) . \tag{5.2.15}$$

We now introduce the completeness relation $\sum_n |\langle v_m|w_n\rangle|^2 = 1$ into the first term, and the unity operator $1 = \sum_n |w_n\rangle\langle w_n|$ after $\ln \mathcal{R}$, thus obtaining from (5.2.12) the *result*

$$\begin{aligned} \mathrm{Tr}(\rho \ln \rho) - \mathrm{Tr}(\mathcal{R} \ln \mathcal{R}) &= \sum_{nm} |\langle v_m|w_n\rangle|^2 \{r_m(\ln r_m - \ln R_n)\} \\ &\geqq \sum_{nm} |\langle v_m|w_n\rangle|^2 (r_m - R_n) \\ &= \sum_m r_m - \sum_n R_n = 0 . \end{aligned} \tag{5.2.16}$$

Here, we have made use of the inequality $\ln x \geqq 1 - 1/x$ [cf. (5.1.7)], i.e. $r_m(\ln r_m - \ln R_n) \geqq r_m - R_n$; furthermore, we have taken into account the relations

$$\sum_n |\langle v_m|w_n\rangle|^2 = \sum_m |\langle v_m|w_n\rangle|^2 = 1$$

(completeness relation for $|w_n\rangle$ and for $|v_m\rangle$) and

$$\sum_n r_m = \mathrm{Tr}\,\rho = 1 , \qquad \sum_m R_n = \mathrm{Tr}\,\mathcal{R} = 1 .$$

The *equals sign* in (5.2.16) holds only when, for all $\langle v_m|w_n\rangle \neq 0$, the eigenvalues are equal, i.e. $r_m = R_n$. Then, from the spectral representation, we have

$$\begin{aligned} \rho = \sum_m r_m|v_m\rangle\langle v_m| &= \sum_{nm} r_m|w_n\rangle\langle w_n|v_m\rangle\langle v_m| \\ &= \sum_{nm} R_n|w_n\rangle\langle w_n|v_m\rangle\langle v_m| \\ &= \sum_n R_n|w_n\rangle\langle w_n| = \mathcal{R} , \end{aligned}$$

[18] The indices m and n denumerate the states $|v_m\rangle$ and $|w_n\rangle$ and thus include the degeneracies of the corresponding eigenvalues r_m and R_n. (The index n naturally has nothing to do with the number of observables in the observation level $\{\mathcal{G}\}$.)

that is, $\eta[\rho]$ assumes its maximum value $\eta[\mathcal{R}]$ with the constraint (5.2.2) *only* for $\rho = \mathcal{R}$.

Equation (5.2.16) proves the *fundamental inequality*,

$$\boxed{\eta[\mathcal{R}_{\{\mathcal{G}\}}] = -k\mathrm{Tr}(\mathcal{R}_{\{\mathcal{G}\}} \ln \mathcal{R}_{\{\mathcal{G}\}}) \geqq \eta[\rho_{\{\mathcal{G}\}}] = -k\,\mathrm{Tr}(\rho_{\{\mathcal{G}\}} \ln \rho_{\{\mathcal{G}\}})} \tag{5.2.17}$$

for all possible ρ which fulfill the constraint (5.2.2) in the observation level $\{\mathcal{G}\}$. The *generalized canonical statistical operator $\mathcal{R}_{\{\mathcal{G}\}}$ yields the maximum uncertainty measure for a given observation level $\{\mathcal{G}\}$*. Therefore, $\mathcal{R}$ contains *only* the information from the chosen observation level, and no more.

The maximum uncertainty measure for a particular observation level $\{\mathcal{G}\}$ is called *the entropy with respect to the observation level $\{\mathcal{G}\}$*

$$\boxed{S_{\{\mathcal{G}\}} \equiv \eta_{\max} = -k\,\mathrm{Tr}(\mathcal{R}_{\{\mathcal{G}\}} \ln \mathcal{R}_{\{\mathcal{G}\}})}\;. \tag{5.2.18}$$

There thus exist *not one entropy*, but rather a *multiplicity of entropies, depending on the observation level chosen.* The index $\{\mathcal{G}\}$ is used to remind us of this. Thermal equilibrium of the system is *not* a condition for this definition of the entropy!

By substituting the expression for $\mathcal{R}_{\{\mathcal{G}\}}$ (5.2.6) in (5.2.18), we obtain an expression for the entropy $S_{\{\mathcal{G}\}}$:

$$\boxed{S_{\{\mathcal{G}\}} = k\sum_{\nu} \lambda_\nu \langle \mathcal{G}_\nu \rangle + k \ln Z_{\{\mathcal{G}\}}}\;. \tag{5.2.19}$$

Employing (5.2.5) or its solution (5.2.9), the entropy can be expressed as a function of the Lagrange multipliers λ_ν or the expectation values $\langle \mathcal{G}_\nu \rangle$ (or as a mixed expression),

$$S_{\{\mathcal{G}\}} = S(\lambda_1, \ldots, \lambda_n) = S(\langle \mathcal{G}_1 \rangle, \ldots, \langle \mathcal{G}_n \rangle)\;.[19] \tag{5.2.20}$$

The *change in the entropy $S_{\{\mathcal{G}\}}$ on changing the values $\langle \mathcal{G}_\nu \rangle$* is obtained by differentiating (5.2.19), keeping in mind the dependence (5.2.9) on the Lagrange multipliers and using (5.2.8); it is

$$dS_{\{\mathcal{G}\}} = k\sum_{\nu} \lambda_\nu d\langle \mathcal{G}_\nu \rangle\;. \tag{5.2.21}$$

[19] The right-hand sides of (5.2.20) are naturally different functions, which however give the same value for the entropy.

If one considers a possible dependence of the observables $\mathcal{G}_\nu = \mathcal{G}_\nu(h_1, \ldots, h_a)$ on external parameters h_α (= fields, volumes, etc.), then (5.2.19) leads to the important general relation

$$\boxed{\begin{aligned} dS_{\{\mathcal{G}\}} &= k \sum_\nu \lambda_\nu \left\{ d\,\mathrm{Tr}(\mathcal{R}_{\{\mathcal{G}\}} \mathcal{G}_\nu) - \sum_\alpha \mathrm{Tr}\left(\mathcal{R}_{\{\mathcal{G}\}} \frac{\partial \mathcal{G}_\nu}{\partial h_\alpha} \right) dh_\alpha \right\} \\ &= k \sum_\nu \lambda_\nu \, \mathrm{Tr}\,(d\mathcal{R}_{\{\mathcal{G}\}} \mathcal{G}_\nu) \; ; \end{aligned}} \tag{5.2.22}$$

[20]

we shall make repeated use of this equation.

5.2.3 Linear Transformations Within an Observation Level

When the expectation values $\langle \mathcal{G}_\nu \rangle$ are given, so are all linear combinations of the corresponding observables. Thus, all the linear combinations of the $\mathcal{G}_\nu$,

$$\mathcal{G}'_\nu = \sum_\mu c_{\nu\mu} \mathcal{G}_\mu \; , \tag{5.2.23}$$

give rise to the same observation level. $\mathcal{R}$ and S must therefore be invariant towards such transformations,

$$\begin{aligned} \mathcal{R}'_{\{\mathcal{G}'\}} &= \frac{\exp\left(-\sum_\nu \lambda'_\nu \mathcal{G}'_\nu\right)}{\mathrm{Tr}\exp\left(-\sum_\nu \lambda'_\nu \mathcal{G}'_\nu\right)} = \mathcal{R}_{\{\mathcal{G}\}} \\ S'[\mathcal{R}'] &= S[\mathcal{R}] \; . \end{aligned} \tag{5.2.24}$$

As a result, the Lagrange multipliers transform contravariantly to (5.2.23)

$$\begin{aligned} \lambda'_\nu &= \sum_\mu C_{\nu\mu} \lambda_\mu \; , \\ \sum_\nu C_{\nu\mu} c_{\nu\rho} &= \delta_{\mu\rho} \; . \end{aligned} \tag{5.2.25}$$

The transformation in (5.2.23) can be used to orthogonalize the observables of an observation level with respect to a scalar product in Liouville space, which can often be of practical importance (cf. Chap. 8).

[20] Note that the observables $\partial \mathcal{G}_\nu / \partial h_\alpha$ do not, in general, belong to the observation level $\{\mathcal{G}\}$.

5.2.4 Extension of the Observation Level

If an observation level $\mathcal{G}_1, \ldots, \mathcal{G}_n$ is extended by including further observables $\mathcal{M}_1, \ldots, \mathcal{M}_s$, the addition of the expectation values $\langle \mathcal{M}_1 \rangle, \ldots, \langle \mathcal{M}_s \rangle$ can only increase the available information about the state of the system, so that the maximum uncertainty measure can only decrease or remain unchanged. The entropy $S_{\{\mathcal{G}, \mathcal{M}\}}$ of the extended observation level $\{\mathcal{G}, \mathcal{M}\}$ is therefore smaller or equal to the entropy $S_{\{\mathcal{G}\}}$ of the original observation level $\{\mathcal{G}\}$,

$$\boxed{S_{\{\mathcal{G}, \mathcal{M}\}} \leqq S_{\{\mathcal{G}\}}} \;. \quad (5.2.26)$$

[21]

This result also follows immediately from the inequality (5.2.17), since the generalized canonical statistical operator of the observation level $\{\mathcal{G}, \mathcal{M}\}$,

$$\mathcal{R}_{\{\mathcal{G}, \mathcal{M}\}} = \frac{\exp\left(-\sum_\nu \lambda_\nu \mathcal{G}_\nu - \sum_\sigma \mu_\sigma \mathcal{M}_\sigma\right)}{\operatorname{Tr}\exp\left(-\sum_\nu \lambda_\nu \mathcal{G}_\nu - \sum_\sigma \mu_\sigma \mathcal{M}_\sigma\right)} \quad (5.2.27)$$

belongs in the set of the $\rho_{\{\mathcal{G}\}}$ with respect to which $\eta[\rho]$ was maximized (Fig. 5.2). The corresponding Lagrange multipliers λ_ν and μ_σ are calculated with (5.2.9) from the given expectation values of the observation level $\{\mathcal{G}, \mathcal{M}\}$ and are, in general, functions of *all* the expectation values,

$$\begin{aligned} \lambda_\nu &= \lambda_\nu(\langle \mathcal{G}_1 \rangle, \ldots, \langle \mathcal{G}_n \rangle, \langle \mathcal{M}_1 \rangle, \ldots, \langle \mathcal{M}_s \rangle) \quad (\nu = 1, \ldots, n) \;, \\ \mu_\sigma &= \mu_\sigma(\langle \mathcal{G}_1 \rangle, \ldots, \langle \mathcal{G}_n \rangle, \langle \mathcal{M}_1 \rangle, \ldots, \langle \mathcal{M}_s \rangle) \quad (\sigma = 1, \ldots, s) \;. \end{aligned} \quad (5.2.28)$$

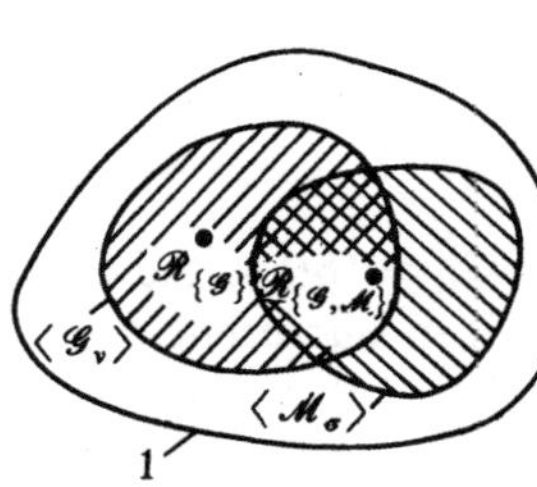

Fig. 5.2. Extension of the observation level $\{\mathcal{G}\}$ by $\{\mathcal{M}\}$. The set of all ρ with $\operatorname{Tr}\rho = 1$ is represented by the outer boundary (1). The left-hand shaded region /// is the set of all ρ with given $\langle \mathcal{G}_\nu \rangle = \operatorname{Tr}(\rho \mathcal{G}_\nu)$. The right-hand shaded region \\\ is the set of all ρ with given $\langle \mathcal{M}_\sigma \rangle = \operatorname{Tr}(\rho \mathcal{M}_\sigma)$. η is a maximum for $\mathcal{R}_{\{\mathcal{G}, \mathcal{M}\}}$

[21] The difference $S_{\{\mathcal{G}\}} - S_{\{\mathcal{G}, \mathcal{M}\}}$ is the information gain resulting from the extension of the observation level.

The equals sign in (5.2.26) holds only for $\mu_\sigma = 0$, owing to the considerations discussed on (p. 54).

If one considers the possible expectation values $\langle \mathcal{M}_\sigma \rangle$ for a set of fixed expectation values $\langle \mathcal{G}_\nu \rangle$, it is found that there is a special set of values $\langle \mathcal{M}_\sigma \rangle_{\{\mathcal{G}\}}$ given by

$$\langle \mathcal{M}_\sigma \rangle_{\{\mathcal{G}\}} = \mathrm{Tr}(\mathcal{R}_{\{\mathcal{G}\}} \mathcal{M}_\sigma) \ . \tag{5.2.29}$$

For these, all the μ_σ are zero, and we have

$$S_{\{\mathcal{G}, \mathcal{M}\}}(\langle \mathcal{G}_1 \rangle, \dots, \langle \mathcal{G}_n \rangle, \langle \mathcal{M}_1 \rangle_{\{\mathcal{G}\}}, \dots, \langle \mathcal{M}_s \rangle_{\{\mathcal{G}\}})$$
$$= S_{\{\mathcal{G}\}}(\langle \mathcal{G}_1 \rangle, \dots, \langle \mathcal{G}_n \rangle) \ . \tag{5.2.30}$$

We will discuss the significance of this result in the following section (see also Fig. 5.4).

5.2.5 A Sufficient Observation Level. Representativity of a Generalized Canonical Statistical Operator

A generalized canonical statistical operator $\mathcal{R}_{\{\mathcal{G}\}}$ belonging to a particular observation level $\{\mathcal{G}\}$ yields *by definition* the correct expectation values belonging to $\{\mathcal{G}\}$,

$$\langle \mathcal{G}_\nu \rangle = \mathrm{Tr}(\mathcal{R}_{\{\mathcal{G}\}} \mathcal{G}_\nu) \ . \tag{5.2.31}$$

The decisive question with regard to its applicability for the description of a physical system Σ (at a particular time) is whether or not it allows the *prediction* of the expectation values of *other* observables $\mathcal{F}$ with sufficient accuracy[22], so that the calculated quantities $\mathrm{Tr}(\mathcal{R}_{\{\mathcal{G}\}} \mathcal{F})$ are in near agreement with the physically relevant expectation values $\langle \mathcal{F} \rangle$[23]:

$$\langle \mathcal{F} \rangle = \mathrm{Tr}(\rho \mathcal{F}) \approx \mathrm{Tr}(\mathcal{R}_{\{\mathcal{G}\}} \mathcal{F}) \ . \tag{5.2.32}$$

Depending on the degree to which this is the case, we call $\mathcal{R}_{\{\mathcal{G}\}}$ more or less *representative with respect to* $\mathcal{F}$. In *macroscopic* systems, both the observables of the observation level, $\mathcal{G}_\nu$, and the observables $\mathcal{F}$ which are to be calculated should be *macroscopic* observables, so that the $\approx$ sign in (5.2.32) is valid within the region of the deviation $\Delta\mathcal{F}$ [cf. (4.1.12)]. Then the "macro state" of the system is correctly described by the (few) observables of the observation level $\{\mathcal{G}\}$. An observation level $\{\mathcal{G}\}$ is thus *sufficient* in view of other observables $\mathcal{F}$ of

[22] An example for such observables $\mathcal{F}$ are the derivatives $\partial \mathcal{G}_\nu / \partial h_\alpha$ which occurred in (5.2.22), i.e. the derivatives of the observables $\mathcal{G}_\nu$ with respect to external parameters h_α.

[23] See also Exercise 18.2.

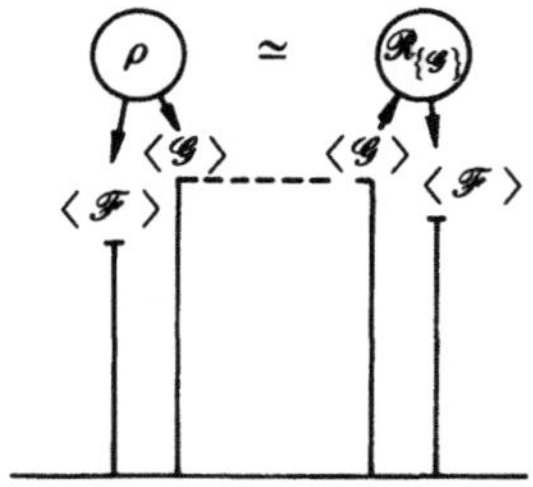

Fig. 5.3. ρ is equivalent to $\mathscr{R}_{\{\mathscr{G}\}}$ with respect to $\mathscr{F}$ when $\mathrm{Tr}(\rho\mathscr{F}) \simeq \mathrm{Tr}(\mathscr{R}_{\{\mathscr{G}\}}\mathscr{F})$

the system, when (5.2.32) holds to a good approximation. The choice of the observation level is then physically reasonable[24].

If, through some other considerations, one knows the statistical operator ρ of the system at the time in question (e.g. via dynamic calculations using the von Neumann equation), then when (5.2.31) and (5.2.32) hold, ρ is *equivalent* (Fig. 5.3) to the generalized canonical statistical operator $\mathscr{R}_{\{\mathscr{G}\}}$.

$$\rho \simeq \mathscr{R}_{\{\mathscr{G}\}} \quad \text{with respect to } \mathscr{F}\ .^{25} \tag{5.2.33}$$

This does *not* mean that ρ must be equal to $\mathscr{R}_{\{\mathscr{G}\}}$! However, as long as one is only interested in the observables $\mathscr{F}$, ρ can be replaced by $\mathscr{R}_{\{\mathscr{G}\}}$:

$$\boxed{\rho \Rightarrow \mathscr{R}_{\{\mathscr{G}\}}}\ .^{26} \tag{5.2.34}$$

The transition (5.2.34), in general, leads to a *loss of information*

$$\boxed{\eta[\rho] \leqq \eta[\mathscr{R}_{\{\mathscr{G}\}}] \equiv S_{\{\mathscr{G}\}}}\ , \tag{5.2.35}$$

which we have already recognized in constructing $\mathscr{R}_{\{\mathscr{G}\}}$ [cf. (5.2.17)].

The preceding discussion may also be interpreted in the reverse sense. For a given observation level $\{\mathscr{G}\}$, i.e. a given generalized canonical statistical operator $\mathscr{R}_{\{\mathscr{G}\}}$, the quantities Tr $(\mathscr{R}_{\{\mathscr{G}\}}\mathscr{F})$ may be calculated. The latter either *agree or do not agree* with the physically relevant expectation values $\langle\mathscr{F}\rangle = \mathrm{Tr}(\rho\mathscr{F})$, whereby for macroscopic observables, 'agreement' is meant in the sense that differences should be negligible compared to the mean square

[24] If this is not the case, the relevant subset $\{\mathscr{G}\}$ must be extended according to Sect. 5.2.4, in a suitable manner, in order to become representative.

[25] As observables, we can also consider operators $\mathscr{F}(t)$ at later times t (Heisenberg formulation). The equivalence of ρ and $\mathscr{R}_{\{\mathscr{G}\}}$ then relates to the temporal dynamics.

[26] This transition, in the language of other treatments of statistical physics, corresponds to a "coarse graining" of the statistical operator.

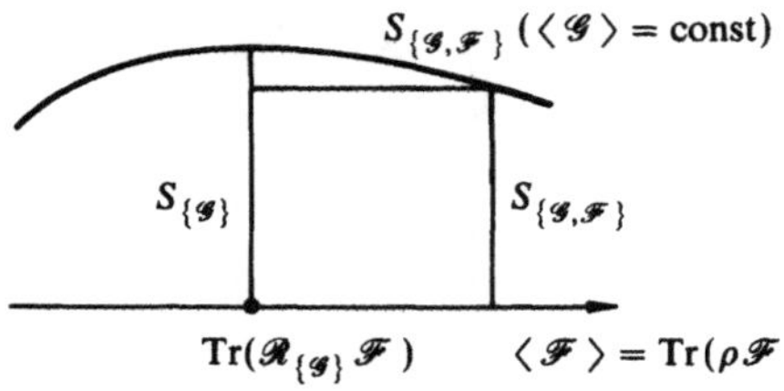

Fig. 5.4. The entropy $S_{\{\mathcal{G},\mathcal{F}\}}$ as a function of $\langle\mathcal{F}\rangle$ for fixed $\langle\mathcal{G}\rangle$. The maximum lies at $\langle\mathcal{F}\rangle = \mathrm{Tr}(\mathcal{R}_{\{\mathcal{G}\}}\mathcal{F})$ and has the value $S_{\{\mathcal{G}\}}$

deviation of the observable $\mathcal{F}$. The given statistical operator of the system therefore belongs (at a particular time) either to the set of ρ which are equivalent to $\mathcal{R}_{\{\mathcal{G}\}}$, or it does not belong to this set. Using this fact, one can arrive at a *classification* of the possible states ρ of a system.

Whether an observation level $\{\mathcal{G}\}$ is *sufficient* with respect to a set of other observables $\mathcal{F}_\sigma$, may be determined by referring to the *entropy* of the observation level $\{\mathcal{G},\mathcal{F}\}$ which has been extended to include the $\mathcal{F}_\sigma$.

Following Sect. 5.2.4, we know that $S_{\{\mathcal{G},\mathcal{F}\}} \leqq S_{\{\mathcal{G}\}}$, where the equals sign is valid for those expectation values $\langle\mathcal{F}_\sigma\rangle$ given by $\langle\mathcal{F}_\sigma\rangle = \mathrm{Tr}(\mathcal{R}_{\{\mathcal{G}\}}\mathcal{F}_\sigma)$ (Fig. 5.4). The degree of deviation of $S_{\{\mathcal{G},\mathcal{F}\}}$ from its maximum value $S_{\{\mathcal{G}\}}$ can thus serve as a *criterion* for determining whether or not the observation level $\{\mathcal{G}\}$ is sufficient for calculating the expectation values $\langle\mathcal{F}_\sigma\rangle$.

If, instead of $\mathcal{F}$, one chooses the observables $\mathcal{F}(t)$ (Heisenberg formulation) [cf. Footnote 25 (p. 59)] at a *later* time ("later" relative to the particular, fixed time of the observation level $\{\mathcal{G}\}$), then the deviation of the entropy $S_{\{\mathcal{G},\mathcal{F}(t)\}}$ of the *multi-temporal* observation level $\{\mathcal{G},\mathcal{F}(t)\}$ from $S_{\{\mathcal{G}\}}$ can be used as a criterion for whether the expectation value at time t, $\langle\mathcal{F}\rangle(t) = \mathrm{Tr}(\rho(t)\mathcal{F}) = \mathrm{Tr}(\rho\mathcal{F}(t))$, agrees with the value $\mathrm{Tr}(\mathcal{R}_{\{\mathcal{G}\}}\mathcal{F}(t))$ calculated from $\mathcal{R}_{\{\mathcal{G}\}}$.

5.2.6 Stationary Generalized Canonical Statistical Operators

The observation level $\{\mathcal{G}\}$ is supposed to describe the state of the system at an arbitrary but fixed time (apart from the multi-temporal observation levels mentioned above). In Chap. 6, we shall compare observation levels at different times, thereby investigating the significance of the generalized canonical statistical operator in treating dynamical processes. However, we now anticipate a special point. For an explicitly time-independent Hamiltonian, when $\rho(t_0) = \mathcal{R}_{\{\mathcal{G}\}}$ and

$$[\mathcal{H},\mathcal{R}_{\{\mathcal{G}\}}] = \mathcal{O} \tag{5.2.36}$$

every explicitly time-independent observable $\mathcal{F}$ of the system possesses time independent expectation values, according to (3.2.21). This *stationary condition* is ensured if $\mathcal{R}_{\{\mathcal{G}\}}$ commutes with $\mathcal{H}$ for arbitrary values of the Lagrange multipliers λ_v; this means that *all* the observables of the observation level must commute with $\mathcal{H}$,

$$[\mathcal{H}, \mathcal{G}_v] = \mathcal{O} , \tag{5.2.37}$$

i.e. they must be constants of the motion.

Exercise 5.4 Let the observation level for a one-dimensional particle be x, p, x^2, p^2. Which relations exist between the expectation values

$$\langle x \rangle, \langle p \rangle, \langle (x - \langle x \rangle)^2 \rangle = (\Delta x)^2 \ , \quad \langle (p - \langle p \rangle)^2 \rangle = (\Delta p)^2$$

and the Lagrange multipliers of the corresponding generalized canonical statistical operator? (*Hint*: use the operators $b = [1/\sqrt{2}\hbar a](ax + \mathrm{i}p)$ and $b^\dagger$ instead of x and p.)

Exercise 5.5 For a system of fermions with a variable number of particles, choose as the observation level all the number operators $\mathcal{N}_j = a_j^\dagger a_j$ of a basis $|v_j\rangle$. Calculate the corresponding entropy $S_{\{\mathcal{N}_j\}}$ as a function of the expectation values.

Exercise 5.6 For two modes of photon field (denoted as 1 and 2), let the expectation values $\langle b_i^\dagger b_k \rangle$ (i, $k = 1$, 2) be given. Calculate the entropy of the radiation field for this observation level as a function of the numbers of particles $\langle \mathcal{N}_1 \rangle$ and $\langle \mathcal{N}_2 \rangle$ and of the degree of coherence j, where

$$j = \frac{\langle b_1^\dagger b_2 \rangle \langle b_2^\dagger b_1 \rangle}{\langle \mathcal{N}_1 \rangle \langle \mathcal{N}_2 \rangle} \ .$$

5.3 Examples of Generalized Canonical Statistical Operators

In the following section, we briefly give some examples of particular observation levels which lead to thermodynamic results, most of which will be well known to the reader. We mention equilibrium statistics in order to show how it fits into the overall scheme of the general considerations.

5.3.1 The Hamiltonian as an Observation Level

If the Hamiltonian

$$\mathcal{H} = \mathcal{H}(h_1, \ldots, h_a) \tag{5.3.1}$$

represents a *sufficient* observation level, we expect that the expectation values of the macroscopic observables of Σ may be calculated with the aid of the *canonical statistical operator*

$$\boxed{\mathcal{R}(\beta, h_1, \ldots, h_a) = \frac{e^{-\beta\mathcal{H}(h_1, \ldots, h_a)}}{\operatorname{Tr} e^{-\beta\mathcal{H}(h_1, \ldots, h_a)}}} \quad . \tag{5.3.2}$$

The c-numbers h_α ($\alpha = 1, \ldots, a$) are here taken to represent external parameters, such as fields, volumes, etc.; the Lagrange multiplier β will prove to be proportional to the reciprocal of the absolute temperature T of Σ. The observation level $\mathcal{H}$ is representative of a system in *thermal equilibrium* (the standard system of thermostatics). All other macroscopic quantities of Σ can be expressed in terms of the $a+1$ independent numbers $\beta, h_1, \ldots, h_a$ (by means of the equations of state). The *internal energy* U is given by

$$U(\beta, h_1, \ldots, h_a) = \langle \mathcal{H} \rangle = \frac{\operatorname{Tr} \mathcal{H} e^{-\beta\mathcal{H}}}{\operatorname{Tr} e^{-\beta\mathcal{H}}} \tag{5.3.3}$$

and the *entropy* S (in the usual sense) according to (5.2.19) by

$$S(\beta, h_1, \ldots, h_a) = -k \operatorname{Tr}(\mathcal{R} \ln \mathcal{R}) = k(\beta U + \ln Z) \ . \tag{5.3.4}$$

If one compares different equilibrium states, which differ only by infinitesimals $d\beta, dh_1, \ldots, dh_a$, the corresponding difference in the internal energy is

$$dU = \operatorname{Tr}(\mathcal{R}\, d\mathcal{H}) + \operatorname{Tr}(\mathcal{H}\, d\mathcal{R}) \tag{5.3.5}$$

and that of the entropy, following (5.2.22), is

$$\boxed{dS = k\beta\left(dU - \sum_\alpha \left\langle \frac{\partial \mathcal{H}}{\partial h_\alpha} \right\rangle dh_\alpha\right) = k\beta \operatorname{Tr}(\mathcal{H}\, d\mathcal{R})} \quad . \tag{5.3.6}$$

In systems with a fixed number of particles[27], the sums in (5.3.5) have the following meanings:

$$\mathrm{D}A = \operatorname{Tr}(\mathcal{R}\, d\mathcal{H}) \tag{5.3.7}$$

$$\mathrm{D}Q = \operatorname{Tr}(\mathcal{H}\, d\mathcal{R}) \ , \tag{5.3.8}$$

which are the infinitesimal amounts of *work* and *heat* added to the system[28]. Thus, (5.3.6) takes on the well-known form

$$dS = \frac{\mathrm{D}Q}{T} \tag{5.3.9}$$

---

[27] In systems which are open to particle exchange, the separation of dU into heat and work terms is somewhat arbitrary and therefore not unambiguously defined in the literature. Relation (5.3.6) is, in contrast, valid even for systems in which particle exchange is permitted.

[28] $\mathrm{D}A$ and $\mathrm{D}Q$ are linear differential forms (Pfaffian forms) and not total differentials. This is expressed by the use of the symbol D.

and β is related to the absolute temperature T by the equation

$$\boxed{\beta=\frac{1}{kT}} \,. \tag{5.3.10}$$

Equations (5.3.5) and (5.3.9) are the quantum-statistical formulation of the *differential First and Second Laws of Thermostatics.*

Equation (5.3.4) implies that the free energy $F=U-TS$ is given by the partition function according to

$$F(\beta, h_1, \ldots, h_a)=-\frac{1}{\beta}\ln Z(\beta, h_1, \ldots, h_a)\,. \tag{5.3.11}$$

As is well known, all the other state functions of the system may be obtained by differentiation from this thermodynamic potential.

The derivatives of the Hamiltonian with respect to the parameters are called the *operators of the generalized forces*

$$\mathcal{M}_\alpha=-\frac{\partial\mathcal{H}}{\partial h_\alpha}\,, \tag{5.3.12}$$

since the work (5.3.7) may be expressed in terms of their expectation values:

$$\mathrm{D}A=\sum_\alpha\left\langle\frac{\partial\mathcal{H}}{\partial h_\alpha}\right\rangle dh_\alpha=-\sum\langle\mathcal{M}_\alpha\rangle dh_\alpha\,. \tag{5.3.13}$$

Examples:

$h_\alpha=$	$\langle\mathcal{M}_\alpha\rangle=\mathrm{Tr}(\mathcal{R}\mathcal{M}_\alpha)$	
volume V	pressure p	
magnetic field $\mathfrak{H}(\mathfrak{r})$	magnetization $\mathfrak{M}(\mathfrak{r})$	(5.3.14)
electric field $\mathfrak{E}(\mathfrak{r})$	polarization $\mathfrak{P}(\mathfrak{r})$.	

Keeping in mind the discussion in Sect. 5.2, we point out that the expectation values $\langle\mathcal{M}_\alpha\rangle$ are *not* contained in the observation level (5.3.1), but rather the values calculated from $\mathcal{R}$ using $\mathrm{Tr}(\mathcal{R}\mathcal{M}_\alpha)$ represent a *prediction* of the expectation values $\langle\mathcal{M}_\alpha\rangle$ which will apply to the physical system [compare Footnote 22 (p. 58)].

If one chooses the Hamiltonian $\mathcal{H}$ and the particle-number operator $\mathcal{N}$ (in a unitary space having a variable number of particles) as the observation level, then the mean energy $\langle\mathcal{H}\rangle$ and the mean number of particles $\langle\mathcal{N}\rangle$ are fixed, and one obtains by finding the extremum of η the socalled *grand canonical statistical operator*

$$\mathscr{R}_{\mathscr{H},\mathscr{N}}=\frac{e^{-\beta\mathscr{H}-\mu\mathscr{N}}}{\operatorname{Tr}e^{-\beta\mathscr{H}-\mu\mathscr{N}}}\,, \tag{5.3.15}$$

(where $-\mu kT$ = chemical potential). For macroscopic systems with large numbers of particles, this operator is, however, *equivalent* to $\mathscr{R}_{\mathscr{H}}$, since the deviation from the expectation value $\langle\mathscr{N}\rangle$ is negligible and $\langle\mathscr{N}\rangle$ is thus practically identical to the fixed particle number N, as is implicit in (5.3.2). We shall not present here a list of the well-known results which follow from (5.3.15). The relation which is obtained from (5.2.22) for the observation level $\{\mathscr{H},\mathscr{N}\}$ becomes Gibbs' fundamental equation.

5.3.2 Partial Hamiltonians as Decomposable Observation Levels

Within a standard system, whose observation level is given by the total Hamiltonian $\mathscr{H}$, there are no subsystems which are adiabatically isolated from one another. If such "walls" exist within a system Σ, one can fix the values of the partial energies $\langle\mathscr{H}_\nu\rangle$ of each subsystem Σ_ν *independently*. For a system with *internal adiabatic separations*, i.e. isolated subsystems, an observation level consisting only of $\mathscr{H}$ is therefore *not sufficient*. It must be extended in such a way that all of the partial energies

$$\mathscr{H}_\nu=\mathscr{H}_\nu(h_1,\ldots,h_a) \tag{5.3.16}$$

enter the observation level. The resulting generalized canonical statistical operator

$$\boxed{\mathscr{R}(\beta_1,\ldots,\beta_n,h_1,\ldots,h_a)=\frac{\exp\left(-\sum_\nu\beta_\nu\mathscr{H}_\nu\right)}{\operatorname{Tr}\exp\left(-\sum_\nu\beta_\nu\mathscr{H}_\nu\right)}} \tag{5.3.17}$$

can then be representative of the overall system Σ.[29]

In the simplest case, such a system with different temperatures may be constructed by placing several systems Σ^ν with different temperatures β^ν in spatially adjacent positions. Each subsystem Σ^ν is then described in *its own unitary space* $\mathfrak{U}^\nu$, in which the corresponding partial Hamiltonian $\mathscr{H}^\nu$ acts. The

[29] In the case that $\sum_\nu\mathscr{H}_\nu=\mathscr{H}$, (5.3.17) with $\beta_1=\beta_2=\ldots=\beta$ becomes the canonical statistical operator (5.3.2). This equality of the Lagrange multipliers β_ν thus indicates *thermal equilibrium* in the overall system Σ, i.e. β has the property of an *empirical temperature*, as is defined in the *Zeroth Law of thermodynamics*.

unitary space $\mathfrak{U}$ of the overall system Σ is the *direct product* (Sect. 3.3)

$$\mathfrak{U}=\mathfrak{U}^1\times\mathfrak{U}^2\times\ldots\mathfrak{U}^\nu\times\mathfrak{U}^n\ . \tag{5.3.18}$$

The Hamiltonians $\mathscr{H}^\nu$ thus commute. The generalized canonical statistical operator (5.3.17) factorizes into

$$\mathscr{R}=\mathscr{R}^1\mathscr{R}^2\ldots\mathscr{R}^\nu\ldots\mathscr{R}^n\ . \tag{5.3.19}$$

Each of the factors $\mathscr{R}^\nu$ is a canonical statistical operator for the subsystem Σ^ν with temperature β^ν. In the case – as we assume here – that each of the partial Hamiltonians $\mathscr{H}^\nu$ of the observation level acts in its own unitary subspace $\mathfrak{U}^\nu$, we term the observation level to be *decomposable.*

When adiabatic isolation exists between the subsystems Σ^ν, then we have

$$\mathscr{H}=\sum_\nu\mathscr{H}^\nu\ .^{30} \tag{5.3.20}$$

$\mathscr{R}$ is then stationary. We have described the trivial case of several completely independent subsystems Σ^ν, each of which is in internal thermal equilibrium at a temperature β^ν.

As we have already emphasized in Sect. 5.2, the concept of the observation level however by no means implies a limitation to thermal-equilibrium states. Instead of (5.3.20), we can permit an interaction between the subsystems Σ^ν,

$$\mathscr{H}=\sum_\nu\mathscr{H}^\nu+\sum_{\nu\mu}\mathscr{H}^{\nu\mu} \tag{5.3.21}$$

(cf. Fig. 3.2b). The requirement of internal adiabatic walls is thus lifted, and they are replaced by *diathermal* walls (no particle exchange) between the Σ^ν. When the interaction $\mathscr{H}^{\nu\mu}$ is sufficiently small, however, then *the observation level $\mathscr{H}^1,\ldots,\mathscr{H}^n$ will still be sufficient* for observables of the Σ^ν. Thus, the subsystems Σ^ν may still be described at each time t by

$$\mathscr{R}_{\{\mathscr{H}^\nu\}}(t)=\frac{\exp\left[-\sum_\nu\beta^\nu(t)\mathscr{H}^\nu\right]}{\operatorname{Tr}\exp\left[-\sum_\nu\beta^\nu(t)\mathscr{H}^\nu\right]} \tag{5.3.22}$$

(compare also Sect. 6.4). *The temperatures $\beta^\nu(t)$ of the subsystems are then functions of the time* (and, as a consequence, so are all the state functions of the Σ^ν). This corresponds to the situation of *heat conduction* between the subsystems. The calculation of the time dependence $\beta^\nu(t)$ is presented in Chap. 16.

[30] Indices written as superscripts refer to subspaces.

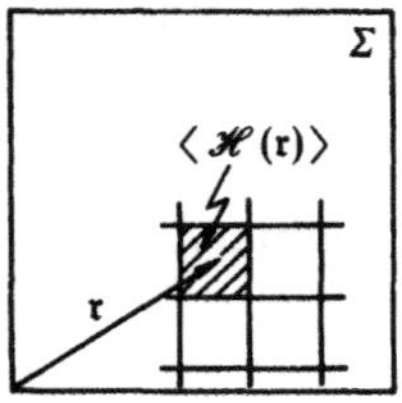

Fig. 5.5. The macroscopic energy density $\mathscr{H}(\mathbf{r})$ as an observation level. The local temperature is given by position-dependent Lagrange multipliers $\beta(\mathbf{r})$

In the case of *spatial* heat conduction, the subsystems are macroscopically small, but microscopically large volume elements, so that the conditions for macroscopic systems as given in Chap. 4 still apply to them. These volume elements can then be described by a position vector r (Fig. 5.5) which appears continuous on a macroscopic scale, thus converting the β^ν to position-dependent temperature parameters $\beta^\nu(\mathbf{r})$; in the formulas, the sums over ν thus become volume integrals. This concept is called *local equilibrium.*

However, the subsystems *need not by any means be spatially separate,* as one sees in the example of spin-lattice relaxation. Here, the energy exchange occurs between the spin system and the lattice (phonons) and is described by time-dependent temperatures $\beta^{\mathrm{S}}(t)$ and $\beta^{\mathrm{L}}(t)$. In this case, also, the observation level is decomposable. The overall system is described in a unitary space $\mathfrak{U}$ which is the direct product space of $\mathfrak{U}^{\mathrm{S}}$ (the space of the spin variables) and of $\mathfrak{U}^{\mathrm{L}}$ (that of the lattice variables).

5.3.3 Partial Hamiltonians as Nondecomposable Observation Levels

The generalized canonical statistical operator (5.3.17) over an observation level $\mathscr{H}_1, \ldots, \mathscr{H}_n$ may also be applied to systems in which partial Hamiltonians $\mathscr{H}_\nu$ act in *the entire unitary space* $\mathfrak{U}$ of the system and *not* only in their own subspaces $\mathfrak{U}^\nu$ of a product space. An example is provided by spin systems, in which the Zeeman energy $\mathscr{H}_{\mathrm{Z}}$, the exchange energy $\mathscr{H}_{\mathrm{Ex}}$, and possibly the crystal field energy $\mathscr{H}_{\mathrm{C}}$ serve as the observation level. These operators do not act in separate unitary spaces, but rather in the common unitary space of all the spins. Another example is an electron gas in a magnetic field with orbital energy $\mathscr{H}_{\mathrm{Orb}}$ and spin energy $\mathscr{H}_{\mathrm{S}}$, which act in the antisymmetric space $\mathfrak{U}^-$ of the fermions. The latter cannot be split up into subspaces of a product space in such a way that the orbital and spin Hamiltonians each act in one subspace.

Nevertheless, even in such cases the generalized canonical statistical operator

$$\mathscr{R}(\beta_1, \ldots, \beta_n, h_1, \ldots, h_a) = \frac{\exp\left[-\sum_\nu \beta_\nu \mathscr{H}_\nu(h_1, \ldots, h_a)\right]}{\operatorname{Tr}\exp\left[-\sum_\nu \beta_\nu \mathscr{H}_\nu(h_1, \ldots, h_a)\right]} \tag{5.3.23}$$

may be representative of the overall system Σ. In this *generalized thermodynamics of a nondecomposable observation level*, Eqns. (5.3.18 and 19) are no longer valid. However, one can easily convince oneself (Exercise 5.7) that the Lagrange multipliers $\beta_1, \ldots, \beta_n$ which occur still have the properties of (inverse) temperatures. The *internal partial energies* U_ν

$$\langle \mathscr{H}_\nu \rangle = \frac{\mathrm{Tr}\left(\mathscr{H}_\nu \exp\left[-\sum_\nu \beta_\nu \mathscr{H}_\nu(h_1, \ldots, h_a)\right]\right)}{\mathrm{Tr}\left(\exp\left[-\sum_\nu \beta_\nu \mathscr{H}_\nu(h_1, \ldots, h_a)\right]\right)}$$

$$= U_\nu(\beta_1, \ldots, \beta_n, h_1, \ldots, h_a) \tag{5.3.24}$$

then in general each depend on *all* the temperatures $\beta_1, \ldots, \beta_n$! By means of a change in only one variable, e.g. in the temperature β_2, one can produce a change for example in the internal energy U_1, although no interaction energy between (1) and (2) exists! One thus obtains in this generalized thermodynamics, in contrast to the usual thermodynamics, for example *mixed* heat capacities

$$C_{\nu\mu} = \frac{\partial U_\nu}{\partial T_\mu} \, . \tag{5.3.25}$$

We shall dispense here with a list of additional, readily derivable formulas of such a generalized thermodynamics. However, we note that, in contrast to the decomposable case, in which an entropy S^ν is defined for each subsystem, *in the nondecomposable case, only a single entropy* $S = -k\mathrm{Tr}(\mathscr{R} \ln \mathscr{R})$ *exists for the whole system* Σ, for which we then find

$$dS = k \sum_\nu \beta_\nu \mathrm{D}Q_\nu = \sum_\nu \frac{\mathrm{D}Q_\nu}{T_\nu} \, . \tag{5.3.26}$$

The partial heats $\mathrm{D}Q_\nu$ have no individual integrating factors; there are no partial entropies.

These results of the generalized thermodynamics of a nondecomposable observation level are practically self-evident following the treatment in Sect. 5.2. In the language of conventional thermodynamics (Sects. 5.3.1 and 5.3.2), they are at first somewhat surprising.

If, in addition to the partial energies $\mathscr{H}_\nu$, there exists also a weak interaction energy $\mathscr{H}_{\nu\mu}$, then, in the nondecomposable case, just as in the decomposable case, the nondecomposable observation level $\mathscr{H}_1, \ldots, \mathscr{H}_n$ will remain sufficient. The temperatures β_ν will become time-dependent functions $\beta_\nu(t)$ (see also Sect. 6.4), which describe the exchange of energy among the subsystems.

5.3.4 Projectors $\{\mathscr{P}_F\}$ as Observation Levels

If the probabilities w_F of all the eigenvalues F of one observable $\mathscr{F}$ are known (Fig. 5.6), the observation level is formed from the *projectors* $\mathscr{P}_F$ which project

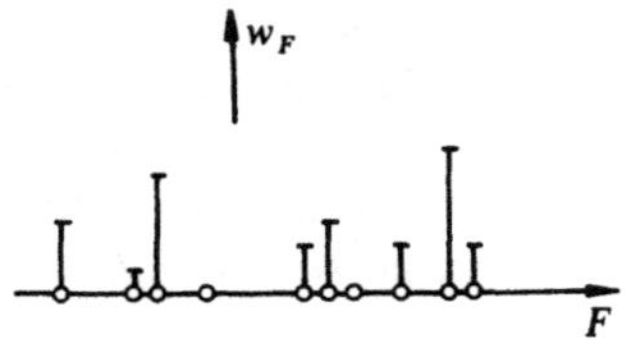

Fig. 5.6. Illustrating the observation level $\{\mathscr{P}_F\}$: the probabilities $w_F = \mathrm{Tr}(\rho\mathscr{P}_F)$ for the eigenvalues of an observable $\mathscr{F}$

onto the eigenspaces $\mathfrak{u}_F$ (2.3.8) and (2.3.9), because the probabilities are the expectation values of these projectors,

$$w_F = \langle\mathscr{P}_F\rangle \ . \tag{5.3.27}$$

The corresponding generalized canonical statistical operator then becomes, according to (5.2.6),

$$\mathscr{R}_{\{\mathscr{P}_F\}} = \exp\left(-\sum_F \lambda_F \mathscr{P}_F\right) .^{31} \tag{5.3.28}$$

It may also be written in a rather different form, if one takes into account the property

$$\mathscr{P}_F\mathscr{P}_{F'} = \delta_{FF'}\mathscr{P}_F \tag{5.3.29a}$$

in expanding the exponential function and also employs

$$\sum_F \mathscr{P}_F = 1 \ . \tag{5.3.29b}$$

We first obtain

$$\exp\left(-\sum_F \lambda_F\mathscr{P}_F\right) = \sum_F \mathrm{e}^{-\lambda_F}\mathscr{P}_F \ . \tag{5.3.30}$$

The resulting relation between w_F and the Lagrange multipliers λ_F is given by

$$w_F = \mathrm{Tr}(\mathscr{R}_{\{\mathscr{P}_F\}}\mathscr{P}_F) = \mathrm{Tr}(\mathrm{e}^{-\lambda_F}\mathscr{P}_F) = \mathrm{e}^{-\lambda_F} d_F \ , \tag{5.3.31}$$

where d_F is the degeneracy of the eigenvalue F. Equation (5.3.28) is thus converted to

$$\boxed{\mathscr{R}_{\{\mathscr{P}_F\}} = \sum_F w_F \frac{\mathscr{P}_F}{d_F}} \ . \tag{5.3.32}$$

[31] The partition function does not occur here, since the observables $\mathscr{P}_F$ and the unit operator are linearly dependent and therefore, specifying all the expectation values w_F already implies the condition $\mathrm{Tr}\,\mathscr{R}_{\{\mathscr{P}_F\}} = 1$.

Due to the idempotence of $\mathscr{P}_F$, the exponential structure has completely vanished! $\mathscr{R}_{\{\mathscr{P}_F\}}$ becomes a sum of *microcanonical statistical operators*. The corresponding entropy is found using (5.2.18) to be

$$\boxed{S_{\{\mathscr{P}_F\}} = -k \sum_F w_F \ln \frac{w_F}{d_F}} \ . \tag{5.3.33}$$

If *one particular* value has been measured, one is led back to the relations (5.1.8 and 5.1.9). Such a precise result cannot, however, be obtained for macroscopic systems. One can assume that the measured values fall in a certain interval δF. In such a case it is reasonable to divide the spectrum of F into intervals δF and to define projectors onto these intervals with respect to an observation level. For these, (5.3.30) holds analogously, so that the respective generalized canonical statistical operator can also be written in the form of (5.3.32). With the special assumption that the probability for a particular interval δF is one, and all others are zero,

$$\begin{aligned} w_{\delta F} &= \sum_{F'} w_{F'} = 1 \quad \text{for } F' \text{ in } \delta F \quad \text{and} \\ w_{F'} &= 0 \quad \text{for } F' \text{ not in } \delta F \ , \end{aligned} \tag{5.3.34}$$

the statistical operator becomes

$$\mathscr{R}_{\delta F} = \frac{\sum_{F'} \mathscr{P}_{F'}}{\sum_{F'} d_{F'}} \quad (F' \text{ in } \delta F) \ . \tag{5.3.35}$$

The entropy

$$S_{\delta F} = k \ln \sum_{F'} d_{F'} = k \ln g_{\delta F} \tag{5.3.36}$$

is given by the *number of states* in the interval δF,

$$g_{\delta F} = \sum_{F'} d_{F'} \quad (F' \text{ in } \delta F) \ . \tag{5.3.37}$$

If one compares the observation level of the Hamiltonian $\mathscr{H}$ with that of $\{\mathscr{P}_{\delta E}\}$, one finds the latter to be more precise. For macroscopic systems, the two are however equivalent, since the standard deviation of $\mathscr{H}$ is negligible and the expectation value $\langle \mathscr{H} \rangle$ is practically identical to the measured value of the energy (Chap. 4)[32]. It is usually more convenient to calculate with the canonical

[32] From (5.3.36), one sees that the number of energy levels in an interval δE, $g_{\delta E} = \exp(S_{\delta E}/k)$, increases exponentially with the particle number N, since the entropy $S_{\delta E} \simeq S_{\mathscr{H}}$ is proportional to N.

statistical operator than with the microcanonical operator, so we shall choose the *canonical statistical operator* (5.3.2) in the following for the *description of thermal equilibrium.* The transition to the grand canonical statistical operator (5.3.15), which is also frequently expedient, is formally easily accomplished.

Exercise 5.7 Show that between two systems Σ^1 and Σ^2, each with non-decomposable observation levels $\mathscr{H}_1^1, \ldots, \mathscr{H}_n^1$ and $\mathscr{H}_1^2, \ldots, \mathscr{H}_n^2$ (with $[\mathscr{H}_\nu^\alpha, \mathscr{H}_\mu^\beta] = 0$), thermal contact is possible by means of a suitable interaction $\mathscr{H}^{12}$, in which for example only the partial energies $\langle \mathscr{H}_1^1 \rangle$ and $\langle \mathscr{H}_1^2 \rangle$ change, but all other partial energies remain constant. Formulate the Zeroth Law for nondecomposable systems with this result; it allows the interpretation of the Lagrange multipliers β_ν as empirical temperatures.

6. The Significance of Generalized Canonical Statistical Operators for Dynamic Processes

The considerations of Chap. 5 concerning the properties of a generalized canonical statistical operator were limited to an arbitrary but fixed time t. In the following sections, especially in Sects. 6.2 to 6.4, we shall make use of the fact that a number of such times can be chosen in succession in the course of a dynamic process.

6.1 The Statistical Operator at the Beginning of a Process

The von Neumann equation (3.2.5) is the differential equation for the time development of the statistical operator $\rho(t)$ (in the Schrödinger formulation) for an *adiabatically isolated* system Σ. The initial-state operator $\rho(t_0)$ can be freely chosen in a mathematical sense, as in any differential equation. It must describe the physical initial state of the system.

We assume that at the time t_0, the (macroscopic) system is described by a *sufficient* observation level $\{\mathscr{G}\}$. The expectation values $\langle\mathscr{G}_\nu\rangle(t_0)$ yield, through maximization of the uncertainty measure η, the generalized canonical statistical operator,

$$\mathscr{R}_{\{\mathscr{G}\}}(t_0)=\frac{\exp\left[-\sum_\nu \lambda_\nu(t_0)\mathscr{G}_\nu(h_\alpha(t_0))\right]}{Z_{\{\mathscr{G}\}}(t_0)}\,,^{1} \tag{6.1.1}$$

which is not necessarily stationary. Since the observation level $\{\mathscr{G}\}$ was chosen to be sufficient, other macroscopic observables $\mathscr{F}$ of the system at time t_0 are also correctly described by $\mathscr{R}_{\{\mathscr{G}\}}(t_0)$. Furthermore, $\mathscr{R}_{\{\mathscr{G}\}}(t_0)$ is supposed to guarantee that the future measured values of macroscopic quantities can be correctly calculated as expectation values of $\mathscr{F}(t)$ (cf. Sect. 5.2.5). We then use $\mathscr{R}_{\{\mathscr{G}\}}(t_0)$ as the initial statistical operator $\rho(t_0)$,

$$\boxed{\rho(t_0)=\mathscr{R}_{\{\mathscr{G}\}}(t_0)}\,, \tag{6.1.2}$$

[1] If the operators $\mathscr{G}_\nu(h_\alpha(t))$ are explicitly time-dependent, we refer to them here at the time t_0 (Schrödinger formulation).

for the dynamical calculation of $\rho(t)$ with the von Neumann equation. The quantities $h_\alpha(t_0)$ which occur in (6.1.1) are the external parameters at the time t_0, such as external fields, volumes, etc. The Lagrange multipliers $\lambda_\nu(t_0)$ may be expressed in principle through the expectation values $\langle\mathscr{G}_1\rangle(t_0)$, $\langle\mathscr{G}_2\rangle(t_0)$, ... [cf. (5.2.9)]. The entropy belonging to the chosen observation level $\{\mathscr{G}\}$ at the time t_0 is given by

$$S_{\{\mathscr{G}\}}(t_0) = -k\,\mathrm{Tr}(\mathscr{R}_{\{\mathscr{G}\}}(t_0)\ln\mathscr{R}_{\{\mathscr{G}\}}(t_0)) \ . \tag{6.1.3}$$

6.2 Entropy Production in Dynamic Processes of Adiabatic Systems

If the system Σ is adiabatically isolated for times $t > t_0$, then the Hamiltonian $\mathscr{H}(h_\alpha(t))$ [or the corresponding Liouville operator $\mathsf{L}(h_\alpha(t))$], which may also be explicitly time-dependent via time-dependent fields $h_\alpha(t)$, yields the time evolution of $\rho(t)$ according to (3.2.6 and 3.2.7):

$$\rho(t) = \mathsf{U}(t, t_0)\mathscr{R}_{\{\mathscr{G}\}}(t_0) \quad (t \geqq t_0) \ . \tag{6.2.1}$$

This statistical operator allows the calculation of the expectation values $\langle\mathscr{F}\rangle(t)$ of observables $\mathscr{F}$ at later times $t > t_0$. The uncertainty measure η belonging to $\mathscr{R}_{\{\mathscr{G}\}}(t_0)$ remains constant in time, according to (5.1.13):

$$\eta(t) = -k\,\mathrm{Tr}\,\rho(t)\ln\rho(t) = S_{\{\mathscr{G}\}}(t_0) \ . \tag{6.2.2}$$

We now ask the question whether at a certain *later* time $t = t_1$ ($> t_0$), the statistical operator dynamically calculated using (6.2.1), $\rho(t_1)$, is equivalent to a generalized canonical statistical operator $\mathscr{R}_{\{\tilde{\mathscr{G}}\}}(t_1)$ over an observation level $\{\tilde{\mathscr{G}}\}$. The observables $\tilde{\mathscr{G}}_\mu(h_\alpha(t_1))$ of this observation level need not be identical to those of the observation level $\{\mathscr{G}\}$, *neither in their number, nor in their meanings.* When the expectation values $\langle\mathscr{F}\rangle(t_1)$ of other macroscopic observables $\mathscr{F}$, calculated on the one hand using $\rho(t_1)$ and on the other hand using $\mathscr{R}_{\{\tilde{\mathscr{G}}\}}(t_1)$, are nearly in agreement,

$$\mathrm{Tr}(\rho(t_1)\mathscr{F}) \approx \mathrm{Tr}(\mathscr{R}_{\{\tilde{\mathscr{G}}\}}(t_1)\mathscr{F}) \ , \tag{6.2.3}$$

then we can make the replacement (for $\mathscr{F}$)

$$\boxed{\rho(t_1) = \mathsf{U}(t_1, t_0)\,\rho(t_0) \Rightarrow \mathscr{R}_{\{\tilde{\mathscr{G}}\}}(t_1) = \frac{\exp\left[-\sum_\mu \lambda_\mu(t_1)\tilde{\mathscr{G}}_\mu(h_\alpha(t_1))\right]}{Z_{\{\tilde{\mathscr{G}}\}}(t_1)}} \ . \tag{6.2.4}$$

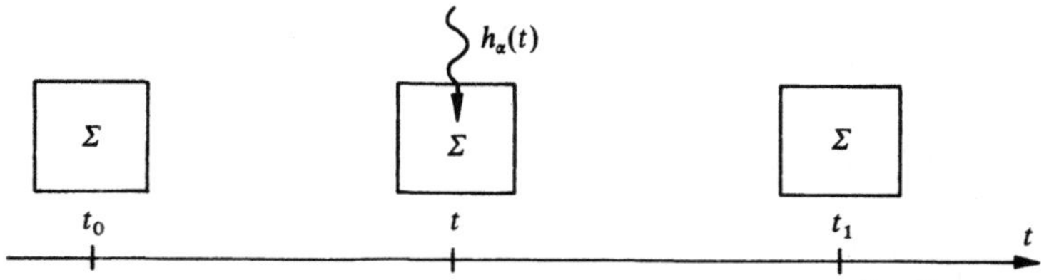

Observation level:	$\{\mathscr{G}\}: \mathscr{G}_\nu(h_\alpha(t_0))$		$\{\tilde{\mathscr{G}}\}: \tilde{\mathscr{G}}_\mu(h_\alpha(t_1))$
External parameters:	$h_\alpha(t_0)$	$h_\alpha(t)$	$h_\alpha(t_1)$
Expectation values:	$\langle \mathscr{G}_\nu \rangle(t_0)$	von Neumann eq. with $\mathscr{H}(h_\alpha(t))$	$\nearrow \langle \tilde{\mathscr{G}}_\mu \rangle(t_1)$
Statistical operator:	$\mathscr{R}_{\{\mathscr{G}\}}(t_0) = \rho(t_0)$ ——	$\rho(t)$ ——	$\rightarrow \rho(t_1) \Rightarrow \mathscr{R}_{\{\tilde{\mathscr{G}}\}}(t_1)$
Lagrange multipliers:	$\lambda_\nu(t_0)$		$\lambda_\mu(t_1)$
Entropy:	$S_{\{\mathscr{G}\}}(t_0)$ =	$\eta(t)$ =	$\eta(t_1) \leqq S_{\{\tilde{\mathscr{G}}\}}(t_1)$

Fig. 6.1. An adiabatic system Σ under the influence of external dynamic fields $h_\alpha(t)$. The von Neumann equation yields the statistical operator $\rho(t_1)$ from the initial value $\rho(t_0) = \mathscr{R}_{\{\mathscr{G}\}}(t_0)$. The transition $\rho(t_1) \Rightarrow \mathscr{R}_{\{\tilde{\mathscr{G}}\}}(t_1)$ causes an *increase in entropy*

For the construction of $\mathscr{R}_{\{\tilde{\mathscr{G}}\}}(t_1)$, we proceed as follows (as in Sect. 5.2.2): from $\rho(t_1)$, we calculate the expectation values at the time t_1

$$\mathrm{Tr}(\rho(t_1))\tilde{\mathscr{G}}_\mu(h_\alpha(t_1)) = \langle \tilde{\mathscr{G}}_\mu \rangle(t_1) \tag{6.2.5}$$

and use the resulting numerical values as constraints in the construction of $\mathscr{R}_{\{\tilde{\mathscr{G}}\}}(t_1)$ through maximization of the uncertainty measure η. The whole procedure which we describe here can best be grasped by examining the overview in Fig. 6.1.

The maximum $S_{\{\tilde{\mathscr{G}}\}}(t_1)$ can, according to (5.2.35), not be smaller than $\eta(t_1) = S_{\{\mathscr{G}\}}(t_0)$. Therefore, following (6.2.2), we find in general for adiabatic processes between observation levels $\{\mathscr{G}\}$ and $\{\tilde{\mathscr{G}}\}$

$$\boxed{S_{\{\mathscr{G}\}}(t_0) \leqq S_{\{\tilde{\mathscr{G}}\}}(t_1)}\ , \tag{6.2.6}$$

a general formulation of the Second Law for the dynamics of adiabatic systems[2]. The statement that the observation level $\{\tilde{\mathscr{G}}\}$ is sufficient to describe the system Σ at the time t_1 means, as we have already mentioned, that macroscopic expectation values $\langle \mathscr{F} \rangle(t_1)$ can be calculated from the values $\langle \tilde{\mathscr{G}}_\mu \rangle(t_1)$. The past, however, i.e. the way in which these expectation values came about before the time t_1, is irrelevant. *This possibility of dispensing with the detailed structure of $\rho(t_1)$ (irrelevance of the past) is thus the cause of the production of entropy (6.2.6).*

[2] This result is independent of whether or not the system Σ is in thermal equilibrium at the times t_0 and t_1.

When the observation level $\{\tilde{\mathscr{G}}\}$ is already sufficient for $\langle\mathscr{F}\rangle(t_1)$ at some earlier time $t_1' < t_1$, then t_1 may be replaced in (6.2.3 through 6.2.6) by the earlier time t_1'.

Whether an observation level is "sufficient" or not is a question which depends upon the physical properties of the system, so that the production of entropy (6.2.6) is not a subjective observation (as it might at first seem), but rather an objective physical fact!

With the possibility of describing the system at the "final" time t_1 by a generalized canonical statistical operator $\mathscr{R}_{\{\tilde{\mathscr{G}}\}}(t_1)$, we have arrived back at a situation similar to that at the initial time t_0, when the state of the system was characterized by $\mathscr{R}_{\{\mathscr{G}\}}(t_0)$. Therefore, we can employ the arguments just given to calculate the dynamics through still later times $t > t_1$, using $\mathscr{R}_{\{\tilde{\mathscr{G}}\}}(t_1)$ as the new initial-state operator in the von Neumann equation. At a later time $t_2 > t_1$, it may again prove to be possible that $\rho(t_2)$ is equivalent to a generalized canonical statistical operator $\mathscr{R}_{\{\tilde{\tilde{\mathscr{G}}}\}}(t_2)$ of an observation level $\{\tilde{\tilde{\mathscr{G}}}\}$, which is then connected to a further entropy production,

$$S_{\{\mathscr{G}\}}(t_0) \leqq S_{\{\tilde{\mathscr{G}}\}}(t_1) \leqq S_{\{\tilde{\tilde{\mathscr{G}}}\}}(t_2) \quad \text{etc} \; . \tag{6.2.7}$$

We thus conclude that *in all possible adiabatic processes, in which the initial and the final states are describable by generalized canonical statistical operators, a decrease in the entropy is impossible.* Under these conditions, there exists no adiabatic process $t_1 \to t_2$ in which the entropy increase of the original process $t_0 \to t_1$ can be reversed, i.e. these dynamical processes are *irreversible.*

Exercise 6.1 A free, one-dimensional particle, with $\mathscr{H} = p^2/2m$, is described at time $t=0$ by the generalized canonical statistical operator over the observation level x, x^2, p, p^2

$$\rho(t=0) = \mathscr{R}_{x,x^2,p,p^2} \quad \text{(compare Exercise 5.4)} \; .$$

a) Determine the generalized canonical statistical operator over the same observation level at a later time $t > 0$ from the expectation values obtained using the von Neumann equation,

$$\langle x\rangle(t),\ \langle x^2\rangle(t),\ \langle p\rangle(t), \text{ and } \langle p^2\rangle(t) \; .$$

b) At a time $t_1 > 0$, make the transition from $\rho(t_1)$ to $\mathscr{R}(t_1)$. What difference does this make in the expectation values of the observable $\mathscr{F} = xp + px$?

c) With $\mathscr{R}(t_1)$ as the new initial-state operator in the von Neumann equation, calculate the expectation values $\langle x\rangle(t_2)$, $\langle x^2\rangle(t_2)$, $\langle p\rangle(t_2)$, and $\langle p^2\rangle(t_2)$ at a time $t_2 > t_1$. Compare them to those obtained directly from $\rho(t=0)$.

6.3 Examples of Entropy Production in Dynamic Adiabatic Processes

After the above discussion, the following examples need no detailed explanation. They should be clear immediately from the accompanying figures.

6.3.1 The Dynamics of an Adiabatic Process in Going from One Thermal Equilibrium to Another

The prime example for thermodynamic considerations is the transition of a system Σ from a state of thermal equilibrium, (0), to another thermal equilibrium state (1) by means of a dynamical adiabatic process. The observation level defined by the Hamiltonian $\mathscr{H}$ is taken to be sufficient at the beginning and at the end of the process. The internal energy

$$\langle\mathscr{H}\rangle(t_0)=U(\beta(t_0),h_\alpha(t_0)) \tag{6.3.1}$$

leads to the canonical statistical operator $\mathscr{R}(t_0)$ with the initial temperature $\beta(t_0)$ of the system. If external forces $h_\alpha(t)$ act on Σ in the time interval $t>t_0$, then the explicitly time-dependent Hamiltonian $\mathscr{H}(h_\alpha(t))$ causes a change in $\rho(t)$ in the von Neumann equation. In the case that $\rho(t_1)$ is equivalent to a new canonical statistical operator at a later time,

$$\rho(t_1)\Rightarrow\mathscr{R}(t_1)=\frac{\exp[-\beta(t_1)\mathscr{H}(h_\alpha(t_1))]}{Z(t_1)}\ , \tag{6.3.2}$$

then the system is again (concerning macroscopic properties) in thermal equilibrium (Fig. 6.2).

The final temperature $\beta(t_1)$ is defined by

$$\mathrm{Tr}(\rho(t_1)\mathscr{H}(h_\alpha(t_1))=\mathrm{Tr}(\mathscr{R}(t_1)\mathscr{H}(t_1))=U(\beta(t_1),h_\alpha(t_1))\ . \tag{6.3.3}$$

The corresponding maximization of the uncertainty measure yields the entropy $S(t_1)$, which must be larger than $S(t_0)$,

$$S(t_0)\leqq S(t_1)\ . \tag{6.3.4}$$

6.3.2 The Dynamics of an Adiabatic Process in Going from Thermal Equilibrium to an Inhibited Equilibrium

The assumption made in the previous section, that $\rho(t_1)$ should be equivalent to a canonical statistical operator, i.e. that at a time t_1, thermal equilibrium again

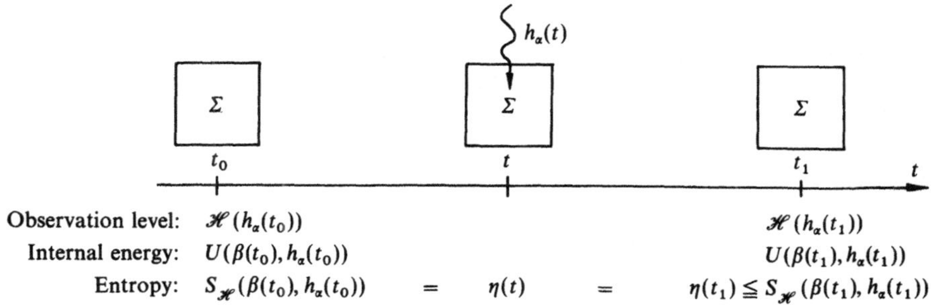

Fig. 6.2. The dynamic transition of an adiabatic system from a thermal-equilibrium state (0): $h_\alpha(t_0)$, $\beta(t_0)$, $U(t_0)$, $S(t_0)$ to another thermal-equilibrium state (1): $h_\alpha(t_1)$, $\beta(t_1)$, $U(t_1)$, $S(t_1)\geq S(t_0)$

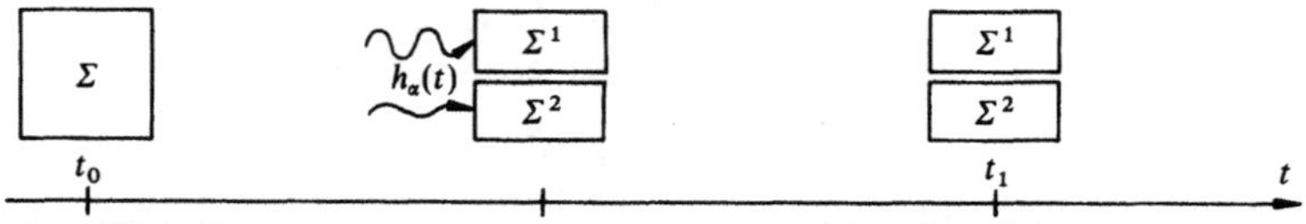

Observation level: $\mathscr{H}(h_\alpha(t_0))$ $\qquad$ $\mathscr{H}^1(h_\alpha(t_1)), \mathscr{H}^2(h_\alpha(t_1))$

Internal energies: $U(\beta(t_0), h_\alpha(t_0))$ $\qquad$ $U^1(\beta^1(t_1), h_\alpha(t_1)), U^2(\beta^2(t_1), h_\alpha(t_1))$

Entropy: $S_{\mathscr{H}}(\beta(t_0), h_\alpha(t_0)) \quad = \quad \eta(t) \quad = \quad \eta(t_1) \leqq S_{\mathscr{H}^1, \mathscr{H}^2}(\beta^1(t_1), \beta^2(t_1), h_\alpha(t_1))$

Fig. 6.3. The dynamic transition of an adiabatic system from a thermal-equilibrium state (0): $h_\alpha(t_0)$, $\beta(t_0)$, $U(t_0)$, $S(t_0)$ to an inhibited equilibrium (1): $h_\alpha(t_1)$, $\beta^1(t_1)$, $\beta^2(t_1)$, $U^1(t_1)$, $U^2(t_1)$, $S(t_1 \geq S(t_0)$

be established, must of course not *necessarily* be fulfilled. As a counterexample, we might consider the case where for times $t > t_0$, an adiabatic wall is inserted into Σ, and different forces act on the two subsystems Σ^1 and Σ^2 (Fig. 6.3).

Then the statistical operator $\rho(t_1)$ can at most be equivalent to a generalized canonical statistical operator with two different temperatures $\beta^1(t_1)$ and $\beta^2(t_1)$ [compare Sect. 5.3.2]:

$$\rho(t_1) \Rightarrow \mathscr{R}_{\mathscr{H}^1, \mathscr{H}^2}(t_1) = \frac{\exp\left[-\sum_\nu \beta^\nu(t_1) \mathscr{H}^\nu(h_\alpha(t_1))\right]}{Z_{\mathscr{H}^1, \mathscr{H}^2}} . \tag{6.3.5}$$

The phrase "insertion of an adiabatic wall" must naturally not be taken literally. It is enough that e.g. an external field acts *only* on one part of the system Σ and, when the interaction between the two parts is weak, practically no energy exchange with the rest of the system occurs up to the time t_1 ("inhibited equilibrium"). Finally, we point out that the observation level $\mathscr{H}_1, \mathscr{H}_2$ may also be *nondecomposable* in the sense of Sect. 5.3.3.

In the present example of a larger observation level $\mathscr{H}^1, \mathscr{H}^2$ at a time t_1, the entropy [from (6.2.6)] can also not decrease,

$$S_{\mathscr{H}}(\beta(t_0), h_\alpha(t_0)) \leqq S_{\mathscr{H}^1, \mathscr{H}^2}(\beta^1(t_1), \beta^2(t_1), h_\alpha(t_1)) . \tag{6.3.6}$$

This statement is not in contradiction to (5.2.26), since that equation requires only that $S_{\mathscr{H}}(t_1) \geqq S_{\mathscr{H}^1 \mathscr{H}^2}(t_1)$.

At this point, we give a nearly trivial, but important explicit example of how a system in thermal equilibrium can acquire two different temperatures by means of dynamics. We consider a magnetic system, which is in a static magnetic field H_0 at time t_0 and is in thermal equilibrium, so that

$$\mathscr{R}_{\mathscr{H}}(t_0) = \frac{\exp - \beta(t_0) \mathscr{H}(H_0)}{Z_0} . \tag{6.3.7}$$

The Hamiltonian $\mathscr{H}$ is composed of the Zeeman energy $\mathscr{H}_Z(H_0) = -\mathscr{M} H_0$ as well as a magnetic-field-independent energy $\mathscr{H}_R$ (e.g. a lattice energy or an exchange energy):

$$\mathscr{H}(H_0) = -\mathscr{M} H_0 + \mathscr{H}_R \quad (\text{for } t_0) \ . \tag{6.3.8}$$

Now, we switch on an additional magnetic field h suddenly, as a step function (Fig. 6.4a), so that the Hamiltonian at a time $t_+ = t_0 + \varepsilon$ (ε sufficiently small) is given by

$$\mathscr{H}(H_0 + h) = -\mathscr{M}(H_0 + h) + \mathscr{H}_R \quad (\text{for } t_+) \ . \tag{6.3.9}$$

From the von Neumann equation, however, it can be shown (in the "sudden approximation") that the statistical operator does not change as a result of a jump in the Hamiltonian [compare Exercise 7.1], so that we have

$$\rho(t_+) = \mathscr{R}_{\mathscr{H}}(t_0) = \frac{\exp[\beta(t_0)\mathscr{M} H_0 - \beta(t_0)\mathscr{H}_R]}{Z_0} \ . \tag{6.3.10}$$

However, we can also write $\rho(t_+)$ as a generalized canonical statistical operator over the observation level $\mathscr{H}_Z(H_0 + h)$, $\mathscr{H}_R$,

$$R_{\mathscr{H}_Z, \mathscr{H}_R}(t_+) = \frac{\exp[+\beta_Z(t_+)\mathscr{M} \cdot (H_0 + h) - \beta_R(t_+)\mathscr{H}_R]}{Z_0} \ , \tag{6.3.11}$$

where one defines the Zeeman temperature β_Z at the time t_+

$$\beta_Z(t_+) = \frac{H_0}{H_0 + h} \beta(t_0) \tag{6.3.12a}$$

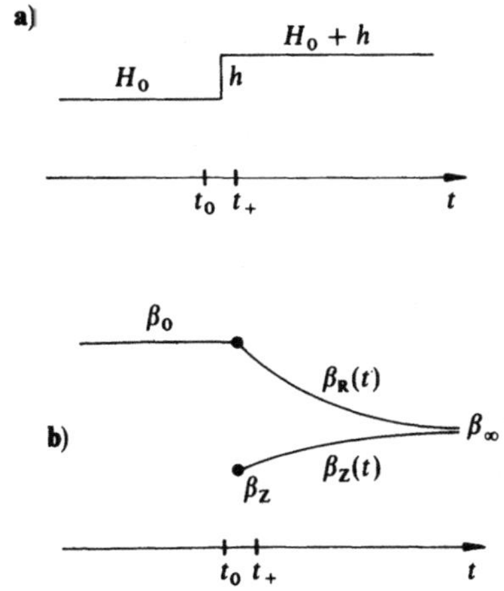

Fig. 6.4. (a) Sudden jump in the field h. **(b)** Different temperatures $\beta_Z(t_+)$ and $\beta_R(t_+)$. With subsequent thermal conduction, $\beta_Z(t)$ and $\beta_R(t)$ converge to a common final temperature β_∞

and, for the remaining part of the Hamiltonian,

$$\beta_R(t_+) = \beta(t_0) \ . \tag{6.3.12b}$$

For the entropies, when the application of the additional field is sudden, the equals sign holds:

$$S_{\mathscr{H}}(\beta(t_0)) = S_{\mathscr{H}_Z, \mathscr{H}_R}(\beta_Z(t_+), \beta_R(t_+)) \ . \tag{6.3.13}$$

An "inhibited" equilibrium with different temperatures can of course not be maintained for long times, since even the smallest interaction gives rise to an exchange of energy (relaxation process), so that after a sufficiently long time t_2, the statistical operator $\rho(t_2)$ for standard systems becomes equivalent to a canonical statistical operator with a common final temperature β_∞. The entropy naturally increases during this process.

6.4 Accompanying Entropy $S_{\{\mathscr{G}\}}(t)$ with Respect to an Observation Level $\{\mathscr{G}\}$

To conclude this chapter, we consider the definition of an entropy $S_{\{\mathscr{G}\}}$ which depends *continuously* on the time t. We begin with the concept of the construction of a generalized canonical statistical operator $\mathscr{R}_{\{\mathscr{G}\}}$ for a system Σ, as developed in Chap. 5. This is carried out at each time t: we use the same observation level $\{\mathscr{G}\}$ at all times $t > t_0$ and calculate the expectation values of the observation level

$$\langle \mathscr{G}_\nu \rangle(t) = \mathrm{Tr}(\mathsf{U}(t, t_0)\rho(t_0) \cdot \mathscr{G}_\nu) \tag{6.4.1}$$

for a particular initial state $\rho(t_0)$. Thus, $\langle \mathscr{G}_\nu \rangle(t)$ for adiabatic systems may be calculated from the dynamic evolution $\mathsf{U}(t, t_0)$. However, we can also allow the system considered, Σ, to be a non-adiabatically-isolated subsystem, if the expectation values $\langle \mathscr{G}_\nu \rangle(t)$ are determined from the Hamiltonian of an enlarged system [compare Sect. 3.3].

These given expectation values $\langle \mathscr{G}_\nu \rangle(t)$ are used as constraints for the maximization of η *at any time* t to obtain a generalized statistical operator

$$\mathscr{R}_{\{\mathscr{G}\}}(t) = \frac{\exp\left[-\sum\limits_\nu \lambda_\nu(t)\mathscr{G}_\nu(h_\alpha(t))\right]}{\mathrm{Tr}\exp\left[-\sum\limits_\nu \lambda_\nu(t)\mathscr{G}_\nu(h_\alpha(t))\right]} \ . \tag{6.4.2}$$

In doing this, we perform a *mapping* of $\rho(t)$ onto $\mathscr{R}_{\{\mathscr{G}\}}(t)$ (see Chap. 18). The observables $\mathscr{G}_\nu$ may themselves be explicitly time-dependent through external parameters $h_\alpha(t)$. Independently of whether the chosen observation level $\{\mathscr{G}\}$ is sufficient for other observables or not, we define (following Schwegler) the *time-dependent* Lagrange multipliers[3]:

$$\lambda_\nu(t) = \lambda_\nu(\langle\mathscr{G}_1\rangle(t), \ldots \langle\mathscr{G}_n\rangle(t), h_1(t), \ldots) \tag{6.4.3}$$

and the *time-dependent entropy*:

$$\begin{aligned} S_{\{\mathscr{G}\}}(t) &= -k\,\mathrm{Tr}(\mathscr{R}_{\{\mathscr{G}\}}(t) \ln \mathscr{R}_{\{\mathscr{G}\}}(t)) \\ &= k \sum_\nu \lambda_\nu(t)\langle\mathscr{G}_\nu\rangle(t) + k \ln Z_{\{\mathscr{G}\}}(t) \end{aligned} \tag{6.4.4}$$

over an arbitrary, but fixed observation level $\{\mathscr{G}\}$. These definitions of the Lagrange multipliers $\lambda_\nu(t)$ and of the entropy $S_{\{\mathscr{G}\}}(t)$ are independent of how far the system is removed from thermodynamic equilibrium. One refers to $\mathscr{R}_{\{\mathscr{G}\}}(t)$ as the accompanying canonical statistical operator and $S_{\{\mathscr{G}\}}(t)$ as the *accompanying entropy with respect to the observation level* $\{\mathscr{G}\}$.

We note that $\mathscr{R}_{\{\mathscr{G}\}}(t)$ does not obey the von Neumann equation, but rather an integro-differential equation given by Robertson, which we shall treat in Sect. 18.2.

The time variation of this accompanying entropy $S_{\{\mathscr{G}\}}(t)$ is given by the relation [from (5.2.22)]

$$\boxed{\begin{aligned} \frac{dS_{\{\mathscr{G}\}}(t)}{dt} &= k \sum_\nu \lambda_\nu(t) \left\{ \frac{d}{dt}\langle\mathscr{G}_\nu\rangle(t) - \sum_\alpha \mathrm{Tr}\left(\mathscr{R}_{\{\mathscr{G}\}} \frac{\partial \mathscr{G}_\nu}{\partial h_\alpha}\right) \dot{h}_\alpha(t) \right\} \\ &= k \sum_\nu \lambda_\nu(t)\,\mathrm{Tr}\left(\frac{d\mathscr{R}_{\{\mathscr{G}\}}}{dt} \mathscr{G}_\nu(h_\alpha(t))\right) \end{aligned}} \quad , \tag{6.4.5}$$

where the equation must be read from right to left, according to the preceding arguments: the expectation values are given by the dynamics and determine the variation of the accompanying entropy. *This entropy can both increase and decrease* (even in the special case of an adiabatically isolated system)!

We summarize these results: equations (6.4.2) through (6.4.4) are valid for any (even nonadiabatic) system Σ, which may be arbitrarily far from equilibrium, and for any observation level, which need not even be representative. This generality of definition of $\lambda_\nu(t)$ and of $S_{\{\mathscr{G}\}}(t)$ is paid for by the lack of a specification of the sign of $\dot{S}_{\{\mathscr{G}\}}(t)$ (cf. Fig. 6.5).

The reader may be surprised that these statements are not in contradiction to the considerations of Sect. 6.2. There, however, the course of the dynamics of

[3] In the example of the partial energies $\mathscr{H}_\nu$ as the observation level, we thus obtain a definition of time-dependent temperatures $\beta_\nu(t)$, even far from equilibrium.

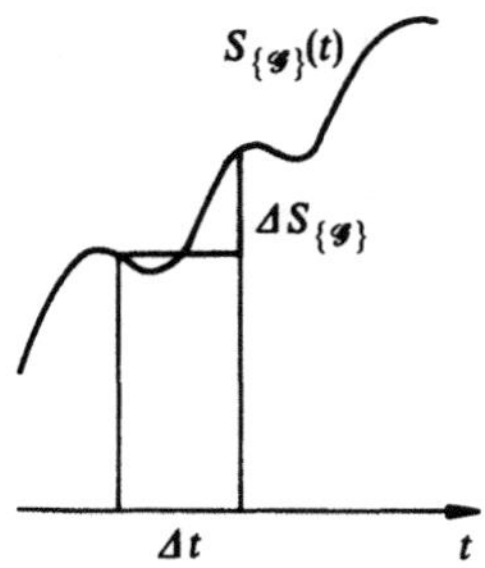

Fig. 6.5. The accompanying entropy $S_{\{\mathcal{G}\}}(t)$. Over a time interval $\Delta t\,(>\tau_c)$, which may be moved at will on the time axis, $\Delta S_{\{\mathcal{G}\}}/\Delta t$ is chosen to be ≥ 0 (6.4.8); this is compatible with $dS_{\{\mathcal{G}\}}/dt \gtreqless 0$

$\rho(t)$ was changed at certain times t_n by the transition

$$\rho(t_n) \Rightarrow \mathcal{R}_{\{\mathcal{G}\}}(t_n) \ . \tag{6.4.6}$$

since new initial conditions were set at these times,

$$\rho(t_0) \xrightarrow{\text{v.N.}} \rho(t_1) \Rightarrow \mathcal{R}_{\{\mathcal{G}\}}(t_1) \xrightarrow{\text{v.N.}} \rho(t_2) \Rightarrow \mathcal{R}_{\{\mathcal{G}\}}(t_2) \rightarrow \dots \ . \tag{6.4.7}$$

This was done under the assumption that the transition (6.4.6) causes no noticeable error in the dynamics of certain observables of interest. Such errors would be important, if the stepwidth $\Delta t = t_n - t_{n-1}$ were allowed to become too small. On a microscopic time scale, Δt must always be large compared to the correlation times of the microscopic motion. The result of Sect. 6.2 for adiabatic processes

$$\frac{\Delta S_{\{\mathcal{G}\}}}{\Delta t} \geqq 0 \tag{6.4.8}$$

with the replacement $\rho(t_n) \rightarrow \mathcal{R}(t_n)$, thus is *only* valid for finite stepwidths $\Delta t > \tau_c$, but for arbitrary times t_n (c.f. Fig. 13.1 and (15.2.16)). The increase of $\Delta S_{\{\mathcal{G}\}}$ over steps Δt is not in contradiction to a temporary decrease of the accompanying entropy $S_{\{\mathcal{G}\}}(t)$ (cf. Fig. 6.5).

Part 2

Response to Time-Dependent External Fields

In this section, we shall treat the reaction of a sample, termed the *response*, to given time-dependent external fields. The goal will be to relate the response to properties of the sample (dynamic materials properties), which are not functions of the particular time dependence of the external fields.

Following a general introduction to the approximate treatment of external perturbations, we shall concentrate on linear response theory; i.e. we shall be interested in the frequently-occurring case in which the response of the sample is described to a good approximation by the linear terms in the amplitudes of the external fields only. For this treatment, we use a scalar product in Liouville space which is particularly suited to the mathematical description of linear response problems in quantum statistics.

Chapter 10 gives an introduction to nonlinear response theory. For simplicity, we restrict the treatment here to responses which are quadratic in the amplitudes of the external fields.

Extremely nonlinear phenomena, which extend far beyond quadratic response theory, have aroused considerable interest in recent times. We shall not treat these problems in general in this book; however, a starting point for the quantum-statistical derivation of nonlinear dissipative equations may be found in Part 4.

7. The Quantum-Statistical Formulation of Response Theory

7.1 Introduction to the Physical Problem

Let the physical system Σ at time t_0 be in *thermal equilibrium*, having the temperature $T = 1/k\beta$, so that it can be described by the canonical statistical operator[1]

$$\mathscr{R}_\beta = \frac{e^{-\beta\mathscr{H}}}{\mathrm{Tr}\, e^{-\beta\mathscr{H}}} \ . \tag{7.1.1}$$

The Hamiltonian $\mathscr{H}$ of the system depends upon constant parameters such as the volume or on temporally constant fields, as described in Part 1. To distinguish them from time-dependent fields, we shall denote them by H_α^0, but will mention them explicitly *only* when required for clarity.

For times $t \geqq t_0$, we allow additional given external *time-dependent* c-number fields[2] $h_\alpha(t)$ to act on the system Σ (Fig. 7.1). They lead to an explicit time dependence of the Hamiltonian for the system Σ:

$$\mathscr{H}(t) = \mathscr{H}(H_1^0 + h_1(t), \ldots) \quad \text{for} \quad t \geqq t_0 \ . \tag{7.1.2}$$

The question of the *dynamic response* of the system Σ to the time-dependent influence of $h_\alpha(t)$ means the calculation of the time dependence of the expectation values $\langle\mathscr{F}\rangle(t)$ of observables $\mathscr{F}$ of the system for times $t \geqq t_0$ (Fig. 7.1). These expectation values $\langle\mathscr{F}\rangle(t) = \mathrm{Tr}(\rho(t)\mathscr{F}(h(t)))$ are determined in the Schrödinger formulation by the von Neumann equation for $\rho(t)$ and a possible explicit time dependence $\mathscr{F}(h(t))$. The function $\langle\mathscr{F}\rangle(t)$ is not given just by the values of the fields $h_\alpha(t)$ at the time t, but rather by the set of all their values $h_\alpha(t')$ in the preceding time interval $t_0 \leqq t' \leqq t$. $\langle\mathscr{F}\rangle(t)$ is a *functional* of the values of *all* the field strengths $h_\alpha(t')$ which occurred up to the time t,

$$\boxed{\langle\mathscr{F}\rangle(t) = F[h_\alpha(t')]} \qquad (t_0 \leqq t' \leqq t) \ . \tag{7.1.3}$$

---

[1] Depending on the particular problem, it may be expedient to use a grand canonical statistical operator (5.3.15) instead of (7.1.1). The formal replacement $\mathscr{H} \to \mathscr{H} + (\mu/\beta)\mathscr{N}$ $(-\mu/\beta = \text{chemical potential}, \mathscr{N} = \text{particle number operator})$ yields the corresponding formulas.

[2] In general, the functions $h_\alpha(t)$ can have an arbitrary physical nature, for example they could be a time-dependent volume. However, since these functions frequently represent fields, particularly electromagnetic fields, we refer to them generically as "fields".

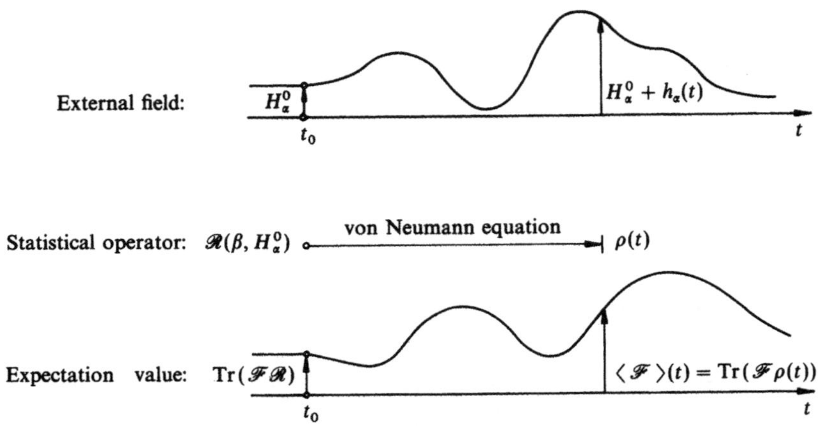

Fig. 7.1. The principle of response theory: the driving field $h_\alpha(t)$ and the response $\langle\mathscr{F}\rangle(t)$

Furthermore, $\langle\mathscr{F}\rangle(t)$ naturally depends upon the statistical operator (7.1.1), i.e. on the initial temperature β and the possible static parameters H_α^0. It is the goal of the present part of this book to find quantum statistical expressions for expansions of the functional (7.1.3).

The functional (7.1.3) for the response requires some interpretation and comment:

The statement that in the calculation of $\langle\mathscr{F}\rangle(t)$ only the past field strengths (for $t' \leqq t$) enter, guarantees that *causality*, a general principle of physics, is obeyed: the response $\langle\mathscr{F}\rangle(t)$ can only be a result of the causes (fields) in the past. The fields which act *after* the time t are arbitrarily variable and thus do not enter into $\langle\mathscr{F}\rangle(t)$.

The system Σ considered is assumed to be *thermally isolated* for all times $t \geqq t_0$. It is described by an explicitly time-dependent Hamiltonian, so that energy is added to the system only in the form of work (cf. Sect. 3.3). This sort of adiabaticity is only an approximation for a physically real system in a particular time interval, since every system enters into energy (heat) exchange with its surroundings after a certain time (cf. Sect. 3.3.4). If one wishes to describe the dynamics of a particular system Σ^1 which is in energetic contact via $\mathscr{H}^{12}$ with another system Σ^2 (reservoir), then it is necessary to consider the combined system Σ obtained from Σ^1 and Σ^2 (Fig. 3.3); this can then again be treated as being adiabatically isolated.

Example: A spin system Σ^S may be driven by a time-dependent magnetic field. However, the spin system is in energetic contact with a lattice (phonons) via the spin-lattice interaction $\mathscr{H}^{SL}$. Although the magnetic field acts directly only on the spins, the lattice also participates in the dynamics through $\mathscr{H}^{SL}$ and thus influences the time dependence of the spin magnetization; this can be treated by considering the combined dynamics of the quantum-statistical total system Σ.

It must be remembered that the quantities $h_\alpha(t)$ refer to *external* fields, which would act even in the absence of the system (i.e. in the absence of matter); for example in the case of electromagnetic fields

$$h_\alpha(t) \to E_i^{\text{ex}}(\mathbf{r}, t) \ , \qquad B_i^{\text{ex}}(\mathbf{r}, t) \ .^3 \tag{7.1.4}$$

However, since matter itself contains a source of electromagnetic fields in the form of charges and currents, these induced fields are superposed on the externally applied fields. It should be remembered that the Maxwell equations for matter refer to the total electromagnetic fields.

The functional (7.1.3) can hardly be specified for arbitrary amplitudes of the external field. However, in the case of small perturbations, one can carry out a *functional expansion*, i.e. an expansion of (7.1.3) in the amplitudes of the field.

To this end, one first expands the Hamiltonian $\mathscr{H}(t)$ in powers of the fields $h_\alpha(t)$,

$$\mathscr{H}(t) = \mathscr{H}|_{h=0} + \sum_\alpha \left.\frac{\partial \mathscr{H}}{\partial h_\alpha}\right|_{h=0} \cdot h_\alpha(t)$$

$$+ \frac{1}{2!} \sum_{\alpha\beta} \left.\frac{\partial^2 \mathscr{H}}{\partial h_\alpha \partial h_\beta}\right|_{h=0} \cdot h_\alpha(t) h_\beta(t) + \dots \tag{7.1.5}$$

with $\mathscr{H}|_{h=0} = \mathscr{H}$, as it enters into (7.1.1). The *linear response theory* consists of representing $\langle \mathscr{F} \rangle(t)$ approximately as a linear functional of $h_\alpha(t')$,

$$\langle \mathscr{F} \rangle(t) = \langle \mathscr{F} \rangle_\beta + \langle \mathscr{F} \rangle_1(t) \ , \tag{7.1.6}$$

or, expressed differently, to obtain the Fourier transform with respect to time, $\langle \mathscr{F} \rangle_1(\omega)$, which is proportional to $h_\alpha(\omega)$ (cf. Sect. 9.1.2); for example, the polarization $\mathfrak{P}(\omega) = \chi^{\text{ex}}(\omega)\, \mathfrak{E}^{\text{ex}}(\omega)$ as the response to an oscillating electric field $\mathfrak{E}^{\text{ex}}(\omega)$.[4] This linear theory is a good approximation for sufficiently small amplitudes of the time-dependent field $h_\alpha(t)$.[5] We emphasize here that the functional expansion of the response is *not* an expansion in the ratio $h_\alpha(t)/H_\alpha^0$, i.e. in the strength of the time-dependent fields $h_\alpha(t)$ relative to the static fields H_α^0. It is quite possible that H_α^0 could be zero, for example. If one includes terms in the

[3] The index α thus represents the components i of a vector *and* the local (spatial) point $\mathbf{r}$.

[4] $\chi^{\text{ex}}(\omega)$ is a material property which in general depends on the geometry of the sample, since it describes the response to the external field in the absence of the sample. In the usual physical nomenclature, this means e.g. that $\chi^{\text{ex}}(\omega)$ for ellipsoids in a homogeneous electric field contains the depolarization factor.

[5] One might at first think that the first term $\langle \mathscr{F} \rangle_1(t)$ of the expansion of $\langle \mathscr{F} \rangle(t)$ around $h_\alpha(t) = 0$ does not include all of the region which is experimentally observed to be linear. This region might contain stronger perturbation amplitudes $h_\alpha(t)$. However, experience shows that linear response theory correctly represents the observed linear properties of materials.

expansion of the functional (7.1.3) beyond the linear approximation in the field amplitude $h_\alpha(t')$, one arrives at *nonlinear response theory* (see Fig. 7.2 and Chap. 10), which, for example, can be applied to problems in nonlinear optics.

7.2 The Mathematical Formulation of the Problem

We shall treat the problem of the quantum statistical dynamics of a system Σ under the action of time-dependent fields in the Schrödinger formulation. An explicitly time *dependent* Liouville operator (3.1.12) corresponds to the Hamiltonian $\mathscr{H}(t)$ (7.1.2). The expansion (7.1.5) is then given by

$$\mathsf{L}(t) = \mathsf{L} + \mathsf{L}_1(t) + \mathsf{L}_2(t) + \ldots \quad (t \geq t_0) \;, \tag{7.2.1}$$

where $\mathsf{L}_\nu(t)$ is thus in the ν-th order in $h_\alpha(t)$. The solution of the von Neumann equation,

$$\dot{\rho}(t) = -\mathrm{i}\mathsf{L}(t)\rho(t) \tag{7.2.2}$$

is obtained from the superoperator $\mathsf{U}(t, t_0)$ for the time evolution, (3.2.6)

$$\rho(t) = \mathsf{U}(t, t_0)\rho(t_0) \;, \tag{7.2.3}$$

which is determined by the differential equation

$$\frac{d}{dt}\mathsf{U}(t, t_0) = -\mathrm{i}\mathsf{L}(t)\mathsf{U}(t, t_0) \quad \text{with} \quad \mathsf{U}(t_0, t_0) = 1 \;. \tag{7.2.4}$$

In order to separate out the unperturbed dynamics in $\mathsf{U}(t, t_0)$ which are given by L,

$$\mathsf{U}_0(t, t_0) = \mathrm{e}^{-\mathrm{i}\mathsf{L}\cdot(t-t_0)} \tag{7.2.5}$$

we make use of the operator identity (3.4.4b):

$$\boxed{\mathsf{U}(t, t_0) = \mathsf{U}_0(t-t_0) - \mathrm{i}\int_{t_0}^{t} \mathsf{U}_0(t-t')(\mathsf{L}_1(t') + \mathsf{L}_2(t') + \ldots)\mathsf{U}(t', t_0)dt'} \;. \tag{7.2.6}$$

It represents a linear integral equation of the second kind for the unknown superoperator $\mathsf{U}(t, t_0)$. Due to the explicit time dependences, $\mathsf{L}_\nu(t)$, the superoperator $\mathsf{U}(t, t_0)$ [in contrast to $\mathsf{U}_0(t-t_0)$] does not depend only on the time difference $t-t_0$.

An approximate solution of the integral equation (7.2.6) for small amplitudes $h_\alpha(t)$ can be obtained from a series expansion,

$$\mathsf{U}(t, t_0) = \mathsf{U}_0(t-t_0) + \mathsf{U}_1(t, t_0) + \mathsf{U}_2(t, t_0) + \ldots \;, \tag{7.2.7}$$

where $\mathsf{U}_\nu(t, t_0)$ symbolizes the contribution of ν-th order in $h_\alpha(t)$. On inserting (7.2.7) into both sides of (7.2.6), and collecting terms of ν-th order[6], we obtain an expression for $\mathsf{U}_\nu(t, t_0)$. For the first terms, this yields

$$\mathsf{U}_1(t, t_0) = -\mathrm{i} \int_{t_0}^{t} \mathsf{U}_0(t-t')\mathsf{L}_1(t')\mathsf{U}_0(t'-t_0)dt' \;, \tag{7.2.8}$$

$$\begin{aligned} \mathsf{U}_2(t, t_0) &= -\mathrm{i} \int_{t_0}^{t} \mathsf{U}_0(t-t')\mathsf{L}_1(t')\mathsf{U}_1(t', t_0)dt' \\ &\quad -\mathrm{i} \int_{t_0}^{t} \mathsf{U}_0(t-t')\mathsf{L}_2(t')\mathsf{U}_0(t'-t_0)dt' \;, \end{aligned} \tag{7.2.9}$$

etc.

The ν-th order, $\mathsf{U}_\nu(t, t_0)$, is thus obtained by inserting all the preceding expressions for $\mathsf{U}_\mu(t, t_0)$ $(\mu = 1, \ldots, \nu-1)$ into the ν-th equation. It contains the field $h_\alpha(t)$ in functional form ν times, i.e. at different times. The time integrals which occur imply that the Neumann series, (7.2.7), can be interpreted in physical language as a "power-series expansion with memory". This is the mathematical expression of the fact stated in Sect. 7.1 that the response does not instantaneously follow the cause $h_\alpha(t)$.

According to (7.2.8), $\mathsf{U}_1(t, t_0)$ is a linear operator-functional of $\mathsf{L}_1(t)$. On substituting (7.2.8) in (7.2.9), $\mathsf{U}_2(t, t_0)$ becomes a quadratic functional of $\mathsf{L}_1(t)$ and a linear functional of $\mathsf{L}_2(t)$,

$$\begin{aligned} \mathsf{U}_2(t, t_0) &= (-\mathrm{i})^2 \int_{t'=t_0}^{t} \mathsf{U}_0(t-t')\mathsf{L}_1(t') \int_{t''=t_0}^{t'} \mathsf{U}_0(t'-t'')\mathsf{L}_1(t'')\mathsf{U}_0(t''-t_0)dt'\,dt'' \\ &\quad -\mathrm{i} \int_{t_0}^{t} \mathsf{U}_0(t-t')\mathsf{L}_2(t')\mathsf{U}_0(t'-t_0)dt' \;. \end{aligned} \tag{7.2.10}$$

Equation (7.2.8) is the starting point of linear response theory (Chap. 9), while (7.2.10) yields quadratic response theory (Chap. 10), etc.

The expansion of the time evolution $\mathsf{U}(t, t_0)$ given by (7.2.8) and (7.2.9) is pictorially illustrated in Fig. 7.2. One readily recognizes the rule for obtaining higher orders, e.g. for $\mathsf{U}_3(t, t_0)$.

---

[6] This procedure is simply a time-dependent perturbation theory. It can be carried out formally most expediently by introducing a parameter λ into (7.2.6). With

$$\mathsf{L}(t) = \mathsf{L} + \lambda \mathsf{L}_1(t) + \lambda^2 \mathsf{L}_2(t) + \ldots \quad \text{and}$$

$$\mathsf{U}(t, t_0) = \mathsf{U}_0(t-t_0) + \lambda \mathsf{U}_1(t, t_0) + \lambda^2 \mathsf{U}_2(t, t_0) + \ldots \;,$$

one obtains immediately from (7.2.6) by comparison of equal powers of λ the ordering in powers of the fields $h_\alpha(t')$. In the result, λ is again set equal to 1, and we obtain (7.2.8) and (7.2.9) for $\mathsf{U}_1(t, t_0)$ and $\mathsf{U}_2(t, t_0)$ as well as those for the higher orders.

On inserting the expansion (7.2.7) in (7.2.3), we obtain the corresponding expansion of the statistical operator

$$\rho(t)=\rho_0(t)+\rho_1(t)+\rho_2(t)+\dots \tag{7.2.11}$$

with

$$\rho_\nu(t)=\mathsf{U}_\nu(t,t_0)\rho(t_0)\ . \tag{7.2.12}$$

In particular, if $\rho(t_0)=\mathscr{R}_\beta$, i.e. it is stationary, as we assumed in Sect. 7.1[7], then

$$\mathsf{L}\mathscr{R}_\beta=0 \quad \text{and thus} \quad \mathsf{U}_0(t-t_0)\mathscr{R}_\beta=\mathscr{R}_\beta\ . \tag{7.2.13}$$

Equations (7.2.8) and (7.2.10) lead in this case to the result

$$\rho_0(t)=\mathscr{R}_\beta\ , \tag{7.2.14}$$

$$\rho_1(t)=-\mathrm{i}\int_{t_0}^{t}\mathsf{U}_0(t-t')\mathsf{L}_1(t')dt'\mathscr{R}_\beta\ , \tag{7.2.15}$$

$$\begin{aligned}\rho_2(t)=(-\mathrm{i})^2&\int_{t'=t_0}^{t}\mathsf{U}_0(t-t')\mathsf{L}_1(t')\int_{t''=t_0}^{t'}\mathsf{U}_0(t'-t'')\mathsf{L}_1(t'')dt'dt''\mathscr{R}_\beta\\ &-\mathrm{i}\int_{t_0}^{t}\mathsf{U}_0(t-t')\mathsf{L}_2(t')dt'\mathscr{R}_\beta\ ,\end{aligned} \tag{7.2.16}$$

etc.

With this result, we have derived the functional expansion of the expectation values mentioned in Sect. 7.1

$$\langle\mathscr{F}\rangle(t)=\mathrm{Tr}(\mathscr{F}\rho(t))=\langle\mathscr{F}\rangle_\beta+\langle\mathscr{F}\rangle_1(t)+\langle\mathscr{F}\rangle_2(t)+\dots \tag{7.2.17}$$

with

$$\langle\mathscr{F}\rangle_\nu(t)=\mathrm{Tr}(\mathscr{F}\rho_\nu(t))\ . \tag{7.2.18}$$

$\langle\mathscr{F}\rangle_\nu(t)$ is the response to the external fields $h_\alpha(t)$ in ν-th order which we had sought, in the case that $\mathscr{F}$ is explicitly time *in*dependent.

[7] In Sect. 9.9, we shall depart from this limitation and allow nonstationary $\rho(t_0)$'s.

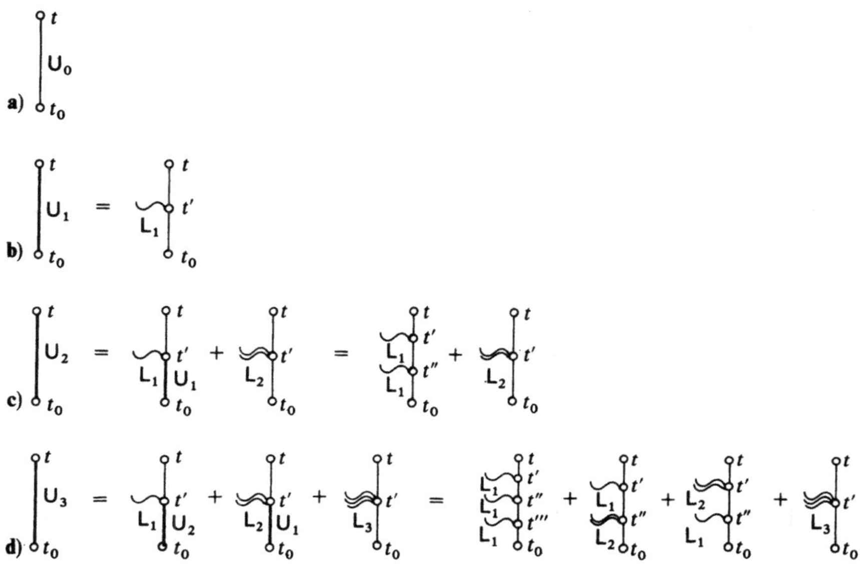

Fig. 7.2a–d. Graphical representation of the expansion of $\mathsf{U}(t, t_0)$ in the driving field

When the observable itself depends on the fields $h_\alpha(t)$, $\mathscr{F}(h_1(t), \ldots)$,[8] then these functions must likewise be expanded in powers of $h_\alpha(t)$,

$$\mathscr{F}(h_1(t), \ldots) = \mathscr{F}|_{h=0} + \sum_\alpha \left.\frac{\partial \mathscr{F}(h)}{\partial h_\alpha}\right|_{h=0} h_\alpha(t) + \ldots , \tag{7.2.19}$$

and included in the appropriate order (cf. Sect. 9.9).

The expression for the *linear response*, (7.2.18) with $\nu = 1$, may in a straightforward manner be brought into a form which can serve as the starting point for further manipulations by means of the definition of a *suitable scalar product* in Liouville space. We shall thus introduce this scalar product and deal with its special properties in the following Chap. 8.

Exercise 7.1 The time-dependent Liouville operator (in the Schrödinger formulation) of a system in which a field is suddenly turned on is given by

$$\mathsf{L}(t) = \mathsf{L} - \mathsf{M}\hat{h}\,\theta(t) ,$$

---

[8] If we describe the action of external electromagnetic fields by 4-component potentials $A_\mu^{\mathrm{ex}}(\mathbf{r}, t)$ (c-numbers) ($\mu = 1, \ldots, 4$), then as is well known, the operators (Schrödinger formulation) for the electric currents, $j_\mu(\mathbf{r}, t) = j_\mu(\mathbf{r}, A_1^{\mathrm{ex}}(\mathbf{r}, t), \ldots, A_4^{\mathrm{ex}}(\mathbf{r}, t))$, are functions of the 4-potentials $A_\mu^{\mathrm{ex}}(\mathbf{r}, t)$ and thus become explicitly time dependent $[h_\alpha(t) \to A_\mu^{\mathrm{ex}}(\mathbf{r}, t),\ \mathscr{F}(h(t)) \to j_\mu(\mathbf{r}, A_1^{\mathrm{ex}}(\mathbf{r}, t), \ldots, A_4^{\mathrm{ex}}(\mathbf{r}, t))]$.

where $\theta(t)$ is the jump function

$$\theta(t) = \begin{cases} 1, t > 0 \\ 0, t < 0 \end{cases} .$$

Calculate the time evolution operator $\mathsf{U}(+0, -0)$ from $t = -0$ to $t = +0$.

Exercise 7.2 a) Let the time-dependent field $h(t)$ in the Liouville operator be given by a δ-pulse:

$$\mathsf{L}(t) = \mathsf{L} - \mathsf{M}\hat{h}\delta(t) .$$

Find the time evolution operator $\mathsf{U}(+0, -0)$ and $\rho(+0)$.

b) Let the Liouville operator $\mathsf{L}(t)$ contain a sum of δ-pulses

$$\mathsf{L}(t) = \mathsf{L} - \mathsf{M} \sum_{\nu=1}^{n} , \hat{h}^{(\nu)} \delta(t - t_\nu) , \quad t_0 < t_1 \ldots < t_n .$$

Determine the statistical operator $\rho(t)$ at a time $t > t_n$, if $\rho(t_0)$ is given by $\mathscr{R}_\beta$.

8. A Scalar Product in the Liouville Space for Linear Response Theory

In Chap. 3, we have already discussed some important properties of the Liouville space $\mathfrak{L}$; these were used in Chap. 7 to formulate the linear response theory. However, up to now, we have not defined the *scalar product* (|) between the elements of $\mathfrak{L}$. Although the mathematical properties of a scalar product (briefly summarized in the following Sect. 8.1) are of course quite general, its actual form can vary, depending on the physical questions to be investigated (cf. Sect. 8.2.1). As we shall see in detail, Mori's definition of the scalar product (Sect. 8.2) is appropriate for all *linear* quantum-statistical problems[1].

8.1 Scalar Products and Projection Operators in Liouville Space

The scalar product in the Liouville space $\mathfrak{L}$ has formally just the same mathematical properties with which we are already familiar from quantum theory in the unitary space of state vectors, $\mathfrak{U}$. The only difference thus lies in the fact that now a complex number corresponds to two elements $\mathscr{F}$ and $\mathscr{G}$ in the Liouville space, which are themselves operators in the unitary space $\mathfrak{U}$. The formal rules are otherwise the same, so that we can limit ourselves here to a brief summary.

The question of the actual form of the scalar product, i.e. the rules for calculating the scalar product from the operators $\mathscr{F}$ and $\mathscr{G}$, will be treated in Sect. 8.2.

8.1.1 Properties of Scalar Products in $\mathfrak{L}$

A complex number a is attributed to pairs of elements $\mathscr{F}, \mathscr{G}, \ldots$ in $\mathfrak{L}$ (i.e. to linear operators in $\mathfrak{U}$ (cf. Chap. 3.1),

$$(\mathscr{F}|\mathscr{G})=a\ .^{2} \tag{8.1.1}$$

[1] In order to avoid confusion, in this book we use curved brackets (|) to denote *only* the Mori scalar product.

[2] We shall use the bra and ket notation, but do not mean it to imply two dual spaces.

Reversing the order leads to the complex conjugate of a, a^*,

$$(\mathscr{G}|\mathscr{F})=(\mathscr{F}|\mathscr{G})^* \ . \tag{8.1.2}$$

Along with the distributive property for addition, the factoring out of a complex number c from the scalar product is of importance. We have

$$(\mathscr{F}|c\mathscr{G})=c(\mathscr{F}|\mathscr{G}) \quad \text{and} \quad (c\mathscr{F}|\mathscr{G})=c^*(\mathscr{F}|\mathscr{G}) \ . \tag{8.1.3}$$

It follows from (8.1.2) that the scalar product of an element with itself must be real. This scalar product is always positive

$$(\mathscr{F}|\mathscr{F})\geqq 0 \ , \tag{8.1.4}$$

where the equals sign is only valid in the case that $\mathscr{F}$ is the null operator $\mathscr{O}$. Two operators $\mathscr{F}$ and $\mathscr{G}$ in $\mathfrak{U}$ are *orthogonal* as elements in $\mathfrak{L}$ if

$$(\mathscr{F}|\mathscr{G})=0 \tag{8.1.5}$$

without one of the factors being the null operator. A consequence of the properties of the scalar product is the Schwarz inequality:

$$|(\mathscr{F}|\mathscr{G})|^2\leqq(\mathscr{F}|\mathscr{F})\cdot(\mathscr{G}|\mathscr{G}) \ . \tag{8.1.6}$$

If a set of elements $|\mathscr{V}_\alpha)$ is orthonormalized,

$$(\mathscr{V}_\alpha|\mathscr{V}_{\alpha'})=\delta(\alpha,\alpha') \ , \tag{8.1.7}$$

and complete,

$$\sum\!\!\!\!\!\!\int_\alpha |\mathscr{V}_\alpha)(\mathscr{V}_\alpha|d\alpha=1 \ , \tag{8.1.8}$$

(where α is discrete or continuous), then the elements $|\mathscr{V}_\alpha)$ form a complete, normalized orthogonal basis in $\mathfrak{L}$, so that each element $|\mathscr{F})$ in $\mathfrak{L}$, i.e. each operator $\mathscr{F}$ in $\mathfrak{U}$, may be written in the form

$$|\mathscr{F})=\sum\!\!\!\!\!\!\int_\alpha |(\mathscr{V}_\alpha)(\mathscr{V}_\alpha|\mathscr{F})d\alpha \ . \tag{8.1.9}$$

8.1.2 Adjoint Operators (Superoperators) in $\mathfrak{L}$

In Sect. 3.1.2, we introduced the linear operators S in $\mathfrak{L}$ (called *superoperators* to distinguish them from the operators $\mathscr{F}$ in $\mathfrak{U}$). In addition to the properties mentioned there, we can now define *adjoint* superoperators S and $\mathsf{S}^\dagger$ using the scalar product:

$$(\mathscr{G}|\mathsf{S}\mathscr{F})=(\mathsf{S}^\dagger\mathscr{G}|\mathscr{F}) \ . \tag{8.1.10}$$

$$\mathsf{S} \text{ is Hermitian in } \mathfrak{L}, \text{ if } \mathsf{S}^\dagger=\mathsf{S} \ ; \tag{8.1.11}$$

$$\mathsf{S} \text{ is unitary in } \mathfrak{L}, \text{ if } \mathsf{S}^\dagger=\mathsf{S}^{-1} \ . \tag{8.1.12}$$

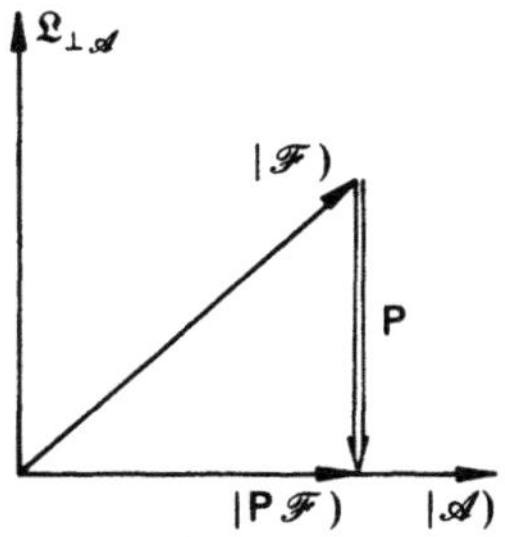

Fig. 8.1. The action of a projection operator $\mathsf{P}=|\mathscr{A})(\mathscr{A}|/(\mathscr{A}|\mathscr{A})$

8.1.3 Projection Operators P in $\mathfrak{L}$

Important superoperators in statistical physics are the projection operators (projectors) in $\mathfrak{L}$. We first define the projector

$$\boxed{\mathsf{P}=\frac{|\mathscr{A})(\mathscr{A}|}{(\mathscr{A}|\mathscr{A})}} \ . \tag{8.1.13}$$

It projects an arbitrary element $\mathscr{F}$ onto the element $\mathscr{A}$ (Fig. 8.1),

$$|\mathsf{P}\mathscr{F})=|\mathscr{A})\frac{(\mathscr{A}|\mathscr{F})}{(\mathscr{A}|\mathscr{A})} \ . \tag{8.1.14}$$

The operator P is linear, Hermitian in $\mathfrak{L}$,

$$\mathsf{P}^{\dagger}=\mathsf{P} \ ,[3] \tag{8.1.15}$$

and idempotent,

$$\mathsf{P}^2=\mathsf{P} \ . \tag{8.1.16}$$

The definition (8.1.13) of the projection operator may be expanded upon by projecting not onto an element $|\mathscr{A})$, but onto a particular subspace $\mathfrak{L}_{\{\mathscr{A}\}}$ in $\mathfrak{L}$. If the *subspace* $\mathfrak{L}_{\{\mathscr{A}\}}$ is spanned by linearly independent elements $\mathscr{A}_\nu$, the following relation holds:

$$\boxed{\mathsf{P}=\sum_{\nu\nu'}|\mathscr{A}_\nu)g_{\nu\nu'}(\mathscr{A}_{\nu'}|} \ . \tag{8.1.17}$$

[3] In Part 4 in this book, we use the symbol P also for linear superoperators which are only idempotent.

Equation (8.1.16) leads to

$$\sum_{\nu'} g_{\nu\nu'}(\mathscr{A}_{\nu'}|\mathscr{A}_{\mu})=\delta_{\nu\mu} \tag{8.1.18}$$

The *metric* $g_{\nu\nu'}$ is thus the inverse of the matrix $(\mathscr{A}_\nu|\mathscr{A}_{\nu'})$. One can easily convince oneself that the representation (8.1.17) of P is independent of the choice of the basis elements $\mathscr{A}_\nu$ spanning the subspace $\mathfrak{L}_{\{\mathscr{A}\}}$.

One naturally has a particularly convenient representation of P when the basis elements $\tilde{\mathscr{A}}_\nu$ are orthonormalized, since then the metric $\tilde{g}_{\nu,\nu'}$ is the unit matrix $\delta_{\nu,\nu'}$, and we have

$$\mathsf{P}=\sum_{\nu}|\tilde{\mathscr{A}}_\nu)(\tilde{\mathscr{A}}_\nu| \; . \tag{8.1.19}$$

The observables $\mathscr{A}_\nu$ which are of interest in physical problems are, however, in general not orthogonal.

8.1.4 The Generation of Orthogonal Elements in $\mathfrak{L}$ Using Projection Operators

The Schmidt orthogonalization method, applied to a linearly independent set of observables $\mathscr{A}, \mathscr{B}, \mathscr{C}, \ldots$ (operators in $\mathfrak{U}$, elements in $\mathfrak{L}$) may be formulated in a very practical and simple manner using projection operators in $\mathfrak{L}$. We begin (arbitrarily) with the observable $\mathscr{A}$ and write the projector onto $\mathscr{A}$,

$$\mathsf{P}_{\mathscr{A}}=\frac{|\mathscr{A})(\mathscr{A}|}{(\mathscr{A}|\mathscr{A})} \; . \tag{8.1.20}$$

Then the elements

$$\mathscr{B}_1=(1-\mathsf{P}_{\mathscr{A}})\mathscr{B} \; , \qquad \mathscr{C}_1=(1-\mathsf{P}_{\mathscr{A}})\mathscr{C} \; , \ldots \tag{8.1.21}$$

are all orthogonal to $\mathscr{A}$, while $\mathscr{B}_1$, $\mathscr{C}_1, \ldots$ may not necessarily be mutually orthogonal (cf. Fig. 8.2). With the projector

$$\mathsf{P}_{\mathscr{B}_1}=\frac{|\mathscr{B}_1)(\mathscr{B}_1|}{(\mathscr{B}_1|\mathscr{B}_1)} \tag{8.1.22}$$

we generate operators in $\mathfrak{L}$ (elements in $\mathfrak{U}$)

$$\mathscr{C}_2=(1-\mathsf{P}_{\mathscr{B}_1})\mathscr{C}_1=(1-\mathsf{P}_{\mathscr{B}_1}-\mathsf{P}_{\mathscr{A}})\mathscr{C} \; , \ldots \; , \tag{8.1.23}$$

which are orthogonal to $\mathscr{A}$ and $\mathscr{B}_1$. Here, we have taken into account the fact that by definition, the orthogonality condition

$$\mathsf{P}_{\mathscr{A}}\mathsf{P}_{\mathscr{B}_1}=0 \tag{8.1.24}$$

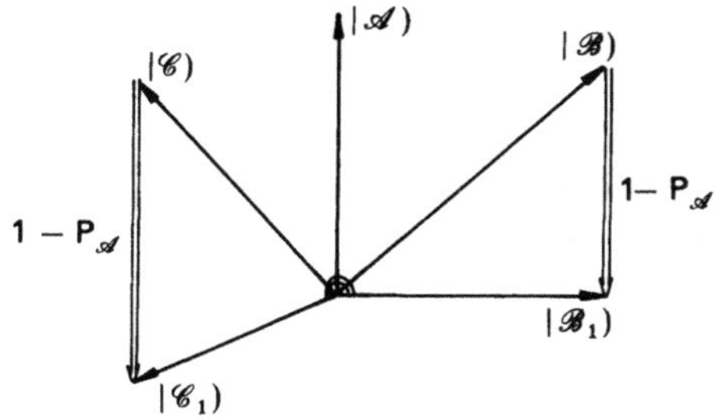

Fig. 8.2. The Schmidt orthogonalization using projection operators

holds. This method may be applied in a corresponding way to further observables $\mathscr{D}$, $\mathscr{E}$, $\mathscr{F}$, ...

Exercise 8.1 A space is spanned by the linearly independent but not orthogonal elements $\mathscr{A}_1$, $\mathscr{A}_2$, $\mathscr{A}_3$. Consider the elements $|\tilde{\mathscr{A}}_\nu)$ defined by determinants,

$$|\tilde{\mathscr{A}}_\nu)=\frac{1}{\sqrt{\Delta_\nu\Delta_{\nu-1}}}\begin{vmatrix}(\mathscr{A}_1|\mathscr{A}_1) & \dots & (\mathscr{A}_{\nu-1}|\mathscr{A}_1) & |\mathscr{A}_1)\\ \vdots & & \vdots & \vdots\\ (\mathscr{A}_1|\mathscr{A}_\nu) & \dots & (\mathscr{A}_{\nu-1}|\mathscr{A}_\nu) & |\mathscr{A}_\nu)\end{vmatrix}$$

with

$$\Delta_\nu=\begin{vmatrix}(\mathscr{A}_1|\mathscr{A}_1) & \dots & (\mathscr{A}_\nu|\mathscr{A}_1)\\ \vdots & & \vdots\\ (\mathscr{A}_1|\mathscr{A}_\nu) & \dots & (\mathscr{A}_\nu|\mathscr{A}_\nu)\end{vmatrix}\;, \qquad \Delta_0=1 \;,$$

and show that they represent an orthonormalized basis for $\nu=1, 2, 3$.

8.2 The Liouville Space with the Mori Scalar Product

8.2.1 Definition of the Mori Scalar Product

As we have already mentioned in Sect. 8.1, the scalar products in $\mathfrak{L}$ may be practically realized in various ways. A particular scalar product is chosen depending on the physical problem at hand.

We mention the following possibilities: definition by means of the trace

$$(\mathscr{F}|\mathscr{G})=\mathrm{Tr}(\mathscr{F}^\dagger\mathscr{G}) \;, \tag{8.2.1}$$

definition according to the thermal average value,

$$(\mathscr{F}|\mathscr{G})=\mathrm{Tr}(\mathscr{R}_\beta\mathscr{F}^\dagger\mathscr{G}) \quad \text{with} \quad \mathscr{R}_\beta=\frac{\mathrm{e}^{-\beta\mathscr{H}}}{Z} \;, \tag{8.2.2}$$

and definition by means of the thermal average of the anticommutator,

$$(\mathcal{F}|\mathcal{G})=\tfrac{1}{2}\mathrm{Tr}(\mathcal{R}_\beta[\mathcal{F}^\dagger,\mathcal{G}]_+)\ . \tag{8.2.3}$$

It can readily be shown that these definitions of the scalar product fulfill the requirements given in Sect. 8.1. In particular, the Liouville operator is Hermitian with these scalar products,

$$\mathsf{L}=\mathsf{L}^\dagger\ . \tag{8.2.4}$$

The following definition of the *Mori scalar product* between operators $\mathcal{F}$ and $\mathcal{G}$ may at first seem to be rather complicated compared to the definitions (8.2.1–3). However, we shall see that it is adequate for *all quantum statistical, linear* response and relaxation problems. We define

$$\boxed{(\mathcal{F}|\mathcal{G})=\frac{1}{\beta}\int_0^\beta \mathrm{Tr}(\mathcal{R}_\beta\mathcal{F}^\dagger\,\mathrm{e}^{-\alpha\mathcal{H}}\mathcal{G}\,\mathrm{e}^{\alpha\mathcal{H}})d\alpha}\ .[4] \tag{8.2.5}$$

Here, $\mathcal{R}_\beta$ is the canonical statistical operator,

$$\mathcal{R}_\beta=\frac{\mathrm{e}^{-\beta\mathcal{H}}}{Z}\ . \tag{8.2.6}$$

This scalar product is *temperature dependent* due to $\beta=1/kT$. Which temperature is meant will be discussed for each particular application.

The definition of the scalar product (8.2.5) is valid for arbitrary operators $\mathcal{F}$ and $\mathcal{G}$, thus also for *time-dependent* operators. We shall see that those scalar products $(\mathcal{F}|\mathcal{G}(t))$ are important, in which $\mathcal{G}(t)$ contains the dynamics in the Heisenberg formulation.

8.2.2 Properties of the Mori Scalar Product

Due to the possibility of cyclic permutation within the trace, (8.2.5) may be written in the form

$$(\mathcal{F}|\mathcal{G})=\frac{1}{\beta}\int_0^\beta \mathrm{Tr}(\mathcal{R}_\beta \mathrm{e}^{\alpha\mathcal{H}}\mathcal{F}^\dagger\mathrm{e}^{-\alpha\mathcal{H}}\mathcal{G})d\alpha\ . \tag{8.2.7}$$

The definition (8.2.5) has the properties of a scalar product required in Sect. 8.1; e.g. we have

$$(\mathcal{F}|c\mathcal{G})=c(\mathcal{F}|\mathcal{G})\quad\text{and}\quad(c\mathcal{F}|\mathcal{G})=c^*(\mathcal{F}|\mathcal{G}) \tag{8.2.8}$$

[4] The expression $(\mathcal{F}^\dagger|\mathcal{G})$ is also referred to as a "canonical correlation function" between $\mathcal{F}$ and $\mathcal{G}$; we use only the term "Mori product".

and

$$(\mathscr{F}|\mathscr{F}) \geqq 0 \quad (=0 \text{ only for } \mathscr{F} = \mathcal{O}) \tag{8.2.9}$$

as well as the following important relations which in part go beyond (8.1.2),

$$\boxed{(\mathscr{F}|\mathscr{G}) = (\mathscr{G}^\dagger|\mathscr{F}^\dagger) = (\mathscr{G}|\mathscr{F})^* = (\mathscr{F}^\dagger|\mathscr{G}^\dagger)^*} \tag{8.2.10}$$

(compare Exercises 8.2b and 8.3).

For Hermitian observables $\mathscr{F}$ and $\mathscr{G}$ in $\mathfrak{U}$, it follows from (8.2.10) that

$$(\mathscr{F}|\mathscr{G}) = (\mathscr{F}|\mathscr{G})^* = \text{real} . \tag{8.2.11}$$

A further important property of the Mori scalar product is that the Liouville operator $\mathfrak{L}$ is Hermitian (cf. Exercise 8.4):

$$\boxed{(\mathscr{F}|\mathsf{L}\mathscr{G}) = (\mathsf{L}\mathscr{F}|\mathscr{G})} \;.^{5} \tag{8.2.12}$$

The Liouville operator L as in (3.1.14) may be introduced into the defining equation (8.2.5):

$$(\mathscr{F}|\mathscr{G}) = \frac{1}{\beta} \int_0^\beta \langle \mathscr{F}^\dagger \cdot e^{-\hbar\alpha\mathsf{L}} \mathscr{G} \rangle_\beta \, d\alpha \; . \tag{8.2.13}$$

One thus obtains, after carrying out the integration, the compact form of the Mori scalar product,

$$\boxed{(\mathscr{F}|\mathscr{G}) = \left\langle \mathscr{F}^\dagger \cdot \frac{1 - e^{-\beta\hbar\mathsf{L}}}{\beta\hbar\mathsf{L}} \mathscr{G} \right\rangle_\beta} \; . \tag{8.2.14}$$

The superoperator $(1 - e^{-\beta\hbar\mathsf{L}})/\beta\hbar\mathsf{L}$ does *not* have a singularity for the eigenvalue zero of L, since, on expanding the exponential function in a power series, we find

$$\frac{1 - e^{-\beta\hbar\mathsf{L}}}{\beta\hbar\mathsf{L}} = 1 - \frac{\beta\hbar\mathsf{L}}{2!} + \frac{(\beta\hbar\mathsf{L})^2}{3!} - \; \dots \; . \tag{8.2.15}$$

With a sufficiently small β, i.e. when the energy differences which occur are small relative to kT, one can limit (8.2.15) to the first term. Then, $(\mathscr{F}|\mathscr{G})$ becomes

[5] This property applies only to the Hamiltonian for which the scalar product is defined by (8.2.5). For a partial Hamiltonian $\mathscr{H}_0$, it does not in general apply, but it can be shown in particular cases in suitable approximations (perturbation theory).

the usual correlation function between $\mathscr{F}^\dagger$ and $\mathscr{G}$:

$$(\mathscr{F}|\mathscr{G}) \approx \langle \mathscr{F}^\dagger \mathscr{G} \rangle_\beta \quad \text{for} \quad \Delta E \ll kT \ . \tag{8.2.16}$$

In the case that $\mathscr{F}$ or $\mathscr{G}$ *commutes* with $\mathscr{H}$, (8.2.16) is exact for all temperatures, as one can see e.g. from (8.2.5),

$$\boxed{\begin{array}{lll} (\mathscr{F}|\mathscr{G}) = \langle \mathscr{F}^\dagger \cdot \mathscr{G} \rangle_\beta \ , & & \text{in case either} \\ [\mathscr{F}, \mathscr{H}] = \mathcal{O} & \text{or} & [\mathscr{G}, \mathscr{H}] = \mathcal{O} \end{array}} \ . \tag{8.2.17}$$

The difference between the Mori scalar product, $(\mathscr{F}|\mathscr{G})$, and the usual correlation function, $\langle \mathscr{F}^\dagger \mathscr{G} \rangle$, is thus due to the quantum-theoretical noncommutativity of the observables $\mathscr{F}$ or $\mathscr{G}$ with the Hamiltonian $\mathscr{H}$.

From the general equation (8.2.14) we can obtain the *Kubo identity*, which permits us to write the thermal expectation value of the *commutator* between $\mathscr{F}^\dagger$ and $\mathscr{G}$ as a Mori product (compare Exercise 8.5):

$$\boxed{(\mathscr{F}|\mathsf{L}\mathscr{G}) = \frac{1}{\beta\hbar} \langle [\mathscr{F}^\dagger, \mathscr{G}] \rangle_\beta} \ . \tag{8.2.18}$$

For explicitly time *in*dependent observables $\mathscr{F}$ and $\mathscr{G}$, we therefore obtain

$$\boxed{(\mathscr{F}|\mathring{\mathscr{G}}) = \frac{\mathrm{i}}{\beta\hbar} \langle [\mathscr{F}^\dagger, \mathscr{G}] \rangle_\beta = -(\mathring{\mathscr{F}}|\mathscr{G})} \ , \tag{8.2.19}$$

i.e. in particular,

$$(\mathscr{F}^\dagger|\mathring{\mathscr{F}}) = 0 \ . \tag{8.2.20}$$

Examples: For the positions $x_{i\nu}$ and the mechanical momenta $p_{i\nu} = m_\nu \mathring{x}_{i\nu}$ of particles ν and ν', the following results are obtained from the commutation relations:

$$\begin{aligned} &(x_{i\nu}|p_{i'\nu'}) = 0 \ , \\ &(p_{i\nu}|p_{i'\nu'}) = m_\nu kT \delta_{ii'} \delta_{\nu\nu'} \ , \\ &(x_{i\nu}|\mathring{p}_{i'\nu'}) = -kT \delta_{ii'} \delta_{\nu\nu'} \ , \\ &(p_{i\nu}|\mathring{p}_{i'\nu'}) = 0 \end{aligned} \tag{8.2.21}$$

and for the spin components, e.g.

$$(s_{x\nu}|\mathring{s}_{y\nu'}) = -kT \langle s_{z\nu} \rangle_\beta \delta_{\nu\nu'} \ . \tag{8.2.22}$$

Equations (8.2.21 and 22) are "equipartition theorems", formulated in quantum-statistical language.

For the scalar product of the unity operator I with an observable $\mathscr{G}$, the defining equation (8.2.5) yields

$$\boxed{(I|\mathscr{G})=\langle\mathscr{G}\rangle} \ , \tag{8.2.23}$$

that is, the thermal expectation value of $\mathscr{G}$ is the Mori product of I and $\mathscr{G}$. Correspondingly,

$$(\mathscr{F}|I)=\langle\mathscr{F}^\dagger\rangle \ . \tag{8.2.24}$$

For the *fluctuations* of the observables $\mathscr{F}$ and $\mathscr{G}$ around the thermal expectation value,

$$\delta\mathscr{F}=\mathscr{F}-\langle\mathscr{F}\rangle I \quad \text{and} \quad \delta\mathscr{G}=\mathscr{G}-\langle\mathscr{G}\rangle I \tag{8.2.25}$$

we obtain finally

$$\boxed{\begin{aligned}(\mathscr{F}|\delta\mathscr{G})&=(\delta\mathscr{F}|\mathscr{G})=(\delta\mathscr{F}|\delta\mathscr{G})\\ &=(\mathscr{F}|\mathscr{G})-\langle\mathscr{F}^\dagger\rangle\langle\mathscr{G}\rangle\end{aligned}} \ . \tag{8.2.26}$$

The equations just derived can be given a geometrical interpretation in the sense of Sect. 8.1. Although the elements $|I)$ and $|\mathscr{G})$ are not, in general, orthogonal to one another, $|I)$ and $|\delta\mathscr{G})$ *are*, since

$$(I|\delta\mathscr{G})=\langle\delta\mathscr{G}\rangle=0 \ . \tag{8.2.27}$$

We can therefore interpret the transition from $|\mathscr{G})$ to $|\delta\mathscr{G})$ as an orthogonalization to $|I)$, which the projector in $\mathfrak{L}$ (superoperator, cf. Sect. 8.1.3),

$$\mathsf{1}-\mathsf{P}_{|I)}=\mathsf{1}-|I)(I| \tag{8.2.28}$$

can carry out (see Fig. 8.3):

$$(\mathsf{1}-\mathsf{P}_{|I)})|\mathscr{G})=|\mathscr{G})-|I)(I|\mathscr{G})=|\delta\mathscr{G}) \ . \tag{8.2.29}$$

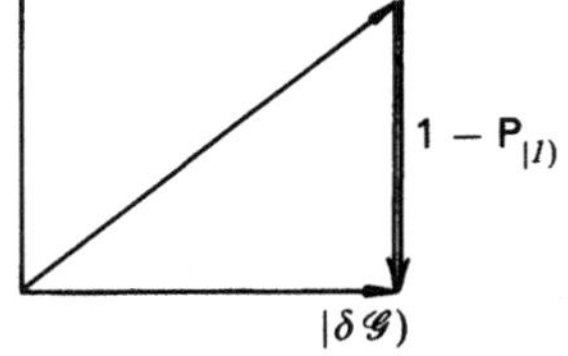

Fig. 8.3. Generation of $\delta\mathscr{G}$ by the projector $\mathsf{1}-\mathsf{P}_{|I)}$

Exercise 8.2 a) Write the Mori product in the energy representation.
b) Show that $(\mathscr{F}|\mathscr{F}) \geqq 0$ holds and that $(\mathscr{F}|\mathscr{F})=0$ only for $\mathscr{F}=\mathcal{O}$.

Exercise 8.3 Prove the validity of properties (8.2.10) of the Mori product.

Exercise 8.4 Demonstrate the Hermiticity of the Liouville operator in the Mori product (8.2.12).

Exercise 8.5 Prove the Kubo identity (8.2.18)

a) by carrying out the integration in the definition of the scalar product;
b) by introducing the generalized statistical operator

$$\mathscr{R}(\beta, \lambda)=\frac{1}{Z(\beta, \lambda)} \exp(-\beta\mathscr{H}-\sum_{\nu}\lambda_\nu\mathscr{F}_\nu^\dagger)$$

and differentiation of

$$\left[\mathscr{H}+\sum_{\nu}\frac{\lambda_\nu}{\beta}\mathscr{F}_\nu^\dagger, \mathscr{R}(\beta, \lambda)\right]=\mathcal{O}$$

with respect to λ_ν.

8.3 The Physical Significance of the Mori Product

8.3.1 Interpretation of the Mori Product as a Linear Variation, $\mathrm{Tr}(d\mathscr{R}\cdot\mathscr{G})$

In Sect. 8.2, we gave a formal definition of the Mori product and became familiar with some of its properties. In order to recognize the physical significance of the Mori products, we shall show that they represent the *variation of the expectation values* $\mathrm{Tr}(\mathscr{R}\mathscr{G})$ due to a variation of the canonical statistical operator

$$\mathscr{R}_\beta(h_1, \ldots) = \frac{\exp[-\beta\mathscr{H}(H_1^0+h_1,\ldots)]}{Z(\beta, h_1, \ldots)} \tag{8.3.1}$$

with the parameters $\beta, h_1, \ldots$ in *linear* approximation,

$$d\mathscr{R} = \sum_\alpha \left.\frac{\partial\mathscr{R}}{\partial h_\alpha}\right|_{h=0} dh_\alpha + \left.\frac{\partial\mathscr{R}}{\partial\beta}\right|_{h=0} d\beta\ . \tag{8.3.2}$$

If we multiply this equation by an arbitrary operator $\mathscr{G}$ and take the trace, $\mathrm{Tr}(d\mathscr{R}\cdot\mathscr{G})$, then the *coefficients of this linear differential form may be written as Mori products*:

$$\boxed{\mathrm{Tr}(d\mathscr{R}\cdot\mathscr{G}) = \sum_\alpha \beta(\delta\mathscr{M}_\alpha|\mathscr{G})dh_\alpha-(\delta\mathscr{H}|\mathscr{G})d\beta}\ . \tag{8.3.3}$$

Here, the operator $\mathscr{G}$ is arbitrary, while

$$\mathscr{M}_\alpha = -\left.\frac{\partial \mathscr{H}}{\partial h_\alpha^*}\right|_{h=0} \tag{8.3.4}$$

represent the operators of the generalized forces at $h_\alpha = 0$; they occur in the differentiation of $d\mathscr{R}$ on applying the differentiation rule for exponential operators, cf. (3.4.11),

$$\frac{\partial \mathscr{R}}{\partial h_\alpha} = -\int_0^\beta \mathscr{R}\, e^{\lambda \mathscr{H}} \frac{\partial \mathscr{H}}{\partial h_\alpha} e^{-\lambda \mathscr{H}} d\lambda + \beta \mathscr{R} \left\langle \frac{\partial \mathscr{H}}{\partial h_\alpha} \right\rangle_\beta . \tag{8.3.5}$$

The second term in (8.3.3) follows from the simpler relation

$$\frac{\partial \mathscr{R}}{\partial \beta} = -\mathscr{R}\,\delta \mathscr{H} . \tag{8.3.6}$$

The operators

$$\delta \mathscr{M}_\alpha = \mathscr{M}_\alpha - \langle \mathscr{M}_\alpha \rangle_\beta 1 , \quad \delta \mathscr{H} = \mathscr{H} - \langle \mathscr{H} \rangle_\beta 1 \tag{8.3.7}$$

are the "fluctuations" which have already been defined; they describe the deviation of the corresponding operators from their expectation values in thermal equilibrium. Following (8.2.26), we can replace the operators $\mathscr{G}$ in (8.3.3) by $\delta\mathscr{G}$,

$$\boxed{\mathrm{T\dot{r}}(d\mathscr{R}\cdot\mathscr{G}) = \sum_\alpha \beta(\delta \mathscr{M}_\alpha | \delta \mathscr{G}) dh_\alpha - (\delta \mathscr{H} | \delta \mathscr{G}) d\beta} . \tag{8.3.8}$$

Relations (8.3.3) and (8.3.8) may be applied to the comparison of neighboring states in equilibrium statistical thermodynamics for time independent operators $\mathscr{G}$. They can also be used in the Heisenberg formulation for time dependent operators $\mathscr{G}(t)$. Equation (8.3.3) in that case yields the variation, for a fixed time t, of $\langle \mathscr{G} \rangle(t)$ on varying the parameters β and h_α in the initial conditions $\rho(t_0) = \mathscr{R}_\beta(h)$.

The fact that the linear differential forms $\mathrm{Tr}(d\mathscr{R}\cdot\mathscr{G})$ may be written as Mori products is the reason that these products in Liouville space provide an appropriate theoretical tool for describing both the static and the dynamic *linear* response of physical systems. Before we clarify this statement by considering some static relations, (Sect. 8.3.3), we make a remark concerning formalism.

8.3.2 A Note on Formal Calculation with Non-Hermitian "Observables"

The physical parameters h (fields, volumes) on which the Hamiltonian depends are, like the temperature β_n, in principle real quantities. In many cases, however (similarly to electrodynamics), it is expedient to choose complex linear combinations for calculations, e.g.

$$h_\alpha \to h_\pm = h_x \pm i h_y = h_\mp^* \ , \tag{8.3.9}$$

or complex Fourier transforms,

$$h_\alpha \to h_i(\mathfrak{k}) = \int h_i(\mathfrak{r})\, e^{-i\mathfrak{k}\mathfrak{r}}\, d^3x \quad (i = x, y, z \text{ or } \pm, z) \ . \tag{8.3.10}$$

In this case, the corresponding generalized forces (8.3.4)

$$\mathcal{M}_\alpha = -\left.\frac{\partial \mathcal{H}}{\partial h_\alpha^*}\right|_{h=0} \tag{8.3.11}$$

become non-Hermitian operators in $\mathfrak{U}$, but in such a way that their Hermitian conjugates $\mathcal{M}_\alpha^\dagger$ also occur. *In a summation over* α, *by convention the adjoint expressions are always included in the sum.* Then the sum

$$\sum_\alpha \mathcal{M}_\alpha\, dh_\alpha^* = \sum_\alpha \mathcal{M}_\alpha^\dagger\, dh_\alpha \tag{8.3.12}$$

remains a Hermitian operator in $\mathfrak{U}$. With this convention, it is thus *also reasonable to refer to non-Hermitian operators as "observables"*, since the corresponding adjoint operator always occurs in the same expression. Therefore, the principle of quantum theory which states that physical observables are described by Hermitian operators in $\mathfrak{U}$ is not violated.

In the first part of this book (Chaps. 1–6), in which only general aspects were considered, we did not make use of this formal possibility. In the following sections, therefore, in keeping with the convention just cited, some formulas will contain additional adjoint symbols or complex-conjugate symbols when compared to the earlier chapters.

8.3.3 The Isothermal Susceptibility

As a first static example of the procedures outlined in Sect. 8.3.1, we choose $\mathcal{G}$ to be an observable $\mathcal{F}(h_1, \ldots)$ which may depend on static fields h_α. The total differential of its expectation value in thermal equilibrium for $h_\alpha = 0$,

$$d\langle \mathcal{F}(h)\rangle = \sum_\alpha \left.\frac{\partial\langle \mathcal{F}(h)\rangle}{\partial h_\alpha}\right|_{h=0} dh_\alpha + \left.\frac{\partial\langle \mathcal{F}(h)\rangle}{\partial \beta}\right|_{h=0} d\beta \tag{8.3.13}$$

contains in its first term the variation with the working parameters h_α and in its second term the variation with the temperature β. Similarly to the nomenclature used in electromagnetism, we denote

$$\boxed{\begin{aligned}\chi^{\mathrm{T}}_{M_\alpha F(h)} &\equiv \left.\frac{\partial\langle\mathscr{F}(h)\rangle}{\partial h_\alpha}\right|_{h=0}\\ &= \mathrm{Tr}\left.\left(\frac{\partial\mathscr{R}}{\partial h_\alpha}\mathscr{F}(h)\right)\right|_{h=0} + \mathrm{Tr}\left.\left(\mathscr{R}\frac{\partial\mathscr{F}(h)}{\partial h_\alpha}\right)\right|_{h=0}\end{aligned}} \tag{8.3.14}$$

as the *isothermal susceptibility*[6] *with respect to the observables* $\mathscr{M}_\alpha$ *and* $\mathscr{F}(h)$. According to (8.2.23) and (8.3.8), it may be expressed using Mori products,

$$\boxed{\chi^{\mathrm{T}}_{M_\alpha F(h)} = \beta(\delta\mathscr{M}_\alpha|\delta\mathscr{F}) + \left(1\left|\frac{\partial F}{\partial h_\alpha}\right.\right)} \; . \tag{8.3.15}$$

In this equation and in the following, we understand the operators *without the argument h*, for example $\mathscr{F}$, $\delta\mathscr{F}$, and $\partial\mathscr{F}/\partial h_\alpha$, *always to apply at* $h_\alpha = 0$(!).

The first term in (8.3.15) represents the *Mori product of the operators for the fluctuations* $\delta\mathscr{M}_\alpha$ and $\delta\mathscr{F}$. We denote it as

$$\boxed{\chi^{\mathrm{T}}_{M_\alpha F} = \mathrm{Tr}\left(\frac{\partial\mathscr{R}}{\partial h_\alpha}\mathscr{F}\right) = \beta(\delta\mathscr{M}_\alpha|\delta\mathscr{F})} \tag{8.3.16}$$

the isothermal susceptibility[7] *with respect to the observables* $\mathscr{M}$ *and* $\mathscr{F}$, in contrast to (8.3.14 and 15). In this equation and in the following, for clarity we *suppress the index* α, in cases where this will not give rise to confusion. The susceptibility χ^{T}_{MF} is Hermitian in the indices M and F, and, when the operators $\mathscr{M}$ and $\mathscr{F}$ are Hermitian, it is real.

The second term in (8.3.15) is the thermal expectation value $\langle\partial\mathscr{F}/\partial h_\alpha\rangle$ taken at the value $h = 0$.[8] In the particular case that the observables $\mathscr{F}$ are independent of the parameters, this term does not occur; for example in spin systems for

[6] In compressible media, this is the isothermal variation of the pressure with volume, i.e. the reciprocal isothermal compressibility.

[7] At high temperature, (8.2.16) implies that this term becomes the usual correlation function

$$(\delta\mathscr{M}|\delta\mathscr{F}) \approx \langle\delta\mathscr{M}^\dagger\cdot\delta\mathscr{F}\rangle_\beta \quad \text{for} \quad \Delta E_n \ll kT \tag{8.3.17}$$

[8] In particular, if one chooses

$$\mathscr{F}(h) \to -\frac{\partial\mathscr{H}}{\partial h^*_{\alpha'}} = \mathscr{M}_{\alpha'}(h) \; , \tag{8.3.18}$$

then the second term in (8.3.15):

$$\left(1\left|\frac{\partial\mathscr{M}_{\alpha'}}{\partial h_\alpha}\right.\right) = -\left\langle\frac{\partial^2\mathscr{H}}{\partial h_\alpha\partial h^*_{\alpha'}}\right\rangle \tag{8.3.19}$$

is likewise Hermitian in α and α'. For applications to magnetism, this term, in an appropriate gauge, represents the diamagnetic susceptibility.

$\mathscr{F} \to \vec{\sigma}$, since the spin operators $\vec{\sigma}$ are independent of the magnetic field. The current operators $j_\mu(\mathrm{r})$, by contrast, are in general dependent on the vector potential $A_\mu^{\mathrm{ex}}(\mathrm{r})$ [see Footnote 8 (p. 89)].

The *indices* α, as already mentioned, *can, in addition to the vector components, also express the position dependence* r. The derivatives are then to be read as functional derivatives[9]. Equation (8.3.11) thus becomes

$$\mathscr{M}_i(\mathrm{r}') = -\left.\frac{\delta\mathscr{H}[h(\mathrm{r})]}{\delta h_i^*(\mathrm{r}')}\right|_{h=0} , \tag{8.3.20}$$

and (8.3.16), with $\mathscr{F} \to \mathscr{M}_i(\mathrm{r}')$, becomes

$$\chi_{ii'}^{\mathrm{T}}(\mathrm{r}, \mathrm{r}') = \beta(\delta\mathscr{M}_i(\mathrm{r})|\delta\mathscr{M}_{i'}(\mathrm{r}')) = \chi_{i'i}^{\mathrm{T}*}(\mathrm{r}', \mathrm{r}) . \tag{8.3.21}$$

In translationally-invariant problems, these quantities depend only on the differences $\mathrm{r}-\mathrm{r}'$.

In the expression which is Fourier transformed in configuration space, instead of r and r', the propagation vectors $\mathfrak{k}$ and $\mathfrak{k}'$ occur. In translationally-invariant problems, the susceptibility *may be reduced* to a function of *one* propagation vector $\mathfrak{k}$. (see Appendix B1).

8.3.4 The Adiabatic Susceptibility

The adiabatic susceptibility contains information about how the thermal expectation values vary with variation of the static fields h_α, when the system is thermally isolated with respect to its environment. As we shall see (Sect. 9.5), this quantity is also of interest for dynamical processes, in particular for the limit in which the frequency ω of the time-dependent driving fields approaches zero.

In order to find an expression for the adiabatic susceptibility in terms of correlation functions, we first examine the quasistatic *heat transfer* DQ, cf. (5.3.8), to a system. Setting $\mathscr{G} = \mathscr{H}$ in (8.3.8), we obtain

$$\boxed{DQ = \mathrm{Tr}(d\mathscr{R}\cdot\mathscr{H}) = \sum_\alpha \beta(\delta\mathscr{M}_\alpha|\delta\mathscr{H})dh_\alpha - (\delta\mathscr{H}|\delta\mathscr{H})d\beta} . \tag{8.3.22}$$

The first term is the expression for the "latent heat" W_α in terms of correlation functions,

$$W_\alpha \equiv \left(\frac{DQ}{\partial h_\alpha}\right)_\beta = \beta(\delta\mathscr{M}_\alpha|\delta\mathscr{H}) = \beta\langle\delta\mathscr{M}_\alpha^\dagger\cdot\delta\mathscr{H}\rangle_\beta . \tag{8.3.23}$$

[9] Instead of the sums over α, integrals over d^3x and sums over the vector components then occur.

The second term is the heat capacity, C_h with constant parameters h,

$$C_h \equiv \left(\frac{DQ}{\partial T}\right)_h = -k\beta^2 \frac{\partial U}{\partial \beta} = k\beta^2 \langle (\delta \mathscr{H})^2 \rangle \ , \tag{8.3.24}$$

i.e. the average squared energy fluctuation.

The condition of *adiabaticity*, $DQ = 0$, leads to the adiabatic temperature variation due to a variation dh,

$$d\beta^{\mathrm{ad}} = \sum_\alpha \beta \frac{(\delta \mathscr{M}_\alpha | \delta \mathscr{H})}{(\delta \mathscr{H} | \delta \mathscr{H})} dh_\alpha \ . \tag{8.3.25}$$

We insert this quantity in the general expression for the complete differential resulting from (8.3.8, 13, 14 and 15):

$$d\langle \mathscr{F}(h) \rangle = \sum_\alpha \left(\beta(\delta \mathscr{M}_\alpha | \delta \mathscr{F}) + \left(1 \middle| \frac{\partial \mathscr{F}}{\partial h_\alpha} \right) \right) dh_\alpha - (\delta \mathscr{H} | \delta \mathscr{F}) d\beta \tag{8.3.26}$$

and obtain a result for the *adiabatic susceptibility with respect to* $\mathscr{M}$ *and* $\mathscr{F}(h)$

$$\boxed{\begin{aligned} \chi^{\mathrm{ad}}_{M_\alpha F(h)} &\equiv \left. \frac{\partial \langle \mathscr{F}(h) \rangle}{\partial h_\alpha} \right|^{\mathrm{ad}}_{h=0} \\ &= \beta \left(\delta \mathscr{M}_\alpha \middle| \left\{ 1 - \frac{|\delta \mathscr{H})(\delta \mathscr{H}|}{(\delta \mathscr{H} | \delta \mathscr{H})} \right\} \delta \mathscr{F} \right) + \left(1 \middle| \frac{\partial \mathscr{F}}{\partial h_\alpha} \right) . \end{aligned}} \tag{8.3.27}$$

The first term is the Mori product of the fluctuations $\delta \mathscr{M}_\alpha$ and $\delta \mathscr{F}$, in which however *only those components play a role, which are perpendicular to* $\delta \mathscr{H}$. If the projector (superoperator) (cf. Sect. 8.1.3) on the subspace $\mathfrak{L}_{\{1, \mathscr{H}\}}$ spanned by the operators 1 and $\mathscr{H}$ or $\delta \mathscr{H}$ is denoted by

$$\mathsf{P}_{\{1, \mathscr{H}\}} = |1)(1| + \frac{|\delta \mathscr{H})(\delta \mathscr{H}|}{(\delta \mathscr{H} | \delta \mathscr{H})} \tag{8.3.28}$$

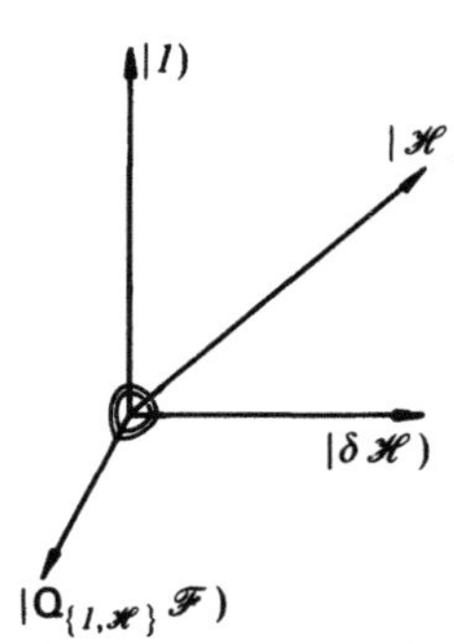

Fig. 8.4. The projector $\mathsf{Q}_{\{1, \mathscr{H}\}}$ in the adiabatic susceptibility

then the operator

$$\mathsf{Q}_{\{1,\mathscr{H}\}} = 1 - \mathsf{P}_{\{1,\mathscr{H}\}} \tag{8.3.29}$$

projects all the components in the direction of *1* and $\mathscr{H}$ *out* (see Fig. 8.4). Thus, the first term of (8.3.27), namely the *adiabatic susceptibility with respect to* $\mathscr{M}$ and $\mathscr{F}$, may be written in the compact form[10]

$$\boxed{\chi_{MF}^{\mathrm{ad}} = \beta(\mathsf{Q}_{\{1,\mathscr{H}\}}\mathscr{M} \,|\, \mathsf{Q}_{\{1,\mathscr{H}\}}\mathscr{F})} \,. \tag{8.3.30}$$

If the scalar product

$$(\delta\mathscr{H} \,|\, \delta\mathscr{F}) = 0 \;, \tag{8.3.31}$$

vanishes, e.g. due to selection rules (cf. Appendix B), then there is no difference between the isothermal and the adiabatic susceptibilities. The difference is also negligible when the heat capacity, corresponding to $(\delta\mathscr{H} \,|\, \delta\mathscr{H})$, is sufficiently large.

Exercise 8.6 For N spins s with the Hamiltonian

$$\mathscr{H} = -\gamma H \sum_i \hat{\sigma}_i^z - \frac{1}{\hbar^2} \sum_{i \neq j} J_{ij} \vec{\hat{\sigma}}_i \vec{\hat{\sigma}}_j$$

calculate, in the high-temperature limit ($\beta \to 0$), the adiabatic susceptibility

$$\chi^{\mathrm{ad}} = \frac{\partial}{\partial h} \langle \mathscr{M} \rangle \bigg|_{h=0}^{\mathrm{ad}} ,$$

where

$$\mathscr{M} = -\frac{\partial \mathscr{H}(h)}{\partial h}\bigg|_{h=0} = \gamma \sum_i \hat{\sigma}_i^z$$

denotes the magnetic moment in the z direction.

[10] The first term originally occurs in the form

$$\beta(\mathscr{M}_\alpha | \mathsf{Q}_{\{1,\mathscr{H}\}} \mathscr{F}) \,.$$

Due to the Hermiticity and idempotence of Q, however, this expression may also be written in the symmetric form (8.3.30).

9. Linear Response Theory

After having introduced the Mori product for linear problems in the previous chapter, we now proceed with the treatment of quantum-statistical dynamics in the presence of time-dependent fields, as described in Chap. 7. In the following chapter, we shall investigate such dynamics in the *linear* approximation, corresponding to Sect. 7.2.

9.1 The Kubo Formula

9.1.1 The Quantum-Statistical Formulation in the Time Domain

Beginning at a particular time t_0, let *time dependent external fields* $h_\alpha(t)$ act on an adiabatically isolated system Σ. The *dynamic response* of the system to these fields, i.e. the time dependence of the expectation values $\langle \mathscr{F} \rangle(t)$ of the observables $\mathscr{F}$ in Σ is to be represented *approximately* as a *linear functional* of the fields $h_\alpha(t)$.

For this purpose, according to the remarks in Chap. 7 we require only the zeroth and first order terms in the expansion of the explicitly time-dependent Hamiltonian $\mathscr{H}(t) = \mathscr{H}(H_1^0 + h_1(t), \ldots)$ in the fields,

$$\mathscr{H}(t) = \mathscr{H} - \sum_\alpha h_\alpha(t) \mathscr{M}_\alpha^\dagger \quad (t \geqq t_0) \ , \tag{9.1.1}$$

where the operators

$$\mathscr{M}_\alpha^\dagger = -\left.\frac{\partial \mathscr{H}(t)}{\partial h_\alpha(t)}\right|_{h(t)=0} \tag{9.1.2}$$

are explicitly time independent. The Liouville operator in first order is then given by

$$\mathsf{L}_1(t) = -\sum_\alpha h_\alpha(t) \mathsf{M}_\alpha^{(\dagger)} \quad (t \geqq t_0) \tag{9.1.3}$$

with the superoperator $\mathsf{M}_\alpha^{(\dagger)}$ defined as

$$\mathsf{M}_\alpha^{(\dagger)}\mathscr{X} = \frac{1}{\hbar}[\mathscr{M}_\alpha^\dagger, \mathscr{X}] \ .^1 \qquad (9.1.4)$$

The integration of the von Neumann equation up to terms of first order in the fields $h_\alpha(t)$ is given by $\mathsf{U}_0(t-t_0)+\mathsf{U}_1(t, t_0)$, where we insert expression (9.1.3) for $\mathsf{L}_1(t)$ into (7.2.8).

For an arbitrary, explicitly time *in*dependent[2] observable $\mathscr{F}$ in Σ, we obtain from (7.2.5, 14, 15, and 17) the time dependence of the expectation value $\langle\mathscr{F}\rangle(t)$ to first order in the fields $h_\alpha(t')$:

$$\langle\mathscr{F}\rangle(t) = \langle\mathscr{F}\rangle_\beta + \mathrm{i}\int_{t_0}^{t}\sum_\alpha h_\alpha(t')\,\mathrm{Tr}(\mathscr{F}\,\mathrm{e}^{-\mathrm{i}\mathsf{L}(t-t')}\mathsf{M}_\alpha^{(\dagger)}\mathscr{R}_\beta)dt' \ . \qquad (9.1.5)$$

Using (3.1.12a, 15, and 16), we can apply the operator $\exp[-\mathrm{i}\mathsf{L}(t-t')]\,\mathsf{M}_\alpha^{(\dagger)}$ to the left-hand side of the expression and let it act on the operator $\mathscr{F}$:

$$\boxed{\langle\mathscr{F}\rangle(t) = \langle\mathscr{F}\rangle_\beta - \mathrm{i}\int_{t_0}^{t}\sum_\alpha h_\alpha(t')\mathrm{Tr}(\mathsf{M}_\alpha^{(\dagger)}\,\mathrm{e}^{\mathrm{i}\mathsf{L}(t-t')}\mathscr{F}\cdot\mathscr{R}_\beta)dt' \quad (t \geqq t_0)} \ . \qquad (9.1.6)$$

In (9.1.5 and 6), it was assumed that the system Σ was in a state of thermal equilibrium at time[3] t_0, describable using the canonical statistical operator $\mathscr{R}_\beta = \exp(-\beta\mathscr{H})/Z$. $\langle\mathscr{F}\rangle_\beta$ is thus the thermal equilibrium value at time t_0.

The Kubo formula (9.1.6) for the linear response $\langle\mathscr{F}\rangle(t)$ of the system Σ is valid for *arbitrary time dependences* $h_\alpha(t')$ of the external fields, as long as their amplitudes permit the linear approximation. This formula therefore *separates the external fields from the internal dynamical properties of the unperturbed system,* which are described by the term $-\mathrm{i}\,\mathrm{Tr}(\mathsf{M}_\alpha^{(\dagger)}\exp[\mathrm{i}\mathsf{L}(t-t')]\mathscr{F}\cdot\mathscr{R}_\beta)$. This term contains only the Liouville operator L of the system *without* the external fields.

---

[1] The superoperators M and $\mathsf{M}^{(\dagger)}$ which are defined by $\mathsf{M}\mathscr{X}=(1/\hbar)[\mathscr{M},\mathscr{X}]$ and $\mathsf{M}^{(\dagger)}\mathscr{X}=(1/\hbar)[\mathscr{M}^\dagger,\mathscr{X}]$ are *not* adjoint to one another in the sense of (8.1.10); we express this fact by enclosing the cross in parentheses.

[2] If an observable $\mathscr{F}(h(t))$ depending on $h_\alpha(t)$ is explicitly time dependent, the following considerations are valid *only* for the observable $\mathscr{F}$, i.e. the first term in the expansion (7.2.19), which is explicitly time independent. For $\langle\mathscr{F}(h)\rangle(t)$ in a linear approximation, we must take account of the "pulling" by the term

$$\sum_\alpha \left.\frac{\partial\mathscr{F}(h(t))}{\partial h_\alpha(t)}\right|_{h(t)=0} h_\alpha(t) \ .$$

This is treated in Sect. 9.9.

[3] The extension to a non-stationary initial state is described in Sect. 9.9.

Since the convolution integral (9.1.6) extends over the time interval $t_0 \leqq t' \leqq t$, the response $\langle \mathscr{F} \rangle(t)$ of the system Σ to the external fields $h_\alpha(t')$ is *retarded* depending upon the internal properties of Σ. The fact that the response $\langle \mathscr{F} \rangle(t)$ occurs *after* its cause $h_\alpha(t')$ is a direct indication of *causality*. The result implied by (9.1.6) that the response to $h_\alpha(t')$ can be represented as a simple *superposition* of $h_\alpha(t')\,\mathrm{Tr}(\mathsf{M}_\alpha^{(\dagger)} \exp[\mathrm{i}\mathsf{L}(t-t')]\,\mathscr{F} \cdot \mathscr{R}_\beta)$ is due to the *linear approximation.*

From now on, we shall write the deviation of the dynamic response $\langle \mathscr{F} \rangle(t)$ from the thermal equilibrium value $\langle \mathscr{F} \rangle_\beta$ in the form

$$\langle \delta\mathscr{F} \rangle(t) = \langle \mathscr{F} \rangle(t) - \langle \mathscr{F} \rangle_\beta \;,^4 \tag{9.1.7}$$

so that the Kubo formula (9.1.6) takes on the form

$$\boxed{\langle \delta\mathscr{F} \rangle(t) = \int_{t_0}^{t} \sum_\alpha h_\alpha(t')\, \varphi_{M_\alpha F}(t-t')\,dt'} \tag{9.1.8}$$

or, with $t_1 = t - t'$,

$$\boxed{\langle \delta\mathscr{F} \rangle(t) = \int_{0}^{t-t_0} \sum_\alpha h_\alpha(t-t_1)\, \varphi_{M_\alpha F}(t_1)\,dt_1} \;. \tag{9.1.9}$$

The internal system property

$$\boxed{\begin{aligned} \varphi_{MF}(t) &= -\mathrm{i}\langle \mathsf{M}^{(\dagger)} \mathrm{e}^{\mathrm{i}\mathsf{L}t} \mathscr{F} \rangle_\beta \\ &= -\frac{\mathrm{i}}{\hbar}\langle [\mathscr{M}^\dagger, \mathrm{e}^{\mathrm{i}\mathsf{L}t} \mathscr{F}] \rangle_\beta \end{aligned}} \tag{9.1.10}$$

is called the *linear response function.* It is the thermal expectation value (β = temperature at the initial time t_0!) of the commutator between $\mathscr{M}^\dagger$ and the operator given by the internal dynamics:

$$\mathscr{F}(t) = \mathrm{e}^{\mathrm{i}\mathsf{L}t} \mathscr{F} \;. \tag{9.1.11}$$

This time dependence is the Heisenberg dynamics determined by L (i.e. *without* fields). For trivial systems, for example independent oscillators or spins, the commutator in (9.1.10) may be easily calculated (see Exercise 9.1). For general, complex systems (e.g. oscillators or spins with interactions), a direct treatment of

---

[4] To avoid confusion, we note that (9.1.7) is *not* the thermal expectation value of the operator $\delta\mathscr{F}(t)$ in (9.1.14). The thermal expectation value of this operator is in fact zero. Instead, (9.1.7) represents, see (7.2.17), the thermal expectation value $\langle \delta\mathscr{F} \rangle(t) = \langle \mathsf{U}^{-1}(t, t_0)\,\delta\mathscr{F} \rangle_\beta$ in the linear approximation, which contains the time dependence of the external fields $h_\alpha(t)$.

(9.1.10) is possible only in rare cases. For this reason, (9.1.10) is rewritten and one searches for *equivalent expressions* to $\varphi_{MF}(t)$, which are more readily evaluated.

One possibility consists of writing the linear response function as a time-dependent Mori product using the Kubo identity (8.2.18):

$$\varphi_{MF}(t) = -\beta(\mathcal{M}|\mathrm{iL}\,\mathrm{e}^{\mathrm{iL}t}\mathcal{F}) \tag{9.1.12}$$

or, alternatively,

$$\boxed{\varphi_{MF}(t) = -\beta(\mathcal{M}|\dot{\mathcal{F}}(t))} \ . \tag{9.1.13}$$

This scalar product is invariant with respect to substitution, cf. (8.2.26):

$$\mathcal{M} \to \delta\mathcal{M} = \mathcal{M} - \langle\mathcal{M}\rangle_\beta \ , \quad \mathcal{F}(t) \to \delta\mathcal{F}(t) = \mathcal{F}(t) - \langle\mathcal{F}\rangle_\beta = \mathrm{e}^{\mathrm{iL}t}\delta\mathcal{F} \ , \tag{9.1.14}$$

so that it may also be written in the form

$$\boxed{\varphi_{MF}(t) = -\beta(\delta\mathcal{M}|\delta\dot{\mathcal{F}}(t))} \ .[5] \tag{9.1.15}$$

For the calculation of the response function (9.1.10), it thus suffices to determine the time dependence of the Mori product of $\delta\mathcal{M}$ and $\delta\mathcal{F}$,

$$\boxed{\Phi_{MF}(t) = \beta(\delta\mathcal{M}|\delta\mathcal{F}(t)) = \beta(\delta\mathcal{M}|\mathrm{e}^{\mathrm{iL}t}\delta\mathcal{F})} \ . \tag{9.1.16}$$

For physical reasons, which we shall encounter in Sect. 9.2.2, $\Phi_{MF}(t)$ is called the *relaxation function.* Its negative time derivative is the response function

$$\boxed{\varphi_{MF}(t) = -\dot{\Phi}_{MF}(t)} \ . \tag{9.1.17}$$

It follows from integration of this relation (see Fig. 9.1), that

$$\boxed{\Phi_{MF}(t) = \Phi_{MF}(0) - \int_0^t \varphi_{MF}(t')dt'} \ , \tag{9.1.18}$$

---

[5] The relation

$$\delta\dot{\mathcal{F}}(t) = \mathrm{iL}\,\mathrm{e}^{\mathrm{iL}t}\delta\mathcal{F} \tag{9.1.15a}$$

holds.

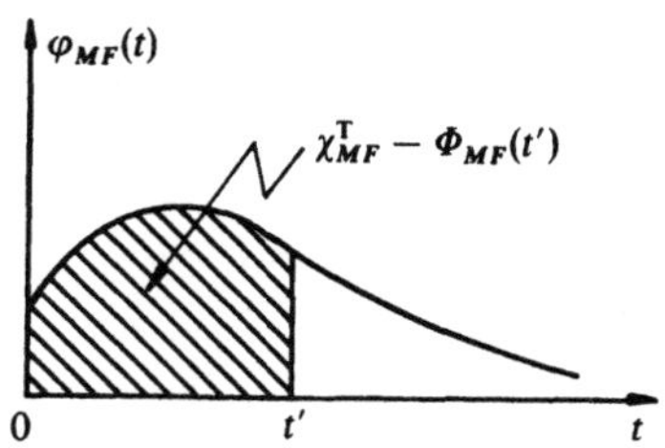

Fig. 9.1. The relation between the response function $\varphi(t)$ and the relaxation function $\Phi(t)$

where the initial value $\Phi_{MF}(0)$ is, according to (9.1.16) and (8.3.16),

$$\boxed{\Phi_{MF}(0) = \chi^{\mathrm{T}}_{MF} = \beta(\delta\mathcal{M}\,|\,\delta\mathcal{F})} \tag{9.1.19}$$

i.e. it is the isothermal susceptibility with respect to the operators $\mathcal{M}$ and $\mathcal{F}$.

Following (9.1.17), the Kubo formula (9.1.8) may also be expressed in terms of the relaxation function $\Phi_{MF}(t)$:

$$\langle\delta\mathcal{F}\rangle(t) = -\int_{t_0}^{t}\sum_{\alpha} h_\alpha(t')\,\dot{\Phi}_{M_\alpha F}(t-t')dt' \ . \tag{9.1.20}$$

After integrating by parts, due to $h_\alpha(t_0)=0$, this relation takes on the form

$$\boxed{\langle\delta\mathcal{F}\rangle(t) = \sum_{\alpha} h_\alpha(t)\,\Phi_{M_\alpha F}(0) - \int_{t_0}^{t}\sum_{\alpha} \dot{h}_\alpha(t')\Phi_{M_\alpha F}(t-t')dt'} \ . \tag{9.1.21}$$

The relaxation functions, owing to their simple mathematical form (9.1.16), are suitable for carrying out adequate approximations for many-particle systems. We shall deal with that problem in the framework of the Mori Theory, which will be treated beginning in Chap. 11.

9.1.2 The Quantum-Statistical Formulation in the Frequency Domain

Since the Kubo formula (9.1.9) is a convolution integral in time, it is reasonable to consider its formulation in terms of a one-sided Fourier (Laplace) transformation. To this end, we multiply (9.1.9) by $\exp(\mathrm{i}zt)$ (z complex, $\mathrm{Im}\,z>0$) and integrate from t_0 to ∞:

$$\begin{aligned}\int_{t_0}^{\infty}\langle\delta\mathcal{F}\rangle(t)\,\mathrm{e}^{\mathrm{i}zt}dt &= \int_{t=t_0}^{\infty}\int_{t_1=0}^{t-t_0}\sum_{\alpha} h_\alpha(t-t_1)\,\varphi_{M_\alpha F}(t_1)\,\mathrm{e}^{\mathrm{i}zt}\,dt_1\,dt \\ &= \int_{t_2=t_0}^{\infty}\int_{t_1=0}^{\infty}\sum_{\alpha} h_\alpha(t_2)\,\mathrm{e}^{\mathrm{i}zt_2}\,\varphi_{M_\alpha F}(t_1)\,\mathrm{e}^{\mathrm{i}zt_1}\,dt_1\,dt_2 \ .\end{aligned} \tag{9.1.22}$$

In going from the first to the second line, we have made the replacement $t_2 = t - t_1$ and have formulated the double integral in the t_1, t_2-plane correspondingly The Kubo formula (9.1.9) then assumes the following form in z:[6]

$$\langle \delta \mathscr{F} \rangle (z) = \sum_{\alpha} h_{\alpha}(z) \chi_{M_{\alpha}F}(z) \ . \tag{9.1.23}$$

If, on the one hand, we assume that the fields $h_{\alpha}(t)$ do not increase unboundedly with increasing time, and because, on the other hand, $|\varphi_{MF}(t)|$ is bounded (cf. Exercise 9.2), we can set $\eta \to +0$ in $z = \omega + i\eta$ after carrying out the integration over time. Then we obtain as a relation between

$$\langle \delta \mathscr{F} \rangle (\omega) = \int_{t_0}^{\infty} \langle \delta \mathscr{F} \rangle (t) \, e^{i\omega t - \eta t} \, dt |_{\eta \to +0} \tag{9.1.24}$$

and

$$h_{\alpha}(\omega) = \int_{t_0}^{\infty} h_{\alpha}(t) \, e^{i\omega t - \eta t} \, dt |_{\eta \to +0} \tag{9.1.25}$$

the *Kubo formula in the frequency domain* as the linear algebraic relation

$$\boxed{\langle \delta \mathscr{F} \rangle (\omega) = \sum_{\alpha} h_{\alpha}(\omega) \, \chi_{M_{\alpha}F}(\omega)} \ .^{7} \tag{9.1.26}$$

It depends on the *linear dynamic susceptibility*

$$\boxed{\chi_{MF}(\omega) = \int_{0}^{\infty} \varphi_{MF}(t) \, e^{i\omega t - \eta t} \, dt |_{\eta \to +0}} \ . \tag{9.1.27}$$

---

[6] This is nothing other than the convolution theorem for the Laplace transformation, whereby however due to (9.1.8 or 9), the integration over time of $\langle \delta \mathscr{F} \rangle (t)$ and $h(t)$ in (9.1.24 and 25) does not begin at $t=0$, but rather at $t = t_0$, while that of $\varphi_{MF}(t)$ in (9.1.27) begins at $t=0$.

[7] If the fields h_{α} are dependent on the *position* $\mathbf{r}$, then relation (9.1.26) is converted by the replacements [see Footnote 3, (p. 85)]

$$h_{\alpha'} \to h_{i'}(\mathbf{r}') \ , \qquad \mathscr{M}_{\alpha'} \to \mathscr{M}_{i'}(\mathbf{r}') \ , \qquad \mathscr{F} \to \mathscr{M}_i(\mathbf{r})$$

into

$$\langle \delta \mathscr{M}_i(\mathbf{r}) \rangle (\omega) = \int \sum_{\mathbf{r}' \ i'} h_{i'}(\mathbf{r}', \omega) \, \chi_{i' i}(\mathbf{r}', \mathbf{r}, \omega) \, d^3 x' \ . \tag{9.1.26a}$$

Here, $\langle \delta \mathscr{M}_i(\mathbf{r}) \rangle (\omega)$ is determined at the position $\mathbf{r}$ by *superposition* of the responses to the fields $h_i(\mathbf{r}', \omega)$. Corresponding equations hold when the *propagation vector* $\mathfrak{k}$ of the spatially Fourier-transformed quantities is used rather than $\mathbf{r}$.

After substituting (9.1.10 or 12) for $\varphi_{MF}(t)$, we can carry out the integration and obtain for the linear dynamic susceptibility the quantum-statistical formulas:

$$\chi_{MF}(\omega)=\left\langle \mathsf{M}^{(\dagger)}\frac{1}{\mathsf{L}+\omega+\mathrm{i}\eta}\mathscr{F}\right\rangle_{\beta}\Bigg|_{\eta\to+0} \tag{9.1.28}$$

$$=\beta\left(\delta\mathscr{M}\left|\frac{\mathsf{L}}{\mathsf{L}+\omega+\mathrm{i}\eta}\delta\mathscr{F}\right.\right)\Bigg|_{\eta\to+0} . \tag{9.1.29}$$

These expressions are also suitable starting points for approximate treatments [cf. for example (11.4.11)]. Various properties of $\chi_{MF}(\omega)$ are given in Sect. 9.4.

Equation (9.1.26) contains the same information as the original Kubo formula (9.1.9). The latter is obtained from (9.1.26) by a back transformation, i.e. by integration over ω. The response function $\varphi_{MF}(t)$ for positive times is obtained from $\chi_{MF}(\omega)$ through

$$\frac{1}{2\pi}\int_{-\infty}^{+\infty}\chi_{MF}(\omega)\,\mathrm{e}^{-\mathrm{i}\omega t}\,d\omega=\begin{cases}\varphi_{MF}(t) & \text{for}\quad t>0\\ 0 & \text{for}\quad t<0\end{cases} . \tag{9.1.30}$$

In summary, we note that the response functions $\varphi_{MF}(t)$, the relaxation functions $\Phi_{MF}(t)$, and the dynamic susceptibilities $\chi_{MF}(\omega)$ may be calculated from one another.

The following paragraphs of this chapter serve the purpose of making the reader familiar with the physical significance of these functions and to illustrate their mathematical properties[8].

At this point, we shall add a few remarks concerning the quantity η which occurs in (9.1.24, 25, and 27–29). A Laplace transformation with a complex variable z [9]

$$\chi_{MF}(z)=\int_{0}^{\infty}\varphi_{MF}(z)\,\mathrm{e}^{\mathrm{i}zt}\,dt\ , \tag{9.1.32}$$

is, as is well known, only defined in the z plane above the line Im z_0, i.e. for Im $z>$ Im z_0.

If the response function $\varphi_{MF}(t)$ is bounded (Exercise 9.2), then one can choose Im $z_0=\eta$ to be positive and arbitrarily small, but in general, it cannot be set

[8] The reader who is not particularly interested in these properties may skip at this point to Chap. 11, where the Mori theory and its approximations for the relaxation function are discussed. He or she should, however, refer occasionally to the later sections of the present chapter.

[9] The relation between z and the complex variable s in the theory of Laplace transformations is

$$z=\mathrm{i}s\ . \tag{9.1.31}$$

equal to zero. We therefore write in (9.1.32)

$$z = \omega + i\eta \quad (\eta > 0,\ \text{arbitrarily small}) \ . \tag{9.1.33}$$

For the operator $\mathsf{L}/(\mathsf{L}+\omega+i\eta)$ which is generated by the integration, we use the δ-function formalism in the limit $\eta \to 0$.

Macroscopic systems (other than model systems, such as independent spins or oscillators with integer energy differences), due to their internal interactions, exhibit quasi-continuous energy spectra. From this fact, it follows for the time behavior of the relaxation functions $\Phi_{MF}(t)$, considering the discussion of Sect. 4.2, that the time axis is divided into two regions: following the physically interesting time region, in which observable relaxation behavior occurs, there is a region in which destructive interference takes place (compare Fig. 4.5 in Sect. 4.2). The rapid, small fluctuations, as well as the unphysical recurrence time, are suppressed in the *thermodynamic limit*, $N \to \infty$, $N/V = \text{const.}$, which, however, can only be taken *after computing the scalar products*. For times which are longer than the region of relaxation behavior (an exact definition and analysis of the relaxation time τ_R is to be found in Part 3), the relaxation functions $\Phi_{MF}(t)$ *take on constant values in the thermodynamic limit, and the response functions fall to zero*:

$$\lim_{\substack{N\to\infty \\ N/V=\text{const}}} \varphi_{MF}(t) = 0 \quad \text{for} \quad t \to \infty \ . \tag{9.1.34}$$

When this decrease is sufficiently rapid, the Laplace transform

$$\chi_{MF}(\omega) = \int_0^\infty \lim_{\substack{N\to\infty \\ N/V=\text{const}}} \varphi_{MF}(t)\, e^{i\omega t}\, dt \tag{9.1.35}$$

exists for these limiting values *even for* $\eta = 0$.

The quantity η thus has *nothing* to do with a physical damping constant! Rather, it allows the Laplace transformation of the quantum-statistical dynamics to be carried out mathematically, i.e. it permits calculations with quantum-statistical expressions, as they occur in (9.1.28 and 29), for systems with a finite number of particles, before taking the thermodynamic limit.

Exercise 9.1 The Hamiltonian of a spin system is

$$\mathscr{H} = -\gamma H \sum_{i=1}^{N} \sigma_i^z \ .$$

Calculate the response functions $\varphi_{\nu\mu}(t)$ for the observables $\mathscr{S}^{\pm} = \sum_i \sigma_i^{\pm}$, $\mathscr{S}^0 = \sum_i \sigma_i^z$.

Exercise 9.2 Using Schwarz's inequality, find an upper bound for the response function $|\varphi_{MF}(t)|$.

9.2 The Physical Interpretation of the Kubo Formula Using Particular Time-Dependent Fields

In order to gain a physical understanding of the functions $\varphi_{MF}(t)$, $\Phi_{MF}(t)$, and $\chi_{MF}(\omega)$ which occur in the Kubo formula, we consider fields $h_\alpha(t)$ with particular time dependences which will allow us to recognize the physical meaning of these functions.

9.2.1 A Pulsed External Field

A physical interpretation of the response function $\varphi_{MF}(t)$ is possible when one allows a short pulse[10] to act upon the system; in the limit of short pulse time, the field pulse is given by

$$h_\alpha(t) = \hat{h}_\alpha \delta(t) . \tag{9.2.1}$$

It follows from (9.1.8) (assuming t_0 is chosen <0):

$$\langle \delta \mathscr{F} \rangle(t) = \sum_\alpha \hat{h}_\alpha \varphi_{M_\alpha F}(t) \quad (t>0) . \tag{9.2.2}$$

Thus, we can interpret the *response function* $\varphi_{MF}(t)$ directly as the dynamic response of the system Σ to a $\delta(t)$ pulse (compare Fig. 9.2). Carrying out such a measurement of $\langle \delta \mathscr{F} \rangle(t)$ with the various components of $\hat{h}_\alpha$ therefore determines $\varphi_{M_\alpha F}(t)$. The results $\langle \delta \mathscr{F} \rangle(t)$ for arbitrary *other* time-dependent fields $h_\alpha(t)$ may then be theoretically predicted within a linear approximation from the Kubo formula (9.1.9).

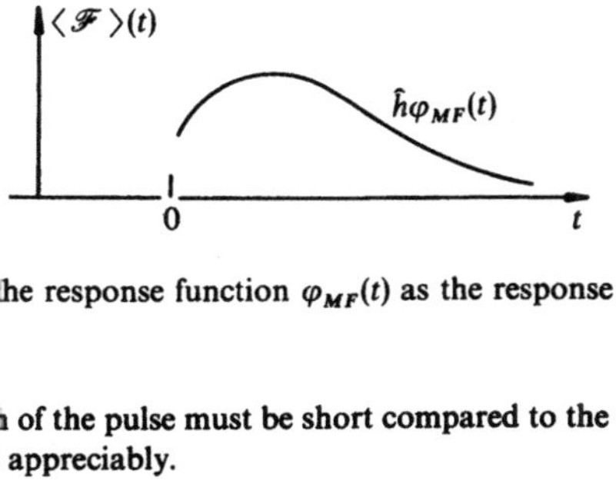

Fig. 9.2. Interpretation of the response function $\varphi_{MF}(t)$ as the response to a δ-function field pulse

[10] More precisely: the length of the pulse must be short compared to the time in which the response function $\varphi_{MF}(t)$ can change appreciably.

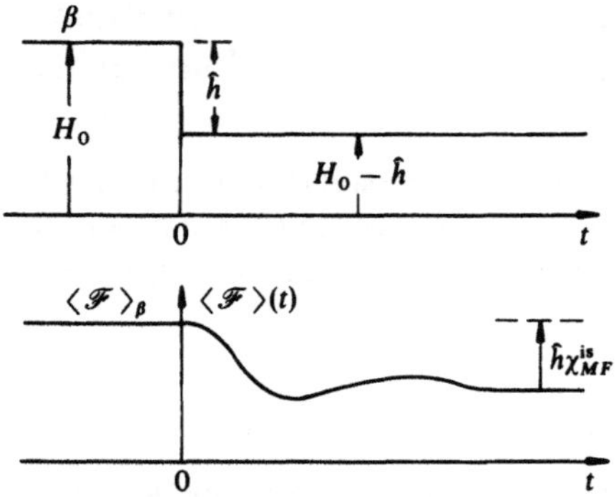

Fig. 9.3. The interpretation of the relaxation function $\Phi_{MF}(t)$

9.2.2 A Sudden Change in the External Field

The significance of $\Phi_{MF}(t)$ as a *relaxation function* may be seen from the following experiment:

Up to the time $t=-0$, let the system Σ be acted upon by a static external field H^0_α and be in thermal equilibrium at the temperature β. Within the short time interval from -0 to $+0$, the field is changed *suddenly* by the amount $\hat{h}_\alpha(>0)$[11], so that for times $t>0$, the external field has the value $H^0_\alpha - \hat{h}_\alpha$ (see Fig. 9.3). The canonical statistical operator *before* the field switching ($t<0$):

$$\mathcal{R}(\beta, H^0_\alpha) = \frac{\exp[-\beta\mathcal{H}(H^0_\alpha)]}{Z(\beta, H^0_\alpha)} \tag{9.2.3}$$

remains unchanged during the rapid switching process ["sudden approximation" of quantum theory (cf. Exercise 7.1)], so that $\mathcal{R}(\beta, H^0_\alpha)$ *also* remains the statistical operator of the system at the time $t=+0$. It thus describes the *nonstationary initial state for times* $t>0$; the dynamics follow the von Neumann equation with the Hamiltonian $\mathcal{H}=\mathcal{H}(H^0_\alpha - \hat{h}_\alpha)$. For $t>0$, we obtain the time dependence of the expectation value $\langle \mathcal{F} \rangle(t)$ of an observable $\mathcal{F}$ in the linear approximation, i.e. relevant to small $\hat{h}_\alpha$, from the Kubo formula (9.1.8) ($t_0<0$) on setting

$$h_\alpha(t) = \begin{cases} 0 & \text{for} \quad t<0 \\ -\hat{h}_\alpha & \text{for} \quad t>0 \end{cases} . \tag{9.2.4}$$

In this manner, we obtain for $t>0$,

$$\langle \mathcal{F} \rangle(t) = \langle \mathcal{F} \rangle_\beta - \sum_\alpha \hat{h}_\alpha \int_0^t \varphi_{M_\alpha F}(t')\,dt' \tag{9.2.5}$$

[11] It is not necessary that $\hat{h}_\alpha$ be small compared to H^0_α (H^0_α may also be zero)!

or, using (9.1.18 and 19),

$$\langle \mathscr{F} \rangle(t) = \langle \mathscr{F} \rangle_\beta + \sum_\alpha \hat{h}_\alpha [\Phi_{M_\alpha F}(t) - \chi^{\mathrm{T}}_{M_\alpha F}] \tag{9.2.6}$$

(cf. Fig. 9.3). Here, corresponding to the derivation of (9.1.8), the right sides of the equations depend on the temperature β and the field H^0_α which applied *before* the sudden field change.

Since this *relaxation process* is described by the function $\Phi_{MF}(t) = \beta(\delta\mathscr{M}|\delta\mathscr{F}(t))$, it is termed the *relaxation function*. At all times $t>0$, no time-variable fields are present; therefore (9.2.6), i.e. the relaxation function $\Phi_{MF}(t)$, provides a direct insight into the *internal dynamics* of the system Σ, determined, for $t>0$, by the Hamiltonian $\mathscr{H}(H^0_\alpha - \hat{h}_\alpha)$, which is time *in*dependent. This is, on the one hand, the reason why we shall treat the relaxation function (within the Mori theory) in detail in Part 3 of this book (Chap. 11 and following). On the other hand, if we know $\Phi_{MF}(t)$, we can use (9.1.20 or 21) to find also the response $\langle \mathscr{F} \rangle(t)$ to other, arbitrary time-dependent external fields $h_\alpha(t)$ within the linear approximation.

At *long times t*, $\Phi_{MF}(t)$ assumes a constant value $\Phi_{MF}(\infty)$ in the thermodynamic limit for systems with internal interactions. Depending on the nature of the system, this limiting value can be finite or zero. Predictions of the limiting value are treated in Sect. 9.5.

9.2.3 An Harmonically Oscillating External Field

Beginning at the initial time $t_0=0$, let the field $h_\alpha(t)$ oscillate harmonically with the frequency ω_0:

$$h_\alpha(t) = \hat{h}_\alpha \sin\omega_0 t \quad (t \geqq 0) \; . \tag{9.2.7}$$

For simplicity, $\mathscr{F}$ and $\mathscr{M}$ are here taken to be Hermitian operators. The response function $\varphi_{MF}(t)$ [(9.1.10)], like the field amplitude, $\hat{h}_\alpha$, is thus *real.* The linear response of the system to the perturbation (9.2.7) is given by (9.1.9) as:

$$\langle \delta\mathscr{F} \rangle(t) = \int_0^t \sum_\alpha \hat{h}_\alpha \sin\omega_0(t-t_1)\varphi_{M_\alpha F}(t_1)\,dt_1 \; . \tag{9.2.8}$$

In the steady state, i.e. for times long compared to transient times, $t \gg \tau_R$, for which according to (9.1.34) the response function $\varphi_{MF}(t)$ is damped out, we can replace the upper integration limit by ∞ and obtain

$$\boxed{\langle \delta\mathscr{F} \rangle(t) = \sum_k \hat{h}_\alpha (\mathrm{Re}\,\chi_{M_\alpha F}(\omega_0)\sin\omega_0 t - \mathrm{Im}\,\chi_{M_\alpha F}(\omega_0)\cos\omega_0 t)} \; . \tag{9.2.9}$$

The real part of the dynamic susceptibility,

$$\mathrm{Re}\,\chi_{MF}(\omega_0) = \int_0^\infty \varphi_{MF}(t)\cos\omega_0 t\,dt \tag{9.2.10}$$

determines the portion of the response which is *in phase* with the external field; the imaginary part,

$$\operatorname{Im}\chi_{MF}(\omega_0)=\int_0^\infty \varphi_{MF}(t)\sin\omega_0 t\,dt \tag{9.2.11}$$

determines the portion which oscillates 180° *out of phase*[12].

Exercise 9.3 Let the relaxation function $\Phi(t)=\Phi(0)\exp(-\Gamma t)$ $(t\geqq 0)$ be known for the observables $\mathscr{F}=\mathscr{M}$. Calculate the response $\langle\delta\mathscr{M}\rangle(t)$ for a harmonically oscillating field $h(t)=\hat{h}\sin\omega_0 t$ which is switched on at time $t=0$.

9.3 Properties of the Response and Relaxation Functions

9.3.1 The Linear Response Function

Applying the quantum statistical formulas (9.1.12) and (9.1.13) for the linear response function:

$$\boxed{\varphi_{MF}(t)=-\beta(\mathscr{M}\,|\,\mathrm{iL}\,\mathrm{e}^{\mathrm{iL}t}\mathscr{F})=-\beta(\mathscr{M}\,|\,\dot{\mathscr{F}}(t))\quad (t\gtreqless 0)} \tag{9.3.1}$$

let us demonstrate some immediately evident properties[13]. We first wish to fix the convention that expression (9.3.1), the response function $\varphi_{MF}(t)$, is defined not only for positive times [in the Kubo formula, (9.1.9), only positive times occur], but also for negative times[14].

The Hermiticity of L in the Mori product implies the following relation:

$$\varphi_{MF}(t)=\beta(\mathrm{iL}\mathscr{M}\,|\,\mathscr{F}(t))=\beta(\dot{\mathscr{M}}\,|\,\mathscr{F}(t)) \ . \tag{9.3.2}$$

From (8.2.10), it also follows that (cf. Exercise 9.4)

$$\varphi_{MF}(t)=-\varphi_{F^\dagger M^\dagger}(-t)=-\varphi^*_{FM}(-t) \ , \tag{9.3.3}$$

thus, in particular,

$$\varphi_{MM^\dagger}(0)=0 \ . \tag{9.3.4}$$

---

[12] This implies, as we shall see in Sect. 9.6, that in the steady state, the average power absorbed by the system Σ from the external field is given by the imaginary part of the linear dynamic susceptibility, or alternatively by the part χ'' (9.4.4).

[13] In all these formulas, following (9.1.15) one can replace $\mathscr{M}$ by $\delta\mathscr{M}$ and $\mathscr{F}(t)$ by $\delta\mathscr{F}(t)=\exp(\mathrm{iL}t)\delta\mathscr{F}$.

[14] In the literature, another definition is also frequently encountered, in which the response function is set equal to zero for negative times, $\varphi_{MF}(t)\Theta(t)$. We believe, however, that the above convention is more expedient for many problems.

For Hermitian operators $\mathcal{M}$ and $\mathcal{F}$ in $\mathfrak{U}$, $\varphi_{MF}(t)$ is *real*, cf. (9.1.10), therefore

$$\varphi_{MF}(t)=\varphi^*_{MF}(t)=-\varphi_{FM}(-t), \quad \text{when } \mathcal{M} \text{ and } \mathcal{F} \text{ are Hermitian} . \tag{9.3.5}$$

In particular, for identical observables, $\mathcal{F}=\mathcal{M}$, the response function is *odd* in time:

$$\varphi_{MM}(t)=-\varphi_{MM}(-t), \quad \text{when} \quad \mathcal{M}=\mathcal{M}^\dagger . \tag{9.3.6}$$

Further symmetry relations, in particular those which follow from time-reversal invariance, are to be found in Appendix B.

If $\varphi_{MF}(t)$ is differentiated w.r.t. time, an additional factor iL appears in the scalar product (9.3.1):

$$\dot{\varphi}_{MF}(t)=-\beta(\mathcal{M}|(\mathrm{iL})^2\mathcal{F}(t))=\varphi_{M\dot{F}}(t)=-\varphi_{\dot{M}F}(t) . \tag{9.3.7}$$

We thus have for the nth derivative

$$\frac{d^n}{dt^n}\varphi_{MF}(t)=-\beta(\mathcal{M}|(\mathrm{iL})^{n+1}\mathcal{F}(t)) , \tag{9.3.8}$$

so that the Taylor expansion of the response function can be written in the form

$$\varphi_{MF}(t)=-\beta\sum_{n=0}^{\infty}(\mathcal{M}|(\mathrm{iL})^{n+1}\mathcal{F})\frac{t^n}{n!} \tag{9.3.9}$$

without making a statement about its region of convergence. We shall meet the derivatives $d^n/dt^n\,\varphi_{MF}(0)$ again in Sect. 9.4.5 as the *moments* of the spectral function.

9.3.2 The Linear Relaxation Function

The time-dependent Mori scalar product over $\delta\mathcal{F}(t)$,

$$\Phi_{MF}(t)=\beta(\delta\mathcal{M}|\delta\mathcal{F}(t))=\beta(\delta\mathcal{M}|\mathrm{e}^{\mathrm{iL}t}\delta\mathcal{F}) \quad (t \gtreqless 0) \tag{9.3.10}$$

is called the *relaxation function* for the physical reasons mentioned in Sect. 9.2.2. This function, in contrast to the response function $\varphi_{MF}(t)$, is *not* invariant with respect to a transformation $\delta\mathcal{M}\to\mathcal{M}$, $\delta\mathcal{F}\to\mathcal{F}$.

From the Hermiticity of L in the Mori product and from (8.2.10), it follows that the property

$$\Phi_{MF}(t)=\Phi_{F^\dagger M^\dagger}(-t)=\Phi^*_{FM}(-t) . \tag{9.3.11}$$

holds.

For Hermitian operators $\mathcal{M}$ and $\mathcal{F}$ in $\mathfrak{U}$, $\Phi_{MF}(t)$ is therefore *real*:

$$\Phi_{MF}(t)=\Phi^*_{MF}(t)=\Phi_{FM}(-t), \quad \text{when } \mathcal{M} \text{ and } \mathcal{F} \text{ are Hermitian} \tag{9.3.12}$$

In particular, for identical observables, $\mathcal{F}=\mathcal{M}$, in this case the relaxation function is *even* in time:

$$\Phi_{MM}(t)=\Phi_{MM}(-t) \quad (\mathcal{M}=\mathcal{M}^{\dagger}) \ . \tag{9.3.13}$$

The relation between the relaxation function $\Phi_{MF}(t)$ and the response function $\varphi_{MF}(t)$ has already been demonstrated in Sect. 9.1.1 (9.1.17, 18, and 19). Relations between $\Phi_{MF}(t)$ and the dynamic susceptibility $\chi_{MF}(\omega)$ are given in Sect. 9.4.2. In Sect. 9.5, the behavior of $\Phi_{MF}(t)$ for long times is discussed in detail in relation to the properties of $\chi_{MF}(\omega=0)$.

Symmetry relations for $\Phi_{MF}(t)$, in particular those which result from time-reversal invariance, are to be found in Appendix B.

Exercise 9.4 Show that for the application of a function of the Liouville operator to an observable $\mathcal{X}$, $f(\mathsf{L})\mathcal{X}=\mathcal{Y}$, the relation $\mathcal{Y}^{\dagger}=f^{*}(-\mathsf{L})\mathcal{X}^{\dagger}$ holds. Specialize $f(\mathsf{L})$ to

$$\mathrm{e}^{\mathrm{i}\mathsf{L}t}, \frac{1}{\mathsf{L}+z}, \delta(\mathsf{L}+\omega) \ .$$

9.4 Properties of the Dynamic Susceptibility

9.4.1 Decomposition of $\chi_{MF}(\omega)$ into Two Hermitian Matrices, $\chi'_{MF}(\omega)$ and $\chi''_{MF}(\omega)$

The matrix of the dynamic susceptibility, $\chi_{MF}(\omega)$ (9.1.29),

$$\boxed{\chi_{MF}(\omega)=\beta\left(\delta\mathcal{M}\left|\frac{\mathsf{L}}{\mathsf{L}+\omega+\mathrm{i}\eta}\delta\mathcal{F}\right)\right|_{\eta\to+0}} \tag{9.4.1}$$

may be decomposed w.r.t. the indices M and F into two Hermitian matrices, *without* presupposing that $\mathcal{F}$ and $\mathcal{M}$ are Hermitian operators:

$$\chi_{MF}(\omega)=\chi'_{MF}(\omega)+\mathrm{i}\chi''_{MF}(\omega) \tag{9.4.2}$$

with

$$\chi'_{MF}(\omega)=\frac{1}{2}(\chi_{MF}(\omega)+\chi^{*}_{FM}(\omega))=\chi'^{*}_{FM}(\omega) \ , \tag{9.4.3}$$

$$\chi''_{MF}(\omega)=\frac{1}{2\mathrm{i}}(\chi_{MF}(\omega)-\chi^{*}_{FM}(\omega))=\chi''^{*}_{FM}(\omega) \ . \tag{9.4.4}$$

Equation (9.4.2) is in general *not* a decomposition into real and imaginary parts. Instead, we see from (9.4.3 and 4) that

$$\chi'_{MF}(\omega)=\frac{1}{2}\,\mathrm{Re}\,(\chi_{MF}(\omega)+\chi_{FM}(\omega))+\frac{\mathrm{i}}{2}\,\mathrm{Im}\,(\chi_{MF}(\omega)-\chi_{FM}(\omega)) \ , \tag{9.4.5}$$

$$\chi''_{MF}(\omega)=\frac{1}{2}\,\mathrm{Im}\,(\chi_{MF}(\omega)+\chi_{FM}(\omega))-\frac{\mathrm{i}}{2}\,\mathrm{Re}\,(\chi_{MF}(\omega)-\chi_{FM}(\omega)) \ . \tag{9.4.6}$$

The matrices $\chi'_{MF}(\omega)$ and $\chi''_{MF}(\omega)$ may, according to (9.3.3), be expressed as integrals over the response function $\varphi_{MF}(t)$ between $-\infty$ and $+\infty$:

$$\chi'_{MF}(\omega)=\frac{1}{2}\int_0^\infty (\varphi_{MF}(t)\mathrm{e}^{\mathrm{i}\omega t}-\varphi_{MF}(-t)\,\mathrm{e}^{-\mathrm{i}\omega t})\,\mathrm{e}^{\eta t}\,dt\,|_{\eta\to+0} \ , \tag{9.4.7}$$

$$\chi''_{MF}(\omega)=\frac{1}{2\mathrm{i}}\int_{-\infty}^{+\infty}\varphi_{MF}(t)\,\mathrm{e}^{\mathrm{i}\omega t-\eta|t|}\,dt\,|_{\eta\to+0} \ . \tag{9.4.8}$$

Applying the following identity:

$$\left.\frac{1}{x+\mathrm{i}\eta}\right|_{\eta\to+0}=\mathrm{Pr}\left(\frac{1}{x}\right)-\mathrm{i}\pi\delta(x) \ , \tag{9.4.9}$$

to the expression (9.4.1) (where the abbreviation “Pr” stands for “principal value”), we find the important quantum-statistical relations

$$\chi'_{MF}(\omega)=+\beta\left(\delta\mathcal{M}\,|\,\mathsf{L}\ \mathrm{Pr}\left(\frac{1}{\mathsf{L}+\omega}\right)\delta\mathcal{F}\right) \ , \tag{9.4.10}$$

$$\begin{aligned}\chi''_{MF}(\omega)&=-\beta\pi(\delta\mathcal{M}\,|\,\mathsf{L}\delta(\mathsf{L}+\omega)\delta\mathcal{F})\\ &=+\beta\pi\omega(\delta\mathcal{M}\,|\,\delta(\mathsf{L}+\omega)\delta\mathcal{F}) \ .\end{aligned} \tag{9.4.11}$$

From (9.4.3 and 4), and using the Hermiticity of L, one can derive the following symmetry relations:

$$\chi'_{MF}(\omega)=+\chi'_{F^\dagger M^\dagger}(-\omega) \ , \tag{9.4.12}$$

$$\chi''_{MF}(\omega)=-\chi''_{F^\dagger M^\dagger}(-\omega) \ . \tag{9.4.13}$$

They can evidently be simplified when the operators $\mathcal{F}$ and $\mathcal{M}$ are also Hermitian. Further symmetry relations, in particular those derived from time-reversal invariance, are to be found in Appendix B.

In the definition (9.4.1) of the dynamic susceptibility $\chi_{MF}(\omega)$, we can separate out the isothermal susceptibility χ^{T}_{MF} [compare (8.3.16)] by applying the relation

$$\frac{\mathsf{L}}{\mathsf{L}+z}=1-\frac{z}{\mathsf{L}+z}\ ; \tag{9.4.14}$$

we thus obtain

$$\boxed{\chi_{MF}(\omega)=\chi^{\mathrm{T}}_{MF}-\beta\left(\delta\mathcal{M}\left|\frac{\omega+\mathrm{i}\eta}{\mathsf{L}+\omega+\mathrm{i}\eta}\delta\mathcal{F}\right)\right|_{\eta\to+0}}\ .^{15} \tag{9.4.15}$$

If we replace $\mathcal{F}$ in this equation by the operator for its derivative, $\dot{\mathcal{F}}=\mathrm{i}\mathsf{L}\mathcal{F}$, we obtain the relation:

$$\chi_{M\dot{F}}(\omega)=\chi^{\mathrm{T}}_{M\dot{F}}-\mathrm{i}\omega\chi_{MF}(\omega)=-\chi_{\dot{M}F}(\omega)\ . \tag{9.4.16}$$

Similar results may of course be obtained for the higher derivatives. From (9.4.4) we find the relation

$$\chi''_{M\dot{F}}(\omega)=-\mathrm{i}\omega\chi''_{MF}(\omega)=-\chi''_{\dot{M}F}(\omega)\ . \tag{9.4.17}$$

9.4.2 Relations Between $\chi_{MF}(\omega)$ and $\varphi_{MF}(t)$ or $\Phi_{MF}(t)$

Equation (9.1.27), which defines the dynamic susceptibility as the one-sided Fourier (Laplace) transform of the response function $\varphi_{MF}(t)$, permits the derivation of some further relations between $\chi_{MF}(\omega)$ and $\varphi_{MF}(t)$ or $\Phi_{MF}(t)$.

We have already shown in (9.1.30) how the response function $\varphi_{MF}(t)$ (for positive times) may be obtained by Fourier transformation of $\chi_{MF}(\omega)$. The decomposition (9.4.11), on the other hand, demonstrates that by employing the properties of the δ function, one can similarly obtain $\varphi_{MF}(t)$ for both positive and negative times t (9.3.1) by Fourier transformation of $\chi''_{MF}(\omega)$ *alone*:

$$\varphi_{MF}(t)=\frac{\mathrm{i}}{\pi}\int_{-\infty}^{+\infty}\chi''_{MF}(\omega)\,\mathrm{e}^{-\mathrm{i}\omega t}\,d\omega\ .^{16,17} \tag{9.4.18}$$

---

[15] For $\omega\neq 0$, one can replace $\omega+\mathrm{i}\eta$ in the numerator simply with ω. However, for $\omega=0$, this leads in general to incorrect results [cf. (9.4.21) and Sect. 9.5].

[16] The fact that the response function can be derived from $\chi''_{MF}(\omega)$ alone is understandable when one considers the Kramers–Kronig relations [see (9.4.28 and 29)], according to which $\chi'_{MF}(\omega)$ is determined by $\chi''_{MF}(\omega)$.

[17] In some treatments of the subject, the starting point for linear response theory is not $\varphi_{MF}(t)$, but instead $\chi''_{MF}(\omega)$, in terms of which all the other response quantities can be derived. Relation (9.4.18) provides the connection to this approach.

Finally, we return to the connection between $\chi_{MF}(\omega)$ and $\Phi_{MF}(t)$. From (9.1.17) and (9.1.27), we find the relation

$$\chi_{MF}(\omega) = -\int_0^\infty \dot{\Phi}_{MF}(t)\, e^{i\omega t - \eta t}\, dt\, |_{\eta \to +0} \ . \tag{9.4.19}$$

If we replace $\Phi_{MF}(t)$ in this equation by $\Phi_{MF}(t) - \Phi_{MF}(\infty)$, we obtain after integration by parts the result:

$$\boxed{\chi_{MF}(\omega) = \Phi_{MF}(0) - \Phi_{MF}(\infty) + i\omega \int_0^\infty [\Phi_{MF}(t) - \Phi_{MF}(\infty)]\, e^{i\omega t}\, dt} \ .^{18} \tag{9.4.20}$$

This is a relation between $\chi_{MF}(\omega)$ and the one-sided Fourier (Laplace) transform of $\Phi_{MF}(t) - \Phi_{MF}(\infty)$. For $\omega \to 0$ we thus find

$$\boxed{\chi_{MF}(\omega = 0) = \Phi_{MF}(0) - \Phi_{MF}(\infty)} \ . \tag{9.4.21}$$

From this result we see that $\Phi_{MF}(\infty)$ *must differ from zero*, if $\chi_{MF}(\omega = 0) \neq \chi^T_{MF}$ [compare (9.1.19)]. We shall discuss the physical significance of this result in Sect. 9.5.

If one integrates the original form of (9.4.19) by parts with $\Phi_{MF}(t)$, the result is

$$\boxed{\chi_{MF}(\omega) = \chi^T_{MF} + (i\omega - \eta) \int_0^\infty \Phi_{MF}(t)\, e^{i\omega t - \eta t}\, dt|_{\eta \to +0}} \ .^{18} \tag{9.4.22}$$

This integral has a pole at $\omega = 0$, provided that $\Phi_{MF}(\infty) \neq 0$.

9.4.3 The Kramers–Kronig Relations

According to (9.4.7) and (9.4.8), $\chi'_{MF}(\omega)$ and $\chi''_{MF}(\omega)$ are determined by the response function $\varphi_{MF}(t)$. If the latter is eliminated from the equations, a relation between $\chi'_{MF}(\omega)$ and $\chi''_{MF}(\omega)$ will be obtained. We shall derive this relation using the theory of functions. To this end, we replace ω by the complex variable z, see (9.1.23, 32 and 33):

$$\chi_{MF}(z) = \int_0^\infty \varphi_{MF}(t)\, e^{izt}\, dt \ . \tag{9.4.23}$$

[18] In (9.4.20), $\Phi_{MF}(t)$ is understood to be in its thermodynamic limit (Sect. 9.1.2), in which $\Phi_{MF}(\infty)$ takes on a constant value; thus, one can set η to zero in the integrand. In contrast, (9.4.22) is generally valid. In it, even when $\Phi_{MF}(t)$ is taken in the thermodynamic limit, the transition $\eta \to +0$ may only be made after carrying out the integration, since $\Phi_{MF}(\infty) \neq 0$ is allowed.

Because $|\varphi_{MF}(t)|$ is bounded (compare Exercise 9.2), and since the integration in (9.4.23) extends over positive times only (a result of causality), we find

$$\boxed{\chi_{MF}(z)=\text{holomorphic} \quad \text{for} \quad \operatorname{Im} z>0} , \tag{9.4.24}$$

i.e. $\chi_{MF}(z)$ contains no singularities in this region of the upper half plane in z (Fig. 9.4).

By employing Cauchy's integral theorem, we obtain an integral expression for $\chi_{MF}(z)$, choosing the path of integration C in the upper half-plane and enclosing the point z:

$$\chi_{MF}(z)=\frac{1}{2\pi \mathrm{i}}\oint_C \frac{\chi_{MF}(\bar{z})}{\bar{z}-z}\,d\bar{z} . \tag{9.4.25}$$

In particular, if we choose the path shown in Fig. 9.5, then $\chi_{MF}(\bar{z})$ vanishes along C_2 if we allow the radius to approach ∞. The only remaining contribution is along C_1 from $-\infty$ to $+\infty$. Setting $z=\omega+\mathrm{i}\eta$, we have

$$\chi_{MF}(\omega+\mathrm{i}\eta)=\frac{1}{2\pi \mathrm{i}}\int_{-\infty}^{+\infty}\frac{\chi_{MF}(\bar{\omega}+\mathrm{i}\bar{\eta})}{\bar{\omega}-\omega-\mathrm{i}(\eta-\bar{\eta})}\,d\bar{\omega} . \tag{9.4.26}$$

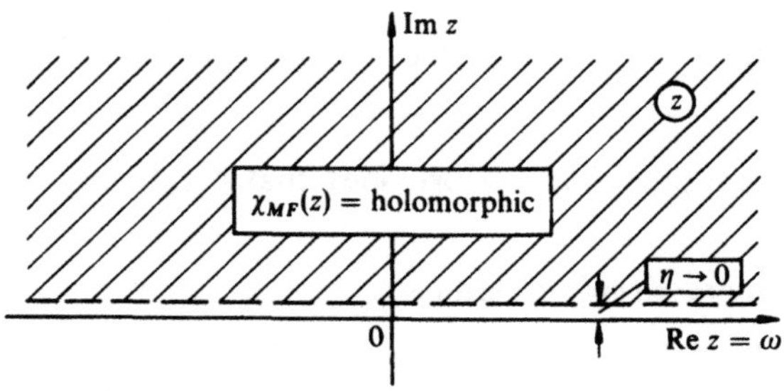

Fig. 9.4. $\chi_{MF}(z)$ in the complex z-plane

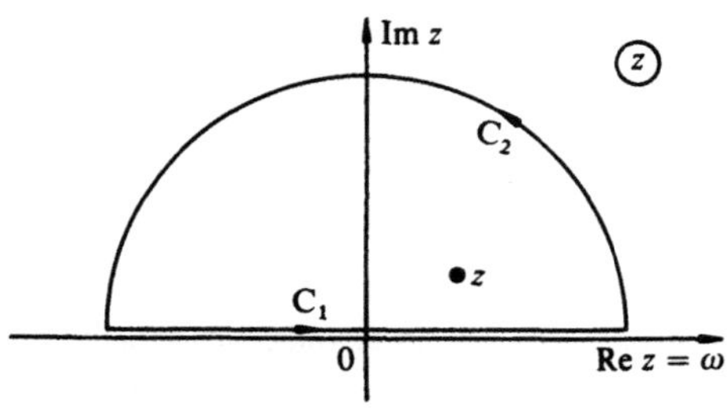

Fig. 9.5. The derivation of the Kramers–Kronig Relations

Using the identity (9.4.9) for $\eta \to \bar{\eta}$, we obtain the result:

$$\boxed{\chi_{MF}(\omega) = \frac{1}{i\pi} \, \mathrm{Pr} \int_{-\infty}^{+\infty} \frac{\chi_{MF}(\bar{\omega})}{\bar{\omega} - \omega} \, d\bar{\omega}} \quad (\omega, \bar{\omega} = \text{real}) \; . \tag{9.4.27}$$

Writing this equation for $\chi^*_{FM}(\omega)$ also, and employing the definitions (9.4.3 and 4), one obtains the *Kramers–Kronig Relations* for the quantities $\chi'_{MF}(\omega)$ and $\chi''_{MF}(\omega)$, which are Hermitian w.r.t. the indices M and F (while the operators $\mathcal{M}$ and $\mathcal{F}$ need not be Hermitian):

$$\chi'_{MF}(\omega) = \frac{1}{\pi} \, \mathrm{Pr} \int_{-\infty}^{+\infty} \frac{\chi''_{MF}(\bar{\omega})}{\bar{\omega} - \omega} \, d\bar{\omega} \; , \tag{9.4.28}$$

$$\chi''_{MF}(\omega) = -\frac{1}{\pi} \, \mathrm{Pr} \int_{-\infty}^{+\infty} \frac{\chi'_{MF}(\bar{\omega})}{\bar{\omega} - \omega} \, d\bar{\omega} \; . \tag{9.4.29}$$

These relations permit us to calculate $\chi'_{MF}(\omega)$ when $\chi''_{MF}(\omega)$ is known for all values of ω, and vice versa.

In the case of Hermitian observables $\mathcal{M}$ and $\mathcal{F}$, for which $\varphi_{MF}(t)$ is real according to (9.3.5), it is reasonable to decompose $\chi_{MF}(\omega)$ into real and imaginary parts [see Sect. 9.2.3 and (9.2.10 and 11)]:

$$\chi_{MF}(\omega) = \mathrm{Re}\, \chi_{MF}(\omega) + \mathrm{i}\, \mathrm{Im}\, \chi_{MF}(\omega) \tag{9.4.30}$$

and to employ the Kramers–Kronig relations between $\mathrm{Re}\, \chi_{MF}(\omega)$ and $\mathrm{Im}\, \chi_{MF}(\omega)$ according to (9.4.27),

$$\mathrm{Re}\, \chi_{MF}(\omega) = \frac{1}{\pi} \, \mathrm{Pr} \int_{-\infty}^{+\infty} \frac{\mathrm{Im}\, \chi_{MF}(\bar{\omega})}{\bar{\omega} - \omega} \, d\bar{\omega} \; , \tag{9.4.31}$$

$$\mathrm{Im}\, \chi_{MF}(\omega) = -\frac{1}{\pi} \, \mathrm{Pr} \int_{-\infty}^{+\infty} \frac{\mathrm{Re}\, \chi_{MF}(\bar{\omega})}{\bar{\omega} - \omega} \, d\bar{\omega} \; . \tag{9.4.32}$$

9.4.4 High-Frequency Behavior of $\chi_{MF}(\omega)$

The dynamic susceptibility $\chi_{MF}(\omega)$ vanishes as a Laplace transform (9.4.23) if z is increased in the upper half-plane ($z = \omega + i\eta$, $\eta > 0$) on an arbitrary path towards infinity. In particular, we have

$$\chi_{MF}(z=\pm\infty+i\eta)=0 \ . \tag{9.4.33}$$

In order to determine the behavior of $\chi_{MF}(\omega)$ in the neighborhood of $z=\infty$, i.e. for *high frequencies* ω, we use a formal expansion in L/z:

$$\frac{\mathsf{L}}{\mathsf{L}+z}=\frac{\mathsf{L}}{z}\frac{1}{1+\frac{\mathsf{L}}{z}}=\sum_{n=1}^{\infty}(-1)^{n+1}\left(\frac{\mathsf{L}}{z}\right)^n \tag{9.4.34}$$

in (9.4.1) and obtain an *asymptotic formula* for $\chi_{MF}(\omega)$ at large z ($\operatorname{Im} z>0$),

$$\boxed{\chi_{MF}(z)=\sum_{n=1}(-1)^{n+1}\frac{\beta(\delta\mathscr{M}\,|\,\mathsf{L}^n\delta\mathscr{F})}{z^n}} \ . \tag{9.4.35}$$

The coefficients $\beta(\delta\mathscr{M}\,|\,\mathsf{L}^n\delta\mathscr{F})$ are, according to (9.3.8), proportional to the $(n-1)$-th derivatives of the response function at time $t=0$, $\varphi_{MF}^{(n-1)}(0)$; these are often relatively easy to calculate for small values of n (cf. also Sect. 9.4.5).

Since, as a rule, only a few terms in the series (9.4.35) are known, we shall derive the remainder and thereby demonstrate that this equation is, in fact, an asymptotic series. We start with the identity

$$\frac{\mathsf{L}}{z+\mathsf{L}}=\frac{\mathsf{L}}{z}-\frac{\mathsf{L}}{z}\frac{\mathsf{L}}{z+\mathsf{L}} \tag{9.4.36}$$

and insert it repeatedly in the last term, for example:

$$\frac{\mathsf{L}}{z+\mathsf{L}}=\frac{\mathsf{L}}{z}-\left(\frac{\mathsf{L}}{z}\right)^2+\left(\frac{\mathsf{L}}{z}\right)^3-\left(\frac{\mathsf{L}}{z}\right)^3\frac{\mathsf{L}}{z+\mathsf{L}} \ . \tag{9.4.37}$$

We thus obtain

$$\chi_{MF}(z)=\sum_{\nu=1}^{n}(-1)^{\nu+1}\frac{\beta(\delta\mathscr{M}\,|\,\mathsf{L}^\nu\delta\mathscr{F})}{z^\nu}+R_n(z) \ , \tag{9.4.38}$$

where the remainder term is given by

$$R_n(z)=(-1)^n\beta\left(\delta\mathscr{M}\,\middle|\,\frac{\mathsf{L}^{n+1}}{z^n(\mathsf{L}+z)}\delta\mathscr{F}\right) \ . \tag{9.4.39}$$

$z^n R_n(z)$ is thus proportional to the dynamic susceptibility $\chi_{MF^{(n)}}(z)$ of the observables $\mathscr{M}$ and $\mathscr{F}^{(n)}=(i\mathsf{L})^n\mathscr{F}$, which, like $\chi_{MF}(z)$, vanishes for $z\to\infty$ ($\operatorname{Im} z>0$). The limiting value $\lim_{z\to\infty} z^n R_n(z)$ is therefore zero for every fixed n, i.e. (9.4.35) [or (9.4.38)] is an asymptotic series for $z\to\infty$ ($\operatorname{Im} z>0$).

9.4.5 The Moments of the Spectral Density Function

The linear response function is determined if the *spectral density function* $\beta(\delta\mathcal{M}|\delta(\mathsf{L}+\omega)\delta\mathcal{F})$ is known. From it, the relaxation function $\Phi_{MF}(t) = \beta(\delta\mathcal{M}|\exp(\mathrm{i}\mathsf{L}t)\delta\mathcal{F})$ may be obtained by Fourier transformation (from $-\infty$ to $+\infty$); multiplying by $-\omega/(z-\omega)$, followed by integration over ω yields the susceptibility $\chi_{MF}(z)$. It is thus useful in a theoretical treatment of the linear response to obtain results for the spectral density function. The n-th moments give a global characterization of the functional form of the spectral density function. They are defined by

$$\boxed{\langle\omega^n\rangle_{MF} \equiv \int_{-\infty}^{+\infty} \beta(\delta\mathcal{M}|\delta(\mathsf{L}+\omega)\delta\mathcal{F})\omega^n\,d\omega} \quad (n=0,1,2,\ldots)\ . \tag{9.4.40}$$

Because of the property of the δ-function, $\delta(x+\omega)\omega^n = (-x)^n\delta(x+\omega)$, the integration may be carried out:

$$\boxed{\langle\omega^n\rangle_{MF} = (-1)^n\beta(\delta\mathcal{M}|\mathsf{L}^n\delta\mathcal{F})} \quad (n=0,1,2,\ldots)\ . \tag{9.4.41}$$

Thus, the n-th moments $\langle\omega^n\rangle_{MF}$ are proportional to the matrix elements of L^n with $\delta\mathcal{M}$ and $\delta\mathcal{F}$.

The zeroth moment is equal to the value of the relaxation function at the time $t=0$, $\Phi_{MF}(0)$, and thus is equal to the isothermal susceptibility,

$$\langle\omega^0\rangle_{MF} = \Phi_{MF}(0) = \chi^{\mathrm{T}}_{MF}\ . \tag{9.4.42}$$

The first moment is proportional to the response function at $t=0$, $\varphi_{MF}(0)$, and the higher moments are proportional to the $(n-1)$-th derivatives w.r.t. time at $t=0$, $\varphi^{(n-1)}_{MF}(0)$,

$$\langle\omega^n\rangle_{MF} = -(\mathrm{i})^n \frac{d^{n-1}}{dt^{n-1}}\varphi_{MF}(t)\bigg|_{t=0} \quad (n=1,2,\ldots)\ . \tag{9.4.43}$$

Numerical calculation of the moments is therefore equivalent to determining the expansion coefficients of the Taylor expansion of $\varphi_{MF}(t)$ in powers of t (9.3.9), or those of the expansion (9.4.35) of $\chi_{MF}(z)$ in powers of $1/z$.

Frequently, the *lowest* moments may be found by calculating the trace. One thereby finds the first terms of the series expansions for $\varphi_{MF}(t)$ and $\chi_{MF}(z)$ and thus the integrals over the spectral density function (9.4.40). The *partial sums*:

$$\varphi_{MF}(t) = -\sum_{n=0}^{N}(-\mathrm{i})^{n+1}\langle\omega^{n+1}\rangle_{MF}\frac{t^n}{n!} \tag{9.4.44}$$

and

$$\chi_{MF}(z) = -\sum_{n=1}^{N} \frac{\langle\omega^n\rangle_{MF}}{z^n} \tag{9.4.45}$$

are usually applicable as approximations only over small ranges of t and z.

More information can be obtained from the moments if one recalls the fact that they are integrals over the spectral density function. The general problem of determining a non-negative function from its given moments is treated in a branch of mathematics known as the "moment problem". In this theory, it is shown how much information about the exact frequency dependence of the function sought is obtainable if only the first N moments are known. With the aid of this theory it is also possible in some cases to find, for example, the frequency regions in which $\chi''_{MF}(\omega)$ differs appreciably from zero, by employing knowledge of only a few of the moments. For a further treatment of these questions, we refer the reader to the specialized mathematical literature[19].

In the definition of the moments, (9.4.40), we can also use $\chi''_{MF}/\pi\omega$ from (9.4.11), instead of the spectral density function:

$$\langle\omega^n\rangle_{MF} = \frac{1}{\pi}\int_{-\infty}^{+\infty} \chi''_{MF}(\omega)\omega^{n-1}\,d\omega \ . \tag{9.4.46}$$

For $n=0$, however, the definitions (9.4.40) and (9.4.46) do *not*, in general, agree. While (9.4.40) yields the isothermal susceptibility for $n=0$ (9.4.42), one finds from the Kramers–Kronig relation

$$\chi'_{MF}(\omega=0) = \frac{1}{\pi}\,\mathrm{Pr}\int_{-\infty}^{+\infty} \frac{\chi''_{MF}(\bar\omega)}{\bar\omega}\,d\bar\omega \ . \tag{9.4.47}$$

We have already pointed out [following (9.4.21)] that $\chi_{MF}(\omega=0)$ need not be equal to χ^{T}_{MF}; the physical significance behind this fact will be discussed in more detail in Sect. 9.5. The apparent discrepancy between (9.4.42) and (9.4.47) can be explained as follows: in going from (9.4.40) to (9.4.46) for $n=0$, we divided by ω, so that $\chi''_{MF}(\omega)/\pi\omega$ differs from $\beta(\delta\mathscr{M}|\delta(\mathsf{L}+\omega)\delta\mathscr{F})$ by an amount $c_{MF}\delta(\omega)$, which just corresponds to the difference between $\chi_{MF}(\omega)$ and χ^{T}_{MF}.

To conclude this section, we mention another problem. It may happen that for a given function $\chi_{MF}(\omega)$, which is, in fact, given by the Laplace transform of a response function, the integrals (9.4.46) do not exist (e.g. in the case of Lorenz lines). Nevertheless, it may still be possible to generate the moments as *limiting*

---

[19]Akhiezer, N.J.: *The Classical Moment Problem*, Edinburgh 1965;
Karlin, S., and Studden, W.J.: *Tschebycheff Systems with Applications in Analysis and Statistics*, New York 1966.

values by extending the definition (9.4.40) or (9.4.46):

$$\langle\omega^n\rangle_{MF} = \lim_{\Delta\to+0} \int_{-\infty}^{+\infty} \beta(\delta\mathscr{M}|\delta(\mathsf{L}+\omega)\delta\mathscr{F}) \left(\frac{e^{-i\omega\Delta}-1}{-i\Delta}\right)^n d\omega$$

$$= \lim_{\Delta\to+0} (-1)^n \beta\left(\delta\mathscr{M}\left|\left(\frac{e^{i\mathsf{L}\Delta}-1}{i\Delta}\right)^n \delta\mathscr{F}\right.\right) \quad (\Delta>0)$$

$$= -i^n \frac{d^{n-1}}{dt^{n-1}} \varphi_{MF}(t)\bigg|_{t\to+0} \quad (n=1,2,\ldots)\,. \tag{9.4.48}$$

This extension then includes cases in which $\varphi_{MF}(t)$ has no derivatives at $t=0$, but instead only limiting values for $t\to+0$ (compare Exercise 9.6).

Exercise 9.5 The relaxation function $\Phi(t)=\Phi^*(t)$ of an Hermitian observable $\mathscr{F}=\mathscr{M}$ is taken to be unknown up to a relaxation time τ_1, but is described by an exponential function $\Phi(t)=\Phi_1\exp(-\Gamma t)$ with $\Gamma\tau_1\ll 1$ for $t\geqq\tau_1$. Discuss $\chi(\omega)$ in the frequency region $\omega\tau_1\ll 1$.

Exercise 9.6 Assume that the imaginary part of a susceptibility ($\mathscr{F}=\mathscr{M}=\mathscr{M}^\dagger$)

$$\chi''(\omega)=\chi^{\mathrm{T}}\frac{\omega\Gamma}{\omega^2+\Gamma^2}$$

is known.

a) Determine $\varphi(t)$ and $\Phi(t)$ for $t\gtrless 0$ and $\chi(\omega)$.
b) Verify the following relation for the expansion coefficients

$$\chi(z)=\sum_{n=1}^{\infty}(i)^n\varphi^{(n-1)}(+0)\frac{1}{z^n}$$

and calculate the moments $\langle\omega^n\rangle$ as the limits in (9.4.48).

Exercise 9.7 A system is described by the Hamiltonian $\mathscr{H}=\sum_{i=1}^N \vec{p}_i^{\,2}/2m + \mathscr{V}(\vec{x}_1,\ldots,\vec{x}_N)$. Determine the 2nd moment $\langle\omega^2\rangle$ of the spectral density function $\beta(n_{\mathfrak{k}}|\delta(\mathsf{L}+\omega)n_{\mathfrak{k}})$, where $n_{\mathfrak{k}}=\int\sum_j\delta(\mathfrak{r}-\vec{x}_j)\exp(-i\mathfrak{k}\mathfrak{r})d^3x=\sum_j\exp(-i\mathfrak{k}\vec{x}_j)$ is the spatial Fourier transform of the particle density.

9.5 The Limit of Slow Field Variation

9.5.1 Properties of the Isolated Susceptibility

For external fields which change very *slowly* (i.e. in the limit $\omega=0$), (9.4.15) yields the following expression for the dynamic susceptibility $\chi_{MF}(\omega=0)$ (which in this limit is called the *isolated susceptibility*, χ^{is}_{MF}):

$$\boxed{\chi^{\mathrm{is}}_{MF} \equiv \chi_{MF}(\omega=0) = \chi^{\mathrm{T}}_{MF} - \beta\left(\delta\mathcal{M}\left|\frac{\mathrm{i}\eta}{\mathsf{L}+\mathrm{i}\eta}\delta\mathcal{F}\right)\right|_{\eta\to+0}} \,. \tag{9.5.1}$$

According to (9.1.27), the isolated susceptibility is the integral over time of the response function (Fig. 9.6),

$$\chi^{\mathrm{is}}_{MF} = \int_0^\infty \varphi_{MF}(t)\mathrm{e}^{-\eta t}\,dt\,\Big|_{\eta\to+0} \,. \tag{9.5.2}$$

We have [compare (9.5.8)]:

$$\chi^{\mathrm{is}}_{MF} = \chi^{\mathrm{is}*}_{FM} \,, \tag{9.5.3}$$

i.e. with (9.4.3, 4),

$$\chi'_{MF}(\omega=0) = \chi^{\mathrm{is}}_{MF} \quad \text{and} \quad \chi''_{MF}(\omega=0) = 0 \,. \tag{9.5.4}$$

It would be incorrect to conclude that the limit $\eta\to 0$ in (9.5.1) causes the second term to vanish and that the isolated susceptibility is therefore equal to the isothermal susceptibility. This is in any case not generally true, because of

$$\lim_{\eta\to 0}\frac{\mathrm{i}\eta}{x+\mathrm{i}\eta} = \begin{cases} 1 & \text{for} \quad x=0\,, \\ 0 & \text{for} \quad x\neq 0\,. \end{cases} \tag{9.5.5}$$

In order to analyze more precisely the properties of the second term in (9.5.1), we insert the decomposition of the unit superoperator in $\mathfrak{L}$ in terms of orthogonal eigenelements of L, $\mathcal{V}_\nu$,

$$1 = \sum_\nu \frac{|\mathcal{V}_\nu)(\mathcal{V}_\nu|}{(\mathcal{V}_\nu|\mathcal{V}_\nu)} \tag{9.5.6}$$

before $\delta\mathcal{F}$. All the $\mathcal{V}_\nu$ for which $\mathsf{L}\mathcal{V}_\nu \neq 0$ holds give vanishing contributions to (9.5.1). This is, however, not generally true for those $\mathcal{V}_{\nu=c}$ which obey

$$\mathsf{L}\mathcal{V}_c = 0 \tag{9.5.7}$$

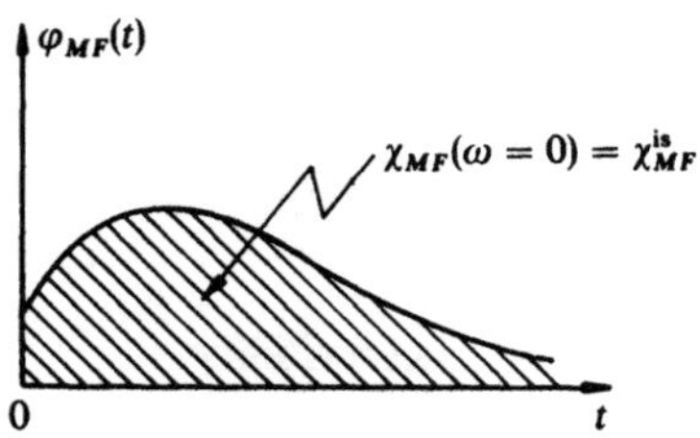

Fig. 9.6. The relation between χ^{is}_{MF} and $\varphi_{MF}(t)$

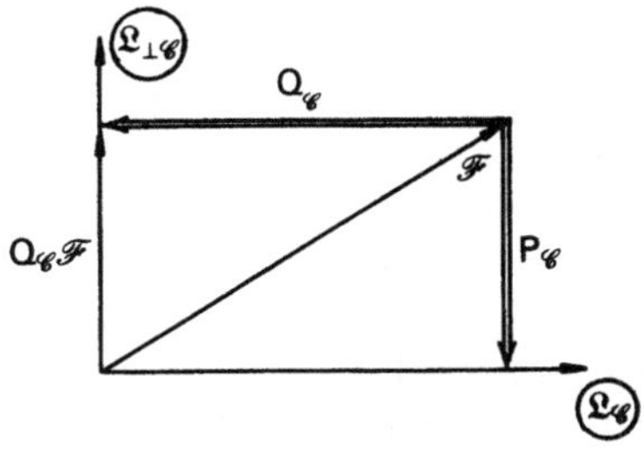

Fig. 9.7. The projection of an observable $\mathscr{F}$ onto the space $\mathfrak{L}_{\mathscr{C}}$ containing the constants of the motion

i.e. which represent *constants of the motion*. We thus obtain the following expression for the isolated susceptibility:

$$\boxed{\begin{aligned}\chi_{MF}^{\mathrm{is}} = \chi_{MF}(\omega=0) = \chi_{MF}^{\mathrm{T}} - \beta\sum_{c}\frac{(\delta\mathscr{M}|\mathscr{V}_c)(\mathscr{V}_c|\delta\mathscr{F})}{(\mathscr{V}_c|\mathscr{V}_c)} \\ = \beta(\mathscr{M}|\mathsf{Q}_{\mathscr{C}}\mathscr{F}) = \beta(\mathsf{Q}_{\mathscr{C}}\mathscr{M}|\mathsf{Q}_{\mathscr{C}}\mathscr{F})\ ,\end{aligned}} \tag{9.5.8}$$

where $\mathsf{Q}_{\mathscr{C}}$ is the superoperator which projects onto the sub(Liouville)space $\mathfrak{L}_{\perp\mathscr{C}}$ which is orthogonal to the space of the constants of the motion (cf. Fig. 9.7).

In a system Σ which is adiabatically isolated there are many constants of the motion, not only the unit operator $\mathit{1}$ and the Hamiltonian $\mathscr{H}$, but also the powers of $\mathscr{H}$, i.e. all projection operators onto the energy eigenspaces [compare (3.1.19)] and possibly also partial Hamiltonians $\mathscr{H}_\nu$ of the system. Fortunately, in many cases one can *neglect* a number of constants of the motion in (9.5.8) for a variety of reasons, e.g. those constants of the motion $\mathscr{V}'_c$ for which the scalar products $(\delta\mathscr{M}|\delta\mathscr{V}'_c)$ or $(\delta\mathscr{F}|\delta\mathscr{V}'_c)$ are zero due to selection rules (Appendix B) or have small numerical values compared to other terms in (9.5.8). This class of constants of the motion includes those for which the scalar products become small in the thermodynamic limit (compare Exercise 9.8).

For observables $\mathscr{F}$ or $\mathscr{M}$ which are *derivatives*, $\mathscr{F}\to\mathrm{i}\mathsf{L}\mathscr{F}$ or $\mathscr{M}\to\mathrm{i}\mathsf{L}\mathscr{M}$, the second term in (9.5.8) vanishes due to the Hermiticity of L and to (9.5.7), so that in these cases, the isolated susceptibility is *always* equal to the isothermal susceptibility, i.e. $\chi_{MF}^{\mathrm{is}} = \chi_{MF}^{\mathrm{T}}$ [cf. also (9.4.16)].

If, for an adiabatically isolated system Σ, the $\mathit{1}$ and the Hamiltonian $\mathscr{H}$ represent the only constants of the motion which need be taken into account, then the *isolated* susceptibility is equal to the *adiabatic* susceptibility, χ_{MF}^{ad} [cf. (8.3.30)]:

$$\chi_{MF}^{\mathrm{is}} = \chi_{MF}^{\mathrm{ad}} = \beta(\mathscr{M}|\mathsf{Q}_{\{\mathit{1},\mathscr{H}\}}\mathscr{F})\ ,^{20} \tag{9.5.9}$$

as is to be expected physically for an adiabatically isolated system.

[20] In (9.5.9), one can in general to a good approximation neglect small interactions of $\mathscr{H}$; this is not true, however, in (9.5.1).

9.5.2 The Physical Significance of the Isolated Susceptibility

Although there is just one $\chi^{\text{is}}_{MF} = \chi_{MF}(\omega = 0)$ associated with each particular Liouville operator L, it may be expedient for the treatment of different problems (i.e. for different ranges of ω) to employ different approximations for L, which then lead to different expressions for χ^{is}_{MF}, each one appropriate to the given problem.

Equation (9.5.9) is only applicable to a physical system Σ (Fig. 9.8) as long as ω is not smaller than the reciprocal isolation time, τ_{is} (compare Sect. 3.3.4) of the system relative to its environment. If there is an *interaction* $\mathscr{H}^{\Sigma\text{B}}$ of Σ with the environment ("bath" = B), then $\mathscr{H}^{\Sigma}$ is no longer a constant of the motion of the extended system $\bar{\Sigma}$. However, for this system there exists a constant of the motion $\bar{\mathscr{H}} = \mathscr{H}^{\Sigma} + \mathscr{H}^{\text{B}} + \mathscr{H}^{\bar{\Sigma}\text{B}}$, so that for $\bar{\Sigma}$ we can assume the following:

$$\chi^{\text{is}(\bar{\Sigma})}_{MF} = \chi^{\text{ad}(\bar{\Sigma})}_{MF} = \chi^{\text{T}(\Sigma)}_{MF} = \beta(\delta\mathscr{M}|\delta\mathscr{F}) \quad \text{for} \quad \omega \ll \frac{1}{\tau_{\text{is}}} \,. \tag{9.5.10}$$

The second equals sign assumes that the bath has an *extremely large heat capacity* (8.3.24); then $(\delta\mathscr{H}|\delta\mathscr{H})$ is so large that the adiabatic susceptibility of $\bar{\Sigma}$ is equal to the isothermal susceptibility of Σ.[21] Σ is then in thermal equilibrium with B, in contrast to the frequency range $\omega > 1/\tau_{\text{is}}$, in which equilibrium is prevented.

From the general relation (9.5.8), one can sometimes obtain results for higher frequencies ω, at which (Σ is now taken to be isolated from the bath) an internal interaction energy $\mathscr{H}^{12}$ in $\mathscr{H}^{\Sigma} = \mathscr{H}^{1} + \mathscr{H}^{2} + \mathscr{H}^{12}$ can be neglected ($\mathscr{H}^{1}$ is assumed to commute with $\mathscr{H}^{2}$). *Within the system Σ, an adiabatic isolation then*

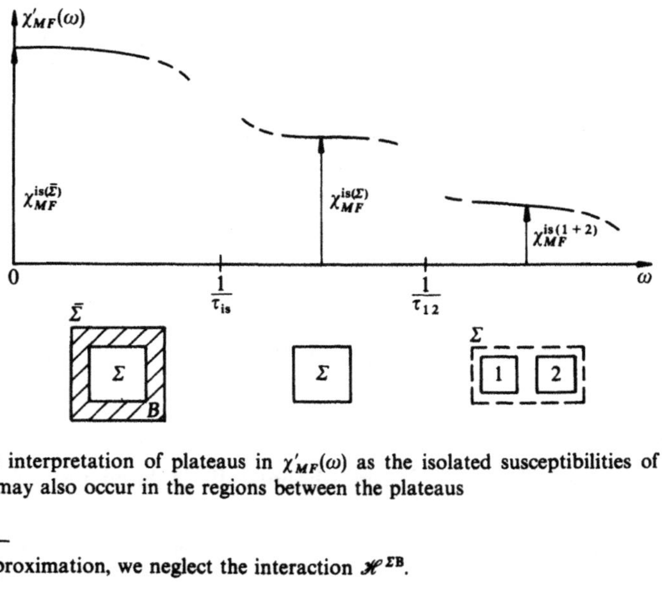

Fig. 9.8. The interpretation of plateaus in $\chi'_{MF}(\omega)$ as the isolated susceptibilities of subsystems. Oscillations may also occur in the regions between the plateaus

---

[21] In this approximation, we neglect the interaction $\mathscr{H}^{\Sigma\text{B}}$.

exists. The partial energies $\mathscr{H}^1$ and $\mathscr{H}^2$ then represent constants of the motion $\mathscr{V}_c$ and yield in (9.5.8) an isolated susceptibility [with respect to (1+2)] which is *equal to the double-adiabatic susceptibility* (cf. Exercise 9.9):

$$\chi_{MF}^{\mathrm{is}(1+2)} = \chi_{MF}^{\mathrm{ad}(1),\mathrm{ad}(2)}|_{\beta_1=\beta_2=\beta} = \beta(\mathscr{M}|\mathrm{Q}_{\{I,\mathscr{H}^1,\mathscr{H}^2\}}\mathscr{F}) \tag{9.5.11}$$

(Fig. 9.8). This value occurs in the system Σ in a frequency range $\omega\tau_{12} > 1$, where τ_{12} is a relaxation time determined by $\mathscr{H}^{12}$. The double-adiabatic susceptibility can also occur in the framework of the generalized thermodynamics of a *nondecomposable observation level* (cf. Sect. 5.3) with the associated properties.

A real physical example, in which steps in $\chi'_{MF}(\omega)$, as indicated schematically in Fig. 9.8, in fact do occur, is the longitudinal relaxation of a spin system (with an oscillating magnetic field $h(t)$ parallel to the static field H^0), where

$$\left.\begin{array}{l} \mathscr{H}^\Sigma \left\{\begin{array}{ll} \mathscr{H}^1 & = \text{Zeeman energy} \\ \mathscr{H}^2 & = \text{isotropic exchange energy} \\ \mathscr{H}^{12} & = \text{dipolar interaction energy} \end{array}\right. \\ \qquad \mathscr{H}^{\mathrm{B}} = \text{lattice energy} \\ \qquad \mathscr{H}^{\Sigma\mathrm{B}} = \text{spin-lattice interaction energy} \end{array}\right\} \tag{9.5.12}$$

($\mathscr{H}^1$ and $\mathscr{H}^2$ form a nondecomposable observation level, while $\mathscr{H}^\Sigma$ and $\mathscr{H}^{\mathrm{B}}$ are decomposable [see Sects. 5.3.2, 3 and Exercise 9.11)].)

A calculation of $\chi'_{MF}(\omega)$ as indicated schematically in Fig. 9.8 is given for a magnetic system in Exercise 13.2.

9.5.3 Plateaus in the Relaxation Function

Equation (9.4.21) showed that in the thermodynamic limit, the value of the relaxation function $\Phi_{MF}(t)$ for long times t is related to the isothermal and the isolated susceptibilities,

$$\boxed{\Phi_{MF}(\infty) = \chi_{MF}^{\mathrm{T}} - \chi_{MF}^{\mathrm{is}}} \ . \tag{9.5.13}$$

This equation can be applied to different approximations for a given Hamiltonian and can be used to explain possible plateaus in $\Phi_{MF}(t)$. To illustrate this, we consider a Hamiltonian $\mathscr{H} = \mathscr{H}^0 + \mathscr{H}'$ and assume that we can calculate the relaxation function $\Phi_{MF}(t)$ up to times $t < \tau$ without taking the internal perturbation $\mathscr{H}'$ into account, i.e. that we can replace $\mathscr{H}$ in the defining formulas by $\mathscr{H}^0$. Thus, for times $t < \tau$, the approximation $\Phi_{MF}(t) \approx \Phi_{MF}^0(t)$ is assumed to hold. If $\Phi_{MF}^0(t)$ now reaches its limiting value $\Phi_{MF}^0(\infty)$ within the time $t < \tau$, for which perturbation theory is applicable, then $\Phi_{MF}(t)$ exhibits a plateau whose height is given by (9.5.13): $\Phi_{MF}^0(0) - \chi_{MF}^{\mathrm{is}(0)} \approx \Phi_{MF}(0) - \chi_{MF}^{\mathrm{is}(0)}$. These considerations may be applied to various time ranges if one decomposes $\mathscr{H}$ in various ways, splitting off smaller and smaller perturbations in order to encompass longer and longer τ. The resulting plateau heights in $\Phi_{MF}(t)$ are therefore described by the isolated susceptibilities corresponding to various approximations for $\mathscr{H}$. The stepped behavior of $\Phi_{MF}(t)$ can, of course, also be shown through Fourier transformation of the susceptibility $\chi'_{MF}(\omega)$ as discussed above.

Exercise 9.8 Express the susceptibility $\beta(\mathscr{M}|\mathfrak{Q}_{\{1,\mathscr{H},\mathscr{H}^2\}}\mathscr{F})$ in terms of derivatives of the equilibrium expectation values $\langle\mathscr{M}^\dagger\rangle_\beta$, $\langle\mathscr{F}\rangle_\beta$, and $\langle\mathscr{H}\rangle_\beta$ w.r.t. β and show that in the thermodynamic limit for a many-body system, the following relations hold:

$$\frac{1}{\langle\mathscr{F}\rangle_\beta}\,\beta(\mathscr{M}|\mathfrak{Q}_{\{1,\mathscr{H},\mathscr{H}^2\}}\mathscr{F})=\frac{1}{\langle\mathscr{F}\rangle_\beta}\,\beta(\mathscr{M}|\mathfrak{Q}_{\{1,\mathscr{H}\}}\mathscr{F})=\frac{1}{\langle\mathscr{F}\rangle_\beta}\,\chi^{\mathrm{ad}}_{MF}\ .$$

Here, $1-\mathfrak{Q}_{\{1,\mathscr{H},\mathscr{H}^2\}}$ is the projection operator onto the subspace $\mathfrak{L}_c$ spanned by the constants of the motion $1, \mathscr{H}, \mathscr{H}^2$.

Exercise 9.9 For the system with $\mathscr{H}=\mathscr{H}_1+\mathscr{H}_2$, prove that the susceptibility

$$\beta(\mathscr{M}_\alpha|\mathfrak{Q}_{\{1,\mathscr{H}_1,\mathscr{H}_2\}}\mathscr{F})$$

can be represented as a double-adiabatic susceptibility $\chi^{\mathrm{ad(1)ad(2)}}_{M_\alpha F}$ in the framework of thermostatics over an observation level $\mathscr{H}_1, \mathscr{H}_2$. Here, $1-\mathfrak{Q}_{\{1,\mathscr{H}_1,\mathscr{H}_2\}}$ again means the projector onto the subspace spanned by $1, \mathscr{H}_1, \mathscr{H}_2$.

Exercise 9.10 Express $\chi^{\mathrm{is}}_{M_\alpha F}$ in the energy representation (cf. Exercise 8.2a) using $\partial/\partial h_\alpha\langle m(h)|\mathscr{F}m(h)\rangle|_{h=0}$, where $|m(h)\rangle$ refers to the eigenvectors of $\mathscr{H}(h)$. For the case

$$\mathscr{F}=\mathscr{M}_\gamma=-\left.\frac{\partial\mathscr{H}(h)}{\partial h_\gamma^*}\right|_{h=0},$$

give $\chi^{\mathrm{is}}_{\alpha\gamma}$ in terms of

$$\mathrm{e}^{-\beta E_m},\quad \left.\frac{\partial^2 E_m(h)}{\partial h_\alpha\partial h_\gamma^*}\right|_{h=0}\quad\text{and}\quad\left\langle\left.\frac{\partial^2\mathscr{H}(h)}{\partial h_\alpha\partial h_\gamma^*}\right|_{h=0}\right\rangle_\beta .$$

Exercise 9.11 A system of N independent ions with spin $s=1$ is described by the spin Hamiltonian

$$\mathscr{H}=\sum_{j=1}^N\left\{-\gamma H\mathfrak{s}_j^z+\frac{D}{\hbar^2}\left\{(\mathfrak{s}_j^z)^2-\frac{2}{3}\hbar^2\right\}+\frac{E}{\hbar^2}\{(\mathfrak{s}_j^x)^2-(\mathfrak{s}_j^y)^2\}\right\}.$$

Calculate the isolated susceptibility χ^{is}_{MM} in the high-temperature approximation ($\beta \to 0$) for

$$\mathcal{M} = -\left.\frac{\partial \mathcal{H}(h)}{\partial h}\right|_{h=0} = \gamma \sum \sigma_j^z \ , \qquad \frac{\partial^2 \mathcal{H}(h)}{\partial h^2} = 0$$

and compare it to the corresponding value of χ^{ad}_{MM}.

9.6 The Work Performed on the System

The *change in the energy* of an adiabatically isolated system Σ is given by the *work performed on the system* when $\mathcal{H}(h_\alpha(t))$ changes with the external parameters $h_\alpha(t)$ (cf. Sect. 3.3.5). Using (9.1.2), we obtain an expression up to second order in the $h_\alpha(t)$,

$$\begin{aligned}\frac{d}{dt}\langle \mathcal{H}(h)\rangle(t) &= \left\langle \left(\frac{\partial \mathcal{H}(h)}{\partial t}\right)_{\text{ex}}\right\rangle \\ &= -\sum_{\alpha'} \dot{h}^*_{\alpha'}(t)\langle \mathcal{M}_{\alpha'}\rangle(t) + \sum_{\alpha\alpha'} \dot{h}^*_{\alpha'}(t) h_\alpha(t) \left\langle \left.\frac{\partial^2 \mathcal{H}}{\partial h^*_{\alpha'}\partial h_\alpha}\right|_{h=0}\right\rangle_\beta \end{aligned} \qquad (9.6.1)$$

or, in integral form,

$$\begin{aligned} A(t,t_0) &= \langle \mathcal{H}(h)\rangle(t) - \langle \mathcal{H}(h)\rangle(t_0) \\ &= -\int_{t_0}^{t} \sum_{\alpha'} \dot{h}^*_{\alpha'}(t')\langle \mathcal{M}_{\alpha'}\rangle(t')dt' + \frac{1}{2}\sum_{\alpha\alpha'} h^*_{\alpha'}(t')h_\alpha(t')\Big|_{t_0}^{t} \left\langle \left.\frac{\partial^2 \mathcal{H}}{\partial h^*_{\alpha'}\partial h_\alpha}\right|_{h=0}\right\rangle_\beta \ . \end{aligned} \qquad (9.6.2)$$

Here, $\langle \mathcal{M}_{\alpha'}\rangle(t) = \langle \mathcal{M}_{\alpha'}\rangle_\beta + \langle \delta\mathcal{M}_{\alpha'}\rangle(t)$ means the response up to first order (9.1.8).

9.6.1 Average Power $\bar{W}(\omega_0)$ in the Harmonic Steady State

We apply relation (9.6.1) to the case that a *harmonically oscillating* field $h_\alpha(t) = \hat{h}_\alpha \sin\omega_0 t$ (with $\hat{h}_\alpha$ real) is acting on the system. In the *steady state*, after transients have died out, $\langle \mathcal{M}_{\alpha'}\rangle$ is given in the linear approximation by (9.2.9), if we replace $\mathcal{F}$ there by $\mathcal{M}_{\alpha'}$. The *time-averaged* power applied to the system $\bar{W}$ is obtained by averaging over the period $2\pi/\omega_0$,

$$\bar{W}(\omega_0) = \sum_{\alpha\alpha'} \hat{h}_\alpha \hat{h}_{\alpha'} \frac{\omega_0}{2} \operatorname{Im} \chi_{\alpha\alpha'}(\omega_0) \ , \qquad (9.6.3)$$

for which, according to (9.4.6), we may also write

$$\boxed{\begin{aligned}\bar{W}(\omega_0) &= \sum_{\alpha\alpha'} \hat{h}_\alpha \hat{h}_{\alpha'} \frac{\omega_0}{2} \chi''_{\alpha\alpha'}(\omega_0) \\ &= \frac{\beta}{2} \pi\omega_0^2 \sum_{\alpha\alpha'} \hat{h}_\alpha \hat{h}_{\alpha'} (\delta\mathcal{M}_\alpha | \delta(\mathsf{L}+\omega_0)\delta\mathcal{M}_{\alpha'}) \geqq 0\end{aligned}} \quad . \tag{9.6.4}$$

Because $(\mathcal{X}|\delta(\mathsf{L}+\omega)\mathcal{X})$ cannot be negative, the time-averaged power applied to the system is found to be *positive*, as long as the temperature β is positive. The change in the energy is not linear within linear response theory, but rather is quadratic in the field amplitude $\hat{h}_\alpha$. Equation (9.6.4) shows that the *absorption of radiation on passing through matter* is determined by $\chi''_{\alpha\alpha'}(\omega_0)$.

9.6.2 The Work Performed, $A(t_1, t_0)$, by a Field Acting from t_0 to t_1

We now apply (9.6.2) to the case of a field $h_\alpha(t)$ with an *arbitrary time dependence* [where α can also include complex conjugate components, i.e. $\mathcal{M}_\alpha$ need not be Hermitian]. We require the work $A(t_1, t_0)$ performed on the adiabatically isolated system Σ when this field acts only between the times t_0 and t_1 and is zero both before t_0 and after t_1 (see Fig. 9.9). Thus, we can allow the integration in (9.6.2) to run formally from $-\infty$ to $+\infty$. Applying the reverse transformation of (9.1.24, 25),

$$\begin{aligned}\langle\delta\mathcal{M}_{\alpha'}\rangle(t') &= \frac{1}{2\pi}\int_{-\infty}^{+\infty} \langle\delta\mathcal{M}_{\alpha'}\rangle(\omega) e^{-i\omega t'} d\omega \ , \\ \dot{h}^*_{\alpha'}(t') &= \frac{i}{2\pi}\int_{-\infty}^{+\infty} h^*_{\alpha'}(\bar{\omega})\bar{\omega}\, e^{+i\bar{\omega}t'} d\bar{\omega}\end{aligned} \tag{9.6.5}$$

in (9.6.2), the integration over time yields $2\pi\delta(\omega-\bar{\omega})$; we then obtain the result

$$A(t_1, t_0) = -\frac{i}{2\pi}\int_{-\infty}^{+\infty} \sum_{\alpha'} h^*_{\alpha'}(\omega)\omega\langle\delta\mathcal{M}_{\alpha'}\rangle(\omega)\, d\omega \ . \tag{9.6.6}$$

For $\langle\delta\mathcal{M}_{\alpha'}\rangle(\omega)$ within linear response theory we use the Fourier-transformed

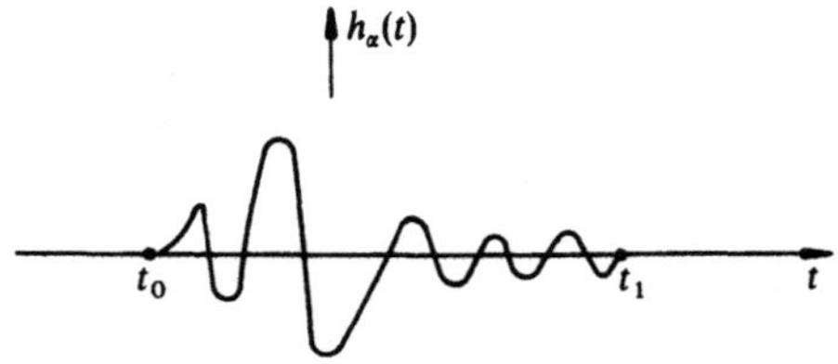

Fig. 9.9. The action of an external field in the time interval from t_0 to t_1

Kubo formula, (9.1.26) (with $\mathscr{F} \to \mathscr{M}_{\alpha'}$). We thus obtain

$$A(t_1, t_0) = -\frac{i}{2\pi} \int_{-\infty}^{+\infty} \sum_{\alpha\alpha'} h_\alpha(\omega) h^*_{\alpha'}(\omega) \omega \chi_{\alpha\alpha'}(\omega)\, d\omega \ . \tag{9.6.7}$$

This expression must be real, according to (9.6.2); it is thus equal to its complex conjugate. On addition and changing the summation indices, and taking (9.4.4) into account, we obtain the final result

$$\boxed{\begin{aligned} A(t_1, t_0) &= \frac{1}{2\pi} \int_{-\infty}^{+\infty} \sum_{\alpha\alpha'} h_\alpha(\omega) h^*_{\alpha'}(\omega) \omega \chi''_{\alpha\alpha'}(\omega)\, d\omega \\ &= \frac{\beta}{2} \int_{-\infty}^{+\infty} \sum_{\alpha\alpha'} h_\alpha(\omega) h^*_{\alpha'}(\omega) \omega^2 (\delta\mathscr{M}_\alpha | \delta(\mathsf{L}+\omega) \delta\mathscr{M}_{\alpha'})\, d\omega \geqq 0 \end{aligned}} \ . \tag{9.6.8}$$

The *work performed*[22] on the system Σ is again determined by $\chi''_{\alpha\alpha'}(\omega)$ and is, for β positive, itself always *positive*.

The fact that work is only *performed on* the system is an *immediate result of the Second Law for dynamic processes* (compare Sect. 6.3.1): at the time t_0, the system Σ was in thermal equilibrium with the internal energy $U_0 = \langle \mathscr{H} \rangle (t_0)$ and the entropy $S_0 = S_0(U_0, h_\alpha = 0)$. When, after the action of the external field at a time $t > t_1$, the system again comes into thermal equilibrium, it has the internal energy $U_1 = \langle \mathscr{H} \rangle (t) = A(t_1, t_0) + U_0$ and the entropy $S_1 = S_1(U_1, h_\alpha = 0)$. Because the state variable $S = S(U, h_\alpha = \text{const.})$ monotonically increases with U, $(\partial S/\partial U)_{h_\alpha} = \beta > 0$, the increase in the internal energy $U_1 - U_0 = A(t_1, t_0)$ implies an increase in the entropy, $S_1 - S_0 > 0$, as required quite generally by the Second Law for adiabatic dynamic processes.

Exercise 9.12 A field $h(t) = h^*(t)$ is increased from $h = 0$ to the value $\hat{h}$ in the time interval $\Delta t = t_1 - t_0$:

$$h(t) = \begin{cases} 0 & t < t_0, \\ \dfrac{t - t_0}{t_1 - t_0} \hat{h} & t_0 \leqq t \leqq t_1 \ , \\ \hat{h} & t > t_1 \ . \end{cases}$$

---

[22] According to (9.1.12), we always have

$$\varphi_{M\mathscr{H}}(t) = 0 \ , \tag{9.6.9}$$

i.e. the application of the linear Kubo formula to the expectation value of the explicitly time-*in*dependent Hamiltonian $\mathscr{H}$ gives no change in the energy $\langle \mathscr{H} \rangle (t) = \langle \mathscr{H} \rangle_\beta$. The work performed appears first with terms quadratic in the external field, but with a coefficient given by the linear susceptibility (cf. Sect. 10.2).

Calculate the work performed on the system

a) in the case of a slow field increase, $\Delta t \to \infty$;

b) in the case of a rapid field increase, $\Delta t \to 0$.

Assume that $\chi^{\mathrm{is}}_{MM} = \chi^{\mathrm{ad}}_{MM}$.

9.7 Relations Between the Fourier-Transformed Time-Dependent Correlation Functions

Along with the time-dependent Mori products $(\mathcal{M}|\mathcal{F}(t))$, the time-dependent *correlation functions* $\langle \mathcal{M}\mathcal{F}(t)\rangle_\beta$ and $\langle [\mathcal{M}, \mathcal{F}]_\pm \rangle_\beta$ are also physically significant. For example, the inelastic scattering cross-section for particles scattered by solids and liquids is determined by the space-time Fourier transform $S(\mathfrak{k}, \omega)$ of the density-density correlation function[23].

In this section, we have collected the relations between the various thermal correlation functions. Their proofs are elementary using the Liouville formalism.

We first note the relation

$$\langle \mathcal{M}\mathcal{F}\rangle_\beta = \langle \mathcal{F} \mathrm{e}^{-\beta\hbar \mathsf{L}} \mathcal{M}\rangle_\beta \ . \tag{9.7.2}$$

For the Fourier transform of $\langle \mathcal{M} \mathrm{e}^{\mathrm{i}\mathsf{L}t} \mathcal{F}\rangle_\beta$ we obtain

$$\int_{-\infty}^{+\infty} \langle \mathcal{M}\mathcal{F}(t)\rangle_\beta \mathrm{e}^{\mathrm{i}\omega t}\, dt = 2\pi \langle \mathcal{M}\cdot\delta(\mathsf{L}+\omega)\mathcal{F}\rangle_\beta \ . \tag{9.7.3}$$

From (9.7.2), it follows that

$$\int_{-\infty}^{+\infty} \langle \mathcal{F}(t)\mathcal{M}\rangle_\beta \mathrm{e}^{\mathrm{i}\omega t}\, dt = 2\pi \mathrm{e}^{\beta\hbar\omega} \langle \mathcal{M}\cdot\delta(\mathsf{L}+\omega)\mathcal{F}\rangle_\beta \ . \tag{9.7.4}$$

By addition or subtraction we obtain for the Fourier transform of the expectation value of the anticommutator $[\mathcal{M}, \mathcal{F}(t)]_+$ or the commutator $[\mathcal{M}, \mathcal{F}(t)]_-$,

---

[23] The space-time correlation function for the electronic spins of ions localized at the lattice positions $\mathfrak{R}_n$ and $\mathfrak{R}=0$ of a solid with discrete translational invariance is the thermal expectation value

$$\langle \mathcal{S}_i(\mathfrak{R}=0, t=0)\, \mathcal{S}_j(\mathfrak{R}_n, t)\rangle_\beta, \quad (i, j = x, y, z) \ .$$

This correlation is accessible through *neutron scattering*, owing to the magnetic interaction between the electronic magnetic moments and those of the neutrons. It is found from quantum scattering theory that the space-time Fourier transform

$$S_{ij}(\mathfrak{k}, \omega) = \sum_n \int_{-\infty}^{+\infty} \langle \mathcal{S}_i(\mathfrak{R}=0, t=0)\, \mathcal{S}_j(\mathfrak{R}_n, t)\rangle_\beta \exp[\mathrm{i}(\mathfrak{k}\mathfrak{R}_n - \omega t)]\, dt \tag{9.7.1}$$

determines the *differential inelastic scattering cross section*. Here, $\hbar\mathfrak{k}$ is the momentum transfer and $\hbar\omega$ is the energy loss of the neutrons. It is shown in the following that $S_{ij}(\mathfrak{k}, \omega)$ may be calculated in terms of the dynamic magnetic susceptibility $\chi''_{ij}(\mathfrak{k}, \omega)$ of the solid.

respectively, the following relation to the Fourier transforms of the correlation functions:

$$\boxed{\int_{-\infty}^{+\infty} \langle [\mathscr{M}, \mathscr{F}(t)]_{\pm} \rangle_{\beta} \mathrm{e}^{\mathrm{i}\omega t} dt = (1 \pm \mathrm{e}^{\beta\hbar\omega}) \int_{-\infty}^{+\infty} \langle \mathscr{M}\mathscr{F}(t) \rangle_{\beta} \mathrm{e}^{\mathrm{i}\omega t} dt} \quad . \tag{9.7.5}$$

The Fourier transform of the Mori product may be written as

$$\int_{-\infty}^{+\infty} (\mathscr{M}|\mathscr{F}(t))\mathrm{e}^{\mathrm{i}\omega t} dt = 2\pi(\mathscr{M}|\delta(\mathsf{L}+\omega)\mathscr{F}) \tag{9.7.6}$$

$$= 2\pi \frac{\mathrm{e}^{\beta\hbar\omega}-1}{\beta\hbar\omega} \langle \mathscr{M}^{\dagger} \cdot \delta(\mathsf{L}+\omega)\mathscr{F} \rangle_{\beta} \ , \tag{9.7.7}$$

where in the second equation we have made use of the expression (8.2.14). Thus, we find the following relation between the Fourier transform of the Mori product and that of the thermal expectation value of the product or of the anti-commutator:

$$\int_{-\infty}^{+\infty} (\mathscr{M}|\mathscr{F}(t))\mathrm{e}^{\mathrm{i}\omega t} dt = \frac{\mathrm{e}^{\beta\hbar\omega}-1}{\beta\hbar\omega} \int_{-\infty}^{+\infty} \langle \mathscr{M}^{\dagger}\mathscr{F}(t) \rangle_{\beta} \mathrm{e}^{\mathrm{i}\omega t} dt \tag{9.7.8}$$

$$= \frac{1}{2\beta E(\beta,\omega)} \int_{-\infty}^{+\infty} \langle [\mathscr{M}^{\dagger}, \mathscr{F}(t)]_{+} \rangle_{\beta} \mathrm{e}^{\mathrm{i}\omega t} dt \ , \tag{9.7.9}$$

where we have introduced the abbreviation

$$E(\beta,\omega) = \hbar\omega\left(\frac{1}{2} + \frac{1}{\mathrm{e}^{\beta\hbar\omega}-1}\right) = \frac{\hbar\omega}{2}\coth\frac{\beta\hbar\omega}{2} \ . \tag{9.7.10}$$

At high temperatures, one obtains the approximation

$$E(\beta,\omega) \approx \frac{1}{\beta} = kT \quad \text{for} \quad kT \gg \hbar\omega \tag{9.7.11}$$

[compare also (8.2.16)].

Finally, from (9.4.11), (9.7.6), and (9.7.9), the following expression for $\chi''_{MF}(\omega)$ as an anticommutator is found immediately:

$$\boxed{\chi''_{MF}(\omega) = \frac{\omega}{4E(\beta,\omega)} \int_{-\infty}^{+\infty} \langle [\delta\mathscr{M}^{\dagger}, \delta\mathscr{F}(t)]_{+} \rangle_{\beta} \mathrm{e}^{\mathrm{i}\omega t} dt} \quad . \tag{9.7.12}$$

With these relations, one can, for example, recalculate the results of inelastic scattering experiments in terms of the dynamic susceptibility from linear response theory.

Exercise 9.13 N spins $s=1/2$ are localized at the positions $\mathfrak{R}_j$. For the spatial Fourier transform

$$\sigma_{\mathfrak{k}}^{\alpha}=\sum_{j=1}^{N}\exp(-\mathrm{i}\mathfrak{k}\mathfrak{R}_j)\sigma_j^{\alpha}\ , \quad \alpha=+,\,-,\,z\ ,$$

assume that $\sigma_{\mathfrak{k}}^{-}(t)=\exp(\mathrm{i}\omega_{\mathfrak{k}}t)\sigma_{\mathfrak{k}}^{-}(0)$ holds.

What is the consequence of this dynamic relation for the equilibrium expectation value $\langle\sum_j\sigma_j^z\rangle_\beta$?

Let the values of $\mathfrak{k}$ be discrete, $\mathfrak{k}=\mathfrak{k}_1,\,\mathfrak{k}_2,\,\ldots,\,\mathfrak{k}_N$, and so chosen that the orthogonality relations $\sum_{\mathfrak{k}}\exp[\mathrm{i}\mathfrak{k}(\mathfrak{R}_i-\mathfrak{R}_j)]=N\delta_{ij}$ are fulfilled.

9.8 The First Fluctuation-Dissipation Theorem

Up to now, we have considered the dynamic susceptibility $\chi_{MF}(\omega)$ as the Fourier transform of the response to external perturbations. The part $\chi''_{MF}(\omega)$, determines the energy absorbed from the external perturbation source (9.6.4 and 8) and thus emphasizes the *dissipative* character of a system. There is, however, a completely different physical phenomenon which is also described by $\chi''_{MF}(\omega)$: the *time-dependent fluctuations* of the observables of an isolated system Σ which *is not subjected to any external fields.* These fluctuations are a result of the *internal* dynamics of the system Σ. In the following, we shall show that $\chi''_{MF}(\omega)$ is related to these fluctuations.

The time dependence of the *fluctuation operator*

$$\delta\mathscr{F}(t)=\mathscr{F}(t)-\langle\mathscr{F}\rangle_\beta 1 \tag{9.8.1}$$

of an explicitly time-*in*dependent observable, i.e. the deviation from the equilibrium value, is given in the Heisenberg formulation by

$$\delta\mathscr{F}(t)=\mathrm{e}^{\mathrm{i}\mathsf{L}t}\delta\mathscr{F}\ , \tag{9.8.2}$$

where $\delta\mathscr{F}$ is the fluctuation at time zero. The thermal expectation value of $\delta\mathscr{F}(t)$ vanishes as a trivial result of the definition (9.8.1),

$$\langle\delta\mathscr{F}(t)\rangle_\beta=0\ . \tag{9.8.3}$$

In a given time interval $0<t<\tau$, we define the operator for the *spectral decomposition of the fluctuation* $\delta\mathscr{F}(t)$[24]:

[24] In the quantum-theoretical sense, $\delta\mathscr{F}(\omega)$ is an observable; more precisely, the observables are the two Hermitian components $(1/2)(\delta\mathscr{F}(\omega)+\delta\mathscr{F}(\omega)^\dagger)$, and $(1/2\mathrm{i})(\delta\mathscr{F}(\omega)-\delta\mathscr{F}(\omega)^\dagger)$, which cannot be determined by measurements of $\delta\mathscr{F}(t)$ at all times t in the interval $0\leqq t\leqq\tau$, since the $\delta\mathscr{F}(t)$ at different times do not, in general, commute.

$$\boxed{\delta\mathscr{F}(\omega) \equiv \frac{1}{\sqrt{2\pi\tau}}\int_0^\tau \delta\mathscr{F}(t)\mathrm{e}^{\mathrm{i}\omega t}\,dt = \frac{1}{\sqrt{2\pi\tau}}\frac{\mathrm{e}^{\mathrm{i}(\mathsf{L}+\omega)\tau}-1}{\mathrm{i}(\mathsf{L}+\omega)}\delta\mathscr{F}} \;.^{25} \tag{9.8.4}$$

Because the thermal expectation value of this operator vanishes, its autocorrelation functions or general correlation functions with other observables $\delta\mathscr{M}(\omega)$ are of interest. For these correlation functions we have from (9.8.4)

$$\langle\delta\mathscr{M}(\omega)^\dagger\cdot\delta\mathscr{F}(\omega)\rangle_\beta = \left\langle\delta\mathscr{M}^\dagger\frac{\tau}{2\pi}\left(\frac{\sin(\mathsf{L}+\omega)\tau/2}{(\mathsf{L}+\omega)\tau/2}\right)^2\delta\mathscr{F}\right\rangle_\beta \;. \tag{9.8.5}$$

At *long times* τ, in the limit $\tau\to\infty$, we obtain a δ-function on the right-hand side:

$$\boxed{\begin{aligned}\langle\delta\mathscr{M}(\omega)^\dagger\cdot\delta\mathscr{F}(\omega)\rangle_\beta &= \langle\delta\mathscr{M}^\dagger\cdot\delta(\mathsf{L}+\omega)\delta\mathscr{F}\rangle_\beta\\ &= \frac{1}{2\pi}\int_{-\infty}^{\infty}\langle\delta\mathscr{M}^\dagger\cdot\delta\mathscr{F}(t)\rangle_\beta\mathrm{e}^{\mathrm{i}\omega t}\,dt\end{aligned}}\;, \tag{9.8.6}$$

i.e. the thermal expectation value of the product $\delta\mathscr{M}(\omega)^\dagger\cdot\delta\mathscr{F}(\omega)$ of the spectral decompositions is the Fourier transform of the time correlation function $\langle\delta\mathscr{M}^\dagger\cdot\delta\mathscr{F}(t)\rangle_\beta$. This quantum-statistical result corresponds to the Wiener–Khintchine theorem.

It follows from (9.8.6) that

$$\boxed{\int_{-\infty}^{+\infty}\langle\delta\mathscr{M}(\omega)^\dagger\cdot\delta\mathscr{F}(\omega)\rangle_\beta\,d\omega = \langle\delta\mathscr{M}^\dagger\cdot\delta\mathscr{F}\rangle_\beta}\;, \tag{9.8.7}$$

so that $\langle\delta\mathscr{M}(\omega)^\dagger\cdot\delta\mathscr{F}(\omega)\rangle_\beta$ may be interpreted as the spectral density of $\langle\delta\mathscr{M}^\dagger\delta\mathscr{F}\rangle_\beta$.

According to (9.7.6 and 7), the left-hand side of (9.8.6) may also be written as a Mori product and can thus be expressed in terms of $\chi''_{MF}(\omega)$:

$$\boxed{\begin{aligned}\langle\delta\mathscr{M}(\omega)^\dagger\cdot\delta\mathscr{F}(\omega)\rangle_\beta &= \frac{\beta\hbar\omega}{\mathrm{e}^{\beta\hbar\omega}-1}(\delta\mathscr{M}\,|\,\delta(\mathsf{L}+\omega)\delta\mathscr{F})\\ &= \frac{\hbar}{\pi}\frac{1}{\mathrm{e}^{\beta\hbar\omega}-1}\chi''_{MF}(\omega)\end{aligned}}\;. \tag{9.8.8}$$

The second equals sign is valid only for $\omega \neq 0$, since it makes use of $\omega\delta(\omega)=0$.

[25] For long times τ (9.8.6), a slightly different definition of $\delta\mathscr{F}(\omega)$, namely

$$\delta\mathscr{F}(\omega) = \sqrt{\frac{\eta}{\pi}}\int_0^\infty \mathrm{e}^{\mathrm{i}(\mathsf{L}+\omega)t-\eta t}\,dt\,\delta\mathscr{F} = \sqrt{\frac{\eta}{\pi}}\frac{\mathrm{i}}{\mathsf{L}+\omega+\mathrm{i}\eta}\delta\mathscr{F} \quad \text{with} \quad \eta\to+0\;,$$

leads to the same result, while being more tractable in many cases.

A similar calculation, making use of the result of Exercise 9.4, yields for

$$\begin{aligned}\langle \delta\mathscr{F}(\omega)\cdot\delta\mathscr{M}(\omega)^\dagger\rangle_\beta &= \langle \delta\mathscr{F}\cdot\delta(\mathsf{L}-\omega)\delta\mathscr{M}^\dagger\rangle_\beta \\ &= \frac{1}{2\pi}\int_{-\infty}^{+\infty}\langle \delta\mathscr{F}\cdot\delta\mathscr{M}^\dagger(t)\rangle_\beta \mathrm{e}^{-\mathrm{i}\omega t}dt \\ &= \frac{-\beta\hbar\omega}{\mathrm{e}^{-\beta\hbar\omega}-1}(\delta\mathscr{F}^\dagger|\delta(\mathsf{L}-\omega)\delta\mathscr{M}^\dagger) \\ &= \frac{-\beta\hbar\omega}{\mathrm{e}^{-\beta\hbar\omega}-1}(\delta\mathscr{M}|\delta(\mathsf{L}+\omega)\delta\mathscr{F}) \\ &= -\frac{\hbar}{\pi}\frac{1}{\mathrm{e}^{-\beta\hbar\omega}-1}\chi''_{MF}(\omega)\ . \end{aligned} \tag{9.8.9}$$

Again, the last equals sign is valid only for $\omega \neq 0$.

If we add (9.8.8 and 9), we obtain for the *thermal expectation value of the anticommutator of the spectral decomposition* (*"fluctuations"*) the following result:

$$\boxed{\begin{aligned}\tfrac{1}{2}\langle[\delta\mathscr{M}(\omega)^\dagger,\delta\mathscr{F}(\omega)]_+\rangle_\beta &= \beta E(\beta,\omega)\,(\delta\mathscr{M}|\delta(\mathsf{L}+\omega)\delta\mathscr{F}) \\ &= \frac{E(\beta,\omega)}{\pi\omega}\chi''_{MF}(\omega)\end{aligned}}\ . \tag{9.8.10}$$

We thus have an expression in terms of $\chi''_{MF}(\omega)$, which on the other hand, according to (9.6.4) or (9.6.8), is a measure of the energy absorption from a time-dependent external field ("dissipation"). This relation is therefore called the *first fluctuation-dissipation theorem.*

This theorem is frequently used for the *time derivatives of operators.* With $\delta\mathscr{M}(t)\to\dot{\mathscr{M}}(t)=\mathrm{i}\mathsf{L}\mathscr{M}(t)$ and $\delta\mathscr{F}(t)\to\dot{\mathscr{F}}(t)=\mathrm{i}\mathsf{L}\mathscr{F}(t)$, we have for the *expectation value of the anticommutator of the spectral decomposition of the time derivatives*, taking account of (9.4.17),

$$\boxed{\begin{aligned}\frac{1}{2}\langle[\dot{\mathscr{M}}(\omega)^\dagger,\dot{\mathscr{F}}(\omega)]_+\rangle_\beta &= \frac{E(\beta,\omega)}{\pi}\omega\chi''_{MF}(\omega) \\ &= \frac{\mathrm{i}E(\beta,\omega)}{\pi}\chi''_{M\dot{F}}(\omega)\end{aligned}}\ . \tag{9.8.11}$$

As an *example* of this theorem, we consider an electrical system Σ having length l. In order that we need not take account of particle exchange with the surroundings, we insert it into a capacitor (see Fig. 9.10). If the charges on the capacitor plates generate an alternating electric field $E^{\mathrm{ex}}(t)$, or a voltage

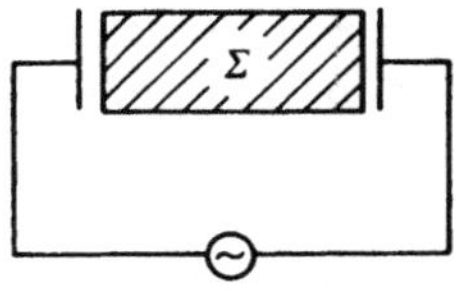

Fig. 9.10. The electric circuit described by (9.8.15)

difference $U^{\mathrm{ex}}(t) = E^{\mathrm{ex}}(t)l$, then the Hamiltonian of Σ is given by

$$\mathscr{H}(t) = \mathscr{H} - E^{\mathrm{ex}}(t)q\sum_\nu x_\nu \ . \tag{9.8.12}$$

Here, a single species of charge carriers q is assumed, having position operators x_ν. Applying the Kubo theory, we then set

$$h(t) \to U^{\mathrm{ex}}(t) \quad \text{and} \quad \mathscr{M} \to \mathscr{X} = \frac{q}{l}\sum_\nu x_\nu \ . \tag{9.8.13}$$

The response $\mathscr{F}$ is represented e.g. by the current:

$$\mathscr{J} = \frac{q}{l}\sum_\nu \dot{x}_\nu = \dot{\mathscr{X}} \ . \tag{9.8.14}$$

With $\langle \mathscr{J} \rangle_\beta = 0$, the Kubo formula (9.1.26) yields

$$\langle \mathscr{J} \rangle(\omega) = U^{\mathrm{ex}}(\omega)\chi_{X\dot{X}}(\omega) \ , \tag{9.8.15}$$

so that the complex resistance (impedance) $Z^{\mathrm{ex}}(\omega)$ is given by

$$Z^{\mathrm{ex}}(\omega) = \frac{1}{\chi_{X\dot{X}}(\omega)} \ . \tag{9.8.16}$$

The energy absorbed [cf. (9.6.8)] is determined by

$$\omega\chi''_{XX}(\omega) = \mathrm{i}\chi''_{X\dot{X}}(\omega) = \mathrm{Re}\,\chi_{X\dot{X}}(\omega) \ , \tag{9.8.17}$$

where we have used (9.4.6) as well as (9.4.16 and 17).

On the other hand, if we use the first fluctuation-dissipation theorem in the form (9.8.11) and set $\mathscr{M} = \mathscr{F} = \mathscr{X}$, we obtain a relation between the correlation of the current fluctuations and $Z^{\mathrm{ex}}(\omega)$:

$$\begin{aligned}\frac{1}{2}\langle[\mathscr{J}(\omega)^\dagger, \mathscr{J}(\omega)]_+\rangle_\beta &= \frac{\mathrm{i}E(\beta,\omega)}{\pi}\chi''_{X\dot{X}}(\omega) = \frac{E(\beta,\omega)}{\pi}\mathrm{Re}\,\chi_{X\dot{X}}(\omega) \\ &= \frac{E(\beta,\omega)}{\pi}\mathrm{Re}\,\frac{1}{Z^{\mathrm{ex}}(\omega)} \ . \end{aligned} \tag{9.8.18}$$

This expression relates the equilibrium expectation value of the squared current fluctuations to the complex resistance $Z^{ex}(\omega)$ defined by (9.8.16). It represents a form of the Nyquist theorem.

Exercise 9.14 Determine the correlation function $\langle[\dot{x}(\omega)^\dagger, \dot{x}(\omega)]_+\rangle_\beta$ of the spectral decomposition of the velocity $\dot{x}(t)$ for a harmonic oscillator with $\mathscr{H} = (p^2/2m) + (m\omega_0^2/2)x^2$

a) by direct calculation of the equilibrium expectation value;
b) by using the first fluctuation-dissipation theorem.

9.9 A Generalization of the Kubo Formula

To conclude this chapter, we shall expand somewhat on the Kubo formula (9.1.8). This generalization consists of two points: on the one hand, we want to permit a *non steady-state* statistical operator at the initial time t_0; however, restricting the treatment to linear deviations from thermal equilibrium. On the other hand, we shall permit the observable of interest for the response, $\mathscr{F}$, to have an *explicit* time dependence, $\mathscr{F}(h_\alpha(t))$, which again will be limited to the case of linearity w.r.t. the external fields.

The statistical operator $\rho(t_0)$ at the initial time t_0 is found as in Sect. 5.2 by maximization of the uncertainty measure over the observation level $\{\mathscr{H}, \mathscr{M}\}$; it will take the form of a generalized canonical statistical operator with (time *in*dependent) Lagrange multipliers β and λ_α,

$$\rho(t_0) = \mathscr{R}_{\{\mathscr{H},\mathscr{M}\}} = \mathscr{R}(\beta, \lambda_\alpha) = \frac{\exp(-\beta\mathscr{H} - \sum\lambda_\alpha\mathscr{M}_\alpha^\dagger)}{Z(\beta, \lambda_1, \ldots)} . \tag{9.9.1}$$

This operator must be representative of the system at the time t_0 (see Sect. 6.1), and could be produced e.g. by a rapid field change just before the time t_0 (cf. Sect. 6.3.2). In the case of small deviations from equilibrium, one must consistently linearize the generalized canonical statistical operator $\rho(t_0)$ with respect to the Lagrange multipliers λ_α, which describe the deviations from thermal equilibrium, following (8.3.5):

$$\rho(t_0) = \mathscr{R}_\beta\left(1 - \frac{1}{\beta}\sum_\alpha\lambda_\alpha\int_0^\beta e^{\alpha'\mathscr{H}}\,\delta\mathscr{M}_\alpha^\dagger\,e^{-\alpha'\mathscr{H}}\,d\alpha'\right) \tag{9.9.2}$$

with

$$\mathscr{R}_\beta = \frac{e^{-\beta\mathscr{H}}}{Z(\beta)} \quad \text{and} \quad \delta\mathscr{M}_\alpha = \mathscr{M}_\alpha - \mathrm{Tr}(\mathscr{R}_\beta\mathscr{M}_\alpha) . \tag{9.9.3}$$

Because the Hamiltonian does not, in general, commute with the operators $\mathscr{M}_\alpha$, we have

$$\mathsf{L}\rho(t_0) \neq \mathcal{O} , \tag{9.9.4}$$

i.e. in contrast to the usual Kubo theory, we cannot begin with (7.2.14 and 15), but rather we must resort to the more general equations (7.2.5 and 8) (Schrödinger formulation):

$$\rho_0(t) = \mathrm{e}^{-\mathrm{i}\mathsf{L}(t-t_0)}\rho(t_0) \ , \tag{9.9.5}$$

$$\rho_1(t) = \mathrm{i}\int_{t_0}^{t}\sum_\alpha \mathrm{e}^{-\mathrm{i}\mathsf{L}(t-t')}h_\alpha(t')\,\mathsf{M}_\alpha^{(\dagger)}\mathrm{e}^{-\mathrm{i}\mathsf{L}(t'-t_0)}\rho(t_0)\,dt' \ . \tag{9.9.6}$$

All the terms which are linear in the fields $h_\alpha(t)$ are taken into account in $\rho_1(t)$. Substituting (9.9.2) in (9.9.5 and 6) yields

$$\rho_0(t) = \mathscr{R}_\beta - \mathrm{e}^{-\mathrm{i}\mathsf{L}(t-t_0)}\frac{\mathscr{R}_\beta}{\beta}\sum_\alpha \lambda_\alpha \int_0^\beta \mathrm{e}^{\alpha'\mathscr{H}}\,\delta\mathscr{M}_\alpha^\dagger\,\mathrm{e}^{-\alpha'\mathscr{H}}\,d\alpha' \ , \tag{9.9.7}$$

$$\rho_1(t) = \mathrm{i}\int_{t_0}^{t}\sum_\alpha \mathrm{e}^{-\mathrm{i}\mathsf{L}(t-t')}h_\alpha(t')\,\mathsf{M}_\alpha^{(\dagger)}\mathscr{R}_\beta\,dt' \ , \tag{9.9.8}$$

where products of type $\lambda_\alpha h_{\alpha'}(t)$ are left out of (9.9.8) in this linear treatment.

We wish to allow the operators $\mathscr{F}$ for the response of interest to have an *explicit* time dependence $\mathscr{F}(h_\alpha(t))$ via the fields $h_\alpha(t)$; we must therefore expand them also. Within the linear theory, however, it is reasonable to retain only the term

$$\mathscr{F}(h(t)) = \mathscr{F} + \sum_\alpha h_\alpha(t)\left.\frac{\partial\mathscr{F}}{\partial h_\alpha}\right|_{h_\alpha(t)=0} \qquad (t \geqq t_0) \tag{9.9.9}$$

(Schrödinger formulation).

Employing (9.9.7 and 8), with $\rho(t) = \rho_0(t) + \rho_1(t)$, we take the expectation value of an explicitly time-dependent observable (9.9.9),

$$\begin{aligned}\langle\mathscr{F}(h)\rangle(t) &= \mathrm{Tr}\,(\rho_0(t)\mathscr{F}) + \sum_\alpha h_\alpha(t)\mathrm{Tr}\left(\rho_0(t)\frac{\partial\mathscr{F}}{\partial h_\alpha}\right) \\ &\quad + \mathrm{Tr}(\rho_1(t)\mathscr{F}) + \sum_\alpha h_\alpha(t)\mathrm{Tr}\left(\rho_1(t)\frac{\partial\mathscr{F}}{\partial h_\alpha}\right) \ .^{26}\end{aligned} \tag{9.9.10}$$

We retain, consistent with the linear treatment, only $\mathscr{R}_\beta$ in the second term in $\rho_0(t)$ and drop the fourth term altogether; thus we write

$$\langle\mathscr{F}(h)\rangle(t) = \mathrm{Tr}(\rho_0(t)\mathscr{F}) + \sum_\alpha h_\alpha(t)\left\langle\frac{\partial\mathscr{F}}{\partial h_\alpha}\right\rangle_\beta + \mathrm{Tr}(\rho_1(t)\mathscr{F}) \ . \tag{9.9.11}$$

[26] Operators without an argument h refer to operators acting at $h_\alpha(t) = 0$ and are thus explicitly time-independent.

Applying the calculational rules (3.1.15 and 17) and the definition of the Mori scalar product (8.2.5), we find from the above

$$\langle \mathscr{F}(h)\rangle(t) = \langle \mathscr{F}\rangle_\beta - \sum_\alpha \lambda_\alpha (\delta \mathscr{M}_\alpha | e^{iL(t-t_0)} \mathscr{F}) + \sum_\alpha h_\alpha(t) \left\langle \frac{\partial \mathscr{F}}{\partial h_\alpha} \right\rangle_\beta - i \int_{t_0}^{t} \sum_\alpha h_\alpha(t') \langle \mathrm{M}_\alpha^{(\dagger)} e^{iL(t-t')} \mathscr{F} \rangle_\beta dt' \qquad (9.9.12)$$

or, expressed in terms of the linear relaxation and response functions defined previously,

$$\boxed{\langle \mathscr{F}(h)\rangle(t) = \langle \mathscr{F}\rangle_\beta - \frac{1}{\beta}\sum_\alpha \lambda_\alpha \Phi_{M_\alpha F}(t-t_0) + \sum_\alpha h_\alpha(t) \left\langle \frac{\partial \mathscr{F}}{\partial h_\alpha} \right\rangle_\beta + \int_{t_0}^{t} \sum_\alpha h_\alpha(t') \varphi_{M_\alpha F}(t-t') dt'} \ . \qquad (9.9.13)$$

As compared to the usual Kubo formula (9.1.8), this generalized formula differs in the second and the third terms. The second term expresses the fact that the system is *not* in a steady state at the initial time t_0, so that there is an additional linear relaxation process. The third term is only present when the observables $\mathscr{F}$ are explicitly time-dependent via $h_\alpha(t)$. Due to the linearization, all four terms are additive. Of course, within the linear approximation one could have intuitively expected that a superposition of four such terms would result; however, we wished to demonstrate the derivation in detail to show where the linear approximation enters.

This generalization of the Kubo formula was given here so that we may employ it later (in Chap. 15).

The initial value $\langle \mathscr{F}(h)\rangle(t_0)$ is, according to (9.9.13), not the thermal expectation value $\langle \mathscr{F}\rangle_\beta$, but rather

$$\langle \mathscr{F}\rangle(t_0) = \langle \mathscr{F}\rangle_\beta - \frac{1}{\beta}\sum_\alpha \lambda_\alpha \chi^{\mathrm{T}}_{M_\alpha F} \ ,^{27} \qquad (9.9.14)$$

since at time t_0, the system is not in thermal equilibrium.

Exercise 9.15 We suppose that from $t=0$ on, an external electromagnetic field acts on a system of particles with electric charge q and the Hamiltonian

$$\mathscr{H} = \sum_i \vec{p}_i^{\,2}/2m + \mathscr{V}(\vec{x}_1, \ldots, \vec{x}_N) \ .$$

[27] At $t = t_0$, $h_\alpha(t)$ is switched on, so that $h_\alpha(t_0) = 0$.

What is the linear response for the current density

$$\vec{j}(\mathbf{r}) = \frac{q}{2}\sum_i (\dot{\vec{x}}_i \delta(\mathbf{r} - \vec{x}_i) + \delta(\mathbf{r} - \vec{x}_i)\dot{\vec{x}}_i)$$

to the external potentials $\mathfrak{A}(\mathbf{r}, t)$ and $\varphi(\mathbf{r}, t)$? Show explicitly that the result is gauge invariant.

10. Quadratic Response Theory

In Chap. 7, we showed in general how the response $\langle\mathcal{F}\rangle(t)-\langle\mathcal{F}\rangle_\beta$ of an observable $\mathcal{F}$ in a system driven by external fields $h_\alpha(t)$ can be expressed as an expansion in powers of the fields. After formulating linear response theory in the preceding chapter, we now turn to that portion of the response which is *quadratic* in the fields $h_\alpha(t)$. This quadratic response, $\langle\mathcal{F}\rangle_2(t)$, is important on the one hand as a correction to the linear response; on the other, its major significance lies in the fact that it describes physical phenomena which do not occur at all in the linear approximation. As we shall see in Sect. 10.3, these include frequency doubling and the appearance of frequency mixing as a response to harmonically oscillating fields $h_\alpha(t)$. We shall first consider the formal description of the response $\langle\mathcal{F}\rangle_2(t)$, and then turn our attention to its physical interpretation.

10.1 The Quadratic Response

10.1.1 Formulation in the Time Domain

We limit ourselves to the case of explicitly time-independent observables $\mathcal{F}$. We then obtain the portion of the expectation value which is quadratic in the fields, $\langle\mathcal{F}\rangle_2(t)$, from the term $\rho_2(t)$ in the expansion (7.2.16) of the statistical operator[1]. In order to write out this term explicitly, we require the expansion of the Hamiltonian, $\mathcal{H}(t)=\mathcal{H}(H_1^0+h_1(t),\ldots)$, not only including the terms of zeroth and first order (9.1.1), but up to the *second-order term.* Using the definition of the explicitly time *in*dependent operator:

$$\mathcal{M}_{\alpha\gamma}=-\left.\frac{\partial^2\mathcal{H}(H_1^0+h_1(t),\ldots)}{\partial h_\alpha^*\,\partial h_\gamma^*}\right|_{h=0}=\mathcal{M}_{\gamma\alpha} \tag{10.1.1}$$

---

[1] In the case that $\mathcal{F}=\mathcal{F}(h)$ is explicitly time-dependent via $h_\alpha(t)$, the additional terms

$$\sum_\alpha h_\alpha(t)\mathrm{Tr}\left\{\left.\frac{\partial\mathcal{F}}{\partial h_\alpha}\right|_{h=0}\rho_1(t)\right\}+\frac{1}{2}\sum_{\alpha\gamma}h_\alpha(t)h_\gamma(t)\,\mathrm{Tr}\left\{\left.\frac{\partial^2\mathcal{F}}{\partial h_\alpha\,\partial h_\gamma}\right|_{h=0}\mathcal{R}_\beta\right\}$$

are to be added [cf. (7.2.19)].

this term is

$$\mathscr{H}_2(t) = -\frac{1}{2}\sum_{\alpha\gamma} \mathscr{M}^{+}_{\alpha\gamma} h_\alpha(t) h_\gamma(t) \ , \tag{10.1.2}$$

so that, employing the superoperator

$$\mathsf{M}^{(\dagger)}_{\alpha\gamma}\mathscr{X} = \frac{1}{\hbar}[\mathscr{M}^{\dagger}_{\alpha\gamma}, \mathscr{X}] \tag{10.1.3}$$

the second-order term in the expansion of $\mathsf{L}(t)$ may be written as:

$$\mathsf{L}_2(t) = -\frac{1}{2}\sum_{\alpha\gamma} \mathsf{M}^{(\dagger)}_{\alpha\gamma} h_\alpha(t) h_\gamma(t) \ . \tag{10.1.4}$$

On inserting $\mathsf{L}_1(t)$ and $\mathsf{L}_2(t)$ in (7.2.16), we then find with (10.1.4), (9.1.3), and (7.2.5) the *quadratic response*:

$$\begin{aligned}\langle \mathscr{F} \rangle_2(t) &= \mathrm{Tr}\,(\mathscr{F}\rho_2(t)) \\ &= \int\limits_{t''=t_0}^{t} \int\limits_{t'=t_0}^{t''} \sum_{\alpha\gamma} h_\alpha(t') h_\gamma(t'')(-\mathrm{i})^2\, \mathrm{Tr}\,(\mathscr{F} \mathrm{e}^{-\mathrm{i}\mathsf{L}(t-t'')} \\ &\quad \times \mathsf{M}^{(\dagger)}_{\gamma}\, \mathrm{e}^{-\mathrm{i}\mathsf{L}(t''-t')}\, \mathsf{M}^{(\dagger)}_{\alpha} \mathscr{R}_\beta)\, dt'\, dt'' \\ &\quad + \sum_{\alpha\gamma} \int\limits_{t'=t_0}^{t} h_\alpha(t') h_\gamma(t') \left(+\frac{\mathrm{i}}{2}\right) \mathrm{Tr}\,(\mathscr{F} \mathrm{e}^{-\mathrm{i}\mathsf{L}(t-t')}\, \mathsf{M}^{(\dagger)}_{\alpha\gamma} \mathscr{R}_\beta) dt' \ .\end{aligned} \tag{10.1.5}$$

If the superoperators which occur within the traces are moved to the left, this expression may be written in the form

$$\boxed{\begin{aligned}\langle \mathscr{F} \rangle_2(t) &= \int\limits_{t''=t_0}^{t} \int\limits_{t'=t_0}^{t''} \sum_{\alpha\gamma} h_\alpha(t') h_\gamma(t'') \varphi_{M_\alpha M_\gamma F}(t''-t', t-t'')\, dt'\, dt'' \\ &\quad + \frac{1}{2} \int\limits_{t'=t_0}^{t} \sum_{\alpha\gamma} h_\alpha(t') h_\gamma(t') \varphi_{M_{\alpha\gamma} F}(t-t')\, dt'\end{aligned}} \ . \tag{10.1.6}$$

Here, we define

$$\varphi_{M_{\alpha\gamma}F}(t) = -\mathrm{i}\,\mathrm{Tr}\,(\mathsf{M}^{(\dagger)}_{\alpha\gamma} \mathrm{e}^{\mathrm{i}\mathsf{L}t} \mathscr{F} \cdot \mathscr{R}_\beta) = -\frac{\mathrm{i}}{\hbar} \langle [\mathscr{M}^{\dagger}_{\alpha\gamma}, \mathscr{F}(t)] \rangle_\beta = \varphi_{M_{\gamma\alpha}F}(t) \tag{10.1.7}$$

to be a linear response function as in (9.1.10), with the observables $\mathscr{M}_{\alpha\gamma}$ and $\mathscr{F}$, while the additional *quadratic response function* $\varphi_{M_\alpha M_\gamma F}(t_1, t_2)$, which is a function of two times, is defined by

$$\boxed{\begin{aligned}\varphi_{M_\alpha M_\gamma F}(t_1, t_2) &= (-\mathrm{i})^2 \mathrm{Tr}\,(\mathsf{M}_\alpha^{(\dagger)} e^{\mathrm{i}\mathsf{L}t_1} \mathsf{M}_\gamma^{(\dagger)} e^{\mathrm{i}\mathsf{L}t_2} \mathscr{F} \cdot \mathscr{R}_\beta) \\ &= \left(-\frac{\mathrm{i}}{\hbar}\right)^2 \langle[\mathscr{M}_\alpha^\dagger, [\mathscr{M}_\gamma^\dagger(t_1), \mathscr{F}(t_1+t_2)]]\rangle_\beta\end{aligned}} \quad . \tag{10.1.8}$$

The significance of the response equation (10.1.6) lies in the fact that the driving fields are explicitly separated out and the response functions represent materials properties, which are determined by the dynamics of the system without external fields, i.e. by L. Thus, if one knows the response functions, either from theory or from experiment, one can predict the response $\langle\mathscr{F}\rangle_2(t)$ to arbitrary time-dependent fields $h_\alpha(t)$. The definition (10.1.8) of the quadratic response function requires some additional remarks.

If we insert the time arguments $t_1 = t'' - t'$ and $t_2 = t - t''$ as given in expression (10.1.6) into (10.1.8), we can also write $\varphi_{M_\alpha M_\gamma F}(t''-t', t-t'')$ in the form

$$\varphi_{M_\alpha M_\gamma F}(t''-t', t-t'') = \left(-\frac{\mathrm{i}}{\hbar}\right)^2 \langle[\mathscr{M}_\gamma^\dagger(t'), [\mathscr{M}_\gamma^\dagger(t''), \mathscr{F}(t)]]\rangle_\beta \tag{10.1.9}$$

which clearly shows the time-translation invariance of the equilibrium expectation values. Therefore, the left side of (10.1.9) may be regarded as a function of two arbitrary differences of the three times t, t', and t''. For this reason, there are differing notations in the literature. We have chosen the time differences in (10.1.8) so as to allow the different behavior as functions of these differences to be described most simply.

The double integral in (10.1.6) extends over the triangle in Fig. 10.1. With the change of variables:

$$\begin{aligned}\bar{t} &= t_1 + t_2 = t - t' \\ \bar{\bar{t}} &= t_2 = t - t''\end{aligned} \tag{10.1.10}$$

we again have a triangular region of integration (Fig. 10.1). We thus have the form

$$\boxed{\begin{aligned}\langle\mathscr{F}\rangle_2(t) = &\int_{\bar{t}=0}^{t-t_0} \int_{\bar{\bar{t}}=\bar{t}}^{t-t_0} \sum_{\alpha\gamma} h_\alpha(t-\bar{t})\, h_\gamma(t-\bar{\bar{t}})\, \varphi_{M_\alpha M_\gamma F}(\bar{t}-\bar{\bar{t}}, \bar{\bar{t}})\, d\bar{t}\, d\bar{\bar{t}} \\ &+ \frac{1}{2} \int_0^{t-t_0} \sum_{\alpha\gamma} h_\alpha(t-t_1)\, h_\gamma(t-t_1) \varphi_{M_{\alpha\gamma} F}(t_1)\, dt_1\end{aligned}} \quad . \tag{10.1.11}$$

This form is useful if one wishes to take the one-sided Fourier (Laplace) transform of $\langle\mathscr{F}\rangle_2(t)$ or to introduce a time-symmetrized form, in which the integration over $\bar{t}$ and $\bar{\bar{t}}$ extends over a square (Sect. 10.1.3).

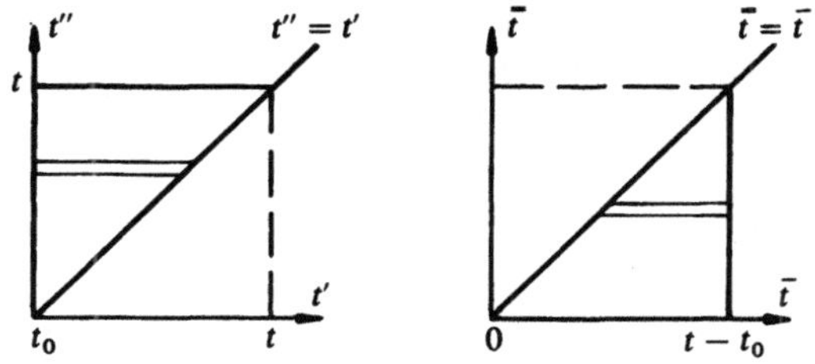

Fig. 10.1. Regions of integration for calculating the quadratic response

10.1.2 Formulation in the Frequency Domain

It is expedient for many problems, as in the linear case, to know the relation between the one-sided Fourier (Laplace) transform of $\langle\mathscr{F}\rangle_2(t)$ and that of the fields $h_\alpha(t)$. In deriving this relation, we shall be led to the Laplace transform of $\varphi_{M_\alpha M_\gamma F}(t_1, t_2)$, a function of two frequencies, which is referred to as a second-order dynamic susceptibility.

In order to calculate the Fourier transform of $\langle\mathscr{F}\rangle_2(t)$, we cannot, as in the linear case, simply apply the convolution theorem for Laplace transforms. We therefore begin with the Fourier (Laplace) transformed fields,

$$h_\alpha(\omega_1) = \int_{t_0}^{\infty} h_\alpha(t) e^{i\omega_1 t - \eta_1 t}\, dt \tag{10.1.12}$$

whose reverse transform is

$$e^{\eta_1 t} \int_{-\infty}^{+\infty} h_\alpha(\omega_1) e^{-i\omega_1 t}\, d\omega_1 = \begin{cases} 2\pi h_\alpha(t) & \text{for} \quad t > t_0 \\ 0 & \text{for} \quad t < t_0 \end{cases} . \tag{10.1.13}$$

Inserting the expression (10.1.13) for the fields into (10.1.11), we can integrate over $\bar{t}$ and $\bar{\bar{t}}$ up to ∞, since $h_\alpha(t-\bar{t})$ vanishes for $\bar{t} > t - t_0$ according to (10.1.13). In this way we obtain from (10.1.11):

$$\begin{aligned} \langle\mathscr{F}\rangle_2(t) = \frac{1}{(2\pi)^2} \iint_{-\infty}^{+\infty} \sum_{\alpha\gamma} d\omega_1\, d\omega_2\, h_\alpha(\omega_1)\, h_\gamma(\omega_2) \\ \times \Bigg\{ \int_{\bar{t}=0}^{\infty} \int_{\bar{\bar{t}}=\bar{t}}^{\infty} \exp[(-i\omega_1 + \eta_1)(t-\bar{t}) + (-i\omega_2 + \eta_2)(t-\bar{\bar{t}})] \\ \times \varphi_{M_\alpha M_\gamma F}(\bar{\bar{t}} - \bar{t}, \bar{t})\, d\bar{t}\, d\bar{\bar{t}} + \tfrac{1}{2} \int_0^{\infty} \exp[(-i\omega_1 + \eta_1)(t - t_1) \\ + (-i\omega_2 + \eta_2)(t - t_1)]\, \varphi_{M_{\alpha\gamma} F}(t_1)\, dt_1 \Bigg\} . \end{aligned} \tag{10.1.14}$$

After replacing the upper limits of integration by ∞, this equation is formally valid also for $t<t_0$, since it then yields the value zero according to (10.1.13). Therefore, in taking the Fourier transform of $\langle\mathscr{F}\rangle_2(t)$,

$$\langle\mathscr{F}\rangle_2(\omega)=\int_{t_0}^{\infty}\exp\left[(\mathrm{i}\omega-\eta_1-\eta_2)t\right]\langle\mathscr{F}\rangle_2(t)\,dt\ , \tag{10.1.15}$$

we can integrate (10.1.14) over t from $-\infty$ to $+\infty$. Employing the representation of the δ-function as a Fourier integral, and with the substitution $\bar{t}-\bar{\bar{t}}\to\bar{t}$, one obtains for the *quadratic* response as a function of frequency the result[2]:

$$\begin{aligned}\langle\mathscr{F}\rangle_2(\omega)=&\frac{1}{2\pi}\iint\limits_{-\infty}^{+\infty}\sum_{\alpha\gamma}h_\alpha(\omega_1)h_\gamma(\omega_2)\chi_{M_\alpha M_\gamma F}(\omega_1,\,\omega_1+\omega_2)\\&\times\delta(\omega-\omega_1-\omega_2)d\omega_1\,d\omega_2\\&+\frac{1}{2\pi}\frac{1}{2}\iint\limits_{-\infty}^{+\infty}\sum_{\alpha\gamma}h_\alpha(\omega_1)h_\gamma(\omega_2)\chi_{M_{\alpha\gamma}F}(\omega_1+\omega_2)\\&\times\delta(\omega-\omega_1-\omega_2)d\omega_1\,d\omega_2\end{aligned} \tag{10.1.16}$$

Here, $\chi_{M_{\alpha\gamma}F}(\omega)$ is the *linear* susceptibility associated with $\varphi_{M_{\alpha\gamma}F}(t)$,

$$\chi_{M_{\alpha\gamma}F}(\omega)=\int_0^{\infty}\varphi_{M_{\alpha\gamma}F}(t)\,\mathrm{e}^{(\mathrm{i}\omega-\eta)t}\,dt\big|_{\eta\to+0}=\chi_{M_{\gamma\alpha}F}(\omega)\ , \tag{10.1.17}$$

while $\chi_{M_\alpha M_\gamma F}(\omega_1,\omega_2)$ represents the *second-order dynamic susceptibility* which depends upon two frequencies,

$$\begin{aligned}\chi_{M_\alpha M_\gamma F}(\omega_1,\,\omega_2)=&\iint\limits_{0}^{\infty}\varphi_{M_\alpha M_\gamma F}(t_1,\,t_2)\\&\times\exp\left[(\mathrm{i}\omega_1-\eta_1)t_1+(\mathrm{i}\omega_2-\eta_2)t_2\right]dt_1\,dt_2\big|_{\eta\to+0}\ .\end{aligned} \tag{10.1.18}$$

As long as the fields $h_\alpha(t)$ do not increase unboundedly, we can consider all the Fourier integrals as in the linear approximation at the limit $\eta\to 0$. However, even in the thermodynamic limit, the limits $\eta\to+0$ in (10.1.18) cannot always be carried out *within* the integrals, due to the fact that, as we shall see in Sect. 10.2, $\varphi_{M_\alpha M_\gamma F}(t_1,t_2)$ can take on a finite value for $t_2\to\infty$ [compare (10.2.19)].

A quantum-statistical expression for the second-order dynamic susceptibility is obtained from the definition (10.1.8) of $\varphi_{M_\alpha M_\gamma F}(t_1,t_2)$ after carrying out the

---

[2] For $\chi_{M_\alpha M_\gamma F}(\omega_1,\,\omega_1+\omega_2)$, $\mathrm{i}\omega_2-\eta_2$ in the definition (10.1.18) is to be replaced by $\mathrm{i}(\omega_1+\omega_2)-(\eta_1+\eta_2)$.

integrations in (10.1.18). The result is:

$$\boxed{\chi_{M_\alpha M_\gamma F}(\omega_1,\omega_2)=\left\langle \mathsf{M}_\alpha^{(\dagger)}\frac{1}{\mathsf{L}+\omega_1+i\eta_1}\mathsf{M}_\gamma^{(\dagger)}\frac{1}{\mathsf{L}+\omega_2+i\eta_2}\mathscr{F}\right\rangle_\beta\Bigg|_{\eta\to+0}} \;. \tag{10.1.19}$$

The susceptibility $\chi_{M_\alpha M_\gamma F}(\omega_1, \omega_2)$ thus contains two denominators, in which the energy differences in the system appear on carrying out a spectral decomposition of L. We shall consider the physical interpretation of the two frequencies in Sect. 10.3.

10.1.3 Symmetrized Expressions

If we exchange the integration variables ω_1 and ω_2 in the response $\langle\mathscr{F}\rangle_2(\omega)$ (10.1.16)[3], and also exchange the summation indices α and γ, we again obtain an expression having the form of (10.1.16), but instead of the susceptibility $\chi_{M_\alpha M_\gamma F}(\omega_1, \omega_1+\omega_2)$, the new expression contains $\chi_{M_\gamma M_\alpha F}(\omega_2, \omega_1+\omega_2)$. Adding it to the original expression (10.1.16) demonstrates that only the sum of the two susceptibilities can be physically relevant. It is therefore expedient to employ a *symmetric form of the response theory* from the beginning; this form leads immediately to the symmetrized second-order dynamic susceptibility. This procedure is used in particular in optics.

The response formula (10.1.6) contains integrals over t' and t'' which are carried out over a triangular region (Fig. 10.1). If the integration variables t' and t'' as well as the summation indices α and γ are exchanged in the expression for $1/2\langle\mathscr{F}\rangle_2(t)$, one can formally integrate over a square (Fig. 10.1),

$$\boxed{\begin{aligned}\langle\mathscr{F}\rangle_2(t)&=\iint_{t_0}^{t}\sum_{\alpha\gamma}h_\alpha(t')h_\gamma(t'')\tilde{\varphi}_{M_\alpha M_\gamma F}(t-t', t-t'')dt'\,dt''\\&\quad+\frac{1}{2}\int_{t_0}^{t}\sum_{\alpha\gamma}h_\alpha(t')h_\gamma(t')\varphi_{M_{\alpha\gamma}F}(t-t')dt' \quad (t>t_0)\end{aligned}} \;, \tag{10.1.20}$$

by introducing the response function

$$\begin{aligned}\tilde{\varphi}_{M_\alpha M_\gamma F}(t-t', t-t'')=\tfrac{1}{2}\{&\Theta(t''-t')\varphi_{M_\alpha M_\gamma F}(t''-t', t-t'')\\&+\Theta(t'-t'')\varphi_{M_\gamma M_\alpha F}(t'-t'', t-t')\} \;.\end{aligned} \tag{10.1.21}$$

The function $\Theta(t)$ here represents a step function:

$$\Theta(t)=\begin{cases}1 & \text{for} \quad t>0\\ 0 & \text{for} \quad t<0 \;.\end{cases} \tag{10.1.22}$$

---

[3] In case $\eta_1 \neq \eta_2$ was chosen, the imaginary parts η_1 and η_2 must also be exchanged, since in (10.1.16), $h_\alpha(\omega_1)$ is defined with η_1 and $h_\gamma(\omega_2)$ with η_2

Owing to the definition (10.1.21), the function $\tilde{\varphi}$ is *symmetric* with respect to exchange of the time arguments $\bar{t}$ and $\bar{\bar{t}}$ *and* the component indices M_α and M_γ:

$$\boxed{\tilde{\varphi}_{M_\alpha M_\gamma F}(\bar{t}, \bar{\bar{t}}) = \varphi_{M_\gamma M_\alpha F}(\bar{\bar{t}}, \bar{t})} \ . \tag{10.1.23}$$

Furthermore, its quantum-statistical definition may be written using (10.1.21) and (10.1.9) as

$$\boxed{\begin{aligned}\tilde{\varphi}_{M_\alpha M_\gamma F}(\bar{t}, \bar{\bar{t}}) = \frac{1}{2}\left(-\frac{\mathrm{i}}{\hbar}\right)^2 \{&\Theta(\bar{t}-\bar{\bar{t}})\langle[\mathcal{M}_\alpha^\dagger(-\bar{t}), [\mathcal{M}_\gamma^\dagger(-\bar{\bar{t}}), \mathcal{F}]]\rangle_\beta \\ &+ \Theta(\bar{\bar{t}}-\bar{t})\langle[\mathcal{M}_\gamma^\dagger(-\bar{\bar{t}}), [\mathcal{M}_\alpha^\dagger(-\bar{t}), \mathcal{F}]]\rangle_\beta\}\end{aligned}} \ . \tag{10.1.24}$$

With the substitution (10.1.10), the response formula (10.1.20) takes on a form corresponding to that of (10.1.11),

$$\boxed{\begin{aligned}\langle\mathcal{F}\rangle_2(t) = &\iint_0^{t-t_0} \sum_{\alpha\gamma} h_\alpha(t-\bar{t}) h_\gamma(t-\bar{\bar{t}}) \tilde{\varphi}_{M_\alpha M_\gamma F}(\bar{t}, \bar{\bar{t}})\, d\bar{t}\, d\bar{\bar{t}} \\ &+ \frac{1}{2} \int_0^{t-t_0} \sum_{\alpha\gamma} h_\alpha(t-t_1) h_\gamma(t-t_1) \varphi_{M_{\alpha\gamma} F}(t_1)\, dt_1 \quad (t > t_0)\end{aligned}} \ . \tag{10.1.25}$$

In this expression, one may set the upper integration limits $t-t_0$ equal to ∞, since $h_\alpha(t)$ vanishes for $t < t_0$.

The form (10.1.25), which is symmetric in the time variables, contains in principle no new features as compared to (10.1.6) or (10.1.11). The situation is different, however, when one considers the formulation in the frequency domain.

Using the Fourier-transformed fields (10.1.13) in (10.1.25), and carrying out a calculation analogous to that of (10.1.14) leading to the Fourier transform (10.1.15), we obtain

$$\boxed{\begin{aligned}\langle\mathcal{F}\rangle_2(\omega) = &\frac{1}{2\pi} \iint_{-\infty}^{+\infty} \sum_{\alpha\gamma} h_\alpha(\omega_1) h_\gamma(\omega_2) \tilde{\chi}_{M_\alpha M_\gamma F}(\omega_1, \omega_2) \delta(\omega-\omega_1-\omega_2)\, d\omega_1\, d\omega_2 \\ &+ \frac{1}{2\pi}\frac{1}{2} \int_{-\infty}^{+\infty} \sum_{\alpha\gamma} h_\alpha(\omega_1) h_\gamma(\omega_2) \chi_{M_{\alpha\gamma} F}(\omega_1+\omega_2) \\ &\times \delta(\omega-\omega_1-\omega_2)\, d\omega_1\, d\omega_2\end{aligned}}$$

(10.1.26)

In this expression, the *symmetric susceptibility* $\tilde{\chi}$ is defined by

$$\boxed{\tilde{\chi}_{M_\alpha M_\gamma F}(\omega_1,\omega_2)=\iint_0^\infty \tilde{\varphi}_{M_\alpha M_\gamma F}(t_1,t_2)\exp[(\mathrm{i}\omega_1-\eta_1)t_1+(\mathrm{i}\omega_2-\eta_2)t_2]\,dt_1\,dt_2} \tag{10.1.27}$$

and, due to (10.1.23), it has the symmetry property

$$\boxed{\tilde{\chi}_{M_\alpha M_\gamma F}(\omega_1,\omega_2)=\tilde{\chi}_{M_\gamma M_\alpha F}(\omega_2,\omega_1)}\ . \tag{10.1.28}$$

We did not carry out the integration over the δ-function in (10.1.26), in order to allow this symmetry to be clearly seen. It guarantees that the expression (10.1.26) remains invariant on interchange of the integration variables ω_1 and ω_2 and of the summation indices α and γ. The quantum-statistical expression for the symmetric susceptibility (10.1.27) is obtained most simply from the relation (10.1.21) between φ and $\tilde{\varphi}$. Substituting (10.1.21) in (10.1.27) and using (10.1.10), we first find

$$\begin{aligned}\tilde{\chi}_{M_\alpha M_\gamma F}(\omega_1,\omega_2)=&\frac{1}{2}\int_{\bar{t}=0}^{\infty}\int_{\bar{\bar{t}}=\bar{t}}^{\infty}\varphi_{M_\alpha M_\gamma F}(\bar{\bar{t}}-\bar{t},\bar{t})\exp[(\mathrm{i}\omega_1-\eta_1)\bar{\bar{t}}+(\mathrm{i}\omega_2-\eta_2)\bar{t}]\,d\bar{t}\,d\bar{\bar{t}}\\&+\frac{1}{2}\int_{\bar{t}=0}^{\infty}\int_{\bar{\bar{t}}=\bar{t}}^{\infty}\varphi_{M_\gamma M_\alpha F}(\bar{\bar{t}}-\bar{t},\bar{t})\exp[(\mathrm{i}\omega_1-\eta_1)\bar{t}\\&+(\mathrm{i}\omega_2-\eta_2)\bar{\bar{t}}]\,d\bar{t}\,d\bar{\bar{t}}\ ,\end{aligned} \tag{10.1.29}$$

from which, by replacing $\bar{\bar{t}}-\bar{t}$ with t_1, or $\bar{\bar{t}}-\bar{t}$ with t_1' and using the relation (10.1.18), we obtain

$$\boxed{\tilde{\chi}_{M_\alpha M_\gamma F}(\omega_1,\omega_2)=\tfrac{1}{2}\{\chi_{M_\alpha M_\gamma F}(\omega_1,\omega_1+\omega_2)+\chi_{M_\gamma M_\alpha F}(\omega_2,\omega_1+\omega_2)\}}\ . \tag{10.1.30}$$

Combining this equation with (10.1.19) leads to the result:

$$\begin{aligned}\tilde{\chi}_{M_\alpha M_\gamma F}(\omega_1,\omega_2)=&\frac{1}{2}\left\langle\left\{\mathsf{M}_\alpha^{(\dagger)}\frac{1}{\mathsf{L}+\omega_1+\mathrm{i}\eta_1}\mathsf{M}_\gamma^{(\dagger)}+\mathsf{M}_\gamma^{(\dagger)}\frac{1}{\mathsf{L}+\omega_2+\mathrm{i}\eta_2}\mathsf{M}_\alpha^{(\dagger)}\right\}\right.\\&\left.\times\frac{1}{\mathsf{L}+\omega_1+\omega_2+\mathrm{i}(\eta_1+\eta_2)}\mathscr{F}\right\rangle_\beta\Bigg|_{\eta\to+0}\ .\end{aligned} \tag{10.1.31}$$

This expression for $\tilde{\chi}$ shows directly the symmetry property (10.1.28). It represents the linear combination of $\chi_{M_\alpha M_\gamma F}(\omega_1,\omega_1+\omega_2)$ and $\chi_{M_\gamma M_\alpha F}(\omega_2,\omega_1+\omega_2)$ which we mentioned at the beginning of this section.

Although the response $\langle\mathscr{F}\rangle_2(\omega)$ contains only the symmetric portion $\tilde{\chi}$ of χ, as seen in (10.1.26), we can, conversely, obtain the unsymmetrized form χ from $\tilde{\chi}$. To this end, one need only realize that Fourier transformation of (10.1.27) allows us to calculate the response function $\tilde{\varphi}$. But, following (10.1.21), it is possible to determine φ from $\tilde{\varphi}$. It follows namely from (10.1.21) for $\bar{t}>\bar{\bar{t}}$ that

$$\tilde{\varphi}_{M_\alpha M_\gamma F}(\bar{t},\bar{\bar{t}})=\tfrac{1}{2}\varphi_{M_\alpha M_\gamma F}(\bar{t}-\bar{\bar{t}},\bar{\bar{t}}) \ , \tag{10.1.32}$$

so that for $t_2=\bar{\bar{t}}>0$ and $t_1=\bar{t}-\bar{\bar{t}}>0$,

$$\varphi_{M_\alpha M_\gamma F}(t_1,t_2)=2\tilde{\varphi}_{M_\alpha M_\gamma F}(t_1+t_2,t_2) \tag{10.1.33}$$

holds and therefore the Fourier transformation (10.1.18) for χ can in principle be carried out.

Exercise 10.1 For a spin system with the Hamiltonian $\mathscr{H}=-\gamma H\sum_j \sigma_j^z$, calculate the response functions $\varphi_{\nu\mu,\lambda}(t_1,t_2)$ of the observables $\mathscr{S}^{\pm}=\sum_j \sigma_j^{\pm}$ and $\mathscr{S}^0=\sum_j \sigma_j^z$.

10.2 The Influence of Energy Entering the System

If fields $h_\alpha(t)$ are permitted to act upon an adiabatically isolated system, work will be performed on the system. In this section, we treat the effects of this energy input on the time dependence of the quadratic response function.

10.2.1 The Behavior of $\varphi_{MMF}(t_1, t_2)$ at Long Times

We first consider the case that the fields $h_\alpha(t)$ act *only* within the time interval from $t=t_0$ to $t=t_1$. Within *linear* response theory, we find for the expectation value of an observable $\mathscr{F}$ for times $t>t_1$ after the fields are switched off, using (9.1.8), the following expression:

$$\langle\mathscr{F}\rangle_1(t)=\langle\mathscr{F}\rangle_\beta+\int_{t_0}^{t_1}\sum_\alpha h_\alpha(t')\varphi_{M_\alpha F}(t-t')\,dt' \ . \tag{10.2.1}$$

At long times, $t\to\infty$, it follows that

$$\lim_{t\to\infty}\langle\mathscr{F}\rangle_1(t)=\langle\mathscr{F}\rangle_\beta \ , \tag{10.2.2}$$

since the linear response function vanishes for $t\to\infty$ according to (9.1.34). The expectation value $\langle\mathscr{F}\rangle_1(\infty)$ is thus given by the equilibrium expectation value at the temperature β. In spite of the possibility of an energy input, there is no change in the statistical operator in (10.2.2) as compared to $\rho(t_0)=\mathscr{R}_\beta$. This is because the energy input, as shown in Sect. 9.6, first appears in the quadratic term in the expansion in terms of the fields and thus plays no role in the linear approximation for $\langle\mathscr{F}\rangle(t)$ [cf. (9.6.9)]. The situation is quite different when we consider the *quadratic response. The energy entering the system can then give rise to detectable effects.* Because the fields are taken to act only between t_0 and t_I, the quadratic response (10.1.6) for times $t > t_I$ is found to be

$$\langle \mathscr{F} \rangle_2(t) = \int_{t''=t_0}^{t_I} \int_{t'=t_0}^{t''} \sum_{\alpha\gamma} h_\alpha(t') h_\gamma(t'') \varphi_{M_\alpha M_\gamma F}(t''-t', t-t'')\, dt'\, dt''$$

$$+\frac{1}{2}\int_{t_0}^{t_I} \sum_{\alpha\gamma} h_\alpha(t') h_\gamma(t') \varphi_{M_{\alpha\gamma} F}(t-t') dt' \ . \tag{10.2.3}$$

In the limit $t \to \infty$, the second term makes no contribution, as may be seen from (10.2.1), and one obtains

$$\lim_{t\to\infty} \langle \mathscr{F} \rangle_2(t) = \int_{t''=t_0}^{t_I} \int_{t'=t_0}^{t''} \sum_{\alpha\gamma} h_\alpha(t')\, h_\gamma(t'')\, \varphi_{M_\alpha M_\gamma F}(t''-t', \infty)\, dt'\, dt'' \ . \tag{10.2.4}$$

This limiting value exists if the system tends towards an equilibrium state as $t \to \infty$. The value of the integral (10.2.4) indicates the extent to which, in the quadratic approximation, the expectation value $\langle \mathscr{F} \rangle(\infty)$ in the final state differs from that in the initial state, $\langle \mathscr{F} \rangle_\beta$, at the initial temperature β. We have to assume that these values may be different and therefore

$$\boxed{\lim_{t_2\to\infty} \varphi_{M_\alpha M_\gamma F}(t_1, t_2) = \varphi_{M_\alpha M_\gamma F}(t_1, \infty) \neq 0} \tag{10.2.5}$$

may hold. This can be seen from the definition (10.1.8) if one chooses $\mathscr{F}$ to be a constant of the motion, $\mathsf{L}\mathscr{F} = 0$, e.g. $\mathscr{F} = \mathscr{H}$. Then the dependence on t_2 is no longer present, and the limit (10.2.5) is generally not equal to zero. We shall discuss a quantum-statistical expression for $\varphi_{M_\alpha M_\gamma F}(t_1, \infty)$ in Sect. 10.2.2.

In contrast to the behavior described by (10.2.5) for $t_2 \to \infty$, we can assume the following behavior of $\varphi_{M_\alpha M_\gamma F}(t_1, t_2)$ in the limit $t_1 \to \infty$:

$$\boxed{\lim_{t_1\to\infty} \varphi_{M_\alpha M_\gamma F}(t_1, t_2) = \varphi_{M_\alpha M_\gamma F}(\infty, t_2) = 0} \ . \tag{10.2.6}$$

This follows from the fact that $\varphi_{M_\alpha M_\gamma F}(t_1, t_2)$ for *fixed* t_2, according to (10.1.8) with

$$\tilde{\mathscr{F}}(t_2) = -\mathrm{i}\mathsf{M}_\gamma^{(\dagger)} e^{\mathrm{i}\mathsf{L}t_2} \mathscr{F} \tag{10.2.7}$$

takes on the form of a linear response function (9.1.10):

$$\varphi_{M_\alpha M_\gamma F}(t_1, t_2) = \varphi_{M_\alpha \tilde{\mathscr{F}}(t_2)}(t_1) \ , \tag{10.2.8}$$

for which (9.1.34) is valid.

10.2.2 $\varphi_{M_\alpha M_\gamma F}(t_1, \infty)$ as a Linear Response Function

In this section, we shall determine the limiting value $\varphi_{M_\alpha M_\gamma F}(t_1, \infty)$ under the condition that *after* switching off the field, for $t \to \infty$ the system can once again be described by a canonical statistical operator,

$$\mathscr{R}(\beta+\Delta\beta, h_\alpha=0)=\frac{\mathrm{e}^{-(\beta+\Delta\beta)\mathscr{H}}}{Z(\beta+\Delta\beta)} \tag{10.2.9}$$

(cf. Sect. 6.3.1). The final temperature $\beta+\Delta\beta$ is then fixed by the expectation value of the energy,

$$\lim_{t\to\infty} \langle \mathscr{H} \rangle(t)=U=\mathrm{Tr}(\mathscr{H}\mathscr{R}(\beta+\Delta\beta)) \ . \tag{10.2.10}$$

In order to determine $\mathscr{R}(\beta+\Delta\beta)$, we interpret $\beta+\Delta\beta$ as a function of U and expand in terms of

$$\Delta U=U-\langle\mathscr{H}\rangle(t_0) \ . \tag{10.2.11}$$

Due to $\langle\mathscr{H}\rangle(t_0)=\mathrm{Tr}(\mathscr{H}\mathscr{R}(\beta))$, this expansion [taking account of (8.3.6) and (8.3.24)] yields

$$\begin{aligned}\mathscr{R}(\beta+\Delta\beta) &= \mathscr{R}(\beta)+\frac{\partial\mathscr{R}}{\partial\beta}\frac{1}{\left.\dfrac{\partial U}{\partial\Delta\beta}\right|_{\Delta\beta=0}}\Delta U+\dots \\ &= \mathscr{R}(\beta)+\mathscr{R}(\beta)\delta\mathscr{H}\frac{1}{\langle\delta\mathscr{H}\cdot\delta\mathscr{H}\rangle}\Delta U+\dots \ .\end{aligned} \tag{10.2.12}$$

On the other hand, due to the dynamics, ΔU is a functional of the fields $h_\alpha(t)$ which acted; its linear term, according to (10.2.2), vanishes for $\mathscr{F}=\mathscr{H}$. The lowest-order term in the expansion for $\mathscr{R}(\beta+\Delta\beta)$ in terms of the fields is therefore obtained from (10.2.12) by inserting the second-order term for ΔU, i.e. using the quadratic response

$$\Delta U=\lim_{t\to\infty}\langle \mathscr{H} \rangle_2(t) \ . \tag{10.2.13}$$

As a result, the quadratic term for the expectation value of an observable $\mathscr{F}$ follows from (10.2.12 and 13):

$$\lim_{t\to\infty}\langle\mathscr{F}\rangle_2(t)=\lim_{t\to\infty}\langle\mathscr{H}\rangle_2(t)\frac{\langle\delta\mathscr{H}\cdot\mathscr{F}\rangle_\beta}{\langle\delta\mathscr{H}\cdot\delta\mathscr{H}\rangle_\beta} \ . \tag{10.2.14}$$

This relation between $\langle\mathscr{F}\rangle_2(\infty)$ and $\langle\mathscr{H}\rangle_2(\infty)$ makes it possible to determine $\varphi_{M_\alpha M_\gamma F}(t_1, \infty)$. Inserting relation (10.2.4), we consider that the result must hold for arbitrary fields $h_\alpha(t)$, so that we first obtain

$$\varphi_{M_\alpha M_\gamma F}(t_1, \infty) = \varphi_{M_\alpha M_\gamma \mathscr{H}}(t_1, \infty) \frac{\langle \delta\mathscr{H} \cdot \mathscr{F} \rangle_\beta}{\langle \delta\mathscr{H} \cdot \delta\mathscr{H} \rangle_\beta} . \tag{10.2.15}$$

However, the response function $\varphi_{M_\alpha M_\gamma \mathscr{H}}(t_1, t_2)$, as seen from (10.1.8) and due to

$$e^{iLt_2}\mathscr{H} = \mathscr{H} \tag{10.2.16}$$

does not depend on t_2 at all, so that, using

$$-iM_\gamma^{(\dagger)} e^{iLt_2}\mathscr{H} = \frac{i}{\hbar}[\mathscr{H}, \mathscr{M}_\gamma^\dagger] = \dot{\mathscr{M}}_\gamma^\dagger \tag{10.2.17}$$

it follows from (10.2.7 and 8) that we can write

$$\varphi_{M_\alpha M_\gamma \mathscr{H}}(t_1, t_2) = \varphi_{M_\alpha \dot{M}_\gamma^\dagger}(t_1) \tag{10.2.18}$$ [4]

as a linear response function. We obtain the overall result from the condition of an equilibrium final state (10.2.9):

$$\boxed{\varphi_{M_\alpha M_\gamma F}(t_1, \infty) = \varphi_{M_\alpha \dot{M}_\gamma^\dagger}(t_1) \frac{\langle \delta\mathscr{H} \cdot \mathscr{F} \rangle_\beta}{\langle \delta\mathscr{H} \cdot \delta\mathscr{H} \rangle_\beta}} . \tag{10.2.19}$$

This limiting value thus appears only when $\langle \delta\mathscr{H} \cdot \mathscr{F} \rangle_\beta \neq 0$. It therefore plays a role, e.g. for translationally invariant systems, when $\mathscr{F}$ contains a spatial Fourier component with $\mathfrak{k} = 0$ [cf. Appendix B]. Furthermore, it can be set to zero due to the denominator $\langle \delta\mathscr{H} \cdot \delta\mathscr{H} \rangle$ when the heat capacity of a subsystem (thermal bath) is very large. Using the example of periodic irradiation by an harmonically oscillating field, one can clearly see the physical effects of the term (10.2.19) (cf. Exercise 10.3).

---

[4] The response which follows from (10.2.3) and (9.6.9) for $t \geqq t_1$,

$$\langle \mathscr{H} \rangle_2(t) = \sum_\gamma \int_{t''=t_0}^{t_1} h_\gamma(t'') \int_{t'=t_0}^{t''} \sum_\alpha h_\alpha(t') \frac{d}{dt''} \varphi M_\alpha M_\gamma^\dagger(t''-t')\, dt'\, dt''$$

may be reformulated making use of

$$\sum_{\alpha\gamma} h_\gamma(t'') h_\alpha(t'') \varphi M_\alpha M_\gamma^\dagger(0) = -\frac{i}{\hbar} \left\langle \left[\sum_\alpha h_\alpha(t'') \mathscr{M}_\alpha^\dagger, \sum_\gamma h_\gamma(t'') \mathscr{M}_\gamma^\dagger \right] \right\rangle = 0$$

and the linear response formula (9.1.8), yielding:

$$\langle \mathscr{H} \rangle_2(t) = \sum_\gamma \int_{t_0}^{t_1} h_\gamma(t'') \frac{d}{dt''} \langle \mathscr{M}_\gamma^\dagger \rangle_1(t'')\, dt'' .$$

Integration by parts with $h_\gamma(t_0) = h_\gamma(t_1) = 0$ shows that the expression for $\langle \mathscr{H} \rangle_2(t)$ agrees with the energy difference of the explicitly time-dependent Hamiltonian, (9.6.2).

10.2.3 Separation of the Response Function into $\varphi_{\parallel}$ and $\varphi_{\perp}$

When the limiting value of $\varphi_{M_\alpha M_\gamma F}(t_1, \infty)$ in (10.2.19) is nonzero, we cannot take the limit $\eta_2 \to +0$ within the integral in calculating the susceptibility (10.1.18). For this reason, it is expedient to *separate off* the portion $\varphi_{M_\alpha M_\gamma F}(t_1, \infty)$ from the response function $\varphi_{M_\alpha M_\gamma F}(t_1, t_2)$, so that the remaining function may be Laplace transformed using $\eta_1 \to +0$ and $\eta_2 \to +0$. This separation can be most simply accomplished by first decomposing the observable $\mathscr{F}$ into the two parts

$$\mathscr{F}_{\parallel} = \delta\mathscr{H} \frac{\langle \delta\mathscr{H} \cdot \mathscr{F} \rangle_\beta}{\langle \delta\mathscr{H} \cdot \delta\mathscr{H} \rangle_\beta} \tag{10.2.20}$$

and

$$\mathscr{F}_{\perp} = \mathscr{F} - \mathscr{F}_{\parallel} \; . \tag{10.2.21}$$

We have chosen the notation $\mathscr{F}_{\parallel}$ because one can consider $\mathscr{F}_{\parallel}$ to be the projection of $\mathscr{F}$ onto $\mathscr{H}$, following (8.1.14). Since, according to (10.1.8), the response function $\varphi_{M_\alpha M_\gamma F}(t_1, t_2)$ is linear in $\mathscr{F}$, the decomposition (10.2.21) of $\mathscr{F}$ leads to the decomposition

$$\varphi_{M_\alpha M_\gamma F}(t_1, t_2) = \varphi_{M_\alpha M_\gamma F_{\perp}}(t_1, t_2) + \varphi_{M_\alpha M_\gamma F_{\parallel}}(t_1, t_2) \; . \tag{10.2.22}$$

The second term on the right is, aside from a factor $\langle \delta\mathscr{H} \cdot \mathscr{F} \rangle_\beta / \langle \delta\mathscr{H} \cdot \delta\mathscr{H} \rangle_\beta$, just the response function $\varphi_{M_\alpha M_\gamma \mathscr{H}}(t_1, t_2)$. Therefore, applying (10.2.18), we obtain an expression in terms of linear response functions:

$$\begin{aligned}\varphi_{M_\alpha M_\gamma F_{\parallel}}(t_1, t_2) &= \varphi_{M_\alpha M_\gamma \mathscr{H}}(t_1) \frac{\langle \delta\mathscr{H} \cdot \mathscr{F} \rangle_\beta}{\langle \delta\mathscr{H} \cdot \delta\mathscr{H} \rangle_\beta} \\ &= \varphi_{M_\alpha \dot{M}_\gamma^\dagger}(t_1) \frac{\langle \delta\mathscr{H} \cdot \mathscr{F} \rangle_\beta}{\langle \delta\mathscr{H} \cdot \delta\mathscr{H} \rangle_\beta} \; .\end{aligned} \tag{10.2.23}$$

If one considers the limit $t_2 \to \infty$ in the decomposition (10.2.22), comparison of (10.2.19) with (10.2.23) shows that $\varphi_{M_\alpha M_\gamma F}(t_1, \infty) = \varphi_{M_\alpha M_\gamma F_{\parallel}}(t_1)$ and thus, after taking the thermodynamic limit, the response function $\varphi_{M_\alpha M_\gamma F_{\perp}}(t_1, t_2)$ tends towards the limiting value

$$\varphi_{M_\alpha M_\gamma F_{\perp}}(t_1, \infty) = 0 \; . \tag{10.2.24}$$

It can therefore be one-sidedly Fourier transformed in (10.1.18) even when the limit $\eta_1 \to +0$ is taken within the integral.

The *separation* (10.2.21) of $\mathscr{F}$ can, as seen from (10.1.18), be transferred to the *susceptibilities*. Inserting (10.2.23) in (10.1.18), we obtain

$$\chi_{M_\alpha M_\gamma F_{\parallel}}(\omega_1, \omega_2) = \chi_{M_\alpha \dot{M}_\gamma^\dagger}(\omega_1) \frac{\mathrm{i}}{\omega_2 + \mathrm{i}\eta_2} \frac{\langle \delta\mathscr{H} \cdot \mathscr{F} \rangle_\beta}{\langle \delta\mathscr{H} \cdot \delta\mathscr{H} \rangle_\beta} \; , \tag{10.2.25}$$

where $\chi_{M_\alpha M_\gamma^\dagger}(\omega_1)$ denotes the *linear* susceptibility (9.1.27) associated with $\varphi_{M_\alpha M_\gamma^\dagger}(t_1)$. The form (10.2.25) demonstrates that $\chi_{M_\alpha M_\gamma F}(\omega_1, \omega_2)$ has a pole at $z_2=0$ for complex $z_2=\omega_2+i\eta_2$.

Because the *symmetrized* response functions (10.1.21) and susceptibilities (10.1.27) are likewise linear in $\mathscr{F}$, they can also be decomposed into two portions using $\mathscr{F}=\mathscr{F}_\perp+\mathscr{F}_\parallel$. Here, $\tilde{\varphi}_{M_\alpha M_\gamma F_\perp}(\bar{t}, \bar{t})$ represents the symmetrized form of $\varphi_{M_\alpha M_\gamma F_\perp}(\bar{t}-\bar{t}, \bar{t})$ [cf. (10.1.32)], so that $\tilde{\chi}_{M_\alpha M_\gamma F_\perp}(\omega_1, \omega_2)$ can also be calculated in the thermodynamic limit taking $\eta_1 \to +0$, $\eta_2 \to +0$ within the integral. From the relation (10.1.30) between $\tilde{\chi}$ and χ and the form (10.2.25) for χ, one can see that $\tilde{\chi}_{M_\alpha M_\gamma F}(\omega_1, \omega_2)$ with complex frequencies $z_1=\omega_1+i\eta_1$ and $z_2=\omega_2+i\eta_2$ has a pole at $z_1+z_2=0$.

Exercise 10.2 A field $h(t)=h^*(t)$ is increased within the time interval $\Delta t=t_1-t_0$ to the value $\hat{h}$:

$$h(t)=\begin{cases} 0\ , & t<0 \\ \dfrac{t-t_0}{t_1-t_0}\hat{h}\ , & t_0<t<t_1\ . \end{cases}$$

What contribution does $\langle \mathscr{F}_\parallel \rangle_2(t_1)$ make to the response $\langle \mathscr{F} \rangle_2(t_1)$ in the limit $\Delta t \to \infty$? Show that the result can be obtained by taking a limiting value of $\chi_{MMF_\parallel}(\omega_1, \omega_2)$. Calculate the total response $\langle \mathscr{F} \rangle_2(t_1)$ for $\Delta t \to \infty$ and express the result in terms of $\tilde{\chi}_{MMF}(\omega_1, \omega_2)$.

10.3 Interpretation Using Time-Dependent Fields

In this section, we wish to choose the fields $h_\alpha(t)$ in a particular way such that the meaning of the two times in the response function $\varphi_{M_\alpha M_\gamma F}(t_1, t_2)$ and of the two frequencies in $\tilde{\chi}_{M_\alpha M_\gamma F}(\omega_1, \omega_2)$ becomes clear. To this end, we make use of the fact that the response functional

$$\langle \mathscr{F} \rangle_2(t)=F_t[h(t')] \tag{10.3.1}$$

is quadratic in the fields and therefore, the response to the sum of two fields

$$h_\alpha(t)=h_\alpha^{(1)}(t)+h_\alpha^{(2)}(t) \tag{10.3.2}$$

is no longer the sum of the responses to the individual fields $h_\alpha^{(1)}(t)$ and $h_\alpha^{(2)}(t)$; instead, an *interference term* occurs:

$$\Delta\langle \mathscr{F} \rangle(t)=F_t[h^{(1)}(t')+h^{(2)}(t')]-F_t[h^{(1)}(t')]-F_t[h^{(2)}(t')]\ . \tag{10.3.3}$$

This term, according to (10.1.6), satisfies

$$\Delta\langle\mathcal{F}\rangle(t)=\int_{t''=t_0}^{t}\int_{t'=t_0}^{t''}\sum_{\alpha\gamma}\{h_\alpha^{(1)}(t')h_\gamma^{(2)}(t'')$$

$$+h_\alpha^{(2)}(t')h_\gamma^{(1)}(t'')\}\varphi_{M_\alpha M_\gamma F}(t''-t',t-t'')\,dt'\,dt''$$

$$+\int_{t_0}^{t}\sum_{\alpha\gamma}h_\alpha^{(1)}(t')h_\gamma^{(2)}(t')\,\varphi_{M_{\alpha\gamma}F}(t-t')\,dt' \tag{10.3.4}$$

or, in symmetrized form (10.1.25),

$$\Delta\langle\mathcal{F}\rangle(t)=2\iint_{0}^{t-t_0}\sum_{\alpha\gamma}h_\alpha^{(1)}(t-\bar{t})\,h_\gamma^{(2)}(t-\bar{\bar{t}})\,\tilde{\varphi}_{M_\alpha M_\gamma F}(\bar{t},\bar{\bar{t}})\,d\bar{t}\,d\bar{\bar{t}}$$

$$+\int_{0}^{t-t_0}\sum_{\alpha\gamma}h_\alpha^{(1)}(t-t_1)h_\gamma^{(2)}(t-t_1)\,\varphi_{M_{\alpha\gamma}F}(t_1)\,dt_1\ . \tag{10.3.5}$$

Here, we have made use of the symmetry of $\varphi_{M_{\alpha\gamma}F}(t_1)$ and of $\tilde{\varphi}_{M_\alpha M_\gamma F}(\bar{t}_1,\bar{t}_2)$ under exchange of α and γ. In the following, we shall investigate $\Delta\langle\mathcal{F}\rangle(t)$ for two δ-function pulses in sequence and for two harmonically-oscillating fields of differing frequencies.

10.3.1 The Superposition of Two Short Pulses

We first choose an overall field (10.3.2) consisting of two short pulses which can be described in the δ-function limit by

$$h_\alpha^{(1)}(t)=\hat{h}_\alpha^{(1)}\delta(t-t^{(1)}) \tag{10.3.6}$$

and

$$h_\alpha^{(2)}(t)=\hat{h}_\alpha^{(2)}\delta(t-t^{(2)})\ .^5 \tag{10.3.7}$$

The order of the times is taken to be

$$t^{(2)}>t^{(1)}>t_0\ . \tag{10.3.8}$$

If we employ (10.3.4), only the term with $h_\alpha^{(1)}(t')h_\gamma^{(2)}(t'')$ is nonvanishing in the region of the integration. By inserting (10.3.6 and 7) in (10.3.4), after carrying out

---

[5] In order to calculate the response $\langle\mathcal{F}\rangle_2(t)=F_t[h^{(1)}(t')]$ to *one* pulse $h^{(1)}(t)$, one cannot immediately go to the δ-function limit, since the portion of the response due to $\varphi_{M_{\alpha\gamma}F}(t)$ would diverge. If one first calculates with a narrow rectangular pulse of width Δt, the divergent portion is proportional to $(\Delta t)^{-1}$, i.e. by taking $\lim_{\Delta t\to 0}\Delta t\langle\mathcal{F}\rangle_2(t)$, one can find the contribution due to $\varphi_{M_{\alpha\gamma}F}(t)$.

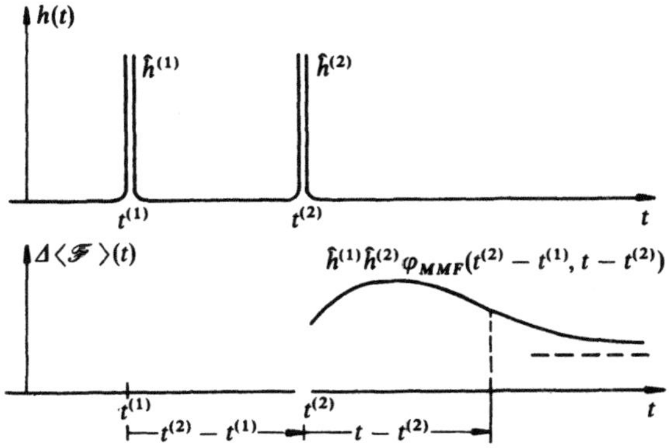

Fig. 10.2. The interpretation of the quadratic response function $\varphi_{MMF}(t^{(2)}-t^{(1)}, t-t^{(2)})$

the integration we obtain:

$$\Delta\langle\mathcal{F}\rangle(t)=\begin{cases}0 & \text{for} \quad t<t^{(2)} \\ \sum\limits_{\alpha\gamma} \hat{h}_\alpha^{(1)}\hat{h}_\gamma^{(2)}\varphi_{M_\alpha M_\gamma F}(t^{(2)}-t^{(1)}, t-t^{(2)}) & \text{for} \quad t>t^{(2)}\end{cases} . \tag{10.3.9}$$

From this expression, we recognize the significance of the time arguments in the response function which depends on two times (cf. Fig. 10.2). The *first* time contains the time difference $t^{(2)}-t^{(1)}$ between the *two δ pulses*, while the *second* argument contains the difference $t-t^{(2)}$ between the *second pulse and the time of observation*. Thus, in principle, by varying the time delay $t^{(2)}-t^{(1)}$ and measuring the resulting response $\Delta\langle\mathcal{F}\rangle(t)$, one can determine the quadratic response function.

10.3.2 The Superposition of Two Harmonically Oscillating Fields

In order to attain an understanding of the *quadratic dynamic susceptibility*, we consider the superposition of two oscillating fields of different frequencies, which are turned on at the time $t_0=0$,

$$h_\alpha^{(1)}(t)=\hat{h}_\alpha^{(1)}\sin\omega_1 t \quad (t\geqq 0) , \tag{10.3.10}$$

$$h_\alpha^{(2)}(t)=\hat{h}_\alpha^{(2)}\sin\omega_2 t \quad (t\geqq 0) . \tag{10.3.11}$$

Let us take $\hat{h}_\alpha^{(1)}$ and $\hat{h}_\gamma^{(2)}$ to be real and $\mathcal{M}_\alpha$, $\mathcal{M}_{\alpha\gamma}$, and $\mathcal{F}$ to be Hermitian operators. In evaluating the interference term $\Delta\langle\mathcal{F}\rangle(t)$ we must allow for the fact that $\varphi_{M_\alpha M_\gamma F}(t_1, t_2)$ possibly tends towards a *finite* value for $t_2\to\infty$, as we saw in Sect. 10.2. For this reason, we employ the decomposition (10.2.21) of the observable $\mathcal{F}$. The interference term $\Delta\langle\mathcal{F}\rangle(t)$ in the response then becomes the sum of the interference terms for the observables $\mathcal{F}_\perp$ and $\mathcal{F}_\parallel$,

$$\Delta\langle\mathcal{F}\rangle(t)=\Delta\langle\mathcal{F}_\perp\rangle(t)+\Delta\langle\mathcal{F}_\parallel\rangle(t) . \tag{10.3.12}$$

The first contribution presents no difficulties. It is most simply calculated using (10.3.5), and writing $\sin\omega_1 t$ as $\mathrm{Re}(\mathrm{i}\exp(-\mathrm{i}\omega_1 t))$. We then obtain:

$$\Delta\langle\mathscr{F}_\perp\rangle(t)=\sum_{\alpha\gamma}\hat{h}_\alpha^{(1)}\hat{h}_\gamma^{(2)}$$

$$\times\,\mathrm{Re}\left\{\iint_0^t \mathrm{e}^{-\mathrm{i}\omega_1(t-\bar{t})}\{\mathrm{e}^{\mathrm{i}\omega_2(t-\bar{\bar{t}})}-\mathrm{e}^{-\mathrm{i}\omega_2(t-\bar{\bar{t}})}\}\,\tilde{\varphi}_{M_\alpha M_\gamma F_\perp}(\bar{t},\bar{\bar{t}})\,d\bar{t}\,d\bar{\bar{t}}\right.$$

$$\left.+\frac{1}{2}\int_0^t \mathrm{e}^{-\mathrm{i}\omega_1(t-t_1)}\{\mathrm{e}^{\mathrm{i}\omega_2(t-t_1)}-\mathrm{e}^{-\mathrm{i}\omega_2(t-t_1)}\}\varphi_{M_{\alpha\gamma}F_\perp}(t_1)\,dt_1\right\}\,. \tag{10.3.13}$$

According to (10.2.6) and (10.2.24), we can make the assumption that the response function $\varphi_{M_\alpha M_\gamma F_\perp}(\bar{t},\bar{\bar{t}})$ decays within a few relaxation times in both time arguments. The linear response function $\varphi_{M_{\alpha\gamma}F_\perp}$ likewise exhibits a similar decay, so that in the steady state in (10.3.13), the upper limits of integration can be replaced by ∞. With the definitions (10.1.27) and (9.1.27) of the susceptibilities, we then obtain for the response:

$$\Delta\langle\mathscr{F}_\perp\rangle(t)=\sum_{\alpha\gamma}\hat{h}_\alpha^{(1)}\hat{h}_\gamma^{(2)}$$

$$\times\,\mathrm{Re}\{\tilde{\chi}_{M_\alpha M_\gamma F_\perp}(\omega_1,\,-\omega_2)\,\mathrm{e}^{-\mathrm{i}(\omega_1-\omega_2)t}$$

$$-\tilde{\chi}_{M_\alpha M_\gamma F_\perp}(\omega_1,\omega_2)\,\mathrm{e}^{-\mathrm{i}(\omega_1+\omega_2)t}+\tfrac{1}{2}\chi_{M_{\alpha\gamma}F_\perp}(\omega_1-\omega_2)\,\mathrm{e}^{-\mathrm{i}(\omega_1-\omega_2)t}$$

$$-\tfrac{1}{2}\chi_{M_{\alpha\gamma}F_\perp}(\omega_1+\omega_2)\,\mathrm{e}^{-\mathrm{i}(\omega_1+\omega_2)t}\}\,. \tag{10.3.14}$$

In this expression, we can also write $\chi_{M_{\alpha\gamma}F}(\omega)$ for the linear susceptibility $\chi_{M_{\alpha\gamma}F_\perp}(\omega)$, since the former is in general zero for a conserved quantity $\mathscr{F}_\parallel$ [cf. (9.6.9)].

In order to calculate the second contribution $\Delta\langle\mathscr{F}_\parallel\rangle(t)$, we use (10.3.4). The second part of that equation makes no contribution due to $\varphi_{M_\alpha M_\gamma F_\parallel}(t)=0$. From (10.2.23), $\varphi_{M_\alpha M_\gamma F_\parallel}$ does not depend on t_2. Making the substitution for t_1 and t_2 as in (10.1.10) and exchanging the order of integration, we can write (10.3.4) in the form:

$$\Delta\langle\mathscr{F}_\parallel\rangle(t)=\sum_{\alpha\gamma}\int_{t_1=0}^{t}\int_{t_2=0}^{t-t_1}\{h_\alpha^{(1)}(t-t_1-t_2)\,h_\gamma^{(2)}(t-t_2)$$

$$+h_\alpha^{(2)}(t-t_1-t_2)h_\gamma^{(1)}(t-t_2)\}\,\varphi_{M_\alpha M_\gamma F_\parallel}(t_1)\,dt_1\,dt_2\,. \tag{10.3.15}$$

The essential point of this equation is that, after insertion of the fields (10.3.10 and 11),

$$\Delta\langle\mathscr{F}_\parallel\rangle(t)=\frac{1}{2}\sum_{\alpha\gamma}\hat{h}_\alpha^{(1)}\hat{h}_\gamma^{(2)}$$

$$\times\,\mathrm{Re}\left\{\int_{t_1=0}^{t}\int_{t_2=0}^{t-t_1}\mathrm{e}^{-\mathrm{i}\omega_1(t-t_1-t_2)}(\mathrm{e}^{\mathrm{i}\omega_2(t-t_2)}\right.$$

$$\left.-\mathrm{e}^{-\mathrm{i}\omega_2(t-t_2)})\,\varphi_{M_\alpha M_\gamma F_\parallel}(t_1)\,dt_1\,dt_2+(\omega_1\leftrightarrow\omega_2,\,\alpha\leftrightarrow\gamma)\right\} \tag{10.3.16}$$

the integrals over t_2 can be evaluated and, for $\omega_1 \neq \omega_2$, only integrals over t_1 with exponential factors, $\exp(\mathrm{i}\omega t_1)$ remain. In the steady state, we can therefore again replace the upper limits of integration t in the integrals over t_1 by ∞. With the definition of $\tilde{\chi}_{M_\alpha M_\gamma F_\parallel}(\omega_1, \omega_2)$ which follows from (10.1.18) and (10.1.30) for $\omega_1 \neq \omega_2$ and $\eta_1, \eta_2 \to +0$:

$$\tilde{\chi}_{M_\alpha M_\gamma F_\parallel}(\omega_1, \omega_2) = \frac{1}{2} \frac{\mathrm{i}}{\omega_1 + \omega_2} \int_0^\infty \{ \mathrm{e}^{\mathrm{i}\omega_1 t_1} \varphi_{M_\alpha M_\gamma F_\parallel}(t_1) + \mathrm{e}^{\mathrm{i}\omega_2 t_1} \varphi_{M_\gamma M_\alpha F_\parallel}(t_1) \} \, dt_1 \;, \tag{10.3.17}$$

the result (10.3.16) for $\Delta \langle \mathscr{F}_\parallel \rangle(t)$ can be expressed in terms of the symmetrized susceptibility $\tilde{\chi}_{M_\alpha M_\gamma F_\parallel}(\omega_1, \omega_2)$. It may be summarized as follows:

$$\begin{aligned} \Delta \langle \mathscr{F}_\parallel \rangle(t) = \sum_{\alpha\gamma} \hat{h}_\alpha^{(1)} \hat{h}_\gamma^{(2)} \operatorname{Re} \{ & \tilde{\chi}_{M_\alpha M_\gamma F_\parallel}(\omega_1, -\omega_2) \mathrm{e}^{-\mathrm{i}(\omega_1 - \omega_2)t} \\ & - \tilde{\chi}_{M_\alpha M_\gamma F_\parallel}(\omega_1, \omega_2) \mathrm{e}^{-\mathrm{i}(\omega_1 + \omega_2)t} \\ & + \tilde{\chi}_{M_\alpha M_\gamma F_\parallel}(\omega_2, -\omega_1) - \tilde{\chi}_{M_\alpha M_\gamma F_\parallel}(\omega_2, \omega_1) \} \;. \end{aligned} \tag{10.3.18}$$

Here, we have made use of the fact that for a complex function, $\operatorname{Re} f(t_1) = \operatorname{Re} f^*(t_1)$. Physically, the response (10.3.18) represents simply the interference term of the energy, $\Delta \langle \mathscr{H} \rangle(t)$ in the steady state, apart from a factor [compare (10.2.20)]. It could also be expressed in terms of the linear susceptibility $\chi_{M_\alpha M_\gamma}(\omega)$ [(10.1.30), (10.2.25), and (9.4.16)]. However, we shall not pursue this point further here, since with the form (10.3.18), we can express both parts of the response (10.3.12) in terms of the overall susceptibility

$$\tilde{\chi}_{M_\alpha M_\gamma F} = \tilde{\chi}_{M_\alpha M_\gamma F_\parallel} + \tilde{\chi}_{M_\alpha M_\gamma F_\perp} \;. \tag{10.3.19}$$

In this manner, we obtain the final result for the interference term due to two harmonically oscillating fields:

$$\boxed{\begin{aligned} \Delta \langle \mathscr{F} \rangle(t) = \sum_{\alpha\gamma} & \hat{h}_\alpha^{(1)} \hat{h}_\gamma^{(2)} \\ & \times \operatorname{Re} \{ \tilde{\chi}_{M_\alpha M_\gamma F}(\omega_1, -\omega_2) \mathrm{e}^{-\mathrm{i}(\omega_1 - \omega_2)t} \\ & - \tilde{\chi}_{M_\alpha M_\gamma F}(\omega_1, \omega_2) \mathrm{e}^{-\mathrm{i}(\omega_1 + \omega_2)t} \\ & + \tfrac{1}{2} \chi_{M_{\alpha\gamma} F}(\omega_1 - \omega_2) \mathrm{e}^{-\mathrm{i}(\omega_1 - \omega_2)t} - \tfrac{1}{2} \chi_{M_{\alpha\gamma} F}(\omega_1 + \omega_2) \mathrm{e}^{-\mathrm{i}(\omega_1 + \omega_2)t} \\ & + \tilde{\chi}_{M_\alpha M_\gamma F_\parallel}(\omega_2, -\omega_1) - \tilde{\chi}_{M_\alpha M_\gamma F_\parallel}(\omega_2, \omega_1) \} \end{aligned}} \;. \tag{10.3.20}$$

In $\Delta \langle \mathscr{F} \rangle(t)$, we thus find two harmonic oscillations at the *sum* and the *difference* frequencies, $\omega_1 + \omega_2$ and $\omega_1 - \omega_2$, as well as an *additional, time-independent term* when $\mathscr{F}_\parallel$ is present. When only a linear coupling to the fields appears in the

Hamiltonian, as is often the case, and the linear susceptibility $\chi_{M_{\alpha\gamma}F}(\omega)$ does not occur, the susceptibility $\tilde{\chi}_{M_\alpha M_\gamma F}(\omega_1, \omega_2)$ can be determined by measuring the complex amplitude at the frequency $\omega_1 + \omega_2$ and variation of the two frequencies.

As may be seen from the formulas which provided the starting point for this calculation, (10.3.4 and 5), the result (10.3.20) can be specialized to the case of *one* oscillating field by taking the limit $\omega_2 \to \omega_1$. The response is then a harmonic oscillation at the *double* frequency $2\omega_1$. The oscillation at the difference frequency 0 is taken up in the constant term; and when the energy absorption is noticeably large, by means of $\mathscr{F}_{\parallel}$ it produces an additional term proportional to the time (heating by irradiation) (cf. Exercise 10.3).

The *total* response to two harmonically oscillating fields therefore consists of a constant and possibly a linearly increasing term in t, as well as harmonic oscillations at the frequencies $2\omega_1$, $2\omega_2$, $\omega_1 + \omega_2$, and $\omega_1 - \omega_2$. In practical investigations of the total response, an analysis at the frequency $\omega_1 + \omega_2$ yields the desired result, $\tilde{\chi}_{M_\alpha M_\gamma F}(\omega_1, \omega_2)$.

Exercise 10.3 An harmonically-oscillating field $h(t) = \hat{h} \sin \omega_0 t$ is turned on at the time $t = 0$ and acts upon a sample. Determine the response $\langle \mathscr{F} \rangle_2(t) = \langle \mathscr{F}_{\parallel} \rangle_2(t) + \langle \mathscr{F}_{\perp} \rangle_2(t)$ in the steady state of oscillation.

Exercise 10.4 The following field acts upon a system: $h_\alpha(t) = \hat{h}_\alpha^{(1)} \delta(t - t^{(1)}) + \hat{h}_\alpha^{(2)} \delta(t - t^{(2)})$, with $t^{(2)} > t^{(1)} > t^{(0)}$. The Liouville operator is given by

$$\mathsf{L}(t) = \mathsf{L} - \sum_\alpha \mathsf{M}_\alpha^{(\dagger)} \{ \hat{h}_\alpha^{(1)} \delta(t - t^{(1)}) + \hat{h}_\alpha^{(2)} \delta(t - t^{(2)}) \} \ .$$

Determine the quadratic response $\langle \mathscr{F} \rangle_2(t)$ for $t > t^{(2)}$ by expanding the exact solution in the amplitudes $\hat{h}$ (cf. Exercise 7.2).

10.4 Concluding Remarks

Additional formal relations for the quadratic dynamic susceptibility, $\chi_{M_\alpha M_\gamma F}(\omega_1, \omega_2)$, may be derived along the lines of those discussed in Chap. 9, for example by transformation of the resolvents. The influence of slowly-varying fields in terms of the isolated quadratic susceptibility could be discussed similarly to Sect. 9.5 by considering the coupling to conserved quantities. Furthermore, one can define a relaxation function $\Phi_{M_\alpha M_\gamma F}(t_1, t_2)$ depending on two times, which physically represents two relaxation processes occurring at different times. However, at present there are few general investigations of such functions which would permit the development of widely applicable quantum-statistical approximation methods for the resonance frequencies and linewidths, comparable to those which we shall meet in Part 3 in the frame of the Mori theory for *linear* response. Therefore, we shall dispense with further manipulations of the quadratic susceptibility.

The formalism for the quadratic response developed in Chap. 10 may also be extended to the third and higher orders, as is done phenomenologically in nonlinear optics. These higher orders become important when the lower orders make no contributions to observable effects, e.g. for symmetry reasons. The corresponding quantum-statistical formulas may be derived from the expansion of $\mathsf{U}(t, t_0)$ [(7.2.6); compare Fig. 7.2].

Finally, we point out that the expansion in the field amplitudes given in Chap. 7 may become entirely meaningless for strong fields. There are approximations which are not limited to the case of weak fields $h_\alpha(t)$, but these will not be treated in the present book.

Part 3

Equations of Motion for Observables in the Case of Small Deviations from Equilibrium

Using the Kubo formalism, we have demonstrated that the response of externally driven systems with linear deviations from thermal equilibrium can be described by *relaxation functions* in the form of Mori scalar products. Their time dependence is determined by the Hamiltonian of the system Σ *without external perturbation.* In the following sections, we shall therefore concern ourselves with the dynamics of systems described by an explicitly time-*in*dependent Liouville operator, L. To this end, we make use of the Heisenberg equation, which has an intuitively clear physical meaning,

$$\dot{\mathscr{G}}(t) = \mathrm{iL}\mathscr{G}(t)$$

and of its formal solution, (11.1.1)

$$\mathscr{G}(t) = \mathrm{e}^{\mathrm{iL}t}\mathscr{G} \ .$$

We take the observables $\mathscr{G}$ to be likewise explicitly time-*in*dependent[1]. In the following, an operator without a time argument will always denote the operator at the initial time $t=0$,

$$\mathscr{G} \equiv \mathscr{G}(0) \ . \tag{11.1.2}$$

Instead of only one observable $\mathscr{G}(t)$, it will often prove expedient to consider an entire set of relevant observables $\mathscr{G}_\mu(t)$ (or of pairs of adjoint, non-Hermitian operators).

Equation (11.1.1), which appears formally to be very simple, can, in general, be solved exactly only for a few elementary systems, owing to the extremely complicated action of $\exp[\mathrm{iL}(t)]$ on $\mathscr{G}$. On the other hand, not all of the many possible observables of the system Σ are of interest, but rather only those observables[2] $\mathscr{G}_\mu$ which are needed for a macroscopic description and whose averaged values show relatively slow variation, in contrast to the rapidly-varying numerous other observables of Σ. This fact will permit us to describe the dynamics of such macroscopic observables by oscillatory and damping terms. The essential idea is to separate out the rapidly varying observables of Σ in the Liouville operator, so that their collective influence can be described as a damping of the motion of $\langle \mathscr{G}_\mu(t) \rangle$. We shall then be in a position to treat linewidths (including their temperature dependences) based on the expressions detailed in the preceding Part 2 (e.g. in $\chi_{\mu\nu}(\omega)$).

---

[1] An extension to explicitly time-dependent operators will be given in Chap. 15.

[2] The operators which were denoted by $\mathscr{M}_\alpha$ in Part 2 and which are acted upon directly by the external fields, as well as the response $\mathscr{F}$, will be subsumed in the following under the notation $\mathscr{G}_\mu$.

11. Exact Integro-Differential Equations for Relaxation Processes

11.1 An Heuristic Introduction to the Langevin–Mori Theory

Before we concern ourselves with the quantum-statistical formalism needed to describe frictional or damping phenomena, let us first make an heuristic remark useful for general understanding.

As we know from mechanics, frictional forces, which for example act on a particle, cannot be directly included in a potential or a Hamiltonian operator. Instead, it is necessary to take the surroundings (of the particle) which cause the friction into consideration; a Hamiltonian for the total system then exists. A standard example is the treatment of the Brownian motion of a particle (e.g. a small oil droplet within a suspension). The mass M of the particle is very large compared to the mass m of a molecule of the liquid ($M/m \gg 1$). This macroscopic particle experiences damping forces due to the viscosity of the liquid. If it is examined under a microscope, one sees that it is subject to a rapid vibrational motion due to random collisions with molecules of the liquid. The liquid thus not only produces a collective damping, but also, the individual liquid molecules give rise to a stochastic residual force $f(t)$ due to collisions; this force, however, vanishes when averaged over time or over the ensemble. The equation of motion

$$M\dot{v}(t) = -\Gamma v(t) + f(t) \tag{11.1.3}$$

is the famous Langevin equation for a particle undergoing Brownian motion. It was shown long ago that the damping Γ is proportional to the integral over the time-dependent correlation function of the residual force $f(t)$:

$$\Gamma \sim \int_0^\infty \langle f f(t) \rangle \, dt \ , \tag{11.1.4}$$

in the case that $f(t)$ for $t > 0$ is nearly *uncorrelated* with $f = f(0)$,

$$\langle f f(t) \rangle \sim \Gamma \delta(t) \ . \tag{11.1.5}$$

At this point, we shall not discuss the theoretical justification of these facts, which we have presented in anticipation of the general treatment of the whole problem to follow. However, we emphasize once more that the equation of motion (11.1.3) is only valid when the mass M of the particles undergoing Brownian motion is

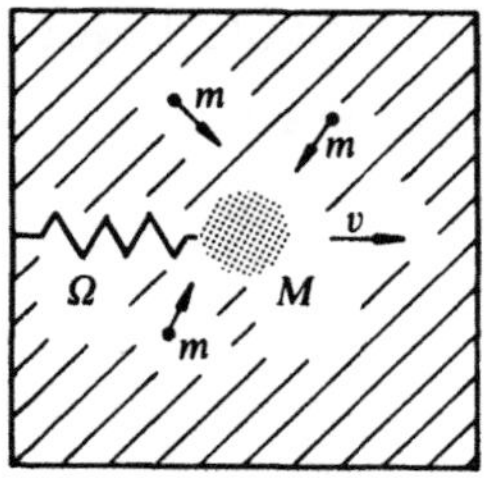

Fig. 11.1. A particle M subject to Brownian motion in a liquid, with a restoring force

much larger than the mass m of the liquid molecules, i.e. when the *average* motion of the particle is *slow* in comparison to the motion of all the other degrees of freedom of the total system.

This Langevin concept of one (or several) slow variables of a system Σ with many other rapid degrees of freedom has proved to be a useful starting point for the statistical and quantum-statistical description of a large number of very diverse systems and problems; in the following, we shall enlarge upon and apply it.

The *macroscopic* observables of a system Σ exhibit in many cases a slowly-varying behavior relative to the dynamics of the other degrees of freedom, so that one is tempted to apply equations of motion of the Langevin type for their decription ($v \rightarrow \mathscr{G}_\mu$). An equation of the Langevin type (11.1.3) is, however, not completely suitable, because the observables are, in general, subject not only to damping forces, but also can oscillate with an eigenfrequency Ω (Fig. 11.1)[3]. We should therefore try to set up an equation of the following type for a (non-Hermitian) observable:

$$\dot{\mathscr{G}}(t) = \mathrm{i}\Omega\mathscr{G}(t) - \Gamma\mathscr{G}(t) + \mathscr{f}(t) \ , \tag{11.1.6}$$

or, rather, for a set of observables $\mathscr{G}_\mu$ we should set up a system of coupled equations of the type (11.1.6).

The goal of the Mori theory, which is described in the following, is to derive such equations as well as quantum-statistical expressions for the quantities Ω, Γ, and $\mathscr{f}(t)$ which occur in them, and to analyze the necessary preconditions corresponding to (11.1.5). In Chap. 16, we present some examples of physical applications which may be illustrative to the study of the general theory.

---

[3] For an undamped harmonic oscillator, the creation operator $\mathscr{b}^\dagger$ in the Heisenberg representation varies according to the equation

$$\dot{\mathscr{b}}^\dagger = \mathrm{i}\Omega\mathscr{b}^\dagger \ .$$

11.2 Mori's Integro-Differential Equations for Operators

11.2.1 Derivation and Interpretation

We consider a set of relevant observables $\{\mathscr{G}\}$ for a system[4] Σ, which are linearly independent and explicitly not time dependent. At the time $t=0$, these operators span a subspace $\mathfrak{L}_{\mathscr{G}}$ of the Liouville space $\mathfrak{L}$. At a later time t, the operators in the Heisenberg formulation with an explicitly time-independent Liouville operator are given by

$$\mathscr{G}_\mu(t)=\mathrm{e}^{\mathrm{i}\mathsf{L}t}\mathscr{G}_\mu \quad (\mathscr{G}_\mu\equiv\mathscr{G}_\mu(0)) \ .^5 \tag{11.2.1}$$

In general, the operators $\mathscr{G}_\mu(t)$ are *not* linear combinations of the $\mathscr{G}_\mu$, but rather also contain components in the space $\mathfrak{L}_{\perp\mathscr{G}}$ which is orthogonal to the space $\mathfrak{L}_{\mathscr{G}}$ (Fig. 11.2).

It is our goal to obtain such an (exact) decomposition of the Liouville operator, which treats the dynamics of the components $\mathsf{Q}\mathscr{G}_\mu(t)$ of $\mathscr{G}_\mu(t)$ which have been rotated out of $\mathfrak{L}_{\mathscr{G}}$ differently from those of the components $\mathsf{P}\mathscr{G}_\mu(t)$ which remain in $\mathfrak{L}_{\mathscr{G}}$. The symbols P and Q represent *time-independent projectors* in $\mathfrak{L}$ which project onto $\mathfrak{L}_{\mathscr{G}}$ or onto $\mathfrak{L}_{\perp\mathscr{G}}$, respectively. As we shall see, the different

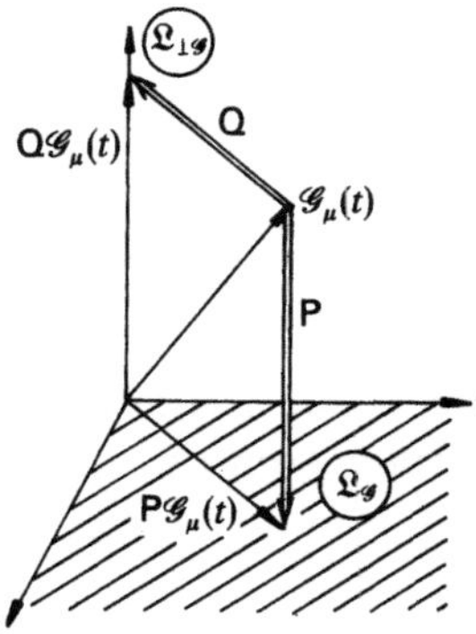

Fig. 11.2. The decomposition of the observables $\mathscr{G}(t)$ in the Mori theory

[4] Instead of using Hermitian operators, it is often expedient to allow non-Hermitian linear combinations of Hermitian observables in the set of $\mathscr{G}_\mu$ [compare Sect. 8.3.2]. It is not necessary for the formal validity of the following expressions that the $\mathscr{G}_\mu^\dagger$ also be included in the set of observables; this is required only in Sect. 11.4.2 and for considering the effects of symmetry operations (e.g. in Appendix B). Physically, it is however reasonable to employ such a set from the beginning, i.e. to begin with an observation level, and to decouple the resulting dynamical equations using symmetry considerations (cf. Sect. 16.3).

[5] We recall the fact that all operators in the Heisenberg representation without time arguments denote the operators at the time $t=0$!

treatment of the motions in $\mathfrak{L}_{\mathscr{G}}$ and in $\mathfrak{L}_{\perp\mathscr{G}}$ leads to equations which under certain conditions take on the form of Langevin equations; these are the Mori equations.

Although the theory which we develop in the following is independent of the actual form of the scalar product in $\mathfrak{L}$ (cf. Sect. 8.2.1), we shall assume the Mori form Eq. (8.2.5) for the scalar product, because the theory is intended precisely for the treatment of *linear dynamic response quantities.* This will become especially clear in Sect. 11.4.2, where we shall see that just for the set of statistical operators $\rho(0)$ which differ *only linearly* from equilibrium, we can find tractable equations for the linear relaxation dynamics of $\langle\mathscr{G}_\mu\rangle(t)$.

The operators $\mathscr{G}_\mu$ of interest do not generally occur in orthonormalized form. Although such a form can always be arrived at owing to the linear independence of the $\mathscr{G}_\mu$, it is not expedient in general considerations to assume that they are orthonormalized. Therefore, the projector P mentioned above, which projects onto $\mathfrak{L}_{\mathscr{G}}$, has the form [cf. (8.1.17)]:

$$\boxed{\mathsf{P}=\sum_{\mu\nu}|\mathscr{G}_\mu)g_{\mu\nu}(\mathscr{G}_\nu|} \ . \tag{11.2.2}$$

The metric $g_{\mu\nu}$ is the inverse matrix of $(\mathscr{G}_\mu|\mathscr{G}_\nu)$ and is given by

$$\sum_{\varrho} g_{\mu\varrho}(\mathscr{G}_\varrho|\mathscr{G}_\nu)=\delta_{\mu\nu} \ . \tag{11.2.3}$$

From (8.2.10), $(\mathscr{G}_\mu|\mathscr{G}_\nu)$ is a Hermitian, positive definite matrix, and the same is true of $g_{\mu\nu}$ according to (11.2.3):

$$(\mathscr{G}_\mu|\mathscr{G}_\nu)=(\mathscr{G}_\nu|\mathscr{G}_\mu)^* \ , \qquad g_{\mu\nu}=g^*_{\nu\mu} \ , \tag{11.2.4}$$

independently of whether the $\mathscr{G}_\mu$ are Hermitian in $\mathfrak{U}$ or not.

The operator Q which projects onto $\mathfrak{L}_{\perp\mathscr{G}}$ is given by

$$\mathsf{Q}=1-\mathsf{P} \ . \tag{11.2.5}$$

Equations (11.2.2 to 5) imply not only the linearity of P and Q, but also the projection properties:

$$\mathsf{P}\mathscr{G}_\mu=\mathscr{G}_\mu \ , \qquad \mathsf{Q}\mathscr{G}_\mu=\mathcal{O} \ , \tag{11.2.6}$$

$$\mathsf{P}^2=\mathsf{P} \ , \quad \mathsf{Q}^2=\mathsf{Q} \quad \text{(idempotence)} \ , \tag{11.2.7}$$

$$\mathsf{P}^\dagger=\mathsf{P} \ , \quad \mathsf{Q}^\dagger=\mathsf{Q} \quad (\text{Hermiticity in } \mathfrak{L}) \ , \tag{11.2.8}$$

$$\mathsf{PQ}=\mathsf{QP}=\mathsf{O} \quad \text{(orthogonality)} \ . \tag{11.2.9}$$

The operators P and Q are naturally invariant with respect to linear transformations of the set $\mathscr{G}_\mu$.

Making use of these introductory remarks, it is simple to bring the Heisenberg equations of motion,

$$\dot{\mathcal{G}}_\mu(t) = e^{iLt}\dot{\mathcal{G}}_\mu \quad (\dot{\mathcal{G}}_\mu = iL\mathcal{G}_\mu \equiv \dot{\mathcal{G}}_\mu(0)) \tag{11.2.10}$$

into the Mori form. To begin, we insert $1 = P + Q$,

$$\dot{\mathcal{G}}_\mu(t) = e^{iLt}P\dot{\mathcal{G}}_\mu + e^{iLt}Q\dot{\mathcal{G}}_\mu \ , \tag{11.2.11}$$

and employ the identity (3.4.8a) in the second term (with $L_1 = QL$ and $L_2 = PL$):

$$e^{iLt} = i\int_0^t e^{iL(t-t')}\,PL\,e^{iQLt'}\,dt' + e^{iQLt} \ , \tag{11.2.12}$$

so that we obtain

$$\dot{\mathcal{G}}_\mu(t) = e^{iLt}P\dot{\mathcal{G}}_\mu + i\int_0^t e^{iL(t-t')}PLe^{iQLt'}\,Q\dot{\mathcal{G}}_\mu\,dt' + e^{iQLt}Q\dot{\mathcal{G}}_\mu \ . \tag{11.2.13}$$

With the definition of P (11.2.2) and $\mathcal{G}_\mu(t) = \exp(iLt)\mathcal{G}_\mu$, we find the result

$$\boxed{\dot{\mathcal{G}}_\mu(t) = \sum_\nu \mathcal{G}_\nu(t)i\Omega_{\nu\mu} - \int_0^t \sum_\nu \mathcal{G}_\nu(t-t')\gamma_{\nu\mu}(t')\,dt' + f_\mu(t)} \ . \tag{11.2.14}$$

These Mori equations form a *system of coupled integro-differential equations for the set of observables* $\mathcal{G}_\mu(t)$. They are, independent of the number and type of observables $\mathcal{G}_\mu$, *exactly* valid; in the derivation from the Heisenberg equations (11.2.10) we have made no approximations.

We shall now discuss the individual terms on the right-hand side of the Mori equations.

The first term contains a generally *temperature-dependent frequency matrix*:

$$\boxed{\begin{aligned} i\Omega_{\nu\mu} &= \sum_\varrho g_{\nu\varrho}(\mathcal{G}_\varrho|\dot{\mathcal{G}}_\mu) = \sum_\varrho g_{\nu\varrho}(\mathcal{G}_\varrho|iL\mathcal{G}_\mu) \\ &= \frac{i}{\beta\hbar}\sum_\varrho g_{\nu\varrho}\langle[\mathcal{G}_\varrho^\dagger, \mathcal{G}_\mu]\rangle_\beta = i\sum_\varrho g_{\nu\varrho}\omega_{\varrho\omega} \end{aligned}} \ .^{6} \tag{11.2.15}$$

If (11.2.14) consisted of this first term only, $\mathcal{G}_\mu(t)$ would comprise a superposition of oscillations (compare Sect. 11.3.1) and would remain in $\mathfrak{L}_\mathcal{G}$. Due to the Hermiticity of L we have

$$\omega_{\lambda\nu} = (\mathcal{G}_\lambda|L\mathcal{G}_\nu) = \omega^*_{\nu\lambda} \tag{11.2.16a}$$

[6] In the second expression, the Mori product was used, and the Kubo identity, (8.2.18), was applied.

or

$$\sum_{\varrho} \Omega_{\lambda\varrho} g_{\varrho\nu} = \sum_{\varrho} g_{\lambda\varrho} \Omega^{*}_{\nu\varrho} \; . \tag{11.2.16b}$$

Further symmetry properties related to time reversal invariance are to be found in Appendix B2. For Hermitian operators $\mathcal{G}_\mu$, $\mathrm{i}\Omega_{\mu\nu}$ is real. In the case of a *single* Hermitian observable, Ω vanishes.

The operators of the third term in (11.2.14),

$$\boxed{f_\mu(t) = \mathrm{e}^{\mathrm{i}\mathsf{QL}t}\mathsf{Q}\dot{\mathcal{G}}_\mu = \mathrm{i}\,\mathrm{e}^{\mathrm{i}\mathsf{QL}t}\mathsf{QL}\mathcal{G}_\mu} \; , \tag{11.2.17}$$

we call the *residual force*[7,8]. *It remains at all times in the space* $\mathsf{L}_{\perp\mathcal{G}}$, since

$$\mathsf{Q}f_\mu(t) = f_\mu(t) \; , \tag{11.2.18}$$

therefore

$$\mathsf{P}f_\mu(t) = \mathcal{O} \tag{11.2.19}$$

holds. All the $f_\mu(t)$ are thus perpendicular to all the $\mathcal{G}_\nu \equiv \mathcal{G}_\nu(0)$,

$$\boxed{(\mathcal{G}_\nu | f_\mu(t)) = 0} \; . \tag{11.2.20}$$

In this sense, they are uncorrelated with one another. It is particularly important that the dynamics of the residual force,

$$\boxed{f_\mu(t) = \mathrm{e}^{\mathrm{i}\mathsf{QL}t} f_\mu = \mathrm{e}^{\mathrm{i}\mathsf{QLQ}t} f_\mu} \tag{11.2.21}$$

with initial operators

$$f_\mu \equiv f_\mu(0) = \mathrm{i}\mathsf{QL}\mathcal{G}_\mu = \dot{\mathcal{G}}_\mu - \sum_\nu \mathcal{G}_\nu \mathrm{i}\Omega_{\nu\mu} \tag{11.2.22}$$

are determined not by the Liouville operator L, but instead by the operator QL or QLQ. Thus, the residual force can vary on an entirely different time scale from that of $\mathsf{P}\mathcal{G}_\mu(t)$, a quantity which corresponds to the average Brownian motion of a macroscopic particle discussed above. This may be seen explicitly in the exactly-solvable model described in Sect. 16.1.

---

[7] If the observables $\mathcal{G}_\mu$ refer, in particular, to momenta, then the entire right-hand side of (11.2.14) represents the operators of the total force, $\dot{\mathcal{G}}_\mu = \mathcal{K}_\mu$. In the literature, $f_\mu(t)$ is often referred to as the "stochastic force". However, such a property is, in spite of (11.2.19) and (11.4.27), in general not *a priori* present. We therefore have chosen the term "residual force".

[8] Due to $\exp(\mathrm{i}\mathsf{QL}t)\mathsf{Q} = \mathsf{Q} + \mathrm{i}\mathsf{QLQ}t + (\mathrm{i}^2/2)\mathsf{QLQLQ}t^2 + \dots$ and $\mathsf{Q}^2 = \mathsf{Q}$, the relation $\exp(\mathrm{i}\mathsf{QL}t)\mathsf{Q} = \mathsf{Q}\exp(\mathrm{i}\mathsf{LQ}t) = \exp(\mathrm{i}\mathsf{QLQ}t)\mathsf{Q}$ is valid.

The integral in the Mori equation (11.2.14) represents a *memory term*, since the integrand depends on the past of all the quantities $\mathcal{G}_\nu(t-t')$, i.e. for $0 \leqq t' \leqq t$. These contributions are weighted by the matrix elements of $\gamma_{\nu\mu}(t')$, which is therefore called the *memory matrix*. It can be found from (11.2.13), taking the Hermiticity of L into account:

$$\boxed{\gamma_{\nu\mu}(t)=\sum_{\varrho} g_{\nu\varrho}(\mathrm{i}\mathsf{L}\mathcal{G}_\varrho|\mathrm{e}^{\mathrm{i}\mathsf{QL}t}\mathsf{Q}\mathrm{i}\mathsf{L}\mathcal{G}_\mu)} \quad . \tag{11.2.23}$$

Due to the idempotence of Q, this operator can be inserted in front of the exponential function [cf. Footnote 8 (p. 176)]; being Hermitian, the latter can also be written in the bra. Employing definition (11.2.21), we can then find the memory matrix in terms of the correlation function of the residual force,

$$\boxed{\gamma_{\nu\mu}(t)=\sum_{\varrho} g_{\nu\varrho}(f_\varrho|f_\mu(t))=\sum_{\varrho} g_{\nu\varrho} l_{\varrho\mu}(t)} \quad . \tag{11.2.24}$$

The properties of $\gamma_{\nu\mu}(t)$ will be summarized in Sect. 11.3.2. Symmetry properties of $l_{\varrho\mu}(t)=(f_\varrho|f_\mu(t))$ due to time-reversal invariance are to be found in Appendix B2.

At the beginning of this section, we indicated that the goal of the present theory is to decompose the dynamics of $\mathcal{G}_\mu(t)$ into parts corresponding to $\mathfrak{L}_{\mathcal{G}}$ and $\mathfrak{L}_{\perp\mathcal{G}}$ (cf. Fig. 11.2). This can be seen from the Mori equation (11.2.13) by projecting it onto $\mathfrak{L}_{\mathcal{G}}$ or onto $\mathfrak{L}_{\perp\mathcal{G}}$. If we multiply it from the left by P or Q, then in the resulting integro-differential equation for $\mathsf{P}\mathcal{G}_\mu(t)$, the residual force $f_\mu(t)$ vanishes because of (11.2.19), while it remains in the equation for $\mathsf{Q}\mathcal{G}_\mu(t)$ as an inhomogeneity; i.e., $\mathsf{P}\mathcal{G}_\mu(t)$ and $\mathsf{Q}\mathcal{G}_\mu(t)$ obey *different dynamics*.

This behavior may be seen still more clearly by writing down expressions for $\mathsf{P}\mathcal{G}_\mu(t)$ and $\mathsf{Q}\mathcal{G}_\mu(t)$. We have

$$\mathsf{P}\mathcal{G}_\mu(t)=\sum_{\nu} \mathcal{G}_\nu \Xi_{\nu\mu}(t) \ , \tag{11.2.25}$$

and

$$\mathsf{Q}\mathcal{G}_\mu(t)=\sum_{\nu} \int_0 f_\nu(t-t')\Xi_{\nu\mu}(t')\,dt' \tag{11.2.26}$$

or, in combined form,

$$\boxed{\mathcal{G}_\mu(t)=\sum_{\nu} \mathcal{G}_\nu \Xi_{\nu\mu}(t)+\sum_{\nu} \int_0^t f_\nu(t-t')\Xi_{\nu\mu}(t')dt'} \quad . \tag{11.2.27}$$

Here, we have introduced an abbreviation for the normalized dynamical correlation function of the observables $\mathcal{G}_\mu(t)$ (compare Sect. 11.4):

$$\boxed{\Xi_{\nu\mu}(t)=\sum_\varrho g_{\nu\varrho}(\mathcal{G}_\varrho|\mathcal{G}_\mu(t))} \ . \tag{11.2.28}$$

The validity of (11.2.25 and 26) follows e.g. from the integro-differenial equations for $\mathsf{P}\mathcal{G}_\mu(t)$ and $\mathsf{Q}\mathcal{G}_\mu(t)$ mentioned above. However, it may also be proved as follows: Equation (11.2.25) results immediately from the definition of P, (11.2.2). For the most simple proof of (11.2.26), we start with the identity (3.4.8b) with $\mathsf{L}_1=\mathsf{LQ}$ and $\mathsf{L}_2=\mathsf{LP}$ and multiply from the left by Q,

$$\mathsf{Q}\,\mathrm{e}^{\mathrm{i}\mathsf{L}t}=\mathsf{Q}\,\mathrm{e}^{\mathrm{i}\mathsf{LQ}t}+\mathrm{i}\int_0^t \mathsf{Q}\,\mathrm{e}^{\mathrm{i}\mathsf{LQ}(t-t')}\mathsf{LP}\,\mathrm{e}^{\mathrm{i}\mathsf{L}t'}dt' \ . \tag{11.2.29}$$

Application to $\mathcal{G}_\mu$ gives $\mathsf{Q}\mathcal{G}_\mu(t)$ on the left-hand side, while the first term on the right-hand side vanishes due to $\mathsf{Q}\mathcal{G}_\mu=0$ [compare (11.2.6)]. Finally, using the relation [cf. Footnote 8 (p. 176)],

$$\mathsf{Q}\,\mathrm{e}^{\mathrm{i}\mathsf{LQ}t}=\mathrm{e}^{\mathrm{i}\mathsf{QL}t}\,\mathsf{Q} \tag{11.2.30}$$

we can shift Q to the end of the integrand, so that on employing (11.2.2) and (11.2.17), the expression (11.2.26), which was to be proved, results.

If, for a given set of $\mathcal{G}_\mu$ with slowly-varying quantities $\mathsf{P}\mathcal{G}_\mu(t)$, the residual force $f_\mu(t)$ is only weakly correlated with f_ϱ, which is the same as saying that $\gamma_{\nu\mu}(t)$ rapidly decays to zero, then in the Mori equation, $\mathsf{P}\mathcal{G}_\nu(t-t')$ may be replaced by $\mathsf{P}\mathcal{G}_\nu(t)$ and taken out of the integral. This qualitative consideration, which we make in anticipation of detailed calculations, shows that under certain conditions, the memory term can become a *damping term*. This important approximation will be treated in detail and more rigorously analyzed in Chap. 13. When the memory matrix lacks the weak correlation mentioned above, this is an indication that the chosen set of observables $\mathcal{G}_\mu$ does not include 'all' the slow observables of the system. It is then necessary to try to extend the set of the $\mathcal{G}_\mu$ in such a way that the thus-extended memory matrix has a shorter correlation time. This extension can be undertaken without difficulty, since the Mori equations hold for an arbitrary set of observables. Some aspects of this procedure will be discussed in the next section.

11.2.2 Choosing a Set of Observables $\mathcal{G}_\mu$

In this section, we shall make some remarks concerning the choice of the set of $\mathcal{G}_\mu$. The derivation of the Mori equation shows that it is valid independently of the number and the nature of the observables $\mathcal{G}_\mu$. However, the separation of $\dot{\mathcal{G}}_\mu(t)$ into the three terms of (11.2.14) depends sensitively on the choice of the set of observables.

The consideration of *two limiting cases* should make this situation more clear.

If we set $\mathsf{P}=1$, i.e. $\mathsf{Q}=0$, then $f_\mu(t)$ becomes $\mathcal{O}$ and $\gamma_{\nu\mu}(t)=0$. The dynamics are then determined *by the Ω-matrix alone*. If we represent $\mathsf{P}=1$ by a complete basis $|\mathscr{V}_\nu)$ in $\mathfrak{L}$ [cf. (3.1.4)], which we assume for simplicity to be orthonormalized,

$$\mathsf{P}=1=\sum_\nu |\mathscr{V}_\nu)(\mathscr{V}_\nu| \;, \tag{11.2.31}$$

we then have

$$\Omega_{\nu\mu}=(\mathscr{V}_\nu|\mathsf{L}\mathscr{V}_\mu) \tag{11.2.32}$$

defining a matrix of large dimensions which can have even a continuous or quasicontinuous spectrum. The relations for $|\mathscr{V}_\mu)$ which follow from the Mori equation (11.2.14),

$$\dot{\mathscr{V}}_\mu(t)=\sum_\nu \mathscr{V}_\nu(t)\mathrm{i}\Omega_{\nu\mu} \tag{11.2.33}$$

then include *all* resonances and linewidths in a complicated manner. Equations (11.2.33) are, of course, none other than the Heisenberg equations of motion,

$$\dot{\mathscr{V}}_\mu(t)=\mathrm{e}^{\mathrm{i}\mathsf{L}t}\mathrm{i}\mathsf{L}\mathscr{V}_\mu \;. \tag{11.2.34}$$

On the other hand, if we set $\mathsf{P}=0$, i.e. $\mathsf{Q}=1$, we can see that in (11.2.13), *only the third term remains*,

$$\dot{\mathscr{G}}_\mu(t)=f_\mu(t) \;, \tag{11.2.35}$$

that is, the residual force then becomes the total force.

One may pose the question as to *which set of observables $\mathscr{G}_\mu$ should be chosen for a given physical problem.* On the basis of the physical situation, we are interested in the dynamics of one or more (macroscopic) observables $\mathscr{G}_\mu$, and we will therefore formulate the Mori integro-differential equation (11.2.14) in terms of these variables. Because the corresponding memory matrix $\gamma_{\nu\mu}(t)$ cannot usually be calculated exactly, we shall require suitable approximations. As we already pointed out above, and as will be detailed in Chap. 13, an approximation is possible when the dynamics of the correlation matrix $\Xi_{\nu\mu}(t)$ are *slow* compared to those of the memory matrix $\gamma_{\nu\mu}(t)$. Such a *separation of time scales* can, however, only be expected when, in the dynamics of $\gamma_{\nu\mu}(t)$, determined by $\exp(\mathrm{i}\mathsf{QLQ}t)$, all of the slow variations are projected out by Q. A general recipe for the choice of such a projector P, however, does *not* exist. In a given case, one can usually find the slow observables which are to be taken into $\mathsf{P}=1-\mathsf{Q}$ by considering the physics of the situation (see Chap. 12).

We can, however, recognize a kind of "especially slow" observables in advance, namely the *constants of the motion* $\mathscr{C}_\alpha$ (e.g. $c\cdot 1, \mathscr{H}, \ldots$ [9]), for which the

[9] Besides $\mathscr{H}$, all the dyads $|u_\mathrm{E}\rangle\langle u_\mathrm{E}|$ formed from energy states represent constants of the motion.

following relation holds:

$$\dot{\mathscr{C}}_\alpha = \mathrm{i}\mathsf{L}\mathscr{C}_\alpha = \mathcal{O} \ . \qquad (11.2.36)$$

If we insert only constants of the motion in the Mori equation, we of course find as a trivial result just the relation (11.2.36). *Nevertheless, consideration of the constants of the motion $\mathscr{C}_\alpha$ is of importance for the treatment of the dynamics of the other physically interesting quantities $\mathscr{G}_\mu$.* In case such conserved quantities are not already contained in the set of $\mathscr{G}_\mu$ as linear combinations, we must *include* them in the choice of the set of observables, in view of later approximations to be made:

$$\mathscr{G}'_\mu = \{\mathscr{G}, \mathscr{C}\} \ , \qquad (11.2.37)$$

assuming that they couple to the $\mathscr{G}_\mu$ through the scalar products

$$(\mathscr{C}_\alpha | \mathscr{G}_\mu) \neq 0 \ . \text{[10]} \qquad (11.2.38)$$

We must choose a projector P' in the Mori equation with *all the* $\mathscr{G}'_\mu$ (11.2.37).

Another, equivalent procedure is the following: if one decomposes $\mathscr{G}_\mu$ into $\mathsf{P}_{\mathscr{C}}\mathscr{G}_\mu$ and $\mathsf{Q}_{\mathscr{C}}\mathscr{G}_\mu$, where $\mathsf{P}_{\mathscr{C}}$ is the projector onto the space $\mathfrak{L}_{\mathscr{C}}$ of the constants of the motion (Fig. 11.3),

$$\mathsf{P}_{\mathscr{C}} + \mathsf{Q}_{\mathscr{C}} = 1 \ , \qquad (11.2.39)$$

then only the components $\mathsf{Q}_{\mathscr{C}}\mathscr{G}_\mu$ vary dynamically. We can thus restrict ourselves to the set of operators

$$\boxed{\tilde{\mathscr{G}}_\mu = \mathsf{Q}_{\mathscr{C}}\mathscr{G}_\mu} \qquad (11.2.40)$$

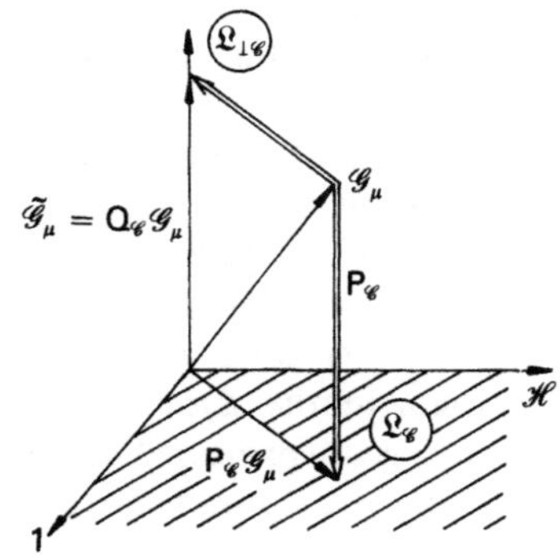

Fig. 11.3. Decomposition of $\mathscr{G}_\mu$ into $\tilde{\mathscr{G}}_\mu$ and constants of the motion

---

[10] In many cases, however, the scalar products $(\tilde{\mathscr{C}}_\alpha | \mathscr{G}_\mu)$ vanish for certain $\mathscr{C}_\alpha$, e.g. due to symmetry selection rules (cf. Appendix B), or in the thermodynamic limit, so that no coupling to the observables of interest exists and therefore these constants of the motion need not be taken into consideration.

It then suffices to formulate the Mori integro-differential equations in terms of the variables $\tilde{\mathscr{G}}_\mu$ with $\tilde{g}_{\mu\varrho}$, the projector $\tilde{\mathsf{P}}=\sum_{\mu\varrho}|\tilde{\mathscr{G}}_\mu)\tilde{g}_{\mu\varrho}(\tilde{\mathscr{G}}_\varrho|$, the frequency matrix $\tilde{\Omega}_{\mu\nu}$, and the memory matrix $\tilde{\gamma}_{\mu\nu}(t)$.

The *equivalence* of the formulation of the Mori equations for the set $\{\mathscr{G}'\}$ to that for the smaller set $\{\tilde{\mathscr{G}}\}$ is seen from the relation

$$\mathsf{P}'=\tilde{\mathsf{P}}+\mathsf{P}_{\mathscr{C}} \tag{11.2.41}$$

and from the property $\mathsf{LP}'=\mathsf{L}\tilde{\mathsf{P}}$, so that the system of equations for $\{\mathscr{G}'\}$ can be decomposed into a system of equations for $\{\tilde{\mathscr{G}}\}$ and into the equations $\dot{\mathscr{C}}_\alpha=\mathcal{O}$.

We mention the frequently-used transformation

$$\boxed{\mathscr{G}_\mu\to\delta\mathscr{G}_\mu=\mathscr{G}_\mu-|\mathit{1})(\mathit{1}|\mathscr{G}_\mu)=\mathscr{G}_\mu-\mathit{1}\langle\mathscr{G}_\mu\rangle_\beta} \tag{11.2.42}$$

as the simplest example of $\mathscr{G}_\mu\to\tilde{\mathscr{G}}_\mu$, in which the constant thermodynamic expectation value $\langle\mathscr{G}_\mu\rangle_\beta$ is subtracted out, so that $\delta\mathscr{G}_\mu$ is always orthogonal to $\mathit{1}$,

$$(\mathit{1}|\delta\mathscr{G}_\mu)=0\ . \tag{11.2.43}$$

The corresponding metric is then inversely proportional to the isothermal static susceptibility [compare (8.3.16)],

$$(\delta\mathscr{G}_\nu|\delta\mathscr{G}_\mu)=\frac{1}{\beta}\chi^{\mathrm{T}}_{\mu\nu}\ .^{11} \tag{11.2.44}$$

In the following, we shall wish to keep these possibilities in mind, but shall continue to use the Mori equations with the notation $\mathscr{G}_\mu$ for general considerations.

11.3 The Frequency and the Memory Matrices

11.3.1 The Eigenelements of the Frequency Matrix

The frequency matrix $\Omega_{\nu\mu}$ defined by (11.2.15) deserves to be considered more closely. Because P projects onto $\mathfrak{L}_{\mathscr{G}}$, $\mathsf{L}\mathscr{G}_\mu$ is in $\mathfrak{L}_{\mathscr{G}}$, i.e. it is a linear combination of the $\mathscr{G}_\mu$. The coefficients which occur in this linear combination are the matrix elements of $\Omega_{\nu\mu}$, since from (11.2.2) and (11.2.15) it immediately follows that

$$\mathsf{PL}\mathscr{G}_\mu=\mathsf{PLP}\mathscr{G}_\mu=\sum_\nu\mathscr{G}_\nu\Omega_{\nu\mu}\ . \tag{11.3.1}$$

---

[11] If it is necessary to project out not only the $\mathit{1}$, but also the Hamiltonian $\mathscr{H}$, one obtains a metric which is reciprocal to the *adiabatic* static susceptibility (8.3.30).

Like L, the operator PLP is Hermitian. The eigenvalue problem for PLP defines eigenvectors $\hat{\mathscr{G}}_\varrho$ in $\mathfrak{L}_{\mathscr{G}}$,

$$\mathsf{PLP}\hat{\mathscr{G}}_\varrho = \Omega_\varrho \hat{\mathscr{G}}_\varrho \tag{11.3.2}$$

with real eigenvalues Ω_ϱ. The orthogonal eigenvectors $\hat{\mathscr{G}}_\varrho$ are thus linear combinations of the $\mathscr{G}_\mu$,

$$\hat{\mathscr{G}}_\varrho = \sum_\nu \mathscr{G}_\nu c_{\nu\varrho} \ . \tag{11.3.3}$$

The columns of the matrix $c_{\nu\varrho}$ are the eigenvectors of the matrix $\Omega_{\nu\mu}$ belonging to the eigenvalue Ω_ϱ:

$$\sum_\mu \Omega_{\nu\mu} c_{\mu\varrho} = \Omega_\varrho c_{\nu\varrho} \ . \tag{11.3.4}$$

It then follows from (11.3.1, 2, and 3) that

$$\mathsf{PLP} \sum_\mu \mathscr{G}_\mu c_{\mu\varrho} = \sum_{\mu\nu} \mathscr{G}_\nu \Omega_{\nu\mu} c_{\mu\varrho} = \Omega_\varrho \sum_\nu \mathscr{G}_\nu c_{\nu\varrho} \ , \tag{11.3.5}$$

from which relation (11.3.4) results owing to the assumed linear independence of the $\mathscr{G}_\mu$. Because the matrix $\Omega_{\nu\mu}$ is, in general, not Hermitian, its eigenvectors $c_{\nu\varrho}$ are not generally orthogonal[12]. The frequency matrix derived from the $\hat{\mathscr{G}}_\mu$ is diagonal; we have

$$\Omega_\varrho = \frac{(\hat{\mathscr{G}}_\varrho | \mathsf{L} \hat{\mathscr{G}}_\varrho)}{(\hat{\mathscr{G}}_\varrho | \hat{\mathscr{G}}_\varrho)} = \frac{1}{\beta\hbar} \frac{\langle [\hat{\mathscr{G}}_\varrho^\dagger, \hat{\mathscr{G}}_\varrho] \rangle_\beta}{(\hat{\mathscr{G}}_\varrho | \hat{\mathscr{G}}_\varrho)} \ . \tag{11.3.6}$$

For the operators $\hat{\mathscr{G}}_\mu(t)$, the Mori equations take on the form

$$\frac{d}{dt} \hat{\mathscr{G}}_\varrho(t) = \mathrm{i}\Omega_\varrho \hat{\mathscr{G}}_\varrho(t) - \int_0^t \sum_\nu \hat{\mathscr{G}}_\nu(t-t') \hat{\gamma}_{\nu\varrho}(t')\, dt' + \hat{f}_\varrho(t) \ . \tag{11.3.7}$$

[12] The matrix $\Omega_{\nu\mu}$ is, according to (11.2.15), the product of the positive definite Hermitian matrix $g_{\nu\varrho}$ and the Hermitian matrix $(\mathscr{G}_\varrho | \mathsf{L}\mathscr{G}_\mu)$; in matrix notation

$$\boldsymbol{\Omega} = \boldsymbol{g}\boldsymbol{\omega} \ .$$

If the eigenvalue equation (11.3.4)

$$\boldsymbol{\Omega c}_\varrho = \boldsymbol{g\omega c}_\varrho = \Omega_\varrho \boldsymbol{c}_\varrho$$

is multiplied from the left by $\boldsymbol{g}^{-1/2}$, we obtain

$$(\boldsymbol{g}^{1/2}\boldsymbol{\omega}\boldsymbol{g}^{1/2})\boldsymbol{g}^{-1/2}\boldsymbol{c}_\varrho = \Omega_\varrho \boldsymbol{g}^{-1/2}\boldsymbol{c}_\varrho \ .$$

Because of the Hermiticity of $\boldsymbol{g}^{1/2}\boldsymbol{\omega}\boldsymbol{g}^{1/2}$, the eigenvectors $\boldsymbol{g}^{-1/2}\boldsymbol{c}_\varrho$ are orthogonal.

The memory matrix $\hat{\gamma}_{\nu\varrho}(t)$ need not be diagonal, of course, so that the $\hat{\mathscr{G}}_\mu(t)$ are mutually coupled by it. If the last two terms in (11.3.7) are negligible, the "normal coordinates"

$$\hat{\mathscr{G}}_\varrho(t) \approx e^{i\Omega_\varrho t}\hat{\mathscr{G}}_\varrho \tag{11.3.8}$$

oscillate at the generally temperature *de*pendent frequency Ω_ϱ (compare Exercise 11.1).

11.3.2 Properties of the Memory Matrix. Dynamic Onsager–Casimir Coefficients

The behavior in time of the memory matrix $\gamma_{\nu\mu}(t)$ is determined by the correlation matrix of the residual force,

$$l_{\varrho\mu}(t) = (f_\varrho | f_\mu(t)) = (f_\varrho | e^{iQLQt} f_\mu) \ . \tag{11.3.9}$$

These expressions are *formally similar to the relaxation functions (9.1.16), if one replaces* L *there by the Hermitian operator* QLQ. We can therefore apply the results which we obtained in Sects. 9.3 and 9.4 for the response function and its Fourier transform, the dynamic susceptibility, directly to $(f_\varrho | f_\mu(t))$, *provided* that we keep in mind that *an additional factor* $-i\beta$L *occurs in the response function*, and that, as in dynamics with L, the relation

$$(e^{iQLQt} f_\mu)^\dagger = e^{iQLQt} f_\mu^\dagger \tag{11.3.10}$$

holds (cf. Exercise 11.2). Here, we suppose that $\mathscr{G}_\nu^\dagger$ is also contained in the set $\mathscr{G}_\nu$.

As in the case of the relaxation functions, the property

$$l_{\varrho\mu}(t) = (f_\varrho | f_\mu(t)) = (f_\mu | f_\varrho(-t))^* = (f_\varrho^\dagger | f_\mu^\dagger(t))^* \tag{11.3.11}$$

follows from the Hermiticity of QLQ in the Mori product and from (8.2.10) and (11.3.10). Therefore, the correlation matrix $l_{\varrho\mu}(t) = (f_\varrho | f_\mu(t))$, and thus also the memory matrix $\gamma_{\nu\mu}(t)$ are *real* when the operators $\mathscr{G}_\mu$ are Hermitian.

In the Taylor expansion of the memory matrix,

$$\gamma_{\nu\mu}(t) = \sum_{n=0}^{\infty} \frac{1}{n!} \gamma_{\nu\mu}^{(n)}(0)\, t^n \tag{11.3.12}$$

the n-th derivatives at time $t=0$ are given, according to (11.2.23), by

$$\boxed{\gamma_{\nu\mu}^{(n)}(0) = (i)^n \sum_\varrho g_{\nu\varrho} (f_\varrho | (QLQ)^n f_\mu)} \ . \tag{11.3.13}$$

These expressions may be represented in terms of $(\mathscr{G}_\varrho | L^m \mathscr{G}_\mu)$, if the definition (11.2.22) of the components f_ϱ of the residual force is inserted. For example, for

$n=0$, we find

$$\gamma_{\nu\mu}(0)=\sum_{\varrho} g_{\nu\varrho}\left\{(\mathscr{G}_{\varrho}|\mathsf{L}^2\mathscr{G}_{\mu})-\sum_{\lambda\sigma} g_{\sigma\lambda}(\mathscr{G}_{\varrho}|\mathsf{L}\,\mathscr{G}_{\sigma})(\mathscr{G}_{\lambda}|\mathsf{L}\,\mathscr{G}_{\mu})\right\}$$

$$=\sum_{\varrho} g_{\nu\varrho}(\mathscr{G}_{\varrho}|\mathsf{L}^2\,\mathscr{G}_{\mu})-\sum_{\sigma}\Omega_{\nu\sigma}\Omega_{\sigma\mu}\ . \tag{11.3.14}$$

The matrix $\gamma_{\nu\mu}(0)$ is therefore called the mean square deviation of L.

The one-sided Fourier transform of the correlation matrix $(f_{\varrho}|f_{\mu}(t))$ is called the (matrix of the) *dynamic Onsager–Casimir coefficients*:

$$\boxed{\begin{aligned}L_{\varrho\mu}(\omega)&\equiv\int_0^{\infty}(f_{\varrho}|\mathrm{e}^{\mathrm{i}(\mathsf{QLQ}+\omega)t-\eta t}f_{\mu})\,dt|_{\eta\to+0}\\&=\left(f_{\varrho}\left|\frac{\mathrm{i}}{\mathsf{QLQ}+\omega+\mathrm{i}\eta}f_{\mu}\right.\right)\Big|_{\eta\to+0}\end{aligned}}\ . \tag{11.3.15}$$

The one-sided Fourier (Laplace) transform of the memory matrix $\gamma_{\nu\mu}(t)$,

$$\boxed{\gamma_{\nu\mu}(\omega)=\sum_{\varrho} g_{\nu\varrho}L_{\varrho\mu}(\omega)}\ , \tag{11.3.16}$$

is thus the product of the reciprocal static susceptibility and the dynamic Onsager–Casimir matrix.

As in Sect. 9.4.1, we decompose the dynamic Onsager–Casimir matrix $L_{\varrho\mu}(\omega)$ into a sum of two matrices which are Hermitian in the indices ϱ and μ:

$$L_{\varrho\mu}(\omega)=L'_{\varrho\mu}(\omega)+\mathrm{i}L''_{\varrho\mu}(\omega) \quad [13] \tag{11.3.17}$$

with

$$\boxed{\begin{aligned}L'_{\varrho\mu}(\omega)&=\tfrac{1}{2}(L_{\varrho\mu}(\omega)+L^{*}_{\mu\varrho}(\omega))=L'^{*}_{\mu\varrho}(\omega)\\&=\frac{1}{2}\int_{-\infty}^{+\infty}(f_{\varrho}|f_{\mu}(t))\,\mathrm{e}^{\mathrm{i}\omega t-\eta|t|}\,dt|_{\eta\to+0}\\&=\pi(f_{\varrho}|\delta(\mathsf{QLQ}+\omega)f_{\mu}) \quad [14,15]\end{aligned}} \tag{11.3.18}$$

[13] This is, in general, not a decomposition into real and imaginary parts [compare Sect. 9.4.1]!

and

$$\boxed{\begin{aligned} L''_{\varrho\mu}(\omega) &= \frac{1}{2\mathrm{i}}(L_{\varrho\mu}(\omega) - L^*_{\mu\varrho}(\omega)) = L''^*_{\mu\varrho}(\omega) \\ &= \int\limits_0^\infty \mathrm{e}^{-\eta t}(f_\varrho|\sin(\mathsf{QLQ}+\omega)t\, f_\mu)\,\mathrm{d}t\Big|_{\eta\to+0} \\ &= \left(f_\varrho\Big|\,\mathrm{Pr}\frac{1}{\mathsf{QLQ}+\omega}f_\mu\right)^{14,15} \end{aligned}} \quad . \tag{11.3.19}$$

From the fact that QLQ is Hermitian in $\mathfrak{L}$ and using (11.3.10 or 11), we find the symmetry relations

$$\boxed{\begin{aligned} L'_{\varrho\mu}(\omega) &= L'_{\mu^\dagger\varrho^\dagger}(-\omega) \\ L''_{\varrho\mu}(\omega) &= -L''_{\mu^\dagger\varrho^\dagger}(-\omega) \end{aligned}} \quad , \tag{11.3.20}$$

which, when the operators $\mathscr{G}_\mu$ are Hermitian in $\mathfrak{U}$, take on the form

$$\begin{aligned} L'_{\varrho\mu}(\omega) &= L'_{\mu\varrho}(-\omega) \ , \\ L''_{\varrho\mu}(\omega) &= -L''_{\mu\varrho}(-\omega) \ . \end{aligned} \tag{11.3.21}$$

Other symmetry properties which are related to time-reversal invariance are listed in Appendix B2.

We shall demonstrate in Sect. 15.2 how the entropy production in irreversible processes is determined by the Onsager–Casimir coefficients and how they can be used to formulate the second fluctuation-dissipation theorem.

For the n-th moment of $L'_{\varrho\mu}(\omega)$, we find from (11.3.18), using the properties of the δ-function

$$\int\limits_{-\infty}^{+\infty} L'_{\varrho\mu}(\omega)\,\omega^n d\omega = (-1)^n \pi(f_\varrho|(\mathsf{QLQ})^n f_\mu) \ , \tag{11.3.22}$$

i.e., the moment is determined by the n-th derivative $\gamma^{(n)}_{\nu\mu}(t=0)$ of the memory matrix at $t=0$ [cf. (11.3.13)]. For general considerations concerning the calculation of moments, see Sect. 9.4.5.

---

[14] Here, we have used the relation $1/(x+\mathrm{i}\eta)|_{\eta\to+0} = \mathrm{Pr}(1/x) - \mathrm{i}\pi\delta(x)$.

[15] The differences compared to (9.4.7, 8, 10, and 11) are due to the additional factor $-\mathrm{i}\beta\mathsf{L}$ in the definition of $\chi_{\varrho\mu}(\omega)$.

The asymptotic expansion of $L_{\varrho\mu}(\omega)$ for *large* ω is obtained as in Sect. 9.4.4, starting with (11.3.15), setting $z=\omega+\mathrm{i}\eta$, and making use of

$$\frac{1}{\mathsf{QLQ}+z}=\sum_{n=0}^{\infty}\frac{(-1)^n}{z}\left(\frac{\mathsf{QLQ}}{z}\right)^n . \tag{11.3.23}$$

The result is

$$L_{\varrho\mu}(z)=\mathrm{i}\sum_{n=0}^{\infty}(-1)^n\frac{(\mathscr{f}_\varrho|(\mathsf{QLQ})^n\,\mathscr{f}_\mu)}{z^{n+1}} . \tag{11.3.24}$$

The following relation between the expansion of $\gamma_{\nu\mu}(z)$ for large z (i.e. at large ω) and that of $\gamma_{\nu\mu}(t)$ for short times t thus holds:

$$\gamma_{\nu\mu}(z)=\sum_{n=0}^{\infty}\frac{\mathrm{i}^{n+1}}{z^{n+1}}\gamma_{\nu\mu}^{(n)}(t=0) . \tag{11.3.25}$$

From (11.3.15), $L_{\varrho\mu}(z)$ is holomorphic in the upper part of the complex z-plane (imaginary part, $\mathrm{Im}\,z>0$). Employing exactly the same arguments as were used in Sect. 9.4.3, we find the Kramers–Kronig relations between $L'_{\varrho\mu}(\omega)$ and $L''_{\varrho\mu}(\omega)$:

$$\begin{aligned} L'_{\varrho\mu}(\omega)&=\frac{1}{\pi}\,\mathrm{Pr}\int_{-\infty}^{+\infty}\frac{L''_{\varrho\mu}(\bar\omega)}{\bar\omega-\omega}d\bar\omega , \\ L''_{\varrho\mu}(\omega)&=-\frac{1}{\pi}\,\mathrm{Pr}\int_{-\infty}^{+\infty}\frac{L'_{\varrho\mu}(\bar\omega)}{\bar\omega-\omega}d\bar\omega . \end{aligned} \tag{11.3.26}$$

These relations permit us to determine the functions $L''_{\varrho\mu}(\omega)$ given the $L'_{\varrho\mu}(\omega)$ and vice versa.

Exercise 11.1 Find the frequency matrix $\Omega^{\mu'\mu}_{\mathfrak{k}\mathfrak{k}}$ for the spin observables of a Heisenberg ferromagnet,

$$\mathscr{S}^\mu_{\mathfrak{k}}=\sum_j \mathrm{e}^{-\mathrm{i}\mathfrak{k}\mathfrak{R}_j}\,\sigma^\mu_j \quad (\mu=\pm)$$

with the Hamiltonian

$$\mathscr{H}=-\gamma H\sum_i\sigma^z_i-\frac{1}{\hbar^2}\sum_{i\neq j}J_{ij}\vec{\sigma}_i\vec{\sigma}_j , \quad J_{ij}=J(\mathfrak{R}_i-\mathfrak{R}_j)\geqq 0 .$$

The positions $\mathfrak{R}_j$ are related to each other by translations, $\mathfrak{a}_1n_1+\mathfrak{a}_2n_2+\mathfrak{a}_3n_3$.

The discrete values $\mathfrak{k}=\mathfrak{k}_1, \ldots, \mathfrak{k}_N$ obey the orthogonality relations

$$\sum_j e^{i(\mathfrak{k}-\mathfrak{k}')\mathfrak{R}_j} = N\delta_{\mathfrak{k}\mathfrak{k}'} \; .$$

What is the limiting value of the frequency matrix for $T \to 0$?

Exercise 11.2 Prove the relation (11.3.10).

11.4 The Integro-Differential Equations for Relaxation Functions

11.4.1 Dynamics of the Correlation Matrix $\Xi_{\nu\mu}(t)$. Relationship to Linear Dynamic Response Theory

We first consider the dynamic correlation matrix of the observables

$$\boxed{\Xi_{\nu\mu}(t) = \sum_{\varrho} g_{\nu\varrho}(\mathcal{G}_\varrho|\mathcal{G}_\mu(t)) = \sum_{\varrho} g_{\nu\varrho}(\mathcal{G}_\varrho|e^{iLt}\mathcal{G}_\mu)} \; , \tag{11.4.1}$$

which was already introduced in Sect. 11.2.1. As a result of (11.2.3), it is given at time $t=0$ by the unit matrix

$$\Xi_{\nu\mu}(0) = \delta_{\nu\mu} \; . \tag{11.4.2}$$

The time dependence of $\Xi_{\nu\mu}(t)$ can be found from the Mori equation (11.2.14) by multiplying it from the left by $(\mathcal{G}_\nu|$ and taking account of the fact that, according to (11.2.20), the residual force $f_\mu(t)$ is always perpendicular to $\mathcal{G}_\nu$:

$$\boxed{\dot{\Xi}_{\nu\mu}(t) = \sum_\lambda \Xi_{\nu\lambda}(t) i\Omega_{\lambda\mu} - \int_0^t \sum_\lambda \Xi_{\nu\lambda}(t-t')\gamma_{\lambda\mu}(t')\,dt'} \; . \tag{11.4.3}$$

This set of coupled integro-differential equations is exact and is valid for an arbitrary set of observables $\mathcal{G}_\nu$; this holds even independently of the actual form of the scalar product in $\mathfrak{L}$. It demonstrates that the dynamics of the correlation function can be related to the frequency matrix $\Omega_{\lambda\mu}$ and the memory matrix $\gamma_{\lambda\mu}(t)$, whose general properties were discussed above.

We introduce the one-sided Fourier transform of $\Xi_{\nu\mu}(t)$:

$$\boxed{\Xi_{\nu\mu}(\omega) = \int_0^\infty \Xi_{\nu\mu}(t) e^{i\omega t - \eta t}\,dt\Big|_{\eta\to+0} = \sum_\varrho g_{\nu\varrho}\left(\mathcal{G}_\varrho \left| \frac{i}{L+\omega+i\eta}\mathcal{G}_\mu\right.\right)_{\eta\to+0}} \; , \tag{11.4.4}$$

for which we obtain the set of *algebraic relations*

$$-\delta_{\nu\mu}-\mathrm{i}\omega\Xi_{\nu\mu}(\omega)=\sum_{\lambda}\Xi_{\nu\lambda}(\omega)(\mathrm{i}\Omega_{\lambda\mu}-\gamma_{\lambda\mu}(\omega)) \quad (11.4.5)$$

or

$$\boxed{\sum_{\lambda}\Xi_{\nu\lambda}(\omega)[\omega\delta_{\lambda\mu}+\Omega_{\lambda\mu}+\mathrm{i}\gamma_{\lambda\mu}(\omega)] = \mathrm{i}\delta_{\nu\mu}} \quad (11.4.6)$$

by applying the convolution theorem to the Mori equation (11.4.3). In matrix notation, this system of equations may be written as

$$\boldsymbol{\Xi}(\omega)[\omega\cdot\mathbf{1}+\boldsymbol{\Omega}+\mathrm{i}\boldsymbol{\gamma}(\omega)] = \mathrm{i}\cdot\mathbf{1} \quad (11.4.7)$$

and has the formal solution:

$$\boxed{\boldsymbol{\Xi}(\omega) = \mathrm{i}[\omega\cdot\mathbf{1}+\boldsymbol{\Omega}+\mathrm{i}\boldsymbol{\gamma}(\omega)]^{-1}}\ . \quad (11.4.8)$$

These results could have also been obtained directly, without explicitly using the Mori equation, by regarding the resolvents:

$$\mathsf{G}(\omega) = \frac{\mathrm{i}}{\mathsf{L}+\omega+\mathrm{i}\eta} \quad (11.4.9)$$

and

$$\mathsf{K}(\omega) = \mathsf{LQ}\frac{\mathrm{i}}{\mathsf{QLQ}+\omega+\mathrm{i}\eta}\mathsf{QL}\ . \quad (11.4.10)$$

As can be readily seen (Exercise 11.4), the following relation holds:

$$\mathsf{PG}(\omega)\mathsf{P} = \frac{\mathrm{i}}{\mathsf{PLP}+\mathrm{i}\mathsf{PK}(\omega)\mathsf{P}+\omega+\mathrm{i}\eta}\mathsf{P}\ ; \quad (11.4.11)$$

it is just (11.4.8) in operator form. In analogy to field theory, the operator $\mathsf{PK}(\omega)\mathsf{P}$ is referred to as the "mass operator". Its matrix elements $(\mathscr{G}_\varrho|\,\mathsf{K}(\omega)\mathscr{G}_\mu)$ are just the Onsager–Casimir coefficients; compare (11.3.15).

We now wish to apply the preceding considerations to *linear response theory*. To this end, we consider the special properties of the Mori product and show how the functions of interest in linear response theory may be calculated from the $\Xi_{\nu\mu}(t)$. In these considerations, we shall not employ the observables $\mathscr{G}_\mu$ themselves, but rather their deviations from thermal equilibrium values, $\delta\mathscr{G}_\mu=\mathscr{G}_\mu-\langle\mathscr{G}_\mu\rangle_\beta$ [compare (11.2.42)]. We then obtain the following results:

From (11.2.44), the isothermal static susceptibility

$$\chi^{\mathrm{T}}_{\nu\varrho} = \beta(\delta\mathscr{G}_\nu|\delta\mathscr{G}_\varrho) \quad (11.4.12)$$

has an inverse matrix given by

$$\sum_\varrho g_{\nu\varrho}\chi^{\mathrm{T}}_{\varrho\mu} = \sum_\varrho \chi^{\mathrm{T}}_{\nu\varrho}g_{\varrho\mu} = \beta\delta_{\nu\mu} \ . \tag{11.4.13}$$

The *relaxation functions* $\Phi_{\nu\mu}(t)=\beta(\delta\mathscr{G}_\nu|\mathscr{G}_\mu(t))$ (9.1.16) are determined by

$$\Phi_{\nu\mu}(t) = \sum_\varrho \chi^{\mathrm{T}}_{\nu\varrho}\Xi_{\varrho\mu}(t) \ , \tag{11.4.14}$$

so that we find for the *response functions* $\varphi_{\nu\mu}(t)=-\dot{\Phi}_{\nu\mu}(t)$ the following expression:

$$\varphi_{\nu\mu}(t) = -\sum_\varrho \chi^{\mathrm{T}}_{\nu\varrho}\dot{\Xi}_{\varrho\mu}(t) \ . \tag{11.4.15}$$

The matrix of the *dynamic susceptibility* $\chi_{\nu\mu}(\omega)$ may be calculated from $\Xi_{\nu\mu}(\omega)$ according to (9.4.22) using

$$\chi_{\nu\mu}(\omega) = \sum_\varrho \chi^{\mathrm{T}}_{\nu\varrho}(\delta_{\varrho\mu} + \mathrm{i}\omega\Xi_{\varrho\mu}(\omega)) \ [16] \tag{11.4.16}$$

or, in matrix notation,

$$\boldsymbol{\chi}(\omega) = \boldsymbol{\chi}^{\mathrm{T}}(\mathbf{1} + \mathrm{i}\omega\boldsymbol{\Xi}(\omega)) \ . \tag{11.4.17}$$

Upon inserting $\boldsymbol{\Xi}(\omega)$ (11.4.8) in this matrix equation, one obtains an expression for $\boldsymbol{\chi}(\omega)$ in terms of the frequency and the memory matrices:

$$\boxed{\boldsymbol{\chi}(\omega) = \boldsymbol{\chi}^{\mathrm{T}}(\boldsymbol{\Omega} + \mathrm{i}\gamma(\omega))(\omega\cdot\mathbf{1} + \boldsymbol{\Omega} + \mathrm{i}\gamma(\omega))^{-1}} \ . \tag{11.4.18}$$

We have thus reduced the determination of the dynamic susceptibility to a calculation of $\boldsymbol{\Omega}$ and of $\gamma(\omega)$, which is more accessible to the application of an approximation method (Chaps. 12 and 13).

11.4.2 Integro-Differential Equations for the Expectation Values $\langle\delta\mathscr{G}_\mu\rangle(t)$

We wish to make the transition from the Mori equations (11.2.14) to equations for the expectation values of the deviations from thermal equilibrium,

$$\delta\mathscr{G}_\mu(t) = \mathscr{G}_\mu(t) - \langle\mathscr{G}_\mu\rangle_\beta \ . \tag{11.4.19}$$

If an arbitrary, non-steady-state operator $\rho(t=0)$ applies at the initial time $t=0$, then for $t>0$, the time dependence of the *expectation values* $\langle\delta\mathscr{G}_\mu\rangle(t)$ *for the*

[16] Owing to conserved quantities, $\Xi_{\nu\mu}(\omega)$ can have a pole at $\omega=0$, so that in general, $\chi_{\nu\mu}(0)\neq\chi^{\mathrm{T}}_{\nu\mu}$ [cf. Sect. 9.5].

relaxation process in the Heisenberg formulation is given by

$$\langle \delta \mathcal{G}_\mu \rangle(t) = \mathrm{Tr}(\rho(t=0)\delta \mathcal{G}_\mu(t)) \ . \tag{11.4.20}$$

From the Mori equations (11.2.14), we find the following inhomogeneous integro-differential equations for these expectation values:

$$\boxed{\frac{d}{dt}\langle \delta \mathcal{G}_\mu \rangle(t) = \sum_\nu \langle \delta \mathcal{G}_\nu \rangle(t)\, \mathrm{i}\Omega_{\nu\mu} - \int_0^t \sum_\nu \langle \delta \mathcal{G}_\nu \rangle (t-t')\gamma_{\nu\mu}(t')dt' + \langle f_\mu \rangle(t)} \ . \tag{11.4.21}$$

We have used here the projection operator

$$\mathsf{P} = \sum_{\nu\mu} g_{\nu\mu} |\delta \mathcal{G}_\nu)(\delta \mathcal{G}_\mu| \ . \tag{11.4.22}$$

If, *at the initial time*, the expectation values are $\langle \mathcal{H} \rangle(0)$ and $\langle \mathcal{G}_\nu \rangle(0)$, then we can use the principle of the maximum uncertainty measure [cf. Sect. 5.2] to obtain the *generalized canonical statistical operator*. Let $\rho(t=0)$ be sufficient in the sense of Sect. 5.2,

$$\rho(t=0, \beta, \lambda) = \frac{\exp\left(-\beta \mathcal{H} - \sum_\nu \lambda_\nu \mathcal{G}_\nu^\dagger\right)}{\mathrm{Tr}\exp\left(-\beta \mathcal{H} - \sum_\nu \lambda_\nu \mathcal{G}_\nu^\dagger\right)} \ .^{17} \tag{11.4.23}$$

Equations (11.4.21) describe the relaxation process of the observables $\langle \delta \mathcal{G}_\mu \rangle(t)$ for $t>0$ starting from the general non-steady-state initial conditions, (11.4.23). The expectation values of the residual force $\langle f_\mu \rangle(t)$ determined from (11.2.21) are, in general, difficult to calculate, so that (11.4.21) in this general form are of limited usefulness.

This difficulty is removed when we restrict our considerations to initial states which deviate only *linearly* from thermal equilibrium,

$$\boxed{\rho(t=0, \beta, \lambda) = \rho(0, \beta, 0) + \sum_\nu \lambda_\nu \left(\frac{\partial \rho}{\partial \lambda_\nu}\right)_{\lambda_\nu = 0}} \ , \tag{11.4.24}$$

for which, corresponding to (8.3.5), we can also write

$$\rho(t=0, \beta, \lambda) = \mathcal{R}_\beta - \mathcal{R}_\beta \frac{1}{\beta} \sum_\nu \lambda_\nu \int_0^\beta \mathrm{e}^{\alpha \mathcal{H}}\, \delta \mathcal{G}_\nu^\dagger \,\mathrm{e}^{-\alpha \mathcal{H}} d\alpha \tag{11.4.25}$$

---

[17] Here, we assume that the set $\mathcal{G}_\nu$ of the Mori theory forms an observation level. The operator $\mathcal{G}_\nu^\dagger$ which is the Hermitian adjoint of $\mathcal{G}_\nu$ then also belongs to the set of the $\{\mathcal{G}\}$ [cf. Sect. 15.1.1].

where

$$\mathscr{R}_\beta = \frac{e^{-\beta\mathscr{H}}}{\mathrm{Tr}\, e^{-\beta\mathscr{H}}} \tag{11.4.26}$$

is the canonical statistical operator for thermal equilibrium. If we use the operator (11.4.25), *which is linear in the* λ_ν, to determine the *expectation values of the residual force* $f_\mu(t)$, *the latter vanish for all times t and for arbitrary values of* λ_μ so long as they are consistent with the linearization:

$$\boxed{\langle f_\mu \rangle(t) = 0} \; . \tag{11.4.27}$$

It is *essential* for this result that the Mori scalar product was assumed in the Mori equations.

Proof of (11.4.27): The expectation value of $f_\mu(t)$ taken using (11.4.25) can be written with the aid of (8.2.7) and (8.2.23) in the form

$$\langle f_\mu \rangle(t) = (1 | f_\mu(t)) + \sum_\nu \lambda_\nu (\delta\mathscr{G}_\nu | f_\mu(t)) \; . \tag{11.4.28}$$

The first term vanishes due to

$$(1 | f_\mu(t)) = (e^{-i\mathsf{QLQ}t} \cdot 1 | f_\mu) = (1 | f_\mu) = i(\mathsf{LQ} \cdot 1 | \mathscr{G}_\mu) = 0 \; , \tag{11.4.29}$$

where we have used $(\delta\mathscr{G}_\mu | 1) = 0$, i.e. $\mathsf{Q} \cdot 1 = 1$. The second term in (11.4.28) is zero because the residual force $f_\mu(t)$ is always perpendicular to $\delta\mathscr{G}_\nu$ [cf. (11.2.19)].

For the set of *generalized canonical statistical operators which are linear in* λ_ν (11.4.25), the expectation value of the residual force thus vanishes, so that (11.4.21) take on the form

$$\boxed{\frac{d}{dt}\langle \delta\mathscr{G}_\mu \rangle(t) = \sum_\nu \langle \delta\mathscr{G}_\nu \rangle(t)\, i\Omega_{\nu\mu} - \int_0^t \sum_\nu \langle \delta\mathscr{G}_\nu \rangle(t-t')\gamma_{\nu\mu}(t')dt'} \; , \tag{11.4.30}$$

a system of closed linear integro-differential equations for the expectation values $\langle \delta\mathscr{G}_\mu \rangle(t)$. The dynamics of these expectation values have thus been related to knowledge of the frequency matrix $\Omega_{\nu\mu}$ and the memory matrix $\gamma_{\nu\mu}(t)$, or of appropriate approximations for them (Chaps. 12 and 13).

Equation (11.4.30) for the expectation values $\langle \delta\mathscr{G}_\mu \rangle(t)$ in a relaxation process has the same form as (11.4.3) for the correlation functions $\Xi_{\nu\mu}(t)$ or the relaxation functions $\Phi_{\nu\mu}(t)$, as must be the case for linear response theory. It is helpful to make this fact explicitly clear by substituting (11.4.25) into (11.4.20) and employing the definition of the Mori scalar product,

$$\langle \delta\mathscr{G}_\mu \rangle(t) = \mathrm{Tr}(\mathscr{R}_\beta \delta\mathscr{G}_\mu(t)) - \sum_\nu \lambda_\nu (\delta\mathscr{G}_\nu | \delta\mathscr{G}_\mu(t)) \; . \tag{11.4.31}$$

The first term is zero because the expectation value of an explicitly time-independent operator is constant in the case of a stationary statistical operator, so that $\mathrm{Tr}(\mathscr{R}_\beta \delta\mathscr{G}_\mu(t)) = \mathrm{Tr}(\mathscr{R}_\beta \mathscr{G}_\mu(t)) - \langle\mathscr{G}_\mu\rangle_\beta = 0$ holds. We thus obtain the expression

$$\boxed{\langle\delta\mathscr{G}_\mu\rangle(t) = -\sum_\nu \lambda_\nu(\delta\mathscr{G}_\nu|\delta\mathscr{G}_\mu(t))}\,, \tag{11.4.32}$$

which makes it clear that $\langle\delta\mathscr{G}_\mu\rangle(t)$ and $\Xi_{\nu\mu}(t)$ both obey the same integro-differential equations.

To conclude, we return to the *initial conditions* defined by $\rho(t=0)$. Considering (11.4.32) at $t=0$, we find

$$\langle\delta\mathscr{G}_\mu\rangle(0) = -\sum_\nu \lambda_\nu(\delta\mathscr{G}_\nu|\delta\mathscr{G}_\mu) = -\frac{1}{\beta}\sum_\nu \lambda_\nu \chi^{\mathrm{T}}_{\nu\mu} \tag{11.4.33}$$

or, solving for λ_ν,

$$\boxed{\lambda_\nu = -\sum_\mu \langle\delta\mathscr{G}_\mu\rangle(0) g_{\mu\nu}}\,. \tag{11.4.34}$$

The *Lagrange multipliers* λ_ν are therefore given by

$$\langle\delta\mathscr{G}_\mu\rangle(0) = \mathrm{Tr}(\rho(0)\mathscr{G}_\mu) - \langle\mathscr{G}_\mu\rangle_\beta\,, \tag{11.4.35}$$

i.e. by the *deviations* of the expectation values $\langle\mathscr{G}_\mu\rangle(0)$ from the equilibrium values $\langle\mathscr{G}_\mu\rangle_\beta$ at the start of the dynamic process. The *linearization in* λ_ν is thus equivalent to a *linearization in the quantities* $\langle\delta\mathscr{G}_\mu\rangle(0)$.

Exercise 11.3 Let a Hermitian observable have the memory function $\gamma(t) = \gamma_0 \exp(-t/\tau_c)$.

Find the normalized relaxation function $\Xi(t)$. Discuss its behavior in particular for $4\gamma_0\tau_c^2 \ll 1$.

Exercise 11.4 Prove relation (11.4.11).

12. Perturbation-Theoretical Treatment of the Frequency and Memory Matrix

For nearly all macroscopic systems, the quantities $\Omega_{\nu\mu}$, $\gamma_{\nu\mu}(t)$, and $f_\mu(t)$ which occur in the Mori theory cannot be precisely determined. One must therefore resort to approximate methods, in particular perturbation theory and Markovian approximations. The latter, which are in part based on the former, will be discussed in Chap. 13. In this chapter, we shall first introduce the perturbation-theoretical treatment of the Mori theory, which is based on the assumption that the time-independent Liouville operator of the isolated system can be decomposed into two parts:

$$\mathsf{L} = \mathsf{L}^0 + \mathsf{L}^1 \ . \tag{12.1.1}$$

Here, we assume further that L^1 is proportional to a smallness parameter[1], ε. In Sect. 12.1, we shall give the formulas which can be derived from a systematic expansion in powers of ε. The possibilities for extending such a perturbation-theoretical approach are discussed in Sect. 12.2.

12.1 The Leading Terms of a Perturbation-Theory Expansion in L^1

12.1.1 A Set of Observables $\{\mathscr{G}\}$ as an Invariant Subspace $\mathfrak{L}_{\{\mathscr{G}\}}$ with Respect to L^0

Beginning with one or several physically relevant observables $\mathscr{G}_1, \mathscr{G}_2, \ldots$, we make use of that invariant subspace $\mathfrak{L}_{\{\mathscr{G}\}}$ of the Liouville space which is transformed back onto itself[2] by application of L^0,

$$\boxed{\mathsf{L}^0 \mathscr{G}_\mu = \sum_\nu \mathscr{G}_\nu \Omega^0_{\nu\mu}} \ . \tag{12.1.2}$$

[1] In this chapter, superscripts 0, 1, 2, ... are used exclusively to indicate the order in the smallness parameter ε.

[2] If necessary, by extending the *a priori* physically interesting observables to a set $\{\mathscr{G}\}$ such that (12.1.2) is obeyed.

The elements $\mathsf{L}^0\mathscr{G}_\mu$ are thus a linear combination of the $\mathscr{G}_\mu$ which span the subspace $\mathfrak{L}_{\{\mathscr{G}\}}$. From (12.1.2) and the definition (11.2.5), it follows that

$$\mathsf{QL}^0\mathscr{G}_\mu = \mathscr{O} . \tag{12.1.3}$$

Due to the precondition (12.1.2), the Ω-matrix (11.2.15) takes on the form

$$\Omega_{\nu\mu} = \Omega^0_{\nu\mu} + \sum_\rho g_{\nu\rho}(\mathscr{G}_\rho | \mathsf{L}^1\mathscr{G}_\mu) \tag{12.1.4}$$

and, for the residual force at $t=0$ (11.2.17), we obtain

$$f_\mu = \mathrm{i}\mathsf{QL}^1\mathscr{G}_\mu . \tag{12.1.5}$$

12.1.2 Perturbation-Theory Expansion of the Scalar Products

Equation (12.1.4), like (12.1.5), does not yet represent an expansion in L^1, since, according to the definition of the Mori product (8.2.14), it contains the total Hamiltonian or the total Liouville operator of the system Σ. In order to obtain a perturbation-theoretical expansion, we must therefore expand the Mori products in the perturbation:

$$(\mathscr{F}|\mathscr{G}) = (\mathscr{F}|\mathscr{G})^0 + (\mathscr{F}|\mathscr{G})^1 + \dots . \tag{12.1.6}$$

If we temporarily extract the smallness parameter from $\mathscr{H}^1$, $\mathscr{H} = \mathscr{H}^0 + \varepsilon\mathscr{H}^1$ (or $\mathsf{L} = \mathsf{L}^0 + \varepsilon\mathsf{L}^1$), then, in (8.2.14) for $(\mathscr{F}|\mathscr{G})$, we must expand not only the canonical statistical operator $\mathscr{R}_\beta = \exp[-\beta(\mathscr{H}^0 + \varepsilon\mathscr{H}^1)]/Z(\varepsilon)$ but also the function of L in powers of ε. Equation (12.1.6) therefore implies

$$(\mathscr{F}|\mathscr{G}) = (\mathscr{F}|\mathscr{G})^{\varepsilon=0} + \frac{\partial}{\partial\varepsilon}(\mathscr{F}|\mathscr{G})|^{\varepsilon=0}\varepsilon + \dots . \tag{12.1.7}$$

$(\mathscr{F}|\mathscr{G})^0$ is *in*dependent of $\mathscr{H}^1$, $(\mathscr{F}|\mathscr{G})^1$ is *linear* in the perturbation operator $\mathscr{H}^1$, etc.

The *zeroth order*, $(\mathscr{F}|\mathscr{G})^0$, is thus the Mori product taken with $\mathscr{H}^0$ alone,

$$\boxed{(\mathscr{F}|\mathscr{G})^0 = \mathrm{Tr}\left(\mathscr{R}^0_\beta\mathscr{F}^\dagger \frac{1-\exp(-\beta\hbar\mathsf{L}^0)}{\beta\hbar\mathsf{L}^0}\mathscr{G}\right)} \qquad (\mathscr{F}, \mathscr{G} = \text{arbitrary}) . \tag{12.1.8}$$

Therefore, $(\mathscr{F}|\mathscr{G})^0$ obeys the equations given in Sect. 8.2.2 if $\mathscr{H}$ is replaced everywhere by $\mathscr{H}^0$. For example, L^0 in $(\mathscr{F}|\mathsf{L}^0\mathscr{G})^0$ is Hermitian. Even the first-order approximation, $(\mathscr{F}|\mathscr{G})^1$, is a complicated expression compared to (12.1.8); its direct evaluation can, however, be avoided in many cases, so that we will not give an explicit formula here (cf. Exercise 12.1).

As a consequence of (12.1.2), the zeroth-order Mori products $(\mathscr{G}_\mu|\mathscr{G}_\nu)^0$ for the operators of the set $\{\mathscr{G}\}$, may be represented in terms of ordinary correlation functions, following (12.1.8):

$$(\mathcal{G}_\mu|\mathcal{G}_\nu)^0 = \sum_\lambda \langle \mathcal{G}_\mu^\dagger \mathcal{G}_\lambda \rangle_\beta^0 \left(\frac{1-\exp(-\beta\hbar\boldsymbol{\Omega}^0)}{\beta\hbar\boldsymbol{\Omega}^0} \right)_{\lambda\nu} . \tag{12.1.9}$$

For the determination of the metric, $g_{\mu\varrho}$, it follows from the orthogonality condition (11.2.3) that

$$\sum_\varrho (g_{\mu\varrho}^0 + g_{\mu\varrho}^1 + \ldots)\,((\mathcal{G}_\varrho|\mathcal{G}_\nu)^0 + (\mathcal{G}_\varrho|\mathcal{G}_\nu)^1 + \ldots) = \delta_{\mu\nu} , \tag{12.1.10}$$

so that $g_{\mu\nu}^0$ is the inverse matrix of (12.1.9),

$$\sum_\varrho g_{\mu\varrho}^0 (\mathcal{G}_\varrho|\mathcal{G}_\nu)^0 = \delta_{\mu\nu} , \tag{12.1.11}$$

while $g_{\mu\nu}^1$ is determined by

$$g_{\mu\nu}^1 = -\sum_{\varrho\lambda} g_{\mu\varrho}^0 (\mathcal{G}_\varrho|\mathcal{G}_\lambda)^1 g_{\lambda\nu}^0 . \tag{12.1.12}$$

The perturbation-theory expansion (12.1.6) of the scalar product implies a corresponding expansion of the projection operators P and Q,

$$\mathsf{P} = \mathsf{P}^0 + \mathsf{P}^1 + \ldots ,^3 \tag{12.1.13}$$

where (for arbitrary $\mathcal{X}$)

$$\mathsf{P}^0|\mathcal{X}) = \sum_{\mu\nu} |\mathcal{G}_\mu)\, g_{\mu\nu}^0 (\mathcal{G}_\nu|\mathcal{X})^0 , \tag{12.1.14}$$

$$\mathsf{P}^1|\mathcal{X}) = \sum_{\mu\nu} |\mathcal{G}_\mu)\, g_{\mu\nu}^1 (\mathcal{G}_\nu|\mathcal{X})^0 + \sum_{\mu\nu} |\mathcal{G}_\mu)\, g_{\mu\nu}^0 (\mathcal{G}_\nu|\mathcal{X})^1 , \tag{12.1.15}$$

etc.

defines the projectors. Using (11.2.16) for the matrix $\Omega_{\mu\nu}^0$, it follows from (12.1.14) that P^0 commutes with L^0, i.e. that

$$\mathsf{P}^0\mathsf{L}^0 = \mathsf{L}^0\mathsf{P}^0 \quad \text{and} \quad \mathsf{Q}^0\mathsf{L}^0 = \mathsf{L}^0\mathsf{Q}^0 \tag{12.1.16}$$

holds.

[3] From the idempotence of $(\mathsf{P}^0 + \mathsf{P}^1 + \ldots)(\mathsf{P}^0 + \mathsf{P}^1 + \ldots) = \mathsf{P}^0 + \mathsf{P}^1 + \ldots$ it follows that

$$\mathsf{P}^0\mathsf{P}^0 = \mathsf{P}^0 , \quad \mathsf{P}^1\mathsf{P}^0 + \mathsf{P}^0\mathsf{P}^1 = \mathsf{P}^1 \tag{12.1.13'}$$

etc.

12.1.3 The Leading Terms of a Perturbation-Theory Expansion of $\Omega_{\nu\mu}$ and $\gamma_{\nu\mu}(t)$

In the perturbation-theoretical expansion of $\Omega_{\nu\mu}$ in $\mathscr{H}^1$,

$$\Omega_{\nu\mu}=\Omega^0_{\nu\mu}+\Omega^1_{\nu\mu}+\Omega^2_{\nu\mu}+\dots \tag{12.1.17}$$

$\Omega^0_{\nu\mu}$ is given by (12.1.2) and the first perturbation correction, $\Omega^1_{\nu\mu}$, according to (12.1.4) by

$$\boxed{\Omega^1_{\nu\mu}=\sum_{\varrho} g^0_{\nu\varrho}(\mathscr{G}_\varrho|\mathsf{L}^1\mathscr{G}_\mu)^0} \ . \tag{12.1.18}$$

Here, the products $(\mathscr{G}_\varrho|\mathsf{L}_1\mathscr{G}_\mu)^0$ follow from (12.1.8) by making the replacement $\mathscr{G}_\mu\to\mathsf{L}^1\mathscr{G}_\mu$. In the zeroth-order scalar product, L^1 is not a Hermitian superoperator. Equation (12.1.18) can*not* be treated according to the Kubo identity (8.2.18). The second-order perturbation term[4] for $\Omega_{\nu\mu}$ follows from (12.1.4):

$$\Omega^2_{\nu\mu}=\sum_{\varrho}\{g^0_{\nu\varrho}(\mathscr{G}_\varrho|\mathsf{L}^1\mathscr{G}_\mu)^1+g^1_{\nu\varrho}(\mathscr{G}_\varrho|\mathsf{L}^1\mathscr{G}_\mu)^0\}\ , \tag{12.1.19}$$

in this expression $g^1_{\nu\varrho}$ is determined by $(\mathscr{G}_\mu|\mathscr{G}_\varrho)^1$ [cf. (12.1.12)].

One can see from (12.1.5) that the *residual force* has the operator L^1 as factor. The expansion

$$f_\mu=f^1_\mu+\dots \tag{12.1.20}$$

is therefore given by the expansion of Q. The first-order term is thus

$$f^1_\mu=\mathrm{i}\mathsf{Q}^0\mathsf{L}^1\mathscr{G}_\mu\ . \tag{12.1.21}$$

The perturbation expansion of the *time evolution operator* for the residual force, $\exp(\mathrm{i}\mathsf{Q}\mathsf{L}t)$, begins with a zeroth-order perturbation term,

$$\mathrm{e}^{\mathrm{i}\mathsf{QL}t}\mathsf{Q}=\mathrm{e}^{\mathrm{i}\mathsf{Q}^0\mathsf{L}^0t}\mathsf{Q}^0+\dots=\mathrm{e}^{\mathrm{i}\mathsf{L}^0t}\mathsf{Q}^0+\dots\ . \tag{12.1.22}$$

In the second equation, we have used the fact that Q^0 commutes with L^0 [compare (12.1.16)] and Q^0 may therefore be removed from the exponent. In the expansion of the time-dependent residual force,

$$f_\mu(t)=\mathrm{e}^{\mathrm{i}\mathsf{QL}t}f_\mu=f^1_\mu(t)+\dots \tag{12.1.23}$$

the first term is then found to be likewise linear in the perturbation:

$$f^1_\mu(t)=\mathrm{i}\mathrm{e}^{\mathrm{i}\mathsf{L}^0t}\mathsf{Q}^0\mathsf{L}^1\mathscr{G}_\mu \tag{12.1.24}$$

[4] The reason that we carry out the expansion for $\Omega_{\nu\mu}$ up to second order may be seen from (12.1.25).

so that in the correlation function $(\mathscr{f}_\varrho|\mathscr{f}_\mu(t))$, the first term $(\mathscr{f}^1_\varrho|\mathscr{f}^1_\mu(t))^0$ is *quadratic* in L^1.

According to the definition (11.2.24) of the *memory function* $\gamma_{\nu\mu}(t)$, its expansion thus also begins with the *second-order* term:

$$\gamma_{\nu\mu}(t)=\gamma^2_{\nu\mu}(t)+\ldots\,, \tag{12.1.25}$$

here, observing that:

$$\gamma^2_{\nu\mu}(t)=\sum_\varrho g^0_{\nu\varrho}(\mathscr{f}^1_\varrho|\mathscr{f}^1_\mu(t))^0\,, \tag{12.1.26}$$

one obtains by inserting (12.1.24) the result

$$\boxed{\gamma^2_{\nu\mu}(t)=\sum_\varrho g^0_{\nu\varrho}(\mathsf{Q}^0\mathsf{L}^1\mathscr{G}_\varrho|e^{i\mathsf{L}^0t}\mathsf{Q}^0\mathsf{L}^1\mathscr{G}_\mu)^0}\,. \tag{12.1.27}$$

Since the projection operator Q^0 is Hermitian in the scalar product $(\,|\,)^0$, it can be left out of *either* the bra *or* the ket owing to its idempotence and the fact that it commutes with L^0.

We make use of the following general relation, which results from the Hermiticity of L in $(\,|\,)$ (arbitrary $\mathscr{F}$ and $\mathscr{G}$):

$$(\mathsf{L}^1\mathscr{F}|\mathscr{G})^0=(\mathscr{F}|\mathsf{L}^1\mathscr{G})^0+(\mathscr{F}|\mathsf{L}^0\mathscr{G})^1-(\mathsf{L}^0\mathscr{F}|\mathscr{G})^1 \tag{12.1.28}$$

to rewrite $\gamma^2_{\nu\mu}(t)$ in the form

$$\begin{aligned}\gamma^2_{\nu\mu}(t)=\sum_\varrho g^0_{\nu\varrho}\{&(\mathscr{G}_\varrho|\mathsf{L}^1e^{i\mathsf{L}^0t}\mathsf{Q}^0\mathsf{L}^1\mathscr{G}_\mu)^0+(\mathscr{G}_\varrho|\mathsf{L}^0e^{i\mathsf{L}^0t}\mathsf{Q}^0\mathsf{L}^1\mathscr{G}_\mu)^1\\ &-(\mathsf{L}^0\mathscr{G}_\varrho|e^{i\mathsf{L}^0t}\mathsf{Q}^0\mathsf{L}^1\mathscr{G}_\mu)^1\}\,.\end{aligned} \tag{12.1.29}$$

This expression at first appears formally to be more complicated than (12.1.27). It can, however, be employed to combine $\gamma^2_{\nu\mu}(t)$ and $\Omega^2_{\nu\mu}$ in approximations (see the example in Sect. 16.3).

Exercise 12.1 Derive the first-order term of the perturbation expansion, $(\mathscr{F}|\mathscr{G})^1$, in terms of expectation values with $\mathscr{R}^0_\beta$. What is found in particular for $(\mathscr{A}_\nu|\mathscr{A}_\mu)^1$, if $\mathsf{L}^0\mathscr{A}_\nu=\Omega^0_\nu\mathscr{A}_\nu$ is valid?

12.2 Extending the Set of Observables in a Manner Appropriate to the Perturbation

In case the terms given in Sect. 12.1 in lowest-order perturbation theory are not sufficiently accurate, it is usually not very helpful to derive higher-order terms using the exact formulas for $\Omega_{\nu\mu}$ and $\gamma_{\nu\mu}(t)$, since the latter quickly become rather complicated. Instead, it is then preferable to *extend* the set of observables $\{\mathscr{G}\}$. While L^0 transforms the subspace $\mathfrak{L}_{\{\mathscr{G}\}}$ back onto itself, according to (12.1.2), this is not true of L^1. However, we can decompose $\mathsf{L}^1\mathscr{G}_\mu$ into a sum of observables $\mathscr{G}_{\bar{\nu}}$ which span a new subspace $\mathfrak{L}_{\{\bar{\mathscr{G}}\}}$ that is invariant with respect to L^0. This subspace, in general, also contains the subspace $\mathfrak{L}_{\{\mathscr{G}\}}$. For all $\mathscr{G}_\mu \in \mathfrak{L}_{\{\mathscr{G}\}}$, the following relations thus hold:

$$\mathsf{L}^1\mathscr{G}_\mu = \sum_{\bar{\nu}} \mathscr{G}_{\bar{\nu}} c_{\bar{\nu}\mu} = \sum_{\nu} \mathscr{G}_\nu c_{\nu\mu} + \sum_{\nu'} \mathscr{G}_{\nu'} c_{\nu'\mu} \ , \tag{12.2.1}$$

where we denote by $\mathscr{G}_{\nu'}$ those basis vectors[5] of $\mathfrak{L}_{\{\bar{\mathscr{G}}\}}$ which are needed *in addition* to the basis vectors[6] $\mathscr{G}_\nu$ of $\mathfrak{L}_{\{\mathscr{G}\}}$.

12.2.1 The Mori Equations for the Extended Set of Observables

In what follows[7], it will prove expedient to choose the additional observables $\mathscr{G}_{\mu'}$ which extend the $\mathscr{G}_\mu$ in $\mathfrak{L}_{\{\bar{\mathscr{G}}\}}$ in such a way that they are *orthogonal* to $\mathfrak{L}_{\{\mathscr{G}\}}$, and to denote the subspace formed by these $\mathscr{G}_{\mu'}$ by $\mathfrak{L}_{\{\mathscr{G}'\}}$. For

$$\mathscr{G}_\mu \in \mathfrak{L}_{\{\mathscr{G}\}} \quad \text{and} \quad \mathscr{G}_{\mu'} \in \mathfrak{L}_{\{\mathscr{G}'\}} \tag{12.2.2}$$

we then have

$$(\mathscr{G}_\mu|\mathscr{G}_{\mu'}) = 0 \quad \text{and} \quad g_{\mu\mu'} = 0 \ . \tag{12.2.3}$$

The property that both $\mathfrak{L}_{\{\mathscr{G}\}}$ and $\mathfrak{L}_{\{\bar{\mathscr{G}}\}}$ are *invariant subspaces* of L^0 implies that

$$\mathsf{L}^0\mathscr{G}_\mu \in \mathfrak{L}_{\{\mathscr{G}\}} \ , \tag{12.2.4}$$

$$\mathsf{L}^0\mathscr{G}_{\bar{\mu}} \in \mathfrak{L}_{\{\bar{\mathscr{G}}\}} \ , \tag{12.2.5}$$

where the expansion coefficients of $\mathsf{L}^0\mathscr{G}_\mu$ are the same as those in (12.1.2),

$$\mathsf{L}^0\mathscr{G}_\mu = \sum_\nu \mathscr{G}_\nu \Omega^0_{\nu\mu} \ . \tag{12.2.6}$$

Furthermore, according to (12.2.1 and 4),

$$\mathsf{L}\mathscr{G}_\mu \in \mathfrak{L}_{\{\bar{\mathscr{G}}\}} \ , \tag{12.2.7}$$

while this is not required for $\mathsf{L}\mathscr{G}_{\mu'}$.

---

[5] One could choose the basis vectors $\mathscr{G}_{\nu'}$ in such a way that they themselves define a subspace which is invariant with respect to L^0. This is, however, *not* done in the following sections.

[6] The whole procedure outlined in this section has practical significance only when $\mathfrak{L}_{\{\bar{\mathscr{G}}\}}$ does not include the total Liouville space $\mathfrak{L}$.

[7] *None* of the equations in Sect. 12.2.1 contains approximations.

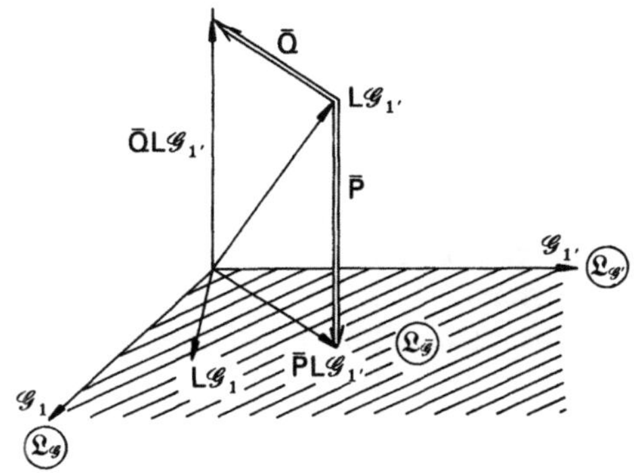

Fig. 12.1. An intuitive picture of the extension of the set of $\mathscr{G}_\mu$

For the *extended set* of observables

$$\mathscr{G}_{\bar\mu}\in\mathfrak{L}_{\{\bar{\mathscr{G}}\}}\ , \tag{12.2.8}$$

we form an enlarged correlation matrix [corresponding to (11.4.1)]:

$$\Xi_{\bar\nu\bar\mu}(t)=\sum_{\bar\varrho}g_{\bar\nu\bar\varrho}(\mathscr{G}_{\bar\varrho}|\mathscr{G}_{\bar\mu}(t))\ , \tag{12.2.9}$$

which satisfies the Mori integro-differential equation [corresponding to (11.4.3)],

$$\dot{\Xi}_{\bar\nu\bar\mu}(t)=\sum_{\bar\lambda}\Xi_{\bar\nu\bar\lambda}(t)\mathrm{i}\Omega_{\bar\lambda\bar\mu}-\int_0^t\sum_{\bar\lambda}\Xi_{\bar\nu\bar\lambda}(t-t')\gamma_{\bar\lambda\bar\mu}(t')\,dt'\ . \tag{12.2.10}$$

The preconditions imply several simplifications in the expressions for $\Omega_{\bar\lambda\bar\mu}$ and $\gamma_{\bar\lambda\bar\mu}(t)$. First, due to (12.2.3), the enlarged projection operator $\bar{\mathsf{P}}=\sum_{\bar\mu\bar\nu}|\mathscr{G}_{\bar\mu})g_{\bar\mu\bar\nu}(\mathscr{G}_{\bar\nu}|$ which projects onto $\mathfrak{L}_{\{\bar{\mathscr{G}}\}}$ can be decomposed into

$$\bar{\mathsf{P}}=\mathsf{P}+\mathsf{P}'\ , \tag{12.2.11}$$

where P projects onto $\mathfrak{L}_{\{\mathscr{G}\}}$ and P' onto $\mathfrak{L}_{\{\mathscr{G}'\}}$. The matrix elements may be grouped into four types:

$$\boxed{\Omega_{\lambda\mu}=\Omega^0_{\lambda\mu}+\sum_{\varrho}g_{\lambda\varrho}(\mathscr{G}_\varrho|\mathsf{L}^1\mathscr{G}_\mu)\ \rightarrow\boldsymbol{\Omega}_{00}}\ , \tag{12.2.12}$$

$$\boxed{\Omega_{\lambda\mu'}=\sum_{\varrho}g_{\lambda\varrho}(\mathsf{L}^1\mathscr{G}_\varrho|\mathscr{G}_{\mu'})\ \rightarrow\boldsymbol{\Omega}_{01}}\ , \tag{12.2.13}$$

$$\boxed{\Omega_{\lambda'\mu}=\sum_{\varrho'}g_{\lambda'\varrho'}(\mathscr{G}_{\varrho'}|\mathsf{L}^1\mathscr{G}_\mu)\ \rightarrow\boldsymbol{\Omega}_{10}}\ , \tag{12.2.14}$$

$$\boxed{\Omega_{\lambda'\mu'}\equiv\sum_{\varrho'}g_{\lambda'\varrho'}(\mathscr{G}_{\varrho'}|\mathsf{L}^0+\mathsf{L}^1|\mathscr{G}_{\mu'})\rightarrow\boldsymbol{\Omega}_{11}}\ . \tag{12.2.15}$$

The right-hand sides indicate a transition (→) to a notation using submatrices

$$\begin{pmatrix} \Omega_{\lambda\mu} & \Omega_{\lambda\mu'} \\ \Omega_{\lambda'\mu} & \Omega_{\lambda'\mu'} \end{pmatrix} \rightarrow \begin{pmatrix} \boldsymbol{\Omega}_{00} & \boldsymbol{\Omega}_{01} \\ \boldsymbol{\Omega}_{10} & \boldsymbol{\Omega}_{11} \end{pmatrix}, \qquad (12.2.16)$$

which will prove to be useful in the following section. $\boldsymbol{\Omega}_{00}$ and $\boldsymbol{\Omega}_{11}$ are square matrices which do not necessarily have the same dimensions; $\boldsymbol{\Omega}_{01}$ and $\boldsymbol{\Omega}_{10}$ are, in general, rectangular matrices.

In the extended set, the residual forces vanish for each observable $\mathcal{G}_\mu \in \mathfrak{L}_{\{\mathcal{G}\}}$ as a result of (12.2.7):

$$\bar{\mathsf{Q}}\mathsf{L}\mathcal{G}_\mu = \mathcal{O} \,. \qquad (12.2.17)$$

However, in general $\bar{\mathsf{Q}}\mathsf{L}\mathcal{G}_{\mu'} \neq \mathcal{O}$, since $\mathsf{L}\mathcal{G}_{\mu'}$ need not lie in $\mathfrak{L}_{\{\mathcal{G}\}}$. This implies the following result for the matrix elements $\gamma_{\bar{\lambda}\bar{\mu}}(t)$ of the enlarged memory matrix:

$$\boxed{\begin{array}{l} \gamma_{\bar{\lambda}\mu}(t) = 0 \\ \gamma_{\lambda\mu'}(t) = 0 \end{array} \rightarrow \begin{pmatrix} \mathbf{0} & \mathbf{0} \\ \mathbf{0} & \gamma_{11}(t) \end{pmatrix}}\,, \qquad \begin{array}{r} (12.2.18) \\ (12.2.19) \end{array}$$

where we have taken the orthogonality condition (12.2.3) into account. Therefore, only the matrix elements $\gamma_{\lambda'\mu'}(t)$ differ from zero. As a result of $\bar{\mathsf{Q}}\mathsf{L}^0\mathcal{G}_{\mu'} = 0$, the associated quantum-statistical expression for $\gamma_{\nu'\mu'}(t)$ is simplified to [corresponding to (11.2.23)]:

$$\boxed{\gamma_{\lambda'\mu'}(t) = \sum_{\varrho'} g_{\lambda'\varrho'}(\bar{\mathsf{Q}}\mathsf{L}^1\mathcal{G}_{\varrho'}|e^{i\bar{\mathsf{Q}}\mathsf{L}t}\bar{\mathsf{Q}}\mathsf{L}^1\mathcal{G}_{\mu'}) \rightarrow \gamma_{11}(t)}\,, \qquad (12.2.20)$$

whose dynamics are thus determined by $\exp(i\bar{\mathsf{Q}}\mathsf{L}t)$.

In the Mori equations (12.2.10), we find a system of *coupled* integro-differential equations for the enlarged correlation matrix:

$$\begin{pmatrix} \Xi_{\nu\mu}(t) & \Xi_{\nu\mu'}(t) \\ \Xi_{\nu'\mu}(t) & \Xi_{\nu'\mu'}(t) \end{pmatrix} \rightarrow \begin{pmatrix} \boldsymbol{\Xi}_{00}(t) & \boldsymbol{\Xi}_{01}(t) \\ \boldsymbol{\Xi}_{10}(t) & \boldsymbol{\Xi}_{11}(t) \end{pmatrix}. \qquad (12.2.21)$$

If one is interested only in the dynamics of the original observables $\mathcal{G}_\mu(t)$, or only in their correlation functions[8] $\Xi_{\nu\mu}(t) \rightarrow \boldsymbol{\Xi}_{00}(t)$, the remaining parts $\boldsymbol{\Xi}_{01}(t)$, $\boldsymbol{\Xi}_{10}(t)$, and $\boldsymbol{\Xi}_{11}(t)$ must be *eliminated* from (12.2.10). This may be most simply carried out

[8] Owing to the orthogonality condition (12.2.3), the part

$$\Xi_{\nu\mu}(t) = \sum_{\bar{\varrho}} g_{\nu\bar{\varrho}}(\mathcal{G}_{\bar{\varrho}}|\mathcal{G}_\mu(t)) = \sum_{\varrho} g_{\nu\varrho}(\mathcal{G}_\varrho|\mathcal{G}_\mu(t))$$

of the enlarged correlation matrix is identical with the smaller correlation matrix of the original set of observables $\{\mathcal{G}\}$.

using the one-sided Fourier-(Laplace-) transformed Mori equations, since they form a system of algebraic equations [compare (11.4.6)]. According to (12.2.12–16), (12.2.18), and (12.2.19), the transformed Mori equations are given by:

$$\begin{pmatrix} \Xi_{00}(z) & \Xi_{01}(z) \\ \Xi_{10}(z) & \Xi_{11}(z) \end{pmatrix} \left[\begin{pmatrix} z\cdot\mathbf{1} & \mathbf{0} \\ \mathbf{0} & z\cdot\mathbf{1} \end{pmatrix} + \begin{pmatrix} \boldsymbol{\Omega}_{00} & \boldsymbol{\Omega}_{01} \\ \boldsymbol{\Omega}_{10} & \boldsymbol{\Omega}_{11}+\mathrm{i}\gamma_{11}(z) \end{pmatrix} \right] = \mathrm{i} \begin{pmatrix} \mathbf{1} & \mathbf{0} \\ \mathbf{0} & \mathbf{1} \end{pmatrix}, \tag{12.2.22}$$

from which we infer the matrix equations

$$\begin{aligned} &\Xi_{00}(z)(z\cdot\mathbf{1}+\boldsymbol{\Omega}_{00})+\Xi_{01}(z)\boldsymbol{\Omega}_{10}=\mathrm{i}\cdot\mathbf{1}\;, \\ &\Xi_{00}(z)\boldsymbol{\Omega}_{01}+\Xi_{01}(z)(z\cdot\mathbf{1}+\boldsymbol{\Omega}_{11}+\mathrm{i}\gamma_{11}(z))=\mathbf{0}\;. \end{aligned} \tag{12.2.23}$$

From these, one finds the solution for the matrix $\Xi_{00}(z)$ of interest:

$$\boxed{\Xi_{00}(z)=\mathrm{i}\left[z\cdot\mathbf{1}+\boldsymbol{\Omega}_{00}+\mathrm{i}\boldsymbol{\Omega}_{01}\frac{\mathrm{i}}{z\cdot\mathbf{1}+\boldsymbol{\Omega}_{11}+\mathrm{i}\gamma_{11}(z)}\boldsymbol{\Omega}_{10}\right]^{-1}}\;. \tag{12.2.24}$$

We see by comparison with (11.4.8) that the correlation matrix $\Xi(z)=\Xi_{00}(z)$ [i.e. comprising the matrix elements $\Omega_{\nu\mu}(z)$ with respect to the original observables $\mathscr{G}_\mu$] is given by the frequency matrix

$$\boxed{\boldsymbol{\Omega}=\boldsymbol{\Omega}_{00}} \tag{12.2.25}$$

and the Fourier transformed memory matrix

$$\boxed{\gamma(z)=\boldsymbol{\Omega}_{01}\frac{\mathrm{i}}{z\cdot\mathbf{1}+\boldsymbol{\Omega}_{11}+\mathrm{i}\gamma_{11}(z)}\boldsymbol{\Omega}_{10}}\;. \tag{12.2.26}$$

These expressions are exact. The frequency matrix $\boldsymbol{\Omega}$ agrees with (12.1.4). The memory matrix $\gamma(z)$ is, to be sure, in a form which differs from the usual expression $\boldsymbol{gL}(z)$ (11.3.16); but the two are in fact identical, as may be verified by explicit calculation.

12.2.2 Perturbation-Theoretical Approximations

The perturbation expansion of $\boldsymbol{\Omega}_{00}$ in $\mathscr{H}^1$ leads, according to (12.2.12), to the expression

$$\boldsymbol{\Omega}_{00}=\boldsymbol{\Omega}_{00}^{0}+\boldsymbol{\Omega}_{00}^{1}+\dots\;, \tag{12.2.27}$$

i.e. to the same expressions as found in (12.1.17 and 12.1.18). If an expansion in powers of $\mathscr{H}^1$ is carried out in (12.2.26), the result is

$$\gamma(z)=(\boldsymbol{\Omega}_{01}^1+\boldsymbol{\Omega}_{01}^2+\dots)\frac{\mathrm{i}}{z\cdot\mathbf{1}+\boldsymbol{\Omega}_{11}^0+\boldsymbol{\Omega}_{11}^1+\dots\mathrm{i}\gamma_{11}^2(z)+\dots}$$

$$\times(\boldsymbol{\Omega}_{10}^1+\boldsymbol{\Omega}_{10}^2+\dots)\ . \qquad (12.2.28)$$

In the *second order* in the perturbation, we find simply

$$\gamma^2(z)=\boldsymbol{\Omega}_{01}^1\frac{\mathrm{i}}{z\cdot\mathbf{1}+\boldsymbol{\Omega}_{11}^0}\boldsymbol{\Omega}_{10}^1\ . \qquad (12.2.29)$$

This expression may be quite naturally rewritten to give (12.1.27). Equation (12.2.28) is clearly particularly suitable for the expansion of $\gamma(z)$ to *higher orders* in $\mathscr{H}^1$; for example, the *third* order may be expressed using only approximations for the enlarged $\boldsymbol{\Omega}$ matrix. A contribution from $\gamma_{11}^2(z)$ occurs only in the *fourth* order. In order to calculate $\boldsymbol{\Omega}_{01}^1, \boldsymbol{\Omega}_{11}^1, \dots, \gamma_{11}^2(z), \dots$, we can again apply the perturbation-theoretical formulas (12.1.18) and (12.1.27), *now with respect to the extended set* $\{\bar{\mathscr{G}}\}$; here, however, one must be aware of the fact that the observables $\mathscr{G}_{\mu'}$ employed, due to the orthogonality condition (12.2.3), can themselves depend upon the perturbation $\mathscr{H}^1$.

The exact equation (12.2.26) is often a better starting point for approximations than (11.2.21) or (11.3.15 and 16), based on the original small set of observables $\mathscr{G}_\mu$. Nevertheless, the calculation of the enlarged $\boldsymbol{\Omega}$ matrices and the inversion in (12.2.24) usually requires considerable effort, particularly in cases where the set of observables $\mathscr{G}_{\mu'}$ is very large.

13. The Transition to Differential Equations with Damping

13.1 One Slow Hermitian Observable

13.1.1 Separation of the Time Scales; Simplified Argument

The memory term in the Mori integro-differential equations (11.4.3) for the expectation values $\langle \mathcal{G}_\mu \rangle(t)$ or the correlation functions $\Xi_{\nu\mu}(t)$ can, under certain conditions, be approximated by a damping term. In order to demonstrate the basic principle most clearly, we shall first consider the simplest case of one Hermitian observable $\mathcal{G} = \mathcal{G}^\dagger$ in a system Σ. We shall then concern ourselves in Sects. 13.2 and 13.3 with the generalization to a set of observables.

In the case of one Hermitian observable, according to (11.2.15) the frequency Ω vanishes, so that the time dependence of the correlation function $\Xi(t) = (\mathcal{G}|\mathcal{G}(t))/(\mathcal{G}|\mathcal{G})$ is determined by the memory function $\gamma(t)$:

$$\dot{\Xi}(t) = -\int_0^t \gamma(t')\Xi(t-t')\,dt' \; . \tag{13.1.1}$$

We can make general statements about $\Xi(t)$ for small times, for which

$$\Xi(t) = 1 - \gamma(0)\cdot\tfrac{1}{2}t^2 + \dots \tag{13.1.2}$$

holds. At longer times, as we shall see in an example in Sect. 16.1, a power law $t^{-\alpha}$ may be asymptotically valid. The exact behavior of $\Xi(t)$ depends upon all the details of the time dependence of $\gamma(t)$. If, however, we consider the case of a slow observable, for which the time scales of $\Xi(t)$ and $\gamma(t)$ are sufficiently different, we shall find that the details of the time dependence of $\gamma(t)$ are no longer important, but instead only the integral $\int_0^\infty \gamma(t)\,dt$ plays a role; furthermore, $\Xi(t)$ can be approximated by an exponential function for all times, in spite of the exact behavior (13.1.2) for $t \to 0$ and a possible asymptotic behavior $t^{-\alpha}$ for $t \to \infty$.

In order that the time scales of $\Xi(t)$ and $\gamma(t)$ be separable, we make the following assumptions concerning the memory function:

i) The memory function $\gamma(t)$ contains a smallness parameter ε^2 as a factor,

$$\gamma(t) = \varepsilon^2 \hat{\gamma}(t) \; .^1 \tag{13.1.3}$$

---

[1] Here, we take $\hat{\gamma}(t)$ to be independent of ε.

This parameter ε^2 can be derived from perturbation theory (cf. Chap. 12) or it can be the general result of a small coupling PLQ of the subspaces $\mathfrak{L}_{\mathscr{G}}$ and $\mathfrak{L}_{\perp\mathscr{G}}$. It can also be due to the denominator $(\mathscr{G}|\mathscr{G})$ in the definition of $\gamma(t)$.

ii) We further assume that the memory function $\gamma(t)$ becomes negligibly small after a characteristic correlation time τ_c (compare Fig. 13.1). We shall consider the possibility of a precise definition of τ_c in Sect. 13.1.2.

When we examine the exact integro-differential equation for $\Xi(t)$ (13.1.1) in view of the assumptions (i) and (ii), we find that assumption (i) implies that the variation $\dot{\Xi}(t)$ is proportional to ε^2, i.e. that $\Xi(t)$ changes slowly. When ε^2 is sufficiently small, $\Xi(t)$ changes only by a very little within a time interval $\Delta t < \tau_c$:

$$\Xi(t-\Delta t) = \Xi(t) + \ \dots \ , \quad \text{for} \quad \Delta t < \tau_c \ ; \tag{13.1.4}$$

this is physically equivalent to a separation of the time scales of $\gamma(t)$ and $\Xi(t)$. On the other hand, requirement (ii) implies that the integration over t' in (13.1.1) extend at most up to τ_c, since $\gamma(t')$ is taken to be negligibly small for $t' > \tau_c$. Together with (13.1.4), this means that we can replace $\Xi(t-t')$ in the integrand of (13.1.1) by $\Xi(t)$. For times $t > \tau_c$, we can then carry out the integral over t' formally up to ∞ instead of to τ_c, and thereby convert the integro-differential equation (13.1.1) into the simple differential equation

$$\dot{\Xi}(t) = -\Gamma\,\Xi(t) \ . \tag{13.1.5}$$

In this equation, the damping constant Γ(or the relaxation time τ_R) is given by the integral over $\gamma(t)$:

$$\boxed{\Gamma = \frac{1}{\tau_R} = \int_0^\infty \gamma(t)\,dt} \ .^2 \tag{13.1.6}$$

In the case of a Hermitian observable, the memory function $\gamma(t)$ is real (11.3.11); therefore, the damping Γ represents just the real part of the Fourier transform of $\gamma(\omega) = \int_0^\infty \gamma(t) \exp(\mathrm{i}\omega t) dt$ at $\omega = 0$, according to (13.1.6), and can thus not become negative [cf. (11.3.18)]. From the microscopic definition, we thus have the property:

$$\Gamma = \Gamma^* = \mathrm{Re}\,\gamma(\omega=0) = \frac{1}{(\mathscr{G}|\mathscr{G})} L'(\omega=0) \geqq 0 \ . \tag{13.1.7}$$

The validity of the transition from the integro-differential equation (13.1.1) to the differential equation (13.1.5) depends on the separation of the time scales

---

[2] The notation $\int_0^\infty \gamma(t) dt$ means that the thermodynamic limit of $\gamma(t)$ [cf. Sect. 4.2] is to be taken *before* the integration over t. Otherwise, the upper limit of the integral must be replaced by a time which is long compared to τ_c but still short compared to the recurrence time of $\gamma(t)$.

(13.1.4), i.e. on the smallness of ε^2 or of $\gamma(t)$. The solution of (13.1.5) together with (13.1.4) thus yields the following condition for the validity of the transition[3]:

$$\boxed{\Gamma\tau_c = \frac{\tau_c}{\tau_R} \ll 1} \quad . \tag{13.1.8}$$

The above considerations show how one arrives at a differential equation with a damping term. However, several questions have not been answered. For example, it is not evident whether one can neglect the time t' in $\Xi(t-t')$ of (13.1.1) in the case that $\gamma(t')$ makes a small contribution for $t' \gg \tau_c$, but still decays only very slowly for $t' \to \infty$. Furthermore, in this connection it is unclear whether one may apply the differential equation (13.1.5) beginning at $t=0$ with the exact initial condition $\Xi(t=0)=1$, or if it is only valid after $t=\tau_c$, as one would expect from the derivation. For these reasons, we shall carry out a mathematically exact treatment of the problem in Sect. 13.1.2, showing that the condition (13.1.8) is sufficient to obtain $\exp(-\Gamma t)$ as a very good approximation to $\Xi(t)$ over the entire time interval $t \geqq 0$, when τ_c is suitably defined.

To conclude this section, we wish to note that the approximate expression (13.1.5) can be obtained formally by replacing the memory function by a δ-function,

$$\gamma(t-t') = 2\delta(t-t')\int_0^\infty \gamma(t')dt' \ , \tag{13.1.9}$$

such as appears in stochastic theories for a Markovian process. The transition from (13.1.1) to (13.1.5) is therefore often termed the Markovian approximation, without referring to a stochastic process.

Another possibility of understanding how the approximate expression (13.1.5) comes about is to use τ_c/τ_R, i.e. the ratio of the time scales, as an expansion parameter. To this end, we start with the definition (13.1.6) of τ_R and write $\Xi(t)$ as a function of t/τ_R; or else we write $\gamma(t)$ as a function of t/τ_c,

$$\Xi(t) = f\left(\frac{t}{\tau_R}\right) , \qquad \gamma(t) = g\left(\frac{t}{\tau_c}\right) . \tag{13.1.10}$$

The integro-differential equation (13.1.1) with respect to the variable $x = t/\tau_R$, after making the substitution $t' = \tau_c x'$, is then given by

$$\frac{d}{dx}f(x) = -\frac{1}{\int_0^\infty g(x')dx'}\int_0^{(\tau_R/\tau_c)x} g(x')f\left(x - \frac{\tau_c}{\tau_R}x'\right)dx' \ . \tag{13.1.11}$$

[3] We have here assumed that $\Gamma \neq 0$. The condition of validity when $\Gamma = 0$ is contained in the exact treatment of Sect. 13.1.2.

This form demonstrates that $f(x)$ can also be considered as a function of the parameter τ_c/τ_R:

$$\frac{\partial}{\partial x} f\left(x, \frac{\tau_c}{\tau_R}\right) = -\frac{1}{\int_0^\infty g(x')dx'} \int_0^{(\tau_R/\tau_c)x} g(x') f\left(x - \frac{\tau_c}{\tau_R}x', \frac{\tau_c}{\tau_R}\right) dx' . \tag{13.1.12}$$

This suggests that when $\tau_c/\tau_R \ll 1$, the following approximation for $f(x, \tau_c/\tau_R)$ is valid:

$$f\left(x, \frac{\tau_c}{\tau_R}\right) = f(x, 0) + \dots . \tag{13.1.13}$$

In this approximation, using (13.1.11) with[4] $\tau_c/\tau_R = 0$, we find

$$\frac{\partial}{\partial x} f(x, 0) = -f(x, 0) , \tag{13.1.14}$$

i.e. [applying the definition (13.1.10)] precisely the differential equation (13.1.5).

13.1.2 Validity of the Approximation

In the previous section, we found the approximate expression $\exp(-\Gamma t)$ for the correlation function $\Xi(t)$ by employing simple arguments. We now want to separate this approximation $\exp(-\Gamma t)$ from the exact solution for $\Xi(t)$ and to investigate the remainder term. We shall discover that the latter cannot be larger than the ratio $2\tau_c/\tau_R$.

In order to find a suitable expression for the remainder term, we first introduce the integral over the memory function with a variable lower limit of integration t:

$$\psi(t) = \int_t^\infty \gamma(t')dt' . \tag{13.1.15}$$

Applying (13.1.6 and 7), we find

$$\psi(0) = \Gamma \geqq 0 . \tag{13.1.16}$$

The integro-differential equation (13.1.1) may then be put in the form

$$\dot{\Xi}(t) = -\Gamma \Xi(t) + \frac{d}{dt}\int_0^t \psi(t-t')\Xi(t')dt' . \tag{13.1.17}$$

We can read this equation formally as a linear differential equation for $\Xi(t)$ with the second term as an inhomogeneity. Integrating with the initial condition

[4] A justification for taking the limit within the integral arises from the results of Sect. 13.1.2.

$\Xi(0)=1$ yields

$$\Xi(t)=e^{-\Gamma t}+\int_0^t dt' e^{-\Gamma(t-t')}\frac{d}{dt'}\int_0^{t'}\psi(t'-t'')\Xi(t'')dt'' , \tag{13.1.18}$$

this represents a decomposition of $\Xi(t)$ into the approximation $\exp(-\Gamma t)$ previously found, and a remainder term $r(t)$:

$$\Xi(t)=e^{-\Gamma t}+r(t) . \tag{13.1.19}$$

The remainder term is computed using integration by parts over t' to be

$$r(t)=\int_0^t \psi(t-t'')\Xi(t'')dt''-\int_0^t dt' \Gamma e^{-\Gamma(t-t')}\int_0^{t'}\psi(t'-t'')\Xi(t'')dt'' . \tag{13.1.20}$$

To estimate the value of the remainder term (13.1.20), we start from the definition of $\Xi(t)$ as a scalar product,

$$\Xi(t)=\frac{1}{(\mathscr{G}|\mathscr{G})}(\mathscr{G}|e^{iLt}\mathscr{G}) \tag{13.1.21}$$

and use the Schwarz inequality,

$$(\mathscr{G}|e^{iLt}\mathscr{G})^2 \leqq (\mathscr{G}|\mathscr{G})(e^{iLt}\mathscr{G}|e^{iLt}\mathscr{G})=(\mathscr{G}|\mathscr{G})^2 , \tag{13.1.22}$$

from which we find an upper bound

$$|\Xi(t)|\leqq 1 . \tag{13.1.23}$$

With this bound, remembering that $\Gamma \geqq 0$, and estimating the integrand in (13.1.20), we find a result for $r(t)$:

$$|r(t)|\leqq \int_0^t |\psi(t-t'')|dt''+\int_0^t dt' \Gamma e^{-\Gamma(t-t')}\int_0^{t'}|\psi(t'-t'')|dt'' . \tag{13.1.24}$$

Changing the variables of integration and employing the following relations,

$$\begin{aligned}&\int_0^{t'}|\psi(t'-t'')|dt''=\int_0^{t'}|\psi(t'')|dt''\leqq\int_0^{t}|\psi(t'')|dt'' \quad \text{for} \quad t\geqq t' ,\\ &\int_0^t \Gamma e^{-\Gamma(t-t')}dt'=1-e^{-\Gamma t}\leqq 1\end{aligned} \tag{13.1.25}$$

we find an expression for the bounds of $r(t)$:

$$\boxed{|r(t)|\leqq 2\int_0^t|\psi(t'')|\,dt''\leqq 2\int_0^\infty|\psi(t'')|\,dt''} . \tag{13.1.26}$$

The deviation of the exact correlation function $\Xi(t)$ from the function $\exp(-\Gamma t)$ is therefore bounded by an integral property of $\psi(t)$, i.e. of $\gamma(t)$.

To interpret the second bound in (13.1.26), we define a correlation time τ_c in the sense of Sect. 13.1.1. If the memory function becomes exactly zero after a time τ_c, this is also true of the function $\psi(t)$, according to its definition (13.1.15). It is therefore reasonable to define the correlation time τ_c in such a way that a rectangle of altitude $|\psi(0)|=\psi(0)=\Gamma$ and of length τ_c has the area $\int_0^\infty |\psi(t)|\,dt$:

$$\tau_c \equiv \int\limits_0^\infty \frac{|\psi(t)|\,dt}{|\psi(0)|} = \frac{1}{\Gamma}\int\limits_0^\infty |\psi(t)|\,dt \ . \tag{13.1.27}$$

If we use (13.1.15) to estimate $\psi(t)$ using $\int_t^\infty |\gamma(t')|\,dt'$ and carry out an integration by parts over t, we obtain the following result in terms of the memory function:

$$\tau_c \leqq \frac{1}{\int\limits_0^\infty \gamma(t)\,dt}\int\limits_0^\infty |\gamma(t)|\,t\,dt \ .[5] \tag{13.1.28}$$

For $\gamma(t)$ positive, the equals sign is valid, so that τ_c represents an average time with $\gamma(t)$ as the weighting function. Using the definition of τ_c, (13.1.27), the remainder term in the simplest approximation (13.1.26) becomes

$$\boxed{r(t) \leqq 2\Gamma\tau_c = 2\frac{\tau_c}{\tau_R}} \ . \tag{13.1.29}$$

Up to this point, all the expressions are valid independently of the particular properties of $\gamma(t)$, i.e. the bounds for $r(t)$ can assume any value and possibly do not yield a reasonable result. However, if $\gamma(t)$ is constructed in such a way that

$$\frac{\tau_c}{\tau_R} \ll 1 \tag{13.1.30}$$

is valid, i.e. the time scales of $\exp(-\Gamma t)$ and $\gamma(t)$ are separated, then according to (13.1.19), $\exp(-\Gamma t)$ is a good approximation for $\Xi(t)$ over the whole time range. The absolute error can never be larger than $2\tau_c/\tau_R$; the correlation function thus lies in the range given by $\exp(-\Gamma t) \pm 2\tau_c/\tau_R$, a band of values having the breadth

[5] There are cases in which $\int_0^\infty |\psi(t)|\,dt$ has a finite value, while $\int_0^\infty |\gamma(t)|\,t\,dt$ diverges. The definition (13.1.27) of τ_c is therefore more appropriate than an apparently more evident one in terms of

$$\frac{1}{\int\limits_0^\infty |\gamma(t)|\,dt}\int\limits_0^\infty |\gamma(t)|\,t\,dt \ .$$

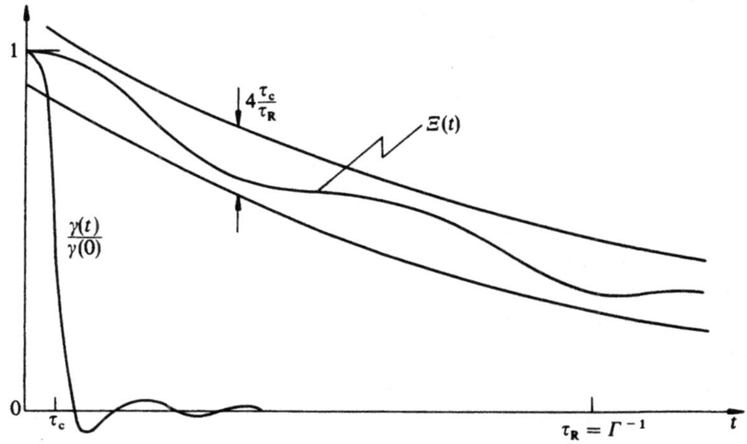

Fig. 13.1. Schematic plot of the memory function $\gamma(t)$ and correlation function $\Xi(t)$ within the bounds $\exp(-\Gamma t) \pm 2\tau_c/\tau_R$

$4\tau_c/\tau_R$ (cf. Fig. 13.1)[6]. For a sufficiently small ratio τ_c/τ_R, $\Xi(t)$ is therefore indistinguishable from $\exp(-\Gamma t)$. The actual value of τ_c/τ_R is known only very rarely in practical problems. However, (13.1.27 and 29) demonstrate that when a smallness parameter ε^2 [cf. (13.1.3)] is present as a factor in $\gamma(t)$ or in $\psi(t)$, the ratio

$$\frac{\tau_c}{\tau_R} = \int_0^\infty |\psi(t)|\,dt \sim \varepsilon^2 \tag{13.1.31}$$

can be made arbitrarily small for $\varepsilon^2 \to 0$, as long as the integral $\int_0^\infty |\psi(t)|\,dt$ does not diverge. If this condition is met, any slowly-decaying contributions to $\gamma(t)$ for $t \to \infty$ influence only the numerical value (13.1.31) of τ_c/τ_R, but not the general result.

The preceding considerations are applicable in the time domain. Employing the definition (13.1.27) of τ_c, we can also estimate the difference between the one-sided Fourier transform of $\Xi(t)$ and $\exp(-\Gamma t)$ and show that the relative error is given by (cf. Exercise 13.1):

$$\left| \frac{\Xi(\omega) - \dfrac{\mathrm{i}}{\omega + \mathrm{i}\Gamma}}{\Xi(\omega)} \right| \leq \frac{\tau_c}{\tau_R} \,. \tag{13.1.32}$$

The general relationships introduced in this section are applied to a concrete example with $\varepsilon^2 = m/M$ in Sect. 16.1.

Exercise 13.1 Prove (13.1.32).

[6] A better, t-dependent bound can be derived from the first inequality in (13.1.26).

13.2 A Set of Slow Observables

13.2.1 Carrying Out the Markovian Approximation

We have shown in Sect. 13.1 that the integro-differential equation for $\Xi(t)$ can be transformed into a differential equation with a damping term. These considerations may be applied without difficulty to a set of observables $\mathcal{G}_\mu$ or to the correlation matrix $\Xi_{\nu\mu}(t)$, so long as the latter varies only little within time intervals of the order of the correlation time of the memory matrix.

In analogy to the case of one Hermitian observable, we shall assume that the memory matrix is proportional to a smallness factor ε^2 and that all the matrix elements $\gamma_{\nu\mu}(t)$ can be neglected for times which are longer than a characteristic correlation time τ_c. In order that the time scales of $\Xi_{\nu\mu}(t)$ and $\gamma_{\nu\mu}(t)$ be separable, we must now however require in addition that the motion of $\Xi_{\nu\mu}(t)$ caused by the matrix $\Omega_{\nu\mu}$ also be slow. We shall therefore assume that the matrix elements $\Omega_{\nu\mu}$ are also proportional to a smallness parameter[7].

With these assumptions, we consider the system of integro-differential equations (11.4.3) for $\Xi_{\nu\mu}(t)$; in matrix notation, they are given by

$$\dot{\boldsymbol{\Xi}}(t) = \boldsymbol{\Xi}(t)\mathrm{i}\boldsymbol{\Omega} - \int_0^t \boldsymbol{\Xi}(t-t')\boldsymbol{\gamma}(t')dt' \ . \tag{13.2.1}$$

This equation shows that $\dot{\boldsymbol{\Xi}}(t)$ is proportional to the smallness parameters mentioned above. We wish to choose these sufficiently small so that all the $\Xi_{\nu\mu}(t)$ are nearly constant within a time interval $\Delta t < \tau_c$:

$$\Xi_{\nu\mu}(t-\Delta t) = \Xi_{\nu\mu}(t) + \ \dots \ . \tag{13.2.2}$$

We may then replace the matrix $\Xi_{\nu\mu}(t-t')$ in the memory term of the integro-differential equation by $\Xi_{\nu\mu}(t)$, since the integration over t' contributes only for $t' < \tau_c$ due to the assumption about $\gamma_{\nu\mu}(t')$. For times $t > \tau_c$, the upper limit of integration t may, for the same reason, be replaced by ∞. In this approximation (Markovian approximation), the system of integro-differential equations (13.2.1) is transformed into the system of linear differential equations:

$$\boxed{\dot{\Xi}_{\nu\mu}(t) = \sum_\sigma \Xi_{\nu\sigma}(t)\mathrm{i}\Lambda_{\sigma\mu}} \ , \tag{13.2.3}$$

[7] One sees from the definitions of $\boldsymbol{\Omega}$ and $\boldsymbol{\gamma}$ that, for example, the case $\boldsymbol{\Omega} \sim \varepsilon$ and $\boldsymbol{\gamma} \sim \varepsilon^2$ can be achieved if $\mathsf{L}\mathcal{G}_\nu$ is proportional to ε and the calculations are limited to the lowest order in perturbation theory (cf. Sect. 12.1).

where the constant matrix $\Lambda_{\nu\mu}$ is determined by

$$\mathrm{i}\Lambda_{\nu\mu} = \mathrm{i}\Omega_{\nu\mu} - \int_0^\infty \gamma_{\nu\mu}(t)\,dt \ . \tag{13.2.4}$$

This evolution matrix $\Lambda_{\nu\mu}$ may be written in another form using the dynamic Onsager–Casimir coefficients $L_{\nu\mu}(\omega)$ at $\omega = 0$ (11.3.16) and defining the abbreviations:

$$\Delta\Omega_{\nu\mu} = -\sum_{\varrho} g_{\nu\varrho} L''_{\varrho\mu}(0) \ , \tag{13.2.5}$$

$$\Gamma_{\nu\mu} = \sum_{\varrho} g_{\nu\varrho} L'_{\varrho\mu}(0) \ . \tag{13.2.6}$$

The result is

$$\boxed{\mathrm{i}\Lambda_{\nu\mu} = \mathrm{i}\Omega_{\nu\mu} - \gamma_{\nu\mu}(\omega=0) = \mathrm{i}(\boldsymbol{\Omega} + \Delta\boldsymbol{\Omega})_{\nu\mu} - \Gamma_{\nu\mu}} \ . \tag{13.2.7}$$

As a result of the definitions (11.2.15) and (11.3.17), the matrices $(\boldsymbol{g}^{-1}(\boldsymbol{\Omega} + \Delta\boldsymbol{\Omega}))_{\nu\mu}$ and $(\boldsymbol{g}^{-1}\boldsymbol{\Gamma})_{\nu\mu}$ are Hermitian.

The formal solution of the system of differential equations (13.2.3), with the initial conditions $\Xi_{\nu\mu}(0) = \delta_{\nu\mu} \approx \Xi_{\nu\mu}(\tau_c)$, is

$$\Xi_{\nu\mu}(t) = (\mathrm{e}^{\mathrm{i}\Lambda t})_{\nu\mu} \ . \tag{13.2.8}$$

If all of the eigenvalues Λ_ϱ of the matrix Λ are different, it may be diagonalized according to a theorem from linear algebra by a similarity transformation $\boldsymbol{S}$:

$$\Lambda_{\nu\mu} = \sum_{\varrho} S_{\nu\varrho} \Lambda_\varrho S^{-1}_{\varrho\mu} \ , \quad (\Lambda_\varrho = \text{complex}) \ , \tag{13.2.9}$$

so that $\boldsymbol{\Xi}(t)$ can be written in the form

$$\Xi_{\nu\mu}(t) = \sum_{\varrho} S_{\nu\varrho} \mathrm{e}^{\mathrm{i}\Lambda_\varrho t} S^{-1}_{\varrho\mu} \ . \tag{13.2.10}$$

The case that two or more eigenvalues are identical can be treated by taking limits in (13.2.9) or (13.2.10).

The solution $\boldsymbol{\Xi}(t) = \exp(\mathrm{i}\Lambda t)$ which we have found is based upon the condition that $\boldsymbol{\Xi}(t-\tau_c)$ can be set $\approx \boldsymbol{\Xi}(t)$, i.e. that the matrix $\exp(-\mathrm{i}\Lambda\tau_c)$ does not differ very much from the unit matrix. We must therefore require that on application to a vector $\boldsymbol{c}$, the deviation of $\exp(-\mathrm{i}\Lambda\tau_c)$ from the unit matrix gives only a small contribution; this is equivalent to requiring that on introducing a vector norm and a matrix norm which is compatible with it[8], the condition

$$\frac{\|(\mathrm{e}^{-\mathrm{i}\Lambda\tau_c}-\mathbf{1})\boldsymbol{c}\|}{\|\boldsymbol{c}\|} \leqq \|\mathrm{e}^{-\mathrm{i}\Lambda\tau_c}-\mathbf{1}\| \ll 1 \tag{13.2.11}$$

is met. Applying the calculational rules for matrix norms, we find

$$\|\mathrm{e}^{-\mathrm{i}\Lambda\tau_c}-\mathbf{1}\| = \left\|\int_0^{\tau_c}\mathrm{e}^{-\mathrm{i}\Lambda t}(-\mathrm{i}\Lambda)dt\right\| \leqq \int_0^{\tau_c}\|\Lambda\|\,\mathrm{e}^{\|\Lambda\|t}dt = \mathrm{e}^{\|\Lambda\|\tau_c}-1 \tag{13.2.12}$$

therefore, condition (13.2.11) is met when

$$\boxed{\|\Lambda\|\,\tau_c = \|\mathrm{i}(\boldsymbol{\Omega}+\Delta\boldsymbol{\Omega})-\boldsymbol{\Gamma}\|\,\tau_c \ll 1} \tag{13.2.13}$$

holds, i.e. when the parameters of $\boldsymbol{\Omega}$ and $\gamma(t)$ are sufficiently small. Equation (13.2.13) is the generalization of relation (13.1.8) which we obtained for a Hermitian observable.

For the error in the Markovian approximation (13.2.8), $\|\Xi_{\text{exact}}(t) - \Xi_{\text{Markov}}(t)\|$, limits can be given in a manner analogous to the method used in the case of one Hermitian observable in Sect. 13.1.2. However, because of the frequency shift $\Delta\boldsymbol{\Omega}$ which occurs in (13.2.7), the result is more difficult to interpret[9]. We shall therefore not treat it here, but shall apply the condition (13.2.13) obtained qualitatively as a criterion for the validity of the Markovian approximation.

The preceding considerations apply to the correlation matrix $\Xi_{\nu\mu}(t)$. However, they are just as valid for the system of equations for the expectation values $\langle\delta\mathscr{G}_\nu\rangle(t)$ in relaxation processes (11.4.30), since this system is identical[10] with that for $\Xi_{\nu\mu}(t)$. With the given assumptions, we thus obtain the following equations for the expectation values $\langle\delta\mathscr{G}_\nu\rangle(t)$:

$$\frac{d}{dt}\langle\delta\mathscr{G}_\nu\rangle(t) = \sum_\mu \mathrm{i}\Lambda_{\mu\nu}\langle\delta\mathscr{G}_\mu\rangle(t) \tag{13.2.14}$$

---

[8] It is reasonable to introduce here the vector norm $\|\boldsymbol{c}\|^2 = \sum_{\nu\mu} c_\nu^*(\mathscr{G}_\nu|\mathscr{G}_\mu)c_\mu$, and, as the compatible matrix norm,

$$\|\boldsymbol{\Lambda}\|^2 = \text{maximum eigenvalue of } \{(\boldsymbol{g}^{-1/2}\boldsymbol{\Lambda}\boldsymbol{g}^{1/2})^\dagger(\boldsymbol{g}^{-1/2}\boldsymbol{\Lambda}\boldsymbol{g}^{1/2})\}, \quad \text{or}$$

$$\|\boldsymbol{\Lambda}\|^2 = \mathrm{Tr}(\boldsymbol{g}^{-1/2}\boldsymbol{\Lambda}\boldsymbol{g}^{1/2})^\dagger(\boldsymbol{g}^{-1/2}\boldsymbol{\Lambda}\boldsymbol{g}^{1/2}).$$

The norm $\|\Lambda\|$ is then invariant with respect to linear transformations of the observables $\mathscr{G}_\nu$ and is associated with $\mathfrak{L}_{\mathscr{G}}$.

[9] The deeper reason for this is that the detuning $\Delta\boldsymbol{\Omega}$ from (13.2.5) is not yet optimally matched to the exact solution (cf. Sect. 13.3), and it can thus strongly influence the error in the case of oscillatory solutions.

[10] In the definition of $\Xi(t)$ and in the projection operator P, the $\mathscr{G}_\mu$ must then be replaced by the $\delta\mathscr{G}_\mu$ [compare (11.4.32)].

with the formal solution

$$\langle\delta\mathcal{G}_\nu\rangle(t)=\sum_\mu(e^{i\Lambda t})_{\mu\nu}\langle\delta\mathcal{G}_\mu\rangle(0)\ . \tag{13.2.15}$$

This can also be found directly from the relationship between the expectation values $\langle\delta\mathcal{G}_\nu\rangle(t)$ and the correlation matrix (11.4.32).

13.2.2 Properties of the Markovian Approximation

The approximation (13.2.8) for $\Xi(t)$ retains two important properties of the exact solution, owing to the microscopic definition (13.2.4) of Λ. First, the Markovian approximation $\Xi(t)$ cannot increase unboundedly for $t\to\infty$. Second, the dynamic susceptibility matrix which can be calculated from $\Xi(t)$ has the property that the "absorption matrix" $\omega\chi''(\omega)$ derived from it [cf. (9.4.4)] is positive semidefinite.

Let us consider the first point in more detail. The time dependence of $\Xi(t)$ is, according to (13.2.10), determined by the eigenvalues Λ_ϱ of the matrix Λ. We now want to demonstrate that due to the definition of Λ, its eigenvalues Λ_ϱ can have no negative imaginary part. To this end we use the fact that [from (13.2.5) and (11.2.15)] the matrix $\boldsymbol{g}^{-1}(\boldsymbol{\Omega}+\Delta\boldsymbol{\Omega})$ is Hermitian, and [from (11.3.18)] the matrix $\boldsymbol{g}^{-1}\boldsymbol{\Gamma}=\boldsymbol{L}'(\omega=0)$ is Hermitian and positive semidefinite. Then, from the eigenvalue equation for Λ, multiplied by $\boldsymbol{g}^{-1}$:

$$\sum_\mu(\boldsymbol{g}^{-1}(\boldsymbol{\Omega}+\Delta\boldsymbol{\Omega}+i\boldsymbol{\Gamma}))_{\nu\mu}c_\mu^\varrho=\Lambda_\varrho\sum_\mu(\boldsymbol{g}^{-1})_{\nu\mu}c_\mu^\varrho\ , \tag{13.2.16}$$

after multiplication by $c_\nu^{\varrho *}$ and summation over ν, due to the conditions

$$\sum_{\nu\mu}c_\nu^{\varrho *}(\boldsymbol{g}^{-1}(\boldsymbol{\Omega}+\Delta\boldsymbol{\Omega}))_{\nu\mu}c_\mu^\varrho=\text{real} \tag{13.2.17}$$

and

$$\sum_{\nu\mu}c_\nu^{\varrho *}L'_{\nu\mu}(0)c_\mu^\varrho\geqq 0 \tag{13.2.18}$$

we obtain for the imaginary part of the eigenvalues Λ_ϱ the following expression:

$$\boxed{\operatorname{Im}\Lambda_\varrho=\frac{\sum_{\nu\mu}c_\nu^{\varrho *}L'_{\nu\mu}(0)c_\mu^\varrho}{\sum_{\nu\mu}c_\nu^{\varrho *}(\boldsymbol{g}^{-1})_{\nu\mu}c_\mu^\varrho}\geqq 0}\ . \tag{13.2.19}$$

Here, we have made use of the fact that

$$\sum_{\nu\mu}c_\nu^{\varrho *}(\boldsymbol{g}^{-1})_{\nu\mu}c_\mu^\varrho=\left(\sum_\nu c_\nu^\varrho\mathcal{G}_\nu\middle|\sum_\mu c_\mu^\varrho\mathcal{G}_\mu\right)>0 \tag{13.2.20}$$

is positive. The result (13.2.19) means, according to (13.2.10), that $\Xi(t)$ can contain no exponentially-increasing contributions and the one-sided Fourier transform $\Xi(z)=\int_0^\infty \Xi(t)\exp(\mathrm{i}zt)\,dt$ is holomorphic in the half plane $\mathrm{Im}\, z>0$ for complex z.

The fact that $\boldsymbol{g}^{-1}\boldsymbol{\Gamma}=\boldsymbol{L}'(0)$ is positive semidefinite also applies to the matrix $\omega\boldsymbol{\chi}''(\omega)$. To see this, we calculate the dynamic susceptibility matrix from (11.4.16). From (13.2.8), we have[11]:

$$\Xi(\omega)=\int_0^\infty \Xi(t)\mathrm{e}^{\mathrm{i}(\omega+\mathrm{i}\eta)t}\,dt=\mathrm{i}[\omega\cdot\mathbf{1}+\boldsymbol{\Lambda}]^{-1}\ .$$

From this, with $\boldsymbol{\chi}(\omega)=\beta\boldsymbol{g}^{-1}[\mathbf{1}+\mathrm{i}\omega\Xi(\omega)]$, it follows that

$$\boldsymbol{\chi}(\omega)=\beta\boldsymbol{g}^{-1}\frac{\boldsymbol{\Lambda}}{\omega\mathbf{1}+\boldsymbol{\Lambda}}=\beta\boldsymbol{g}^{-1}\frac{\boldsymbol{\Omega}+\Delta\boldsymbol{\Omega}+\mathrm{i}\boldsymbol{\Gamma}}{\omega\mathbf{1}+\boldsymbol{\Omega}+\Delta\boldsymbol{\Omega}+\mathrm{i}\boldsymbol{\Gamma}}\ . \tag{13.2.21}$$

This approximation for $\boldsymbol{\chi}(\omega)$ is thus derived from the exact formula (11.4.18) for $\boldsymbol{\chi}(\omega)$ by replacing $\boldsymbol{\gamma}(\omega)$ in it by $\boldsymbol{\gamma}(\omega=0)$. Taking the anti-Hermitian part of (13.2.21), we find, owing to $\boldsymbol{g}=\boldsymbol{g}^\dagger$,

$$\begin{aligned}\frac{1}{2\mathrm{i}}\{\boldsymbol{\chi}(\omega)-\boldsymbol{\chi}(\omega)^\dagger\}&=\frac{\beta}{2\mathrm{i}}\frac{1}{\omega\cdot\mathbf{1}+\boldsymbol{\Lambda}^\dagger}\{(\omega\cdot\mathbf{1}+\boldsymbol{\Lambda}^\dagger)\boldsymbol{g}^{-1}\boldsymbol{\Lambda}\\&\quad-\boldsymbol{\Lambda}^\dagger\boldsymbol{g}^{-1}(\omega\cdot\mathbf{1}+\boldsymbol{\Lambda})\}\frac{1}{\omega\mathbf{1}+\boldsymbol{\Lambda}}\\&=\frac{\beta\omega}{2\mathrm{i}}\frac{1}{\omega\cdot\mathbf{1}+\boldsymbol{\Lambda}^\dagger}(\boldsymbol{g}^{-1}\boldsymbol{\Lambda}-\boldsymbol{\Lambda}^\dagger\boldsymbol{g}^{-1})\frac{1}{\omega\cdot\mathbf{1}+\boldsymbol{\Lambda}}\ .\end{aligned} \tag{13.2.22}$$

In the expression for $\boldsymbol{\Lambda}=\boldsymbol{\Omega}+\Delta\boldsymbol{\Omega}+\mathrm{i}\boldsymbol{\Gamma}$, the matrices $\boldsymbol{g}^{-1}(\boldsymbol{\Omega}+\Delta\boldsymbol{\Omega})$ and $\boldsymbol{g}^{-1}\boldsymbol{\Gamma}=\boldsymbol{L}'(0)$ are Hermitian, so that

$$\omega\boldsymbol{\chi}''(\omega)=\beta\omega^2\frac{1}{\omega\cdot\mathbf{1}+\boldsymbol{\Lambda}^\dagger}\boldsymbol{L}'(0)\frac{1}{\omega\cdot\mathbf{1}+\boldsymbol{\Lambda}} \tag{13.2.23}$$

results. Therefore, for an arbitrary vector c_ν with $b_\nu=\sum_\lambda(1/(\omega\mathbf{1}+\boldsymbol{\Lambda}))_{\nu\lambda}c_\lambda$, the following relation holds:

$$\sum_{\nu\mu}c_\nu^*\,\omega\chi''_{\nu\mu}(\omega)c_\mu=\beta\omega^2\sum_{\nu\mu}b_\nu^*\,L'_{\nu\mu}(0)b_\mu\geqq 0\ . \tag{13.2.24}$$

This means that in the Markovian approximation, the fundamental relation of positive energy absorption in linear response [cf. (9.6.8)] is fulfilled.

[11] In the case that one of the values in $\mathrm{Im}\,\Lambda_\varrho$ is zero, the limit $\eta\to 0$ may be taken only after carrying out the integration.

Exercise 13.2 A magnetic system has the Hamiltonian $\mathscr{H}=\sum_{\nu=1}^{3}\mathscr{H}_\nu+\mathscr{H}^1 = \mathscr{H}^0+\mathscr{H}^1$, where $\mathscr{H}_1=-\mathscr{M}H$ represents the Zeeman energy. The partial Hamiltonian operators $\mathscr{H}_\nu$ mutually commute: $[\mathscr{H}_\nu, \mathscr{H}_\mu]=\mathcal{O}$. The perturbation $\mathscr{H}^1$ is considered to be small (for example $\mathscr{H}_2=$ the lattice; $\mathscr{H}_3=$ isotropic exchange; $\mathscr{H}^1=$ spin-lattice and dipole interactions.)

Determine the susceptibility $\chi_{MM}(\omega)$ using the projection operator onto the space $\delta\mathscr{H}_1$, $\delta\mathscr{H}_2$, $\delta\mathscr{H}_3$ in perturbation theory and the Markovian approximation for the memory matrix. What is the expression for $\chi_{MM}(\omega)$ in the frequency range $\Gamma_2 \ll \omega \ll \Gamma_3$, in the case that the eigenvalues of the damping matrix, $\Gamma_1<\Gamma_2<\Gamma_3$, allow the existence of such a range?

13.3 Modification of the Approximation Due to Rapid Oscillations

In many cases, one can find a set of observables $\mathscr{G}_\mu$ such that the memory matrix is still proportional to a smallness parameter, but this is no longer true of the $\boldsymbol{\Omega}$ matrix. The variation of the correlation matrix $\Xi(t)$ within time intervals of the order of the correlation time of the memory matrix, τ_c, can no longer be considered to be small. Nevertheless, the method described in Sect. 13.2 can still be applied by first splitting off a "rapidly" varying portion $\exp(\mathrm{i}\boldsymbol{\Omega}t)$ from $\Xi(t)$. The results obtained, however, raise some questions which must be regarded as still not completely answered; this is the reason why the following considerations, which in fact include those of Sect. 13.2, were not placed at the beginning of that section.

13.3.1 Principle

The procedure can be most clearly illustrated by considering the special case of one non-Hermitian observable. This case occurs when one wishes to treat a set of two observables $\mathscr{G}$ and $\mathscr{G}^\dagger$ which, for symmetry reasons, are not correlated in the Mori equations. The system of integro-differential equations for the matrix $\Xi_{\nu\mu}(t)$ then is decomposed into two uncoupled equations for the diagonal elements $\Xi_{11}(t)=(\mathscr{G}|\mathscr{G}(t))/(\mathscr{G}|\mathscr{G})$ and $\Xi_{22}(t)=(\mathscr{G}^\dagger|\mathscr{G}^\dagger(t))/(\mathscr{G}|\mathscr{G})$. These are identical to the equations which one obtains with the individual projectors onto $\mathscr{G}$ or $\mathscr{G}^\dagger$ (see example in Sect. 16.3).

We therefore start with *one* equation

$$\dot{\Xi}(t)=\Xi(t)\mathrm{i}\Omega-\int_0^t \Xi(t-t')\gamma(t')dt' \qquad (13.3.1)$$

for the function $\Xi(t)=(\mathscr{G}|\mathscr{G}(t))/(\mathscr{G}|\mathscr{G})$. The frequency

$$\Omega=[1/(\mathscr{G}|\mathscr{G})][1/\beta\hbar]\langle[\mathscr{G}^\dagger,\mathscr{G}]\rangle_\beta$$

is not generally zero, in contrast to Sect. 13.1. We assume the memory function again to be proportional to a smallness parameter and to become negligibly small at times $t' > \tau_c$. Setting

$$\Xi(t) = e^{i\Omega t}\xi(t) \ , \tag{13.3.2}$$

we find from (13.3.1) the following result for the function sought, $\xi(t)$:

$$\dot{\xi}(t) = -\int\limits_0^t \xi(t-t')e^{-i\Omega t'}\gamma(t')dt' \ . \tag{13.3.3}$$

This is an equation to which we can apply the methods of the Markovian approximation as in Sect. 13.1. The only difference is that the new integral kernel $\exp(-i\Omega t)\gamma(t)$ is complex. For this reason, when $\int_0^\infty \exp(-i\Omega t)\gamma(t)dt = \gamma(\omega = -\Omega)$ is decomposed into real and imaginary parts using the Onsager–Casimir coefficients:

$$\mathrm{Im}\int\limits_0^\infty e^{-i\Omega t}\gamma(t)dt = \frac{1}{(\mathcal{G}|\mathcal{G})}L''(-\Omega) \equiv -\Delta\Omega \ , \tag{13.3.4}$$

$$\mathrm{Re}\int\limits_0^\infty e^{-i\Omega t}\gamma(t)dt = \frac{1}{(\mathcal{G}|\mathcal{G})}L'(-\Omega) \equiv \Gamma \ , \tag{13.3.5}$$

we obtain for $\xi(t)$ the differential equation

$$\dot{\xi}(t) = (i\Delta\Omega - \Gamma)\xi(t) \tag{13.3.6}$$

or, by back transformation of this equation, the result for $\Xi(t)$:

$$\boxed{\dot{\Xi}(t) = \{i(\Omega + \Delta\Omega) - \Gamma\}\Xi(t)} \ . \tag{13.3.7}$$

The only condition presupposed by this equation is that $\xi(t-t')$ can be taken $\approx \xi(t)$ for $t' < \tau_c$, i.e. that

$$|i\Delta\Omega - \Gamma|\tau_c \ll 1 \tag{13.3.8}$$

is valid. There is no longer a condition on $|\Omega|\tau_c$. If, however, $|\Omega|\tau_c \ll 1$, then $\exp(-i\Omega t)$ can be set equal to 1 in the integrals (13.3.5 and 6), and we arrive back at the result of Sect. 13.2. In the case of one observable, it is not difficult to see that the damping Γ from (11.3.18) cannot be negative and that the imaginary part of the susceptibility, similarly to (13.2.23), can likewise not take on negative values.

A more exact investigation shows that it is in principle more favorable to define the frequency shift $\Delta\Omega$ somewhat differently. If we take instead of (13.3.2)

$$\Xi(t) = e^{-i(\Omega + \Delta\Omega)t}\xi(t) \tag{13.3.9}$$

with a still-undetermined $\Delta\Omega$, then we can carry out the Markovian approximation in the equation for $\xi(t)$ corresponding to (13.3.3), requiring that there be no more oscillatory contributions to $\xi_{\text{Markov}}(t)$. In this way, we obtain

$$\dot{\Xi}(t)=\{\mathrm{i}(\Omega+\Delta\Omega)-\Gamma\}\Xi(t) \ , \tag{13.3.10}$$

in which $\Delta\Omega$ is implicitly determined by the equation

$$\Delta\Omega=-\operatorname{Im}\int_0^\infty \gamma(t)\mathrm{e}^{-\mathrm{i}(\Omega+\Delta\Omega)t}\,dt=-\operatorname{Im}\gamma(\omega=-\Omega-\Delta\Omega) \tag{13.3.11}$$

and the damping is to be calculated from

$$\Gamma=\operatorname{Re}\int_0^\infty \gamma(t)\,\mathrm{e}^{-\mathrm{i}(\Omega+\Delta\Omega)t}\,dt \ . \tag{13.3.12}$$

If we expand $\Delta\Omega$ and Γ in terms of the smallness parameter of $\gamma(t)$, the first nonvanishing terms give just the previous results (13.3.4 and 5). The meaning of the approximation (13.3.10) can also be seen from the one-sided Fourier transform $\Xi(\omega)=\int_0^\infty \Xi(t)\exp(\mathrm{i}\omega t)\,dt$. Comparison of the result of the Markovian approximation and the exact formula (11.4.8) shows that the Markovian expression for $\Xi(t)$ arises through approximation of the function $\gamma(\omega)$ by its value at the frequency $\omega=-\Omega-\Delta\Omega$, at which $\Xi(\omega)$ has a pole if $\operatorname{Re}\gamma(\omega)$ is neglected (13.3.11)[12]. With the values of $\Delta\Omega$ and Γ determined by (13.3.10 and 11), one can apply the results of Sect. 13.1.2 to obtain a bound for $|\Xi_{\text{exact}}(t)-\Xi_{\text{Markov}}(t)|$. With the definition

$$\psi(t)=\int_t^\infty \gamma(t')\mathrm{e}^{-\mathrm{i}(\Omega+\Delta\Omega)t'}\,dt' \tag{13.3.13}$$

we obtain from (13.3.9) the exact result for $\xi(t)$:

$$\dot{\xi}(t)=-\Gamma\xi(t)+\frac{d}{dt}\int_0^t \psi(t-t')\xi(t')\,dt' \ . \tag{13.3.14}$$

From this, using $|\Xi_{\text{exact}}(t)-\Xi_{\text{Markov}}(t)|=|\xi_{\text{exact}}(t)-\xi_{\text{Markov}}(t)|$ as in Sect. 13.1.2, it follows that

$$|\Xi_{\text{exact}}(t)-\Xi_{\text{Markov}}(t)|<2\int_0^\infty|\psi(t)|\,dt\leq 2\int_0^\infty|\gamma(t)|t\,dt \ . \tag{13.3.15}$$

[12] In order to determine exactly whether a pole is present, one would have to know the analytic continuation of $\gamma(\omega)$ in the lower ω-half plane; this is, however, in the thermodynamic limit not in general given by the integral $\int_0^\infty \gamma(t)\exp(\mathrm{i}\omega t)\,dt$.

13.3.2 Formulation Using Matrices

The procedure which we have described by the example of one non-Hermitian variable will now be extended to a *set* of observables. For the memory matrix $\gamma(t)$, we make the assumption that the matrix elements are proportional to ε^2 and become negligible for $t > \tau_c$. Starting with the expression for the matrix $\Xi(t)$:

$$\Xi(t) = \xi(t)\,e^{i\Omega t} \tag{13.3.16}$$

and applying the same arguments which led us from (13.3.2) to (13.3.7), we go from the integro-differential equation

$$\dot{\Xi}(t) = \Xi(t) i\boldsymbol{\Omega} - \int_0^t \Xi(t-t')\gamma(t')dt' \tag{13.3.17}$$

to the differential equation

$$\dot{\Xi}(t) = \Xi(t) i\Lambda \;, \quad t > 0 \;. \tag{13.3.18}$$

Here, the matrix $i\Lambda$ is given by

$$i\Lambda = i\boldsymbol{\Omega} - \int_0^\infty e^{-i\Omega t'}\gamma(t')\,dt' \;. \tag{13.3.19}$$

In the special case $\|\boldsymbol{\Omega}\|\tau_c \ll 1$, $i\Lambda$ becomes equivalent to expression (13.2.4), which we have already discussed. Equation (13.3.18), with Λ defined by (13.3.19), has produced physically reasonable results in many cases. There are, however, several points which indicate that the right-hand side of the definition of Λ (13.3.19) cannot yet be considered as the optimal form:

i) In contrast to the case discussed in Sect. 13.2, the microscopic definition (13.3.19) of Λ by itself is not sufficient to guarantee that its eigenvalues cannot have negative imaginary parts and that therefore an unbounded increase of the solution $\Xi(t) = \exp(i\Lambda t)$ is not permitted as $t \to \infty$.

ii) If we integrate (13.3.18) using the initial condition $\Xi(0) = \mathbf{1}$, it is no longer certain that the absorption matrix $\omega\boldsymbol{\chi}''(\omega) = (\omega^2/2)\{\boldsymbol{\chi}\Xi(\omega) + \Xi(\omega)^\dagger\boldsymbol{\chi}\}$ is positive semidefinite.

iii) The approximation method with $\Xi(0) = \mathbf{1}$ violates the symmetry which derives from the Hermiticity of the Liouville operator (9.3.11):

$$\boldsymbol{g}\Xi(-t)^\dagger\boldsymbol{g}^{-1} = \Xi(t) \;. \tag{13.3.20}$$

Thus, if the considerations which led to (13.3.18 and 19) are applied to the case of negative times, we obtain on substitution of t by $-t$:

$$\frac{d}{dt}\Xi(-t) = \Xi(-t)\left\{-i\boldsymbol{\Omega} + \int_0^{-\infty} e^{-i\Omega t'}\gamma(t')\,dt'\right\} \;, \quad t > 0 \;. \tag{13.3.21}$$

From this, using the relations (11.2.16) and (11.3.11),

$$\begin{aligned} \gamma(t)^\dagger &= g^{-1}\gamma(-t)g \ , \\ \Omega^\dagger &= g^{-1}\Omega g \end{aligned} \tag{13.3.22}$$

and taking the Hermitian conjugate, for $g\Xi(-t)^\dagger g^{-1}$ we obtain the differential equation

$$\begin{aligned} \frac{d}{dt} g\Xi(-t)^\dagger g^{-1} &= \left\{ \mathrm{i}\Omega - \int_0^\infty \gamma(t'')\,\mathrm{e}^{-\mathrm{i}\Omega t''}\,dt'' \right\} g\Xi(-t)^\dagger g^{-1} \\ &= \mathrm{i}\tilde{\Lambda} g\Xi(-t)^\dagger g^{-1} \quad (t>0) \ . \end{aligned} \tag{13.3.23}$$

Its solution with $\Xi(0)=\mathbf{1}$ does not agree with the solution for $\Xi(t)$ from (13.3.18), since Λ and $\tilde{\Lambda}$ are, in general, different.

We shall discuss these difficulties by making use of a simple example. The condition for validity of (13.3.18) is, analogously to (13.3.8),

$$\left\| \int_0^\infty \mathrm{e}^{-\mathrm{i}\Omega t}\gamma(t)\,dt \right\| \tau_c \ll 1 \ .^{13} \tag{13.3.24}$$

We shall see that this condition limits the parameters which occur in the problem in such a way that the eigenvalues of Λ have no negative imaginary parts; furthermore, the difficulties (ii) and (iii) can be eliminated by making a minor change in the initial condition $\Xi(0)=\mathbf{1}$. To what extent these considerations can be generalized will not be further discussed here.

13.3.3 Discussion Based on the Damped Harmonic Oscillator

As a model system, we consider a harmonic oscillator $\Sigma^{(1)}$ with

$$\mathscr{H}^{(1)} = \frac{p^2}{2m} + \frac{m}{2}\omega_0^2 x^2 \ , \tag{13.3.25}$$

which is coupled via weak interaction $\mathscr{H}^{(12)}$ to another system $\Sigma^{(2)}$. The overall Hamiltonian in $\mathfrak{U}=\mathfrak{U}^1 \times \mathfrak{U}^2$ is thus given by:

$$\mathscr{H} = \mathscr{H}^{(1)} + \mathscr{H}^{(12)} + \mathscr{H}^{(2)} \ . \tag{13.3.26}$$

---

[13] Using the first norm given in Sect. 13.2, we find $\|\exp(\mathrm{i}\Omega t)\| = 1$, so that from the integral equation for $\xi(t)$:

$$\xi(t+\tau) = \xi(t) + \int_t^{t+\tau} \xi(t')\exp(\mathrm{i}\Omega t')(\mathrm{i}\Lambda - \mathrm{i}\Omega)\exp(-\mathrm{i}\Omega t')\,dt' \ ,$$

we obtain the bound

$$\|\xi(t)^{-1}\cdot\xi(t+\tau) - 1\| \leq \exp(\|\mathrm{i}\Lambda - \mathrm{i}\Omega\|\tau) - 1 = \exp\left(\left\| \int_0^\infty \exp(-\mathrm{i}\Omega t')\gamma(t')\,dt' \right\| \tau\right) - 1.$$

Here, we assume that $\mathscr{H}^{(12)}$ does not depend on the momentum p, so that $p = mi\mathsf{L}x$ is fulfilled. We choose as our set of observables

$$\mathscr{G}_1 = x \ , \qquad \mathscr{G}_2 = p \ .^{14} \tag{13.3.27}$$

The Liouville operator $\mathsf{L}^{(1)} + \mathsf{L}^{(2)}$, which corresponds to $\mathscr{H}^{(1)} + \mathscr{H}^{(2)}$, does not lead out of the space spanned by x and p, so that $\mathsf{QL}\mathscr{G}_\nu = \mathsf{QL}^{(12)}\mathscr{G}_\nu$ and therefore the memory matrix $\gamma_{\nu\mu}(t)$ contains a smallness parameter determined by $\mathsf{L}^{(12)}$. The oscillator frequency ω_0 is an independent parameter, which has nothing to do with the smallness parameter.

To demonstrate the theory of Sect. 13.3.2, we need to know the structures of the matrices $\boldsymbol{g}$, $\boldsymbol{\Omega}$, $\boldsymbol{\gamma}(t)$, and $\boldsymbol{\Lambda}$.

Employing the Kubo identity $(\mathscr{F}|\mathsf{L}\mathscr{G}) = (1/\beta\hbar)\langle[\mathscr{F}^\dagger, \mathscr{G}]\rangle_\beta$ and the commutator of x and p, and introducing the abbreviation

$$\frac{1}{\beta(x|x)} = m\alpha^2 \tag{13.3.28}$$

we find for the matrices $\boldsymbol{g}$ and $\mathrm{i}\boldsymbol{\Omega}$:

$$\boldsymbol{g} = \beta\begin{pmatrix} m\alpha^2 & 0 \\ 0 & \dfrac{1}{m}\end{pmatrix} ; \qquad \mathrm{i}\boldsymbol{\Omega} = \begin{pmatrix} 0 & -m\alpha^2 \\ \dfrac{1}{m} & 0\end{pmatrix} . \tag{13.3.29}$$

Due to $\mathsf{Q}\mathrm{i}\mathsf{L}x = (1/m)\mathsf{Q}p = \mathcal{O}$, only the matrix element $\gamma_{pp}(t)$ in the memory matrix is nonzero; from (11.3.11), it is real:

$$\boldsymbol{\gamma}(t) = \begin{pmatrix} 0 & 0 \\ 0 & \gamma_{pp}(t)\end{pmatrix} . \tag{13.3.30}$$

The frequency α is a renormalized oscillator frequency, as can be seen from the case where the interaction is set to zero; then, from the equation of motion for p, $x = (-1/m\omega_0^2)\mathrm{i}\mathsf{L}p$, and the Kubo identity, we find the relation:

$$\frac{1}{\beta(x|x)^{(0)}} = m\omega_0^2 \ . \tag{13.3.31}$$

The exact Mori integro-differential equations for the operators $x(t)$ and $p(t)$ are then

$$\begin{aligned} \frac{d}{dt}x(t) &= \frac{1}{m}p(t) \ , \\ \frac{d}{dt}p(t) &= -m\alpha^2 x(t) - \int_0^t p(t-t')\gamma_{pp}(t')dt' + f_p(t) \ . \end{aligned} \tag{13.3.32}$$

---

[14] For the equilibrium expectation values, we assume $\langle x\rangle_\beta = 0$ and $\langle p\rangle_\beta = 0$, so that $\delta\mathscr{G}_\nu = \mathscr{G}_\nu$.

The matrix $\mathrm{i}\Lambda$, (13.3.19), which we require for making the transition to the differential equations

$$\dot{\Xi}(t)=\Xi(t)\mathrm{i}\Lambda \quad (t>0) \tag{13.3.33}$$

may be found readily by taking into account the fact that from (13.3.29), $\boldsymbol{\Omega}^2=\alpha^2\cdot\mathbf{1}$, and thus

$$\mathrm{e}^{-\mathrm{i}\boldsymbol{\Omega}t}=\mathbf{1}\cos\alpha t-\mathrm{i}\frac{\boldsymbol{\Omega}}{\alpha}\sin\alpha t \tag{13.3.34}$$

holds. Using the abbreviations

$$\Gamma=\int\limits_0^\infty \gamma_{pp}(t)\cos\alpha t\,dt=\mathrm{Re}\int\limits_0^\infty \gamma_{pp}(t)\,\mathrm{e}^{\mathrm{i}\alpha t}\,dt=\frac{1}{(p|p)}L'_{pp}(\alpha)\geqq 0 \tag{13.3.35}$$

and

$$\Delta\alpha=\int\limits_0^\infty \gamma_{pp}(t)\sin\alpha t\,dt=\frac{1}{(p|p)}L''_{pp}(\alpha) \tag{13.3.36}$$

we then find from (13.3.29 and 30) an expression for $\mathrm{i}\Lambda$:

$$\mathrm{i}\Lambda=\begin{pmatrix} 0 & -m\alpha(\alpha+\Delta\alpha) \\ \dfrac{1}{m} & -\Gamma \end{pmatrix}. \tag{13.3.37}$$

The equations for the variables $\langle x\rangle(t)$ and $\langle p\rangle(t)$,

$$\begin{aligned}\frac{d}{dt}\langle x\rangle(t)&=\langle x\rangle(t)\mathrm{i}\Lambda_{xx}+\langle p\rangle(t)\mathrm{i}\Lambda_{px}=\frac{1}{m}\langle p\rangle(t)\ ,\\ \frac{d}{dt}\langle p\rangle(t)&=\langle x\rangle(t)\mathrm{i}\Lambda_{xp}+\langle p\rangle(t)\mathrm{i}\Lambda_{pp}=-\langle x\rangle(t)m\alpha(\alpha+\Delta\alpha)-\langle p\rangle(t)\Gamma\end{aligned} \tag{13.3.38}$$

represent the usual dynamics of a damped harmonic oscillator with the renormalized frequency $\sqrt{\alpha(\alpha+\Delta\alpha)}$.

After having carried out the Markovian approximation, we next discuss the points (i), (ii), and (iii) from the previous section; finally, we show some possible solutions for them.

Problem (i): We first examine the eigenvalues $\mathrm{i}\Lambda_\nu$ of the matrix $\mathrm{i}\Lambda$. With (13.3.37), we obtain from

$$\det|\mathrm{i}\Lambda-\mathrm{i}\Lambda_\nu|=0 \tag{13.3.39}$$

the two values

$$i\Lambda_{1,2} = -\frac{\Gamma}{2} \pm \frac{1}{2}\sqrt{\Gamma^2 - 4\alpha(\alpha + \Delta\alpha)} \ . \tag{13.3.40}$$

This formula shows that the imaginary part of Λ_1 becomes negative when

$$\frac{\Delta\alpha}{\alpha} + 1 < 0 \ . \tag{13.3.41}$$

This case cannot be excluded, owing to the definition of $\Delta\alpha$, (13.3.36). The inequality (13.3.41) may be fulfilled whenever a sufficiently sharp resonance line occurs in the neighborhood of α in the Onsager–Casimir coefficient (11.3.19):

$$L''_{pp}(\omega) = \left(f_p \middle| \Pr \frac{1}{\mathsf{QLQ} + \omega} f_p \right) \ . \tag{13.3.42}$$

Solution: Taking the validity criterion (13.3.24) into account, inequality (13.3.41) cannot be fulfilled. Since the inequality [using the first norm in Footnote 8 (p. 212)]

$$\left\| \int_0^{\tau_c} e^{-i\Omega t} \gamma(t)\, dt \right\| \leqq \int_0^{\tau_c} \| e^{-i\Omega t} \| \, \| \gamma(t) \| \, dt = \int_0^{\tau_c} |\gamma_{pp}(t)| \, dt \leqq \gamma_{pp}(0)\tau_c \tag{13.3.43}$$

is obeyed and $\gamma(t)$ is assumed to be negligibly small for $t > \tau_c$, we require instead of the inequality (13.3.24) that the condition:

$$\gamma_{pp}(0)\tau_c^2 \ll 1 \tag{13.3.44}$$

be met. For $\Delta\alpha$, using (13.3.36) and replacing the upper limit of integration ∞ by τ_c, we find

$$\left| \frac{\Delta\alpha}{\alpha} \right| = \left| \int_0^{\tau_c} \gamma_{pp}(t) \frac{\sin \alpha t}{\alpha} dt \right| < \int_0^{\tau_c} |\gamma_{pp}(t)| \, t \, dt < \gamma_{pp}(0) \frac{\tau_c^2}{2} \ , \tag{13.3.45}$$

so that the condition (13.3.44) yields

$$\left| \frac{\Delta\alpha}{\alpha} \right| \ll 1 \ . \tag{13.3.46}$$

With this, we are certain that the eigenvalues (13.3.40) have positive imaginary parts[15]: $\operatorname{Im} \Lambda_{1,2} \geqq 0$.

---

[15] The result (13.3.46) does not mean that $\Delta\alpha$ can be neglected in calculating the eigenvalues (13.3.40). For that purpose, it is important whether or not $\Delta\alpha$ has a noticeable effect on $\Xi(t)$ within the relaxation time.

Problem (ii): We next investigate the absorption matrix (9.4.4):

$$\omega\chi''(\omega)=\frac{\omega^2}{2}\{\chi\Xi(\omega)+\Xi(\omega)^\dagger\chi\}\ ,$$

which follows from the solution of (13.3.33). Using the initial condition $\Xi(0)=\mathbf{1}$, similarly to (13.2.22) we obtain

$$\omega\chi''(\omega)=-\frac{\omega^2}{2}\frac{1}{\omega+\Lambda^\dagger}\chi^{1/2}\{\chi^{1/2}\mathrm{i}\Lambda\chi^{-1/2}+(\chi^{1/2}\mathrm{i}\Lambda\chi^{-1/2})^\dagger\}\chi^{1/2}\frac{1}{\omega+\Lambda}\ . \tag{13.3.47}$$

The matrix $-\{\chi^{1/2}\mathrm{i}\Lambda\chi^{-1/2}+(\chi^{1/2}\mathrm{i}\Lambda\chi^{-1/2})^\dagger\}$ is calculated from (13.3.29) with $\chi=\beta g^{-1}$ and using (13.3.37); the result is

$$-\{\chi^{1/2}\mathrm{i}\Lambda\chi^{-1/2}+(\chi^{1/2}\mathrm{i}\Lambda\chi^{-1/2})^\dagger\}=\begin{pmatrix}0 & \Delta\alpha\\ \Delta\alpha & 2\Gamma\end{pmatrix}\ . \tag{13.3.48}$$

It has the eigenvalues $\Gamma\pm\sqrt{\Gamma^2+(\Delta\alpha)^2}$. Due to the fact that $\Gamma-\sqrt{\Gamma^2+(\Delta\alpha)^2}$ is negative, the absorption matrix from (13.3.47) is no longer positive semidefinite as it should be.

Possible Solution: We change the initial condition for (13.3.33) to $\Xi(0)=A$ with an as-yet unknown matrix A. We then find from the solution to (13.3.33),

$$\Xi(t)=A\,\mathrm{e}^{\mathrm{i}\Lambda t}\quad(t>0) \tag{13.3.49}$$

with

$$\chi(\omega)=\chi\{\mathbf{1}+\mathrm{i}\omega\Xi(\omega)\}=\chi\left\{\mathbf{1}+\mathrm{i}\omega A\frac{\mathrm{i}}{\omega+\Lambda}\right\} \tag{13.3.50}$$

the following expression [instead of (13.3.47)]:

$$\omega\chi''(\omega)=-\frac{\omega^2}{2}\frac{1}{\omega+\Lambda^\dagger}\chi^{1/2}\{\chi^{1/2}A\mathrm{i}\Lambda\chi^{-1/2}+(\chi^{1/2}A\mathrm{i}\Lambda\chi^{-1/2})^\dagger\}\chi^{1/2}\frac{1}{\omega+\Lambda}\ , \tag{13.3.51}$$

(with the additional condition $\chi A=(\chi A)^\dagger$). We thus have a solution to this problem, if we can find a matrix A which forces the matrix $-\{\chi^{1/2}A\mathrm{i}\Lambda\chi^{-1/2}+(\chi^{1/2}A\mathrm{i}\Lambda\chi^{-1/2})^\dagger\}$ to be positive semidefinite.

Problem (iii): Let us consider the symmetry (13.3.20) of the approximate solutions. Equation (13.3.33) holds for $\Xi(t)$, with the matrix $\mathrm{i}\Lambda$ given by (13.3.37).

The matrix $\boldsymbol{g}\Xi(-t)^{\dagger}\boldsymbol{g}^{-1}(t>0)$ obeys the differential equation (13.3.23) with $\mathrm{i}\tilde{\Lambda}$. We can calculate $\mathrm{i}\tilde{\Lambda}$ from (13.3.23) using (13.3.34, 35, and 36), with the result:

$$\mathrm{i}\tilde{\Lambda}=\begin{pmatrix} 0 & -m\alpha^2 \\ \dfrac{1}{m}\left(1+\dfrac{\Delta\alpha}{\alpha}\right) & -\Gamma \end{pmatrix} . \tag{13.3.52}$$

Thus, $\mathrm{i}\tilde{\Lambda}$ differs from $\mathrm{i}\Lambda$ (13.3.37), again just through terms in $\Delta\alpha/\alpha$. The solutions of (13.3.33),

$$\Xi(t)=\mathrm{e}^{\mathrm{i}\Lambda t} \quad (t>0) \tag{13.3.53}$$

and of (13.3.23),

$$\boldsymbol{g}\Xi(-t)^{\dagger}\boldsymbol{g}^{-1}=\mathrm{e}^{\mathrm{i}\tilde{\Lambda}t} \quad (t>0) \tag{13.3.54}$$

with the initial condition $\Xi(0)=\mathbf{1}$ are therefore different and thus violate the symmetry condition (13.3.20).

Possible Resolution: We integrate both of the above differential equations again using the initial condition $\Xi(0)=\boldsymbol{A}$. $\boldsymbol{A}$ may then be chosen in such a way as to make the two solutions identical. This is possible because the matrices $\mathrm{i}\Lambda$ and $\mathrm{i}\tilde{\Lambda}$ have the same eigenvalues, according to (13.3.37 and 52), and therefore can be transformed into one another by a similarity transformation. There is thus a matrix $\boldsymbol{A}$ with the property that

$$\boldsymbol{A}\mathrm{i}\Lambda\boldsymbol{A}^{-1}=\mathrm{i}\tilde{\Lambda} \tag{13.3.55}$$

or

$$\boldsymbol{A}\mathrm{e}^{\mathrm{i}\Lambda t}\boldsymbol{A}^{-1}=\mathrm{e}^{\mathrm{i}\tilde{\Lambda}t} , \tag{13.3.56}$$

such that

$$\Xi(t)=\boldsymbol{A}\mathrm{e}^{\mathrm{i}\Lambda t}=\mathrm{e}^{\mathrm{i}\tilde{\Lambda}t}\boldsymbol{A} \quad (t>0) \tag{13.3.57}$$

fulfills both the differential equation (13.3.33):

$$\dot{\Xi}(t)=\Xi(t)\mathrm{i}\Lambda \tag{13.3.58}$$

and at the same time the differential equation (13.3.23):

$$\dot{\Xi}(t)=\mathrm{i}\tilde{\Lambda}\Xi(t) . \tag{13.3.59}$$

To find $\boldsymbol{A}$ explicitly, we need only solve the system of linear equations

$$\boldsymbol{A}\mathrm{i}\Lambda=\mathrm{i}\tilde{\Lambda}\boldsymbol{A} \tag{13.3.60}$$

for the matrix elements $A_{\nu\mu}$, inserting (13.3.37 and 52). The result is

$$\boldsymbol{A}=\begin{pmatrix} \frac{\alpha}{\alpha+\Delta\alpha}(m\Gamma A_{21}+A_{22}) & -m^2\alpha^2 A_{21} \\ A_{21} & A_{22} \end{pmatrix} . \qquad (13.3.61)$$

Considering the symmetry condition (13.3.20), it is reasonable to require that

$$\boldsymbol{g}\boldsymbol{A}^{\dagger}\boldsymbol{g}^{-1}=\boldsymbol{A} , \qquad (13.3.62)$$

or, equivalently,

$$\boldsymbol{A}^{\dagger}\boldsymbol{\chi}=\boldsymbol{\chi}\boldsymbol{A} . \qquad (13.3.63)$$

This leads to $A_{21}=0$ and $A_{22}^{*}=A_{22}$. A more precise specification of A_{22} is not possible, since there is always an undetermined factor in the solutions of the differential equations (13.3.58 and 59). The latter must be chosen so that $\boldsymbol{A}$ is as nearly as possible equal to the unit matrix. One such possibility is to set $A_{22}=1$, so that

$$\boldsymbol{A}=\begin{pmatrix} \frac{\alpha}{\alpha+\Delta\alpha} & 0 \\ 0 & 1 \end{pmatrix} . \qquad (13.3.64)$$

With the matrix $\boldsymbol{A}$ found in this manner, the matrix (13.3.51) is also positive semidefinite, since

$$-\{\boldsymbol{\chi}^{1/2}\boldsymbol{A}\mathrm{i}\boldsymbol{\Lambda}\boldsymbol{\chi}^{-1/2}+(\boldsymbol{\chi}^{1/2}\boldsymbol{A}\mathrm{i}\boldsymbol{\Lambda}\boldsymbol{\chi}^{-1/2})^{\dagger}\}=\begin{pmatrix} 0 & 0 \\ 0 & 2\Gamma \end{pmatrix} \qquad (13.3.65)$$

is a positive semidefinite matrix.

By taking the symmetry condition $\boldsymbol{g}\boldsymbol{\Xi}(-t)^{\dagger}\boldsymbol{g}^{-1}=\boldsymbol{\Xi}(t)$ into account in the approximation, we were thus led to the additional result that the absorption matrix behaves "correctly", i.e. it is positive semidefinite. The deviation of the required initial condition $\boldsymbol{A}$ from a unit matrix is understandable, since the differential equations for $\boldsymbol{\Xi}(t)$ are valid only for $t>\tau_c$; $\boldsymbol{A}$ thus represents the initial conditions extrapolated to $t=0$. In the case of a single observable, this is unimportant, since a small deviation of the initial condition from 1 leaves the properties of the approximation unaffected.

The problems demonstrated here taking the example of a damped harmonic oscillator are symptomatic of the case in which a set of observables $\mathscr{G}_\nu$ is treated in the Markovian-like approximation of Sect. 13.3.2 for $\|\boldsymbol{\Omega}\|\tau_c>1$. We shall, however, let our case rest with the above discussion of the special example of the harmonic oscillator, where we have shown a possible way out of the difficulties.

14. Time Derivatives as a Special Set of Observables

Our considerations in Chaps. 11 through 13 were relevant to arbitrary sets of observables $\mathcal{G}_\nu$; only with reference to certain approximations in Chap. 13 did we require particular properties of the observables. In this chapter, we wish to consider a special set of observables, which will prove to have some additional aspects. We shall deal with a set $\mathcal{G}_\mu$ which is given by taking the time derivatives of an observable $\mathcal{G}$. Since these derivatives can be represented by applying powers of iL to $\mathcal{G}$, this set of observables has a close mathematical relationship to the dynamics $\exp(\mathrm{iL}t)\mathcal{G}$ of the observable $\mathcal{G}$, as one might infer from the Taylor series for $\exp(\mathrm{iL}t)\mathcal{G}$ in terms of t. Our treatment within the Mori theory leads us to a branch of mathematics which has been thoroughly investigated and which is described in the literature on continued fractions and the method of moments[1]. The purpose of this chapter is to point out these relationships.

14.1 Specialization of the Mori Integro-Differential Equations

14.1.1 The Space $\mathfrak{L}_{\{\mathcal{G}\}}$ Spanned by the Derivatives

We begin with a set $\mathcal{G}_\mu$ which consists of an observable[2] $\mathcal{G}$ and its $n-1$ time derivatives,

$$\mathcal{G}_1=\mathcal{G}\ , \qquad \mathcal{G}_2=\mathrm{iL}\mathcal{G}\ , \qquad \dots , \qquad \mathcal{G}_n=(\mathrm{iL})^{n-1}\mathcal{G} \tag{14.1.1}$$

these are assumed to form linearly independent operators. Of course, instead of the $\mathcal{G}_\mu$, we could use n other linearly independent linear combinations of the $\mathcal{G}_\mu$. The essential aspect of such a set is that the time dependence of the elements of the memory matrix is given by a *single* function. Furthermore, it offers the possibility of finding continued-fraction expressions for the matrix elements of the dynamic susceptibility.

Regarding the Markovian approximation, the choice of a set of the form (14.1.1) is reasonable if $(\mathrm{iL})^n\mathcal{G}$ depends almost linearly on the $\mathcal{G}_\mu$, i.e. if, when we

---

[1] Jones, W.B., and Thron, W.J.: *Continued Fractions, Analytic Theory and Applications* (London 1980)
Wall, H.S.: *Analytic Theory of Continued Fractions*, New York 1967.

[2] This observable is taken to be defined so that $\langle\mathcal{G}\rangle_\beta=0$, i.e. $\delta\mathcal{G}=\mathcal{G}$. It may also be non-Hermitian [compare example in Sect. 16.3].

neglect a perturbation L^1 in L, the observable $(\mathrm{i}\mathsf{L}^0)^n\mathscr{G}$ becomes a linear combination of $(\mathrm{i}\mathsf{L}^0)^\nu\mathscr{G}$ $(\nu=0,\ldots,n-1)$.[3] In this case, due to $\mathsf{Q}^0\mathsf{L}^0\mathscr{G}_\nu=\mathcal{O}$, the memory matrix vanishes quadratically with L^1, so that for a sufficiently small perturbation, one of the conditions for the approximations given in Chap. 13 is fulfilled.

14.1.2 The Mori Equations for Time Derivatives

We first wish to write down the Mori integro-differential equations for the expectation values $\langle\mathscr{G}_\mu\rangle(t)$, since they illustrate the relationships most clearly. As we shall see, all the equations except one take on a trivial form.

The matrix $\Omega_{\nu\mu}$ has a simple structure, since for $\mu=1,\ldots,n-1$,

$$(\mathscr{G}_\varrho|\mathrm{i}\mathsf{L}\mathscr{G}_\mu)=(\mathscr{G}_\varrho|\mathscr{G}_{\mu+1}) \tag{14.1.2}$$

holds, and thus, from the definition of $g_{\nu\mu}$ as the inverse matrix of $(\mathscr{G}_\nu|\mathscr{G}_\mu)$, it follows that

$$\begin{aligned}\mathrm{i}\Omega_{\nu\mu}&=\sum_\varrho g_{\nu\varrho}(\mathscr{G}_\varrho|\mathrm{i}\mathsf{L}\mathscr{G}_\mu)=\sum_\varrho g_{\nu\varrho}(\mathscr{G}_\varrho|\mathscr{G}_{\mu+1})\\&=\delta_{\nu,\mu+1}\quad\text{for}\quad\mu=1,\ldots,n-1\ .\end{aligned} \tag{14.1.3}$$

The memory matrix $\gamma_{\nu\mu}(t)$ is only nonzero for $\mu=n$, since $\mathrm{i}\mathsf{L}\mathscr{G}_\mu$ lies in the space spanned by $\mathscr{G}_\mu$ for $\mu=1,\ldots,n-1$ and therefore

$$\mathsf{Q}\mathrm{i}\mathsf{L}\mathscr{G}_\nu=0\quad(\nu=1,\ldots,n-1) \tag{14.1.4}$$

applies. From the definition of $\gamma_{\nu\mu}(t)$, (11.2.23), it thus follows that

$$\gamma_{\nu\mu}(t)=g_{\nu n}(\mathscr{G}_n|\mathsf{L}\mathsf{Q}e^{\mathrm{i}\mathsf{Q}\mathsf{L}\mathsf{Q}t}\mathsf{Q}\mathsf{L}\mathscr{G}_n)\delta_{\mu n}=g_{\nu n}(f_n|f_n(t))\delta_{\mu n}\ . \tag{14.1.5}$$

The Mori equations (11.4.30), with the expressions (14.1.3) for $\Omega_{\nu\mu}$ and (14.1.5) for $\gamma_{\nu\mu}$, are given by:

$$\begin{aligned}&\frac{d}{dt}\langle\mathscr{G}_1\rangle(t)=\langle\mathscr{G}_2\rangle(t)\\&\frac{d}{dt}\langle\mathscr{G}_2\rangle(t)=\langle\mathscr{G}_3\rangle(t)\\&\ldots\\&\frac{d}{dt}\langle\mathscr{G}_{n-1}\rangle(t)=\langle\mathscr{G}_n\rangle(t)\\&\frac{d}{dt}\langle\mathscr{G}_n\rangle(t)=\sum_\nu\langle\mathscr{G}_\nu\rangle(t)\mathrm{i}\Omega_{\nu n}-\sum_\nu\int_0^t\langle\mathscr{G}_\nu\rangle(t')g_{\nu n}(f_n|f_n(t-t'))dt'\ .\end{aligned} \tag{14.1.6}$$

[3] The elements $(\mathsf{L}^0)^\nu\mathscr{G}$ $(\nu=0,\ldots,n-1)$ then span a subspace $\mathfrak{L}_{\mathscr{G}}$ which is invariant with respect to L^0 as described in Sect. 12.1.

If the expectation values $\langle \mathscr{G}_\mu \rangle(t)$ for $\mu=2,\ldots,n$ are expressed with the aid of the first $n-1$ equations in terms of time derivatives of $\langle \mathscr{G}_1 \rangle(t)$, then the last equation of (14.1.6) represents an n-th order integro-differential equation for the expectation value $\langle \mathscr{G}_1 \rangle(t)=\langle \mathscr{G} \rangle(t)$,

$$\frac{d^n}{dt^n}\langle \mathscr{G} \rangle(t)=\sum_\nu \frac{d^{\nu-1}}{dt^{\nu-1}}\langle \mathscr{G} \rangle(t)\Omega_{\nu n}$$

$$-\sum_\nu \int_0^t \frac{d^{\nu-1}}{dt'^{\nu-1}}\langle \mathscr{G} \rangle(t')g_{\nu n}(f_n|f_n(t-t'))\,dt' \ . \tag{14.1.6a}$$

14.1.3 Orthogonal Observables

In some cases, in particular for the explicit calculation of the matrix $\{\omega+\boldsymbol{\Omega}+\mathrm{i}\gamma(\omega)\}^{-1}$, it is expedient to begin with orthogonal linear combinations $\hat{\mathscr{G}}_\mu$ of the observables (14.1.1). Formally, this may be most simply achieved by introducing the projection operators

$$\mathsf{P}_\nu=1-\mathsf{Q}_\nu=\frac{|\hat{\mathscr{G}}_\nu)(\hat{\mathscr{G}}_\nu|}{(\hat{\mathscr{G}}_\nu|\hat{\mathscr{G}}_\nu)} \quad (\nu=1,\ldots,n) \ , \tag{14.1.7}$$

which are defined by

$$\begin{aligned}
&\hat{\mathscr{G}}_1=\mathscr{G} \ ,\\
&\hat{\mathscr{G}}_2=\mathsf{Q}_1\mathsf{L}\mathscr{G} \ ,\\
&\ldots\\
&\hat{\mathscr{G}}_{\mu+1}=\mathsf{Q}_\mu\ldots\mathsf{Q}_1\mathsf{L}^\mu\mathscr{G} \ ,\\
&\ldots\\
&\hat{\mathscr{G}}_n=\mathsf{Q}_{n-1}\ldots\mathsf{Q}_1\mathsf{L}^{n-1}\mathscr{G} \ .
\end{aligned} \tag{14.1.8}$$

Then the $\hat{\mathscr{G}}_\nu$ are linear combinations of the $\mathscr{G}_\mu$ $(\mu\leqq\nu)$ and are orthogonal,

$$(\hat{\mathscr{G}}_\nu|\hat{\mathscr{G}}_\mu)=(\hat{\mathscr{G}}_\nu|\hat{\mathscr{G}}_\nu)\delta_{\nu\mu} \ . \tag{14.1.9}$$

Because of their method of construction, the $\hat{\mathscr{G}}_\nu$ obey a simple recursion formula (14.1.14). To derive it, one need only realize that, from (14.1.8), the difference $\hat{\mathscr{G}}_\nu-\mathsf{L}\hat{\mathscr{G}}_{\nu-1}$ is a polynomial of degree $\nu-2$ in L applied to $\mathscr{G}$. However, due to $\mathsf{Q}_\mu\ldots\mathsf{Q}_1=1-\sum_{\varrho=1}^{\mu}\mathsf{P}_\varrho$ from (14.1.8), every power of $\mathsf{L}^\mu\mathscr{G}$ can be represented by a linear combination of $\hat{\mathscr{G}}_1$ to $\hat{\mathscr{G}}_{\mu+1}$, so that

$$\hat{\mathscr{G}}_\nu-\mathsf{L}\hat{\mathscr{G}}_{\nu-1}=\sum_{\varrho=1}^{\nu-1}\hat{\mathscr{G}}_\varrho c_{\varrho\nu} \tag{14.1.10}$$

must hold. Combining this with the orthogonality condition (14.1.9), using the Hermiticity of L, it follows that

$$(\hat{\mathscr{G}}_\mu|\mathsf{L}\hat{\mathscr{G}}_{\nu-1})=(\hat{\mathscr{G}}_{\nu-1}|\mathsf{L}\hat{\mathscr{G}}_\mu)=0 \quad \text{for} \quad \mu>\nu \ , \tag{14.1.11}$$

$$(\hat{\mathscr{G}}_\nu|\mathsf{L}\hat{\mathscr{G}}_{\nu-1})=(\hat{\mathscr{G}}_{\nu-1}|\mathsf{L}\hat{\mathscr{G}}_\nu)=(\hat{\mathscr{G}}_\nu|\hat{\mathscr{G}}_\nu) \ . \tag{14.1.12}$$

The coefficients $c_{\varrho\nu}$ in (14.1.10):

$$c_{\varrho\nu}=\frac{1}{(\hat{\mathscr{G}}_\varrho|\hat{\mathscr{G}}_\varrho)}\{(\hat{\mathscr{G}}_\nu|\hat{\mathscr{G}}_\nu)\delta_{\varrho\nu}-(\hat{\mathscr{G}}_\varrho|\mathsf{L}\hat{\mathscr{G}}_{\nu-1})\} \tag{14.1.13}$$

can therefore, according to (14.1.11 and 12), be nonzero only for $\varrho=\nu-1$ and $\varrho=\nu-2$, so that we finally obtain from (14.1.11–13) the relation

$$\hat{\mathscr{G}}_\nu=\mathsf{L}\hat{\mathscr{G}}_{\nu-1}-\frac{(\hat{\mathscr{G}}_{\nu-1}|\mathsf{L}\hat{\mathscr{G}}_{\nu-1})}{(\hat{\mathscr{G}}_{\nu-1}|\hat{\mathscr{G}}_{\nu-1})}\hat{\mathscr{G}}_{\nu-1}$$
$$-\frac{(\hat{\mathscr{G}}_{\nu-1}|\hat{\mathscr{G}}_{\nu-1})}{(\hat{\mathscr{G}}_{\nu-2}|\hat{\mathscr{G}}_{\nu-2})}\hat{\mathscr{G}}_{\nu-2} \quad (\nu=2,3\ldots;\mathscr{G}_0=0) \ . \tag{14.1.14}$$

It can be used to calculate the orthogonal $\hat{\mathscr{G}}_\nu$ recursively.

As the next step, we wish to write the equations of motion for the expectation values $\langle\hat{\mathscr{G}}_\mu\rangle(t)$ and compare them to (14.1.6). For this purpose, we require the frequency matrix $\hat{\Omega}_{\nu\mu}$ and the memory matrix $\hat{\gamma}_{\nu\mu}(t)$.

The $\hat{\mathscr{G}}_\mu$ are orthogonal, so the matrix $\hat{g}_{\nu\mu}$ is diagonal; from this, it follows from (14.1.11 and 12) that the matrix $\hat{\Omega}_{\nu\mu}$ is tridiagonal, having nonzero matrix elements on the main diagonal and the two adjacent diagonals:

$$\hat{\boldsymbol{\Omega}}=\begin{vmatrix} \hat{\Omega}_{11} & \hat{\Omega}_{12} & & & & \mathbf{0} \\ \hat{\Omega}_{21} & \hat{\Omega}_{22} & \hat{\Omega}_{23} & & & \\ & \hat{\Omega}_{32} & \hat{\Omega}_{33} & \hat{\Omega}_{34} & & \\ & & \cdot & \cdot & \cdot & \\ & & & \cdot & \cdot & \cdot \quad \hat{\Omega}_{n-1,n} \\ \mathbf{0} & & & & \hat{\Omega}_{n,n-1} & \hat{\Omega}_{nn} \end{vmatrix} \ . \tag{14.1.15a}$$

These matrix elements are given by the expressions

$$\hat{\Omega}_{\mu\mu}=\frac{1}{(\hat{\mathscr{G}}_\mu|\hat{\mathscr{G}}_\mu)}(\hat{\mathscr{G}}_\mu|\mathsf{L}\hat{\mathscr{G}}_\mu)$$
$$\hat{\Omega}_{\mu+1,\mu}=1 \tag{14.1.15b}$$
$$\hat{\Omega}_{\mu-1,\mu}=\frac{(\hat{\mathscr{G}}_\mu|\hat{\mathscr{G}}_\mu)}{(\hat{\mathscr{G}}_{\mu-1}|\hat{\mathscr{G}}_{\mu-1})} \ .$$

In calculating the memory matrix

$$\hat{\gamma}_{\nu\mu}(t) = \frac{1}{(\hat{\mathscr{G}}_\nu|\hat{\mathscr{G}}_\nu)}(\hat{f}_\nu|\hat{f}_\mu(t)) \tag{14.1.16}$$

it should be kept in mind that, from (14.1.14),

$$\mathsf{Q}_{\mu+1} \ldots \mathsf{Q}_1 \mathrm{i}\mathsf{L}\hat{\mathscr{G}}_\mu = \mathcal{O} \quad \text{for} \quad \mu = 1, \ldots, n-1 \tag{14.1.17}$$

vanishes and therefore, of the residual forces

$$\hat{f}_\mu = \mathsf{Q}\mathrm{i}\mathsf{L}\hat{\mathscr{G}}_\mu = \mathsf{Q}_n \ldots \mathsf{Q}_1 \mathrm{i}\mathsf{L}\hat{\mathscr{G}}_\mu \tag{14.1.18}$$

only $\hat{f}_n$ is nonzero. The memory matrix (14.1.16) thus consists of the single matrix element:

$$\hat{\gamma}_{\nu\mu}(t) = \frac{1}{(\hat{\mathscr{G}}_n|\hat{\mathscr{G}}_n)}(\hat{f}_n|\hat{f}_n(t))\delta_{\nu n}\delta_{n\mu} \,. \tag{14.1.19}$$

The system of equations [as in (11.4.30)] which follows from $\hat{\Omega}_{\nu\mu}$ and $\hat{\gamma}_{\nu\mu}(t)$:

$$\frac{d}{dt}\langle\hat{\mathscr{G}}_1\rangle(t) = \langle\hat{\mathscr{G}}_1\rangle(t)\mathrm{i}\hat{\Omega}_{11} + \langle\hat{\mathscr{G}}_2\rangle(t)\mathrm{i}\hat{\Omega}_{21}$$

$$\ldots$$

$$\frac{d}{dt}\langle\hat{\mathscr{G}}_\mu\rangle(t) = \langle\hat{\mathscr{G}}_{\mu-1}\rangle(t)\mathrm{i}\hat{\Omega}_{\mu-1,\mu} + \langle\hat{\mathscr{G}}_\mu\rangle(t)\mathrm{i}\hat{\Omega}_{\mu\mu} + \langle\hat{\mathscr{G}}_{\mu+1}\rangle(t)\mathrm{i}\hat{\Omega}_{\mu+1,\mu} \tag{14.1.20}$$

$$\ldots$$

$$\frac{d}{dt}\langle\hat{\mathscr{G}}_n\rangle(t) = \langle\hat{\mathscr{G}}_{n-1}\rangle(t)\mathrm{i}\hat{\Omega}_{n-1,n} + \langle\hat{\mathscr{G}}_n\rangle(t)\mathrm{i}\hat{\Omega}_{nn} - \int_0^t \langle\hat{\mathscr{G}}_n\rangle(t')\hat{\gamma}_{nn}(t-t')dt'$$

has, by contrast to (14.1.6), the simplifying property that the integral kernel in the last equation couples only to $\langle\hat{\mathscr{G}}_n\rangle(t)$. On the other hand, the preceding $n-1$ equations have a more complicated 3-term structure related to the tri-diagonality of the $\hat{\Omega}$ matrix. Altogether, it is still just a question of linear combinations of system (14.1.1).

14.2 A Continued-Fraction Expression for the Correlation Function $\Xi(\omega)$

14.2.1 Exact Description

For the chosen orthogonal set $\hat{\mathscr{G}}_\mu$, the matrix $\hat{\Omega}_{\nu\mu}$, as seen from (14.1.15), is tridiagonal, and the memory matrix (14.1.16) has only one single matrix element,

$\hat{\gamma}_{nn}$. These properties make it possible to compute the matrix elements of the correlation matrix $\hat{\Xi}_{\nu\mu}(\omega)$ (11.4.4) recursively from

$$\sum_{\lambda} \hat{\Xi}_{\nu\lambda}(\omega)\,\{\omega\delta_{\lambda\mu}+\hat{\Omega}_{\lambda\mu}+\mathrm{i}\hat{\gamma}_{\lambda\mu}(\omega)\}=\mathrm{i}\delta_{\nu\mu}\ . \tag{14.2.1}$$

We shall carry this out explicitly for the matrix element[4]

$$\hat{\Xi}_{11}=\frac{1}{(\mathscr{G}|\mathscr{G})}\left(\mathscr{G}\left|\frac{\mathrm{i}}{\mathrm{L}+\omega+\mathrm{i}\eta}\mathscr{G}\right.\right)=\Xi(\omega) \tag{14.2.2}$$

and demonstrate that it may be given in terms of a simple continued-fraction expression.

With (14.1.15 and 16), the system of equations (14.2.1) for $\nu=1$ specializes to

$$\begin{aligned}
&\hat{\Xi}_{11}(\omega)\{\omega+\hat{\Omega}_{11}\}+\hat{\Xi}_{12}(\omega)\hat{\Omega}_{21}=\mathrm{i}\\
&\hat{\Xi}_{12}(\omega)\{\omega+\hat{\Omega}_{22}\}+\hat{\Xi}_{11}(\omega)\hat{\Omega}_{12}+\hat{\Xi}_{13}(\omega)\hat{\Omega}_{32}=0\\
&\dots\\
&\hat{\Xi}_{1n}(\omega)\{\omega+\hat{\Omega}_{nn}\}+\hat{\Xi}_{1,n-1}(\omega)\hat{\Omega}_{n-1,n}+\hat{\Xi}_{1n}(\omega)\mathrm{i}\hat{\gamma}_{nn}(\omega)=0\ .
\end{aligned} \tag{14.2.3}$$

From the first of these equations, we obtain

$$\hat{\Xi}_{11}=\frac{\mathrm{i}}{\omega+\hat{\Omega}_{11}+\dfrac{\hat{\Xi}_{12}}{\hat{\Xi}_{11}}\hat{\Omega}_{21}}\ , \tag{14.2.4}$$

while from the following ones, we have

$$\begin{aligned}
&\frac{\hat{\Xi}_{12}}{\hat{\Xi}_{11}}=-\frac{\hat{\Omega}_{12}}{\omega+\hat{\Omega}_{22}+\dfrac{\hat{\Xi}_{13}}{\hat{\Xi}_{12}}\Omega_{32}}\\
&\frac{\hat{\Xi}_{13}}{\hat{\Xi}_{12}}=-\frac{\hat{\Omega}_{23}}{\omega+\hat{\Omega}_{33}+\dfrac{\hat{\Xi}_{14}}{\hat{\Xi}_{13}}\hat{\Omega}_{43}}\\
&\dots\\
&\frac{\hat{\Xi}_{1n}}{\hat{\Xi}_{1,n-1}}=-\frac{\hat{\Omega}_{n-1,n}}{\omega+\hat{\Omega}_{nn}+\mathrm{i}\hat{\gamma}_{nn}}\ .
\end{aligned} \tag{14.2.5}$$

[4] The calculation of the other matrix elements $\hat{\Xi}_{\nu\mu}(\omega)$ yields no new aspects. We therefore dispense with their further discussion.

If the expressions which result are inserted one after the other into the preceding equations, one obtains a continued fraction for $\hat{\Xi}_{11}(\omega)$:

$$\frac{1}{(\mathscr{G}|\mathscr{G})}\left(\mathscr{G}\left|\frac{\mathrm{i}}{\mathrm{L}+\omega+\mathrm{i}\eta}\mathscr{G}\right.\right)=\hat{\Xi}_{11}(\omega)$$

$$=\frac{\mathrm{i}}{\omega+\hat{\Omega}_{11}+\dfrac{\hat{\Xi}_{12}(\omega)}{\hat{\Xi}_{11}(\omega)}\hat{\Omega}_{21}}$$

$$=\frac{\mathrm{i}}{\omega+\hat{\Omega}_{11}-\dfrac{\hat{\Omega}_{12}\hat{\Omega}_{21}}{\omega+\hat{\Omega}_{22}+\dfrac{\hat{\Xi}_{13}(\omega)}{\hat{\Xi}_{12}(\omega)}\hat{\Omega}_{32}}}$$

$$=\frac{\mathrm{i}}{\omega+\hat{\Omega}_{11}-\hat{\Omega}_{12}\hat{\Omega}_{21}\dfrac{1}{\omega+\hat{\Omega}_{22}-\hat{\Omega}_{23}\hat{\Omega}_{32}\dfrac{1}{\omega+\hat{\Omega}_{33}\ldots-\hat{\Omega}_{n-1,n}\hat{\Omega}_{n,n-1}\dfrac{1}{\omega+\hat{\Omega}_{nn}+\mathrm{i}\hat{\gamma}_{nn}(\omega)}}}}\,. \tag{14.2.6}$$

The coefficients $\hat{\Omega}_{12}\hat{\Omega}_{21},\ldots$ which occur in this expression may be given in terms of quotients of the quantities $(\hat{\mathscr{G}}_\nu|\hat{\mathscr{G}}_\nu)$ as seen from (14.1.15):

$$\hat{\Omega}_{12}\hat{\Omega}_{21}=\frac{(\hat{\mathscr{G}}_2|\hat{\mathscr{G}}_2)}{(\hat{\mathscr{G}}_1|\hat{\mathscr{G}}_1)}\,,\qquad \hat{\Omega}_{23}\hat{\Omega}_{32}=\frac{(\hat{\mathscr{G}}_3|\hat{\mathscr{G}}_3)}{(\hat{\mathscr{G}}_2|\hat{\mathscr{G}}_2)}\,,\ldots \tag{14.2.7}$$

and are therefore positive.

If n is allowed to tend towards ∞ in the above formula, we obtain an infinite continued fraction. Such continued fractions of the type (14.2.6) are well known in the mathematical literature and are closely related to the problem of moments and to questions of function theory. In this connection, we refer to textbooks on continued fractions (cf. p. 226, Footnote 1).

If one only wishes to generate the continued fraction (14.2.6), this can be most simply done by repeated application of the identity (11.4.11) to the resolvents

$$\frac{1}{\mathsf{L}+\omega+\mathrm{i}\eta}\,,\quad \frac{1}{\mathsf{Q}_1\mathsf{L}\mathsf{Q}_1+\omega+\mathrm{i}\eta}\,,\quad \frac{1}{\mathsf{Q}_2\mathsf{Q}_1\mathsf{L}\mathsf{Q}_1\mathsf{Q}_2+\omega+\mathrm{i}\eta}\quad \text{etc.}\,.$$

One must only realize that owing to (14.1.14) and (14.1.8), the following expressions hold:

$$
\begin{aligned}
&\mathsf{Q}_1\mathsf{L}\mathscr{G}=\hat{\mathscr{G}}_2 \;, \\
&\mathsf{Q}_2\mathsf{Q}_1\mathsf{L}\mathsf{Q}_1\mathsf{L}\mathscr{G}=\mathsf{Q}_2\mathsf{Q}_1\mathsf{L}^2\mathscr{G}=\hat{\mathscr{G}}_3 \;, \\
&\mathsf{Q}_3\mathsf{Q}_2\mathsf{Q}_1\mathsf{L}\mathsf{Q}_2\mathsf{Q}_1\mathsf{L}\mathsf{Q}_1\mathsf{L}\mathscr{G}=\mathsf{Q}_3\mathsf{Q}_2\mathsf{Q}_1\mathsf{L}^3\mathscr{G}=\hat{\mathscr{G}}_4 \;.
\end{aligned}
\tag{14.2.8}
$$

We wished, however, to show how the continued fraction (14.2.6) fits into the general Mori theory for a set of observables.

14.2.2 Neglecting the Memory Matrix

We turn to the investigation of some approximations for the continued fraction (14.2.6). The above derivation using the theory for a set of observables $\mathscr{G}_\mu$ shows that all of the approximations described in Chap. 13 can, in principle, be applied, if the necessary preconditions are met.

The simplest approximation for the continued-fraction expression consists of leaving off[5] the matrix element $\hat{\gamma}_{nn}(\omega)$. In this case, poles occur on the real ω-axis (cf. Fig. 14.1a). Therefore, as can be seen from the definition of $\Xi_{11}(\omega)$, we must consider Ξ_{11} for complex frequencies $z=\omega+\mathrm{i}\eta$ in the upper z-half plane, and then take the limit towards the real axis, $\lim_{\eta\to 0}$. The expression (14.2.6) in the variable z, neglecting $\gamma_{nn}(z)$, gives a continued fraction for $\Xi_{11}(z)$ which can be written as the quotient of an $(n-1)$-th degree polynomial divided by an n-th degree polynomial in z:

$$
\hat{\Xi}_{11}(z)=\mathrm{i}\cfrac{1}{z+\hat{\Omega}_{11}-\cfrac{\hat{\Omega}_{12}\hat{\Omega}_{21}}{z+\hat{\Omega}_{22}\dots \atop \qquad -\cfrac{\hat{\Omega}_{n-1,n}\hat{\Omega}_{n,n-1}}{z+\hat{\Omega}_{nn}}}}=\mathrm{i}\frac{A_n(z)}{B_n(z)}{}^{6}
\tag{14.2.9}
$$

where

$$
A_n(z)=z^{n-1}+\dots\;,\qquad B_n(z)=z^n+\dots\;.
\tag{14.2.10}
$$

[5] Terminating the continued fraction (14.2.6) after a partial denominator $\omega+\hat{\Omega}_{\nu\nu}$, $(\nu<n)$, is also included in this approximation, since this would be the same as choosing n smaller.

[6] By using complete induction, one can show that the polynomials $A_n(z)$ and $B_n(z)$ obey the following recursion formula for $n\geqq 2$:

$$
\begin{aligned}
&A_n=(z+\hat{\Omega}_{nn})A_{n-1}-\hat{\Omega}_{n-1,n}\hat{\Omega}_{n,n-1}A_{n-2}\;,\quad A_0=0\;,\quad A_1=1\;,\\
&B_n=(z+\hat{\Omega}_{nn})B_{n-1}-\hat{\Omega}_{n-1,n}\hat{\Omega}_{n,n-1}B_{n-2}\;,\quad B_0=1\;,\quad B_1=z+\hat{\Omega}_{11}\;.
\end{aligned}
$$

The approximation consists in neglecting the entire memory matrix in $\hat{\Xi}_{11}(z) = \mathrm{i}\{z+\hat{\boldsymbol{\Omega}}+\mathrm{i}\hat{\gamma}(z)\}_{11}^{-1}$, so that for $\omega+\mathrm{i}\eta=z$,

$$\frac{A_n(z)}{B_n(z)}=(z+\hat{\boldsymbol{\Omega}})_{11}^{-1} \tag{14.2.11}$$

results.

The most important property of the approximation (14.2.9) is the fact that its power-series expansion in z^{-1},

$$\mathrm{i}\frac{A_n(z)}{B_n(z)}=\mathrm{i}\sum_{\varrho=0}^{\infty} c_\varrho z^{-\varrho-1}=\mathrm{i}\sum_{\varrho=0}^{\infty}(\hat{\boldsymbol{\Omega}}^\varrho)_{11}(-1)^\varrho z^{-\varrho-1} \tag{14.2.12}$$

is identical to the expansion of the exact function,

$$\hat{\Xi}_{11}(z)=\frac{\mathrm{i}}{(\mathscr{G}|\mathscr{G})}\left(\mathscr{G}\left|\frac{1}{z+\mathsf{L}}\mathscr{G}\right.\right)=\frac{\mathrm{i}}{(\mathscr{G}|\mathscr{G})}\sum_{\varrho=0}^{\infty}(\mathscr{G}|\mathsf{L}^\varrho\mathscr{G})(-1)^\varrho z^{-\varrho-1} \tag{14.2.13}$$

up to and including the term $\mathrm{i}c_{2n-1}z^{-2n}$:

$$(\hat{\boldsymbol{\Omega}}^\varrho)_{11}=\frac{1}{(\mathscr{G}|\mathscr{G})}(\mathscr{G}|\mathsf{L}^\varrho\mathscr{G}) \quad \text{for} \quad \varrho\leqq 2n-1\ . \tag{14.2.14}$$

It can thus be used as an approximation at high frequencies.

Proof of (14.2.14): We start with the definition of the matrix $\hat{\Omega}_{\nu\mu}$,

$$\hat{\Omega}_{\nu\mu}=\frac{1}{(\hat{\mathscr{G}}_\nu|\hat{\mathscr{G}}_\nu)}(\hat{\mathscr{G}}_\nu|\mathsf{L}\hat{\mathscr{G}}_\mu) \tag{14.2.15}$$

and of the projection operator

$$\mathsf{P}=\sum_{\nu=1}^{n}|\hat{\mathscr{G}}_\nu)\frac{1}{(\hat{\mathscr{G}}_\nu|\hat{\mathscr{G}}_\nu)}(\hat{\mathscr{G}}_\nu|\ . \tag{14.2.16}$$

These definitions yield immediately the equations

$$\sum_\sigma \hat{\Omega}_{\nu\sigma}\hat{\Omega}_{\sigma\mu}=\frac{1}{(\hat{\mathscr{G}}_\nu|\hat{\mathscr{G}}_\nu)}(\hat{\mathscr{G}}_\nu|\mathsf{LPL}\hat{\mathscr{G}}_\mu)=\frac{1}{(\hat{\mathscr{G}}_\nu|\hat{\mathscr{G}}_\nu)}(\hat{\mathscr{G}}_\nu|(\mathsf{PLP})^2\hat{\mathscr{G}}_\mu) \tag{14.2.17}$$

or

$$(\boldsymbol{\Omega}^\varrho)_{11}=\frac{1}{(\mathscr{G}|\mathscr{G})}(\mathscr{G}|(\mathsf{PLP})^\varrho\mathscr{G})\ . \tag{14.2.18}$$

On the other hand, since all the powers of $\mathsf{L}^\sigma\mathscr{G}$ for $\sigma\leqq n-1$ lie in the space spanned by the set $\mathscr{G}_\nu$, for $\sigma\leqq n-1$ the equation

$$(\mathsf{PLP})^\sigma\mathscr{G}=\mathsf{L}^\sigma\mathscr{G}\ , \tag{14.2.19}$$

is valid; from it, using the Hermiticity of PLP, for $\varrho=\sigma+\sigma'+1$, with $\sigma,\sigma'\leqq n-1$, we find

$$(\boldsymbol{\Omega}^{\varrho})_{11}=\frac{1}{(\mathscr{G}|\mathscr{G})}(\mathscr{G}|(\mathsf{PLP})^{\sigma}\mathsf{L}(\mathsf{PLP})^{\sigma'}\mathscr{G})=\frac{1}{(\mathscr{G}|\mathscr{G})}(\mathsf{L}^{\sigma}\mathscr{G}|\mathsf{L}\mathsf{L}^{\sigma'}\mathscr{G})$$

$$=\frac{1}{(\mathscr{G}|\mathscr{G})}(\mathscr{G}|\mathsf{L}^{\sigma+\sigma'+1}\mathscr{G}) \ . \tag{14.2.20}$$

The conjecture (14.2.14) is thus proved.

The moments of the real part of $\hat{\Xi}_{11}(z)$ for Im $z\to 0$ (cf. Sect. 9.4.5) are closely connected to the power-series expansion in z^{-1}. For the real part of the approximation (14.2.9), which comprises a sum of δ-functions, the first $2n$ moments are equal to the exact values,

$$\int\limits_{-\infty}^{\infty}\omega^{\varrho}\,\mathrm{Re}\,\hat{\Xi}_{11}(\omega)\,d\omega=\pi\frac{(-1)^{\varrho}}{(\mathscr{G}|\mathscr{G})}(\mathscr{G}|\mathsf{L}^{\varrho}\mathscr{G})\quad\text{for}\quad\varrho=0,\dots,2n-1 \ . \tag{14.2.21}$$

The moments, conversely, can be used to obtain the approximation (14.2.9) for $\hat{\Xi}_{11}(z)$.

To see this property of expression (14.2.9), we use a decomposition of $A_n(z)/B_n(z)$ in partial fractions. It can be found from (14.2.11) with the spectral decomposition of the Hermitian matrix $\hat{\boldsymbol{g}}^{-1/2}\hat{\boldsymbol{\Omega}}\hat{\boldsymbol{g}}^{1/2}$, keeping in mind the relations

$$(z+\hat{\boldsymbol{\Omega}})^{-1}_{11}=\{\hat{\boldsymbol{g}}^{-1/2}(z+\hat{\boldsymbol{g}}^{-1/2}\hat{\boldsymbol{\Omega}}\hat{\boldsymbol{g}}^{1/2})^{-1}\hat{\boldsymbol{g}}^{1/2}\}_{11}$$

$$=(z+\hat{\boldsymbol{g}}^{-1/2}\hat{\boldsymbol{\Omega}}\hat{\boldsymbol{g}}^{1/2})^{-1}_{11} \ . \tag{14.2.22}$$

We then obtain the decomposition in the form:

$$\hat{\Xi}_{11}(z)=\mathrm{i}\frac{A_n(z)}{B_n(z)}=\mathrm{i}\sum_{\nu=1}^{n}\frac{M_\nu}{z+\Omega_\nu}\ , \tag{14.2.23}$$

in which Ω_ν represents the eigenvalues of the matrix $\hat{\boldsymbol{g}}^{-1/2}\hat{\boldsymbol{\Omega}}\hat{\boldsymbol{g}}^{1/2}$ or those of $\hat{\boldsymbol{\Omega}}$ and $\boldsymbol{\Omega}$ and the M_ν are real and positive[7]. The values Ω_ν and M_ν can be determined directly from the moments. Comparing the expansion of (14.2.23) in powers of z^{-1} with the expansion (14.2.12), and using (14.2.14), we find a system

[7] Using the recursion formulas for the polynomials $A_n(z)$ and $B_n(z)$, one may show that they can have no common roots. It follows that in (14.2.23) all n eigenvalues Ω_ν must be different and none of the coefficients M_ν can be zero.

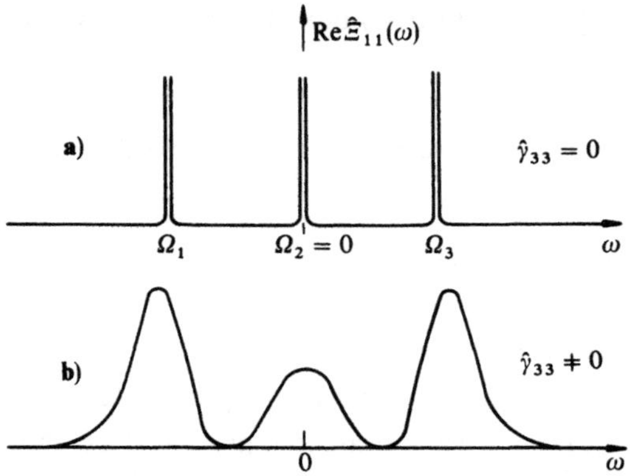

Fig. 14.1. a,b. The spectral function Re $\Xi(\omega)$ for a continued-fraction expansion with $n=3$. **(a)** with neglect of the memory function $\hat{\gamma}_{33}(\omega)$; **(b)** in the Markovian approximation

of equations for the Ω_ν and the M_ν:

$$\begin{aligned}
&\sum_\nu M_\nu = 1\\
&\sum_\nu M_\nu \Omega_\nu = \frac{1}{(\mathscr{G}|\mathscr{G})}(\mathscr{G}|\mathsf{L}\mathscr{G})\\
&\dots\\
&\sum_\nu M_\nu \Omega_\nu^{2n-1} = \frac{1}{(\mathscr{G}|\mathscr{G})}(\mathscr{G}|\mathsf{L}^{2n-1}\mathscr{G}) \ .
\end{aligned} \tag{14.2.24}$$

Now consider the real part of (14.2.23) for $z=\omega+\mathrm{i}\eta$ in the limit $\eta\to+0$; due to $1/(x+\mathrm{i}\eta)=\mathrm{Pr}(1/x)-\mathrm{i}\pi\delta(x)$, it can be cast in the following form (compare Fig. 14.1a):

$$\mathrm{Re}\,\hat{\Xi}_{11}(\omega+\mathrm{i}\eta)=\pi\sum_\nu M_\nu\,\delta(\omega+\Omega_\nu) \ . \tag{14.2.25}$$

The moments of this expression are given by $(-1)^\varrho\pi\sum_\nu M_\nu\Omega_\nu^\varrho$, so that the property (14.2.21) follows from (14.2.24). Beginning with the moments, one can obtain from (14.2.23) via M_ν and Ω_ν the approximation (14.2.9).

14.2.3 The Markovian Approximation

In the chosen set of observables $\hat{\mathscr{G}}_\mu$, the memory matrix $\hat{\gamma}_{\nu\mu}(\omega)$ contains only the single matrix element $\hat{\gamma}_{nn}(\omega)$. Therefore, the Markovian approximation described in Sect. 13.2 for a set of slow observables becomes particularly simple. Formally, according to (13.2.7), the approximation means that the memory matrix $\hat{\gamma}_{\nu\mu}(\omega)$ is to be replaced by its value at $\omega=0$. This is reasonable for the range $|\omega|<1/\tau_c$, in which $\hat{\gamma}_{\nu\mu}(\omega)$ can be considered to be constant, as long as the condition (13.2.13) is fulfilled. In order to make this condition intuitively clear, we intensify it and require that

$$\|\mathrm{i}\hat{\boldsymbol{\Omega}}\|\tau_{\mathrm{c}}+\|\hat{\boldsymbol{\gamma}}(\omega=0)\|\tau_{\mathrm{c}}\ll 1 \ . \tag{14.2.26}$$

This then means that all the eigenvalues of the matrix $\hat{\Omega}_{\nu\mu}$ must lie in the frequency range $|\omega|<1/\tau_{\mathrm{c}}$ and, at the same time, the matrix element $|\hat{\gamma}_{nn}(0)|$ must be sufficiently small compared to τ_{c}^{-1}. When these conditions are met, we find from (14.2.6) with $\hat{\gamma}_{nn}(\omega)=\hat{\gamma}_{nn}(\omega=0)$ as an approximation for $\hat{\Xi}_{11}(\omega)$ the continued-fraction expression

$$\hat{\Xi}_{11}(\omega)=\cfrac{\mathrm{i}}{\omega+\hat{\Omega}_{11}-\hat{\Omega}_{12}\hat{\Omega}_{21}\cfrac{1}{\omega+\hat{\Omega}_{22}-\hat{\Omega}_{23}\hat{\Omega}_{32}\cfrac{1}{\omega+\dots-\hat{\Omega}_{n-1,n}\hat{\Omega}_{n,n-1}\cfrac{1}{\omega+\hat{\Omega}_{nn}+\mathrm{i}\hat{\gamma}_{nn}(0)}}}} \ . \tag{14.2.27}$$

It is at this point only applicable in the frequency range $|\omega|<1/\tau_{\mathrm{c}}$. As in Sect. 14.2.2, the particular feature of this expression is that it is a good approximation even at high frequencies, $\omega\to\infty$, owing to the special character of the $\hat{\boldsymbol{\Omega}}$ matrix. It can thus be interpreted as an interpolation between the low and high frequency regions.

To illustrate the behavior of (14.2.27) at high frequencies, we use an analogous procedure as for (14.2.11). For the continued fraction (14.2.27) in terms of the complex variable z, we have

$$\cfrac{\mathrm{i}}{z+\hat{\Omega}_{11}-\hat{\Omega}_{12}\hat{\Omega}_{21}\cfrac{1}{z+\hat{\Omega}_{22}-\hat{\Omega}_{23}\hat{\Omega}_{32}\cfrac{1}{z+\dots-\hat{\Omega}_{n-1,n}\hat{\Omega}_{n-1,n}\cfrac{1}{z+\hat{\Omega}_{nn}+\mathrm{i}\hat{\gamma}_{nn}(0)}}}}$$

$$=\mathrm{i}(z+\hat{\boldsymbol{\Omega}}+\mathrm{i}\hat{\boldsymbol{\gamma}}(0))_{11}^{-1} \ , \tag{14.2.28}$$

so that the power-series expansion in z^{-1} is

$$\mathrm{i}(z+\hat{\boldsymbol{\Omega}}+\mathrm{i}\hat{\boldsymbol{\gamma}}(0))_{11}^{-1}=\mathrm{i}\sum_{\varrho=0}^{\infty}\{(\hat{\boldsymbol{\Omega}}+\mathrm{i}\hat{\boldsymbol{\gamma}}(0))^{\varrho}\}_{11}(-1)^{\varrho}z^{-\varrho-1} \ . \tag{14.2.29}$$

The $\hat{\boldsymbol{\Omega}}$ matrix is tridiagonal, and the $\hat{\gamma}$ matrix has only one nonzero element, $\hat{\gamma}_{nn}$; therefore, for $\sigma \leqq n-1$, we find

$$\begin{aligned}&\{(\hat{\boldsymbol{\Omega}}+\mathrm{i}\hat{\gamma}(0))^\sigma\}_{1\nu}=(\hat{\boldsymbol{\Omega}}^\sigma)_{1\nu} \quad \text{and}\\&\{(\hat{\boldsymbol{\Omega}}+\mathrm{i}\hat{\gamma}(0))^\sigma\}_{\nu 1}=(\hat{\boldsymbol{\Omega}}^\sigma)_{\nu 1}\ .\end{aligned} \tag{14.2.30}$$

It follows from these relations and (14.2.20) that

$$\{(\hat{\boldsymbol{\Omega}}+\mathrm{i}\hat{\gamma}(0))^\varrho\}_{11}=\frac{1}{(\mathscr{G}|\mathscr{G})}(\mathscr{G}|\mathsf{L}^\varrho\mathscr{G}) \quad \text{for} \quad \varrho=0,\dots,2n-2$$

and, using (14.2.7) and (14.1.19),

$$\begin{aligned}\{(\hat{\boldsymbol{\Omega}}+\mathrm{i}\hat{\gamma}(0))^{2n-1}\}_{11}&=(\hat{\boldsymbol{\Omega}}^{2n-1})_{11}\\&\quad+\mathrm{i}\hat{\Omega}_{12}\hat{\Omega}_{23}\dots\hat{\Omega}_{n-1,n}\hat{\gamma}_{nn}(0)\Omega_{n,n-1}\dots\hat{\Omega}_{32}\hat{\Omega}_{21}\\&=\frac{1}{(\mathscr{G}|\mathscr{G})}(\mathscr{G}|\mathsf{L}^{2n-1}\mathscr{G})+\mathrm{i}\frac{1}{(\mathscr{G}|\mathscr{G})}\left(\hat{f}_n\middle|\frac{\mathrm{i}}{\mathsf{QLQ}+\mathrm{i}\eta}\hat{f}_n\right)\ .\end{aligned} \tag{14.2.31}$$

The power-series expansion of the approximation (14.2.28) in powers of z^{-1} thus agrees with the expansion of the exact function $\hat{\Xi}_{11}(\omega)$ (14.2.13) up to and including the terms $\mathrm{i}c_{2n-2}z^{-2n+1}$.

Similarly to (14.2.9), the approximation (14.2.27) for $\hat{\Xi}_{11}(z)$ can be written as a polynomial of $(n-1)$-th degree, $A_n(z)$, divided by a polynomial of n-th degree, $B_n(z)$. The coefficients which occur in this quotient are now, in general, complex. The zeros of the polynomial $B_n(z)$ are, from (14.2.28), given by the eigenvalues of the matrix $-\boldsymbol{\Lambda}=-(\hat{\boldsymbol{\Omega}}+\mathrm{i}\hat{\gamma}(0))$; they therefore, according to the general results of Sect. 13.2, are to be found in the lower z-semiplane. If all the eigenvalues are different[8], we can carry out a decomposition of $\hat{\Xi}_{11}(z)$ in partial fractions as for (14.2.23), and can find a system of equations analogous to (14.2.24) for the coefficients which occur. From the result (13.2.23), we see that approximation (14.2.28) has the property that the real part of $\hat{\Xi}_{11}(z)$ is never negative (cf. Fig. 14.1b).

The approximation described in this section, as a special case of that in Sect. 13.2, is valid so long as the conditions $|\hat{\gamma}_{nn}(0)|\tau_c \ll 1$ and $\|\hat{\boldsymbol{\Omega}}\|\tau_c \ll 1$ are fulfilled. If the latter condition no longer holds, one can attempt to apply the considerations in Sect. 13.3. There are also other possibilities. For example, one can limit the frequency in $\hat{\Xi}_{11}(\omega)$ to a particular range and approximate $\hat{\gamma}_{nn}(\omega)$ in this range by a constant value, which may differ from $\hat{\gamma}_{nn}(0)$.

---

[8] In the case $n=2$, in particular for $\mathscr{G}=\mathscr{G}^\dagger$ and thus $\hat{\Omega}_{11}=0$, $\hat{\Omega}_{22}=0$, one can see that in contrast to (14.2.23), the eigenvalues may now be degenerate.

Exercise 14.1 Calculate the coefficients of the continued fraction (14.2.6) for

$$\Xi(z) = \frac{1}{(p_\lambda|p_\lambda)}\left(p_\lambda\left|\frac{\mathrm{i}}{\mathsf{L}+z}\right.p_\lambda\right),$$

where p_λ represents the momentum of the λ-th particle in a harmonic chain, $\mathscr{H} = \sum_{\nu=1}^{N} p_\nu^2/2m + k/2\sum_{\nu=1}^{N}(x_{\nu+1} - x_\nu)^2$, with $x_{N+1} = x_1$, and n is chosen to be $n = N$ (odd).

What is the result for $N \to \infty$?

15. Dynamic Onsager–Casimir Coefficients as Linear Response Functions for Generalized Forces

Having treated a special set of observables in Chap. 14, we now return to the general considerations of Chap. 11. There, we developed integro-differential equations for the correlation functions $\Xi_{\nu\mu}(t)$ i.e. for the relaxation functions $\Phi_{\nu\mu}(t)$, which permit us with the aid of the Kubo formalism to describe the linear response of a system to external time dependent fields $h_\alpha(t)$. In this chapter, these two theories will be combined. We shall obtain a generalization of the Mori integro-differential equations (11.4.30) for the expectation values $\langle \delta \mathcal{G}_\mu \rangle (t)$ of a set of observables $\{\mathcal{G}\}$; it will contain the explicitly time-dependent external fields directly. The system Σ is again taken to be adiabatically isolated.

The overall driving "forces"[1] which occur in these generalized equations may be interpreted as time *de*pendent Lagrange multipliers $\lambda_\nu(t)$ of an accompanying generalized canonical statistical operator $\mathcal{R}_{\{\mathcal{G}(h)\}}(t)$, such as we have already met in Sect. 6.4. We shall investigate the accompanying time-dependent entropy $S_{\{\mathcal{G}(h)\}}(t)$ with regard to entropy production in adiabatic systems.

15.1 The Integro-Differential Equations for the Expectation Values in Externally Driven Systems

15.1.1 The Set $\{\mathcal{G}\}$ in the Mori Projection Operator

Let a (macroscopic) system Σ be in a non-stationary state $\rho(0)$ at time $t = 0$; according to Sects. 5.2 and 6.1, $\rho(0)$ is given by maximization of the uncertainty measure with the constraints $G_\mu = \langle \mathcal{G}_\mu \rangle (0)$ (sufficient observation level):

$$\rho(0) = \mathcal{R}_{\{\mathcal{G}\}}(0) = \frac{\exp(-\sum \lambda_\mu \mathcal{G}_\mu^\dagger)}{Z(\lambda_1 \ldots)} = \frac{\exp(-\beta \mathcal{H} - \sum \lambda'_\mu \mathcal{G}_\mu^\dagger)}{Z} , \tag{15.1.1}$$

so that if the expectation values G_μ are known, then the Lagrange multipliers λ_μ are fixed at the time $t = 0$ by the equations:

$$\mathrm{Tr}(\mathcal{G}_\mu \mathcal{R}_{\{\mathcal{G}\}}(0)) = G_\mu . \tag{15.1.2}$$

[1] These are not to be confused with the "forces" $\langle \mathcal{M}_\alpha \rangle$ from (8.3.4).

The Hamiltonian $\mathscr{H} = \mathscr{G}_0$ is to be included in the set of observables $\mathscr{G}_\mu$ in all cases. We set

$$\lambda'_\mu = \lambda_\mu - \beta\delta_{\mu 0} \quad (\mu = 0, 1, 2, \dots) ,$$

where β is an average initial temperature which we choose to be λ_0. The remaining observables (or at least some of them) do not commute with $\mathscr{H}$ and express the non-stationary character of $\rho(0)$. Examples are partial Hamiltonians $\mathscr{H}_\nu$ with Lagrange multipliers $\Delta\beta_\nu = \beta_\nu - \beta$, which represent the deviations of the temperature β_ν of the subsystems from β.

The system is taken to be not too far from equilibrium at $t = 0$; we can thus consider the Lagrange multipliers which contribute to the deviations from a stationary state as small parameters, i.e. we use only the linear expansion of (15.1.1) [compare (9.9.2)]:

$$\rho(0) = \mathscr{R}_\beta \left\{ 1 - \frac{1}{\beta} \sum_\mu \lambda'_\mu \int_0^\beta \mathrm{e}^{\alpha\mathscr{H}}\, \delta\mathscr{G}^\dagger_\mu \mathrm{e}^{-\alpha\mathscr{H}}\, d\alpha \right\} . \tag{15.1.3}$$

At times $t > 0$, we allow external, time dependent fields $h_\nu(t)$ to act on Σ (in contrast to a pure relaxation process; Sect. 11.4.2); we take $h_\nu(0) = 0$. Due to these fields, we must allow for an explicit time dependence of the operators, $\mathscr{G}_\mu(h(t))$ (Schrödinger formulation). Within the framework of a theory of linear deviations from thermal equilibrium it suffices to consider only linear terms in the fields,

$$\mathscr{G}_\mu(h(t)) = \mathscr{G}_\mu + \sum_\nu h_\nu(t) \frac{\partial \mathscr{G}_\mu}{\partial h_\nu} .^2 \tag{15.1.4}$$

The most significant example is the explicitly time dependent Hamiltonian operator $\mathscr{H}(h(t))$ in the linear approximation in the fields,

$$\mathscr{H}(h(t)) = \mathscr{H} + \sum_\nu h_\nu(t) \frac{\partial \mathscr{H}}{\partial h_\nu} = \mathscr{H} - \sum_\nu h_\nu(t) \mathscr{M}^\dagger_\nu . \tag{15.1.5}$$

In the Mori theory, we however require a *projection operator* P *onto a set* $\{\mathscr{G}\}$ of explicitly time-independent observables. We therefore need to make some comments about the choice of the set $\{\mathscr{G}\}$. As pointed out in Sect. 11.4.2, this set initially contains all the observables which occur in $\rho(0)$, including the Hamiltonian operator $\mathscr{H}$. Furthermore, the derivatives of the Hamiltonian $\mathscr{H}(h)$ with respect to the external fields h_ν are *likewise* to be included in the set $\{\mathscr{G}\}$; they were denoted above by $\mathscr{M}^\dagger_\nu$ and are, by definition, explicitly time independent. Further observables can be included as required, in particular for certain approximations. All together, the set $\{\mathscr{G}\}$ may contain only linearly independent

[2] According to Footnote 26 (p. 145), operators without an argument, h, indicate that they are to be taken at $h(t) = 0$, i.e. $\mathscr{G}_\mu$ and $\partial\mathscr{G}_\mu/\partial h_\nu$ are explicitly time independent.

operators $\mathscr{G}_\mu$. With this definition of $\{\mathscr{G}\}$, we can allow the sums in (15.1.3 and 15.1.5) to run over all the observables of the set $\{\mathscr{G}\}$, if we set the additional Lagrange multipliers and fields to zero (e.g. $\lambda'_0=0$, $h_0(t)=0$). We shall make use of this convention in the following. Let the projection operator P be defined by $\delta\mathscr{G}_\mu$.

15.1.2 The Derivation of Generalized Mori Equations for the Expectation Values $\langle\mathscr{G}_\mu\rangle(t)$ in an Externally Driven System

In Sect. 9.9, we derived a generalized, linear Kubo formula (9.9.13) for the expectation values $\mathrm{Tr}\{\rho(t)\mathscr{G}_\mu(h(t))\} = \langle\mathscr{G}_\mu(h)\rangle(t)$ of explicitly time dependent operators $\mathscr{G}_\mu(h(t))$; it allows for a linearized, non-stationary initial state $\rho(0)$, corresponding to (15.1.3). This formula shows that with the condition of linearization about the equilibrium values, the following relation holds for the expectation values of (15.1.4):

$$\langle\mathscr{G}_\mu(h)\rangle(t) = \langle\mathscr{G}_\mu\rangle(t) + \sum_\nu h_\nu(t)\left\langle\frac{\partial\mathscr{G}_\mu}{\partial h_\nu}\right\rangle_\beta . \tag{15.1.6}$$

It therefore suffices to formulate a theory for $\langle\mathscr{G}_\mu\rangle(t)$, since, according to (15.1.6), we can obtain $\langle\mathscr{G}_\mu(h)\rangle(t)$ by adding a term $\sum_\nu h_\nu(t)\langle\partial\mathscr{G}_\mu/\partial h_\nu\rangle_\beta$.

Equation (9.9.13) for $\langle\mathscr{G}_\mu\rangle(t)$ reads:

$$\langle\mathscr{G}_\mu\rangle(t) = \langle\mathscr{G}_\mu\rangle_\beta - \frac{1}{\beta}\sum_\nu \lambda'_\nu\Phi_{\nu\mu}(t) + \sum_\nu\int_0^t h_\nu(t')\varphi_{\nu\mu}(t-t')dt' . \tag{15.1.7}$$

This represents a superposition of the relaxation from the presumed non-stationary initial state (15.1.3) and the retarded influence of the external fields $h_\nu(t')$.

In order to obtain a system of integro-differential equations for the expectation values $\langle\mathscr{G}_\mu\rangle(t)$ of the explicitly time-independent set $\{\mathscr{G}\}$ for a system Σ which is under the influence of time-dependent fields, we differentiate (15.1.7) with respect to time, insert the Mori equation (11.4.3) for $\Phi_{\nu\mu}(t)$, and apply (15.1.7) once more to eliminate $\Phi_{\nu\mu}$. The result of such a calculation (15.1.10 to 12) is

$$\boxed{\begin{aligned}&\frac{d}{dt}\langle\mathscr{G}_\mu\rangle(t)\\ &\quad=\sum_\nu\left\{\langle\mathscr{G}_\nu\rangle(t)-\langle\mathscr{G}_\nu\rangle_\beta-\sum_\rho h_\rho(t)\chi^{\mathrm{T}}_{\rho\nu}\right\}\mathrm{i}\Omega_{\nu\mu}\\ &\quad\quad-\sum_\nu\int_0^t\left\{\langle\mathscr{G}_\nu\rangle(t')-\langle\mathscr{G}_\nu\rangle_\beta-\sum_\rho h_\rho(t')\chi^{\mathrm{T}}_{\rho\nu}\right\}\gamma_{\nu\mu}(t-t')dt'\end{aligned}} \ . \tag{15.1.8}$$

This *closed system of integro-differential equations for the dynamics of the expectation values* $\langle \mathcal{G}_\mu \rangle(t)$ *is valid in linear theory for an arbitrary, isolated system with a non-stationary initial state (15.1.3) under the action of an external field* $h_\rho(t)$ *for times* $t > 0$. In the special case that no external fields are present, $h_\rho(t) = 0$, this system of equations is, of course, identical to the Mori equations, (11.4.30).

The constant Lagrange multipliers λ_ν do not enter (15.1.8) explicitly. They may be *freely chosen* within the validity of the linear theory and determine the non-stationary initial state. From (15.1.3 or 15.1.7), the relation

$$\langle \mathcal{G}_\mu \rangle(0) - \langle \mathcal{G}_\mu \rangle_\beta = -\frac{1}{\beta}\sum_\nu \lambda'_\nu \chi^{\mathrm{T}}_{\nu\mu} , \tag{15.1.9}$$

is valid; we already know this equation from the Mori theory (11.4.33).

Proof of the Integro-Differential Equation (15.1.8) for Externally Driven Systems. We substitute (9.1.17) into (15.1.7) and employ for brevity the usual abbreviation $f*g = g*f$ for a convolution $\int_0^t f(t')g(t-t')\,dt'$. Differentiating (15.1.7) with respect to time, we then obtain

$$\frac{d}{dt}\langle \mathcal{G}_\mu \rangle(t) = \sum_\nu \left\{ -\frac{\lambda'_\nu}{\beta}\dot{\Phi}_{\nu\mu}(t) - h_\nu * \dot{\Phi}_{\nu\mu} \right\} . \tag{15.1.10}$$

For the relaxation functions $\Phi_{\nu\mu}(t)$ we use the Mori equation, (11.4.3):

$$\frac{d}{dt}\langle \mathcal{G}_\mu \rangle(t) = -\sum_{\nu\lambda}\left(\frac{\lambda'_\nu}{\beta}\Phi_{\nu\lambda}(t) + \dot{h}_\nu * \Phi_{\nu\lambda} \right)(\mathrm{i}\Omega_{\lambda\mu} - (*\gamma_{\lambda\mu})) , \tag{15.1.11}$$

where, in the last term, we have made use of the associative property of two convolutions. In order to eliminate $-\sum_\nu(\lambda'_\nu/\beta\,\Phi_{\nu\lambda}(t) + \dot{h}_\nu * \Phi_{\nu\lambda})$, we return to (15.1.7) and carry out an integration by parts of the convolution term, $-h_\nu * \dot{\Phi}_{\nu\mu} = -\dot{h}_\nu * \Phi_{\nu\mu} + h_\nu(t)\chi^{\mathrm{T}}_{\nu\mu}$, remembering the fact that $h_\nu(0) = 0$. Equation (15.1.7) then takes on the form

$$\langle \mathcal{G}_\mu \rangle(t) = \langle \mathcal{G}_\mu \rangle_\beta + \sum_\nu \left\{ -\frac{\lambda'_\nu}{\beta}\Phi_{\nu\mu}(t) - \dot{h}_\nu * \Phi_{\nu\mu} + h_\nu(t)\chi^{\mathrm{T}}_{\nu\mu} \right\} . \tag{15.1.12}$$

Substitution of (15.1.12) into (15.1.11) yields (15.1.8), q.e.d.

The terms of (15.1.8) in curly brackets have an intuitive *physical interpretation*:

In the simplest case of a relaxation process (i.e. without time-dependent fields), the ordinary Mori equation (11.4.30) shows that the time variation of $\langle \mathcal{G}_\mu \rangle(t)$ is produced by the instantaneous deviation from thermal equilibrium, $\langle \mathcal{G}_\nu \rangle(t') - \langle \mathcal{G}_\nu \rangle_\beta$. This means, expressed intuitively, that the "goal" of the relaxation at each instant is the equilibrium value.

In the case of a system which is driven by external fields $h_\nu(t')$, from (15.1.8), the quantity $\langle \mathcal{G}_\nu \rangle_\beta + \sum_\varrho h_\varrho(t') \chi^{\mathrm{T}}_{\varrho\nu}$ is the instantaneous "goal"; it now shows an explicit time dependence[3] due to the fields $h_\varrho(t')$. This quantity can be interpreted in terms of the equation of state $G_\nu(h_\varrho, \beta)$ with $h_\varrho = h_\varrho(t)$, by using a canonical statistical operator $\mathcal{R}_{\mathcal{H}(h(t))}$ and linearizing it in $h_\varrho(t)$:

$$\mathcal{R}_{\mathcal{H}(h(t))} = \frac{\exp\left[-\beta\left(\mathcal{H} - \sum_\varrho h_\varrho(t)\mathcal{G}^\dagger_\varrho\right)\right]}{Z(\beta, h(t))}$$

$$= \mathcal{R}_\beta \left\{ 1 + \sum_\varrho h_\varrho(t) \int_0^\beta \mathrm{e}^{\alpha\mathcal{H}}\, \delta\mathcal{G}^\dagger_\varrho \mathrm{e}^{-\alpha\mathcal{H}}\, d\alpha \right\} + \cdots \qquad (15.1.13)$$

Then, in the linear approximation, the following relation holds:

$$\mathrm{Tr}(\mathcal{R}_{\mathcal{H}(h(t))}\mathcal{G}_\nu) = \langle \mathcal{G}_\nu \rangle_\beta + \sum_\varrho h_\varrho(t) \chi^{\mathrm{T}}_{\varrho\nu} \,. \qquad (15.1.14)$$

This expectation value is just the above mentioned instantaneous "goal", which has a time dependence via the fields $h_\varrho(t)$.

It is useful to rewrite (15.1.8) in terms of the expectation values $\langle \mathcal{G}_\mu(h) \rangle(t)$ of the explicitly time-dependent operators $\mathcal{G}_\mu(h(t))$, using relation (15.1.6). One obtains[4]

$$\begin{aligned} &\frac{d}{dt}\langle \mathcal{G}_\mu(h)\rangle(t) - \sum_\nu \dot{h}_\nu(t) \left\langle \frac{\partial \mathcal{G}_\mu}{\partial h_\nu} \right\rangle_\beta \\ &\quad = \sum_\nu \left\{ \langle \mathcal{G}_\nu(h)\rangle(t) - \langle \mathcal{G}_\nu\rangle_\beta - \sum_\varrho h_\varrho(t)\chi^{\mathrm{T}}_{\varrho\nu(h)} \right\} \mathrm{i}\Omega_{\nu\mu} \\ &\quad\quad - \sum_\nu \int_0^t \left\{ \langle \mathcal{G}_\nu(h)\rangle(t') - \langle \mathcal{G}\rangle_\beta - \sum_\varrho h_\varrho(t')\chi^{\mathrm{T}}_{\varrho\nu(h)} \right\} \gamma_{\nu\mu}(t-t')\,dt' \end{aligned} \,. \qquad (15.1.15)$$

The physical interpretation of the generalized Mori equations (15.1.8 or 15.1.15) is discussed in Sect. 15.1.3 under a somewhat different aspect, which will

---

[3] Instead of the isothermal "goal", an adiabatic "goal" could also be introduced, if the terms with $\Omega_{0\mu}$ and $\gamma_{0\mu}(t)$ in (15.1.8) are eliminated by using

$$\omega_{0\mu} = \sum_\nu \chi^{\mathrm{T}}_{0\nu}\Omega_{\nu\mu} = 0 \quad \text{and} \quad l_{0\mu}(t) = \sum_\nu \chi^{\mathrm{T}}_{0\nu}\gamma_{\nu\mu}(t) = 0 \,.$$

[4] Here, $\chi^{\mathrm{T}}_{\varrho\nu(h)}$ is the *total* susceptibility defined in (8.3.14), $\chi^{\mathrm{T}}_{\varrho\nu} + \langle \partial\mathcal{G}_\nu / \partial h_\varrho \rangle_\beta$, relative to the observables $\mathcal{G}_\varrho$ and $\mathcal{G}_\nu(h)$.

directly illustrate the connection to the thermodynamics of linear, irreversible processes.

15.1.3 Time-Dependent Lagrange Multipliers $\lambda_\nu(t)$ for the Accompanying Generalized Canonical Statistical Operator $\mathscr{R}_{\{\mathscr{G}(h)\}}$ as Generalized Forces

The expectation values $\langle \mathscr{G}_\mu(h) \rangle(t)$ at the time t may be used for the construction of a *temporally accompanying, generalized canonical statistical operator* $\mathscr{R}_{\{\mathscr{G}(h)\}}(t)$ [compare Sect. 6.4] over the observation level $\mathscr{G}_\mu(h(t))$. By maximizing the uncertainty measure while taking account of the constraints, which are given at an arbitrary time t by the expectation values $\langle \mathscr{G}_\mu(h) \rangle(t) = G_\mu(t)$, we construct the accompanying canonical statistical operator over the observation level $\{\mathscr{G}(h)\}$ according to (6.4.2); the result is:

$$\mathscr{R}_{\{\mathscr{G}(h)\}}(t) = \frac{\exp\left[-\sum_\mu \lambda_\mu(t)\mathscr{G}_\mu^\dagger(h(t))\right]}{Z(\lambda_\mu(t), h_\nu(t))} . \tag{15.1.16}$$

The numerical values of $G_\mu(t)$ fix the *time-dependent Lagrange multipliers* $\lambda_\mu(t)$ in (15.1.16) *at every time* t through the definition

$$G_\mu(t) = \mathrm{Tr}\{\mathscr{R}_{\{\mathscr{G}(h)\}}(t)\mathscr{G}_\mu(h(t))\} . \tag{15.1.17}$$

The total Hamiltonian $\mathscr{H}(h(t))$ is always contained in the set of the $\mathscr{G}_\mu(h(t))$. We temporarily split it out in (15.1.16), together with β, in order to carry out an expansion about the equilibrium value,

$$\mathscr{R}_{\{\mathscr{G}(h)\}}(t) = \frac{\exp[-\beta\mathscr{H}(h(t)) - \sum \lambda'_\mu(t)\mathscr{G}_\mu^\dagger(h(t))]}{Z(\beta, \lambda'_\mu(t), h_\nu(t))} . \tag{15.1.18}$$

With (15.1.5) and

$$\lambda'_\mu(t) = \lambda_\mu(t) - \beta\delta_{\mu 0} \quad (\mu = 0, 1, 2, \ldots) , \tag{15.1.19}$$

we obtain, in a consistent linear expansion about the equilibrium values,

$$\mathscr{R}_{\{\mathscr{G}(h)\}}(t) = \mathscr{R}_\beta\left\{1 + \sum_\nu h_\nu(t)\int_0^\beta e^{\alpha\mathscr{H}}\,\delta\mathscr{M}_\nu^\dagger e^{-\alpha\mathscr{H}}\,d\alpha \right.$$
$$\left. - \sum_\mu \frac{\lambda'_\mu(t)}{\beta}\int_0^\beta e^{\alpha\mathscr{H}}\,\delta\mathscr{G}_\mu^\dagger e^{-\alpha\mathscr{H}}\,d\alpha\right\} \tag{15.1.20a}$$

or, combining the observables again,

$$\mathscr{R}_{\{\mathscr{G}(h)\}} = \mathscr{R}_\beta\left\{1 - \sum_\mu (\lambda'_\mu(t) - \beta h_\mu(t))\frac{1}{\beta}\int_0^\beta e^{\alpha\mathscr{H}}\,\delta\mathscr{G}_\mu^\dagger e^{-\alpha\mathscr{H}}\,d\alpha\right\} . \tag{15.1.20b}$$

If we take the expectation values $\langle \mathscr{G}_\mu \rangle(t)$, using the last equation and keeping in mind (15.1.6 and 17), we obtain a system of equations for the time-dependent Lagrange multipliers $\lambda'_\mu(t)$ in the form

$$\langle \mathscr{G}_\mu \rangle(t) = \langle \mathscr{G}_\mu \rangle_\beta - \sum_\nu \{\lambda'_\nu(t) - \beta h_\nu(t)\} \frac{1}{\beta} \chi^{\mathrm{T}}_{\nu\mu} \tag{15.1.21}$$

or, using (11.4.13),

$$\boxed{\lambda'_\nu(t) = \beta h_\nu(t) - \sum_\mu \langle \delta \mathscr{G}_\mu \rangle(t) g_{\mu\nu} \quad (\mu, \nu = 0, 1, 2, \dots)} \; . \tag{15.1.22}$$

For $t = 0$, (15.1.21) naturally becomes identical to (15.1.9), or conversely, *the accompanying generalized canonical statistical operator* $\mathscr{R}_{\{\mathscr{G}(h)\}}(t)$ *arises continuously* at $t > 0$ out of the initial generalized canonical statistical operator, $\mathscr{R}_{\{\mathscr{G}\}}(0)$. *In this process, those Lagrange multipliers which are zero at* $t = 0$ *can continuously* acquire finite values[5].

One should notice in particular that in the case of a *system which is not externally driven*, the fields $h_\nu(t)$ naturally are not present in (15.1.22); however, *nevertheless, time-dependent Lagrange multipliers* $\lambda_\nu(t)$ result, which in this case are generated solely by $\langle \delta \mathscr{G}_\mu \rangle(t)$.

If the computed time-dependent Lagrange multipliers (15.1.21) are inserted into the general integro-differential equations (15.1.8), with consideration of (11.2.15 and 11.2.24) and (11.4.13), these assume the compact form

$$\boxed{\begin{aligned} \frac{d}{dt}\langle \mathscr{G}_\mu \rangle(t) = &-\sum_\nu \lambda_\nu(t)(\mathscr{G}_\nu | \mathrm{i}\mathsf{L}\mathscr{G}_\mu) \\ &+ \sum_\nu \int_0^t \lambda_\nu(t - t')(\not{f}_\nu | \not{f}_\mu(t'))\,dt' \end{aligned}} \;.^{6} \tag{15.1.23}$$

Its one-sided Fourier (Laplace) transform becomes a system of algebraic equations,

$$\boxed{\left(\frac{d}{dt}\langle \mathscr{G}_\mu \rangle\right)(\omega) = -\sum_\nu \lambda_\nu(\omega)\{(\mathscr{G}_\nu | \mathrm{i}\mathsf{L}\mathscr{G}_\mu) - L_{\nu\mu}(\omega)\}} \; . \tag{15.1.24}$$

---

[5] Since the Hamiltonian also belongs to the set $\{\mathscr{G}\}$, (15.1.22), for example, allows one to calculate the variation of the temperature $\beta(t) = \lambda_0(t) = \beta + \lambda'_0(t)$.

[6] In (15.1.23), the part with $\lambda_0(t)$ drops out because of $(\mathsf{L}\mathscr{G}_0| = (\mathsf{L}\mathscr{H}| = 0$; thus, one can write $\lambda_\nu(t)$ instead of $\lambda'_\nu(t)$ in this equation. It is therefore invariant with respect to linear transformations of the $\mathscr{G}_\mu$.

We see that the *time-dependent Lagrange multipliers* $\lambda_\nu(t)$, or their Fourier transforms $\lambda_\nu(\omega)$, act like fields in the Kubo formula. They are for this reason called "forces" in the thermodynamics of linear irreversible processes, and the derivatives $(d/dt\langle\mathscr{G}_\mu\rangle)(\omega)$ which they generate are called "currents". The dynamic Onsager–Casimir coefficients $L_{\nu\mu}(\omega)$ have, *like* $(\mathscr{G}_\nu|\mathrm{i}\mathrm{L}\mathscr{G}_\mu)$, *the properties of a "response function"*, analogous to the dynamic susceptibility $\chi_{\nu\mu}(\omega)$ in the Kubo theory, where these properties, however, are defined with respect to external fields.

Using the same considerations which led us to (15.1.15), we can rewrite (15.1.23) in terms of the expectation values of explicitly time-dependent observables $\mathscr{G}_\mu(h(t))$,

$$\boxed{\begin{aligned}&\frac{d}{dt}\langle\mathscr{G}_\mu(h)\rangle(t)-\sum_\nu\dot{h}_\nu(t)\left\langle\frac{\partial\mathscr{G}_\mu}{\partial h_\nu}\right\rangle_\beta\\&\quad=-\sum_\nu\lambda_\nu(t)(\mathscr{G}_\nu|\mathrm{i}\mathrm{L}\mathscr{G}_\mu)+\sum_\nu\int_0^t\lambda_\nu(t-t')(f_\nu|f_\mu(t'))\,dt'\end{aligned}}\ , \tag{15.1.25}$$

in which the Lagrange multipliers $\lambda_\nu(t)$ obey the relations from (15.1.21):

$$\begin{aligned}-\frac{1}{\beta}\sum_\nu\{\lambda_\nu(t)-\beta\delta_{\nu0}\}\chi^{\mathrm{T}}_{\nu\mu}&=\langle\mathscr{G}_\mu\rangle(t)-\langle\mathscr{G}_\mu\rangle_\beta-\sum_\nu h_\nu(t)\chi^{\mathrm{T}}_{\nu\mu}\\&=\langle\mathscr{G}_\mu(h)\rangle(t)-\langle\mathscr{G}_\mu\rangle_\beta-\sum_\nu h_\nu(t)\chi^{\mathrm{T}}_{\nu\mu(h)}\ .\end{aligned} \tag{15.1.26}$$

In the phenomenological treatment of irreversible processes, the "forces" are usually denoted by $X_\nu(t)$ instead of $\lambda_\nu(t)$, and the "currents" by $J_\mu(t)$ instead of $d/dt\langle\mathscr{G}_\mu\rangle(t)$. The relations (15.1.23 or 24), which we have justified here on a quantum-statistical basis, are known in that field as the "phenomenological linear Ansatz". However, since in a phenomenological description, detailed results for the dynamic Onsager–Casimir coefficients $L_{\nu\mu}(\omega)$ and the frequency matrix $(\mathscr{G}_\nu|\mathrm{i}\,\mathrm{L}\mathscr{G}_\mu)$ are hardly available, these quantities are introduced without precise knowledge of their structures. The socalled "reciprocity relations" for the Onsager–Casimir coefficients are of great importance in this method; they are based upon time reversal behavior. These additional "selection rules" are not used in the present discussion, like other symmetry properties, because they play no role in the general context, although knowledge of them simplifies the computation of certain coefficients [see Appendix B]. In phenomenological irreversible thermodynamics, Markovian *approximations* are frequently used in the "linear Ansatz" (15.1.23 or 24); i.e. $L_{\nu\mu}(\omega)$ is taken as $L_{\nu\mu}(0)$, whereby (15.1.23) become differential equations,

$$\frac{d}{dt}\langle\mathscr{G}_\mu\rangle(t)=-\sum_\nu\lambda_\nu(t)\{(\mathscr{G}_\nu|\mathrm{i}\mathrm{L}\mathscr{G}_\mu)-L_{\nu\mu}(\omega=0)\}\ . \tag{15.1.23a}$$

Exercise 15.1 Two parts of an adiabatic system Σ possess the instantaneous temperatures $\beta_1(t)$ and $\beta_2(t)$. How do the partial energies $\langle \mathscr{H}_1 \rangle(t)$ and $\langle \mathscr{H}_2 \rangle(t)$ vary, if $\mathscr{H} = \mathscr{H}_1 + \mathscr{H}_2$ and $[\mathscr{H}_1, \mathscr{H}_2] \neq 0$?

15.2 The Irreversible Entropy Production in Linear Dynamic Processes

15.2.1 The Accompanying Entropy $S_{\{\mathscr{G}(h)\}}(t)$

The temporally accompanying, generalized canonical statistical operator $\mathscr{R}_{\{\mathscr{G}(h)\}}(t)$ over an observation level $\{\mathscr{G}(h)\}$, which was employed in (15.1.16), yields at every time t the maximum uncertainty measure, the *accompanying entropy* over this subset, as we have shown in Sect. 6.4:

$$
\begin{aligned}
S_{\{\mathscr{G}(h)\}}(t) &= -k\,\mathrm{Tr}(\mathscr{R}_{\{\mathscr{G}(h)\}} \ln \mathscr{R}_{\{\mathscr{G}(h)\}}) \\
&= k \sum_{\mu} \lambda_{\mu}^{*}(t) \langle \mathscr{G}_{\mu}(h) \rangle (t) + k \ln Z_{\{\mathscr{G}(h)\}}(t) \ .
\end{aligned}
\tag{15.2.1}
$$

Its time variation is given by

$$
\begin{aligned}
\frac{d}{dt} S_{\{\mathscr{G}(h)\}}(t) = k \sum_{\mu} \lambda_{\mu}^{*}(t) \Bigg\{ & \frac{d}{dt} \langle \mathscr{G}_{\mu}(h) \rangle (t) \\
& - \sum_{\nu} \left\langle \frac{\partial \mathscr{G}_{\mu}(h)}{\partial h_{\nu}} \right\rangle_{\lambda} (t) \dot{h}_{\nu}(t) \Bigg\} \ .
\end{aligned}
\tag{15.2.2}
$$

These equations are exact and are valid even far from thermal equilibrium[7].

We wish to investigate these relations in particular for *linear* processes in adiabatic systems Σ, on which work can be performed. As we have already shown in Sect. 15.1, the generally time-dependent expectation value $\langle \partial \mathscr{G}_{\mu}(h)/\partial h_{\nu} \rangle_{\lambda}(t)$ is simplified in the linear approximation to the equilibrium expectation value $\langle \partial \mathscr{G}_{\mu}/\partial h_{\nu} \rangle_{\beta}$, because it is multiplied by $\dot{h}_{\nu}(t)$. Using (15.1.6), we can then rewrite (15.2.2) in the form:

$$
\boxed{\frac{d}{dt} S_{\{\mathscr{G}(h)\}}(t) = k \sum_{\mu} \lambda_{\mu}^{*}(t) \frac{d}{dt} \langle \mathscr{G}_{\mu} \rangle (t)} \ ,
\tag{15.2.3}
$$

where the Lagrange multipliers ("forces") are given from (15.1.22) by

$$
\lambda_{\mu}(t) - \beta \delta_{\mu 0} = \beta h_{\mu}(t) - \sum_{\nu} \langle \delta \mathscr{G}_{\nu} \rangle (t) g_{\nu\mu} \ .
\tag{15.2.4}
$$

---

[7] The abbreviation $\langle \mathscr{F} \rangle_{\lambda}$ used in (15.2.2) means $\mathrm{Tr}\,(\mathscr{R}_{\{\mathscr{G}(h)\}} \mathscr{F})$.

The entropy variation (15.2.3) can be expressed in terms of the forces $\lambda_\mu(t)$ by using (15.1.23); the resulting form is

$$\boxed{\frac{d}{dt}S_{\{\mathcal{G}(h)\}}(t)=k\sum_{\mu\nu}\lambda_\mu^*(t)\int_0^t\lambda_\nu(t-t')(\dot{\mathcal{G}}_\nu|\dot{\mathcal{G}}_\mu(t'))\,dt'}\quad, \tag{15.2.5}$$

since the term $\sum_{\mu\nu}\lambda_\mu^*(t)\lambda_\nu(t)(\mathcal{G}_\nu|\mathrm{i}\,\mathrm{L}\mathcal{G}_\mu)$ makes no contribution owing to $S=S^*$. The entropy variation can also be expressed in terms of the deviations $\langle\delta\mathcal{G}_\mu\rangle(t)$ of the expectation values $\langle\mathcal{G}_\mu\rangle(t)$ from their thermal equilibrium values $\langle\mathcal{G}_\mu\rangle_\beta$, and of the fields $h_\nu(t)$, by making use of relation (15.2.4) for the forces:

$$\boxed{\begin{aligned}\frac{d}{dt}S_{\{\mathcal{G}(h)\}}(t)=&\;k\beta\sum_\mu h_\mu^*(t)\frac{d}{dt}\langle\mathcal{G}_\mu\rangle(t)\\&-k\sum_{\nu\mu}\langle\delta\mathcal{G}_\mu\rangle(t)^*g_{\mu\nu}^*\frac{d}{dt}\langle\mathcal{G}_\nu\rangle(t)\end{aligned}}\quad. \tag{15.2.6}$$

In integral form, this equation is

$$\begin{aligned}S_{\{\mathcal{G}(h)\}}(t)&+\frac{k}{2}\sum_{\nu\mu}\langle\delta\mathcal{G}_\mu\rangle(t)^*g_{\mu\nu}^*\langle\delta\mathcal{G}_\nu\rangle(t)-k\beta\int_0^t\sum_\mu h_\mu^*(t')\frac{d}{dt'}\langle\mathcal{G}_\mu\rangle(t')\,dt'\\&=S_{\{\mathcal{G}\}}(0)+\frac{k}{2}\sum_{\nu\mu}\langle\delta\mathcal{G}_\mu\rangle(0)^*g_{\mu\nu}^*\langle\delta\mathcal{G}_\nu\rangle(0)\\&=S_{\mathcal{H}}(\beta)+k\beta\langle\delta\mathcal{H}\rangle(0)\ ,\end{aligned} \tag{15.2.7}$$

where $S_{\mathcal{H}}(\beta)$ represents the equilibrium entropy. This form for $S_{\{\mathcal{G}(h)\}}(t)$ can naturally also be obtained by expanding $\mathcal{R}_{\{\mathcal{G}(h)\}}(t)$ about $\mathcal{R}_{\mathcal{H}}(\beta)$ (Exercise 15.2).

We have already pointed out in Sect. 6.4 that the temporally accompanying entropy $S_{\{\mathcal{G}(h)\}}(t)$ of an adiabatically isolated system can either increase or decrease, so that the expressions (15.2.2, 3, 5, 6) can have either positive or negative values.

In the following, we consider particular physical situations for which we obtain results for the behavior of certain expressions, derived from $S_{\{\mathcal{G}(h)\}}(t)$. We shall see that, based on the forces $\lambda_\mu(t)$, these results are determined by the component $L'_{\nu\mu}(\omega)$ of the Onsager–Casimir coefficients, in analogy to the role of $\chi''_{\nu\mu}(\omega)$ in the Kubo theory for the work performed by external fields.

15.2.2 Significance of the Onsager–Casimir Coefficients $L'_{\nu\mu}(\omega)$ for Entropy Production

We first consider the action of a *harmonically oscillating field* $h_\nu(t)=\hat{h}_\nu\sin\omega_0 t$ ($\mathcal{G}_\nu=\mathcal{G}_\nu^\dagger$) on the system Σ, which is in thermal equilibrium at the time $t=0$. In the steady state, i.e. for times $t > \tau_R$, we can replace the upper limit of integration by ∞ in the third term of (15.1.7). The solution [compare (9.2.9)]

$$\langle \delta \mathscr{G}_\mu \rangle(t) = \sum_\nu \hat{h}_\nu \{\mathrm{Re}\, \chi_{\nu\mu}(\omega_0) \sin \omega_0 t - \mathrm{Im}\, \chi_{\nu\mu}(\omega_0) \cos \omega_0 t\} \tag{15.2.8}$$

could be inserted into (15.2.7) to give an explicit expression for the entropy $S_{\{\mathscr{G}(h)\}}(t)$ as a function of the time t in the steady state (cf. Exercise 15.2). However, a more interesting quantity is the time-averaged entropy production in the system, $\bar{\dot{S}}_{\{\mathscr{G}(h)\}}$, which we obtain from (15.2.3) by averaging over the period $2\pi/\omega_0$:

$$\bar{\dot{S}}_{\{\mathscr{G}(h)\}} = k\beta \sum_{\nu\mu} \hat{h}_\nu \hat{h}_\mu \frac{\omega_0}{2} \chi''_{\nu\mu}(\omega_0) \geqq 0 \ . \tag{15.2.9}$$

Comparing with (9.6.4), we see that the relation

$$\bar{\dot{S}}_{\{\mathscr{G}(h)\}} = \frac{1}{T} \bar{W} \geqq 0 \tag{15.2.10}$$

is fulfilled, i.e. for a harmonically oscillating field, the time-averaged entropy production in an adiabatically isolated system in the steady state is equal to the power input divided by the absolute temperature.

The external field amplitudes $\hat{h}_\mu$ enter (15.2.10). We can naturally formulate the average entropy production on the basis of the forces $\lambda_\mu(t)$. From (15.2.8), $\langle \delta \mathscr{G}_\mu \rangle(t)$ is harmonic in the steady state, so that as a consequence of (15.2.4), $\lambda_\mu(t)$ also oscillates harmonically, $\lambda_\mu(t) - \beta\delta_{\mu 0} = \mathrm{Re}\, \hat{\lambda}_\mu(\exp - i\omega_0 t)$. If this function is inserted into (15.2.5) (with the upper limit of integration as ∞), we obtain after averaging over time

$$\bar{\dot{S}}_{\{\mathscr{G}(h)\}} = \frac{k}{2} \sum_{\nu\mu} \hat{\lambda}_\nu \hat{\lambda}_\mu^* L'_{\nu\mu}(\omega_0) = k\frac{\pi}{2} \sum_{\nu\mu} \hat{\lambda}_\nu \hat{\lambda}_\mu^* (f_\nu | \delta(\mathsf{QLQ} + \omega_0) f_\mu) \geqq 0 \ . \tag{15.2.11}$$

We see that the average entropy production can be described by the portion $L'_{\nu\mu}(\omega_0)$ of the dynamic Onsager–Casimir coefficients, if we make use of the forces $\hat{\lambda}_\nu$.

We now want to reconsider the situation which we have already treated in Sect. 9.6.2 (see Fig. 9.9), applying the more general initial condition (15.1.3): in the time interval between $t_0 = 0$ and t_1, an arbitrary external field $h_\nu(t)$ acts on a system. We wish to calculate the change in entropy[8], $S_{\{\mathscr{G}(h)\}}(\infty) - S_{\{\mathscr{G}\}}(0)$. Here, we assume that the values $\langle \mathscr{G}_\mu \rangle(\infty)$ and $S_{\{\mathscr{G}(h)\}}(\infty)$ exist in the thermodynamic

[8] In Sect. 9.6.2, the energy input to the system is $\langle \mathscr{H} \rangle(t_1) - \langle \mathscr{H} \rangle(0) = \langle \mathscr{H} \rangle(\infty) - \langle \mathscr{H} \rangle(0)$, because no external driving fields act at times $t > t_1$. A corresponding relation does *not* hold for the entropy, since at times $t > t_1$, relaxation processes with accompanying production of entropy take place.

limit. From (15.2.3), we then obtain for $\eta \to +0$:

$$S_{\{\mathcal{G}(h)\}}(\infty)-S_{\{\mathcal{G}\}}(0)=k\int_0^\infty \sum_\mu \lambda_\mu^*(t)\frac{d}{dt}\langle \mathcal{G}_\mu\rangle(t)\mathrm{e}^{-\eta t}dt\ . \tag{15.2.12}$$

Using the formula for the reverse Laplace transformation,

$$\frac{1}{2\pi}\int_{-\infty}^{\infty}\mathrm{e}^{-\mathrm{i}\omega t}\lambda_\mu(\omega)d\omega=\begin{cases}0 & \text{for}\quad t<0\\ \mathrm{e}^{-\eta t}\lambda_\mu(t) & \text{for}\quad t>0\end{cases} \tag{15.2.13}$$

we first convert (15.2.12) to the form

$$S_{\{\mathcal{G}(h)\}}(\infty)-S_{\{\mathcal{G}\}}(0)=\frac{k}{2\pi}\int_{-\infty}^{\infty}\sum_\mu \lambda_\mu^*(\omega)\left(\frac{d}{dt}\langle \mathcal{G}_\mu\rangle\right)(\omega)d\omega\ . \tag{15.2.14}$$

Inserting the relation (15.1.24), we eliminate the contribution from $(\mathcal{G}_\nu|\mathrm{i}\,\mathsf{L}\mathcal{G}_\mu)$ and from $L''_{\nu\mu}(\omega)$ in the double sum, and, using (11.3.18), we obtain:

$$\boxed{\begin{aligned}S_{\{\mathcal{G}(h)\}}(\infty)-S_{\{\mathcal{G}(h)\}}&=\frac{k}{2\pi}\int_{-\infty}^{\infty}\sum_{\mu\nu}\lambda_\mu^*(\omega)\lambda_\nu(\omega)L'_{\nu\mu}(\omega)d\omega\\ &=\frac{k}{2}\int_{-\infty}^{\infty}\sum_{\mu\nu}\lambda_\mu^*(\omega)\lambda_\nu(\omega)(f_\nu|\delta(\mathsf{QLQ}+\omega)f_\mu)\geqq 0\end{aligned}}\ . \tag{15.2.15}$$

The fact that this difference is positive was already shown by the considerations of Sect. 6.2. Formula (15.2.15) again demonstrates that the entropy production can be given as a function of the forces $\lambda_\mu(\omega)$ and the terms $L'_{\nu\mu}(\omega)$ from the dynamic Onsager–Casimir coefficients.

Finally, we shall consider (15.2.5) in the case that $\lambda_\mu(t)$ changes slowly compared to the correlation function $(f_\nu|f_\mu(t'))$ of the residual forces, so that we can carry out a Markovian approximation, i.e. we can set $\lambda_\nu(t-t')\approx\lambda_\nu(t)$. Keeping in mind (11.3.18), we obtain a result in terms of the $L'_{\nu\mu}(0)$ components of the Onsager–Casimir coefficients at the frequency $\omega=0$:

$$\boxed{\begin{aligned}\frac{d}{dt}S_{\{\mathcal{G}(h)\}}(t)&=k\sum_{\mu\nu}\lambda_\mu^*(t)\lambda_\nu(t)L'_{\nu\mu}(0)\\ &=k\pi\sum_{\mu\nu}\lambda_\mu^*(t)\lambda_\nu(t)(f_\mu|\delta(\mathsf{QLQ})f_\mu)\geqq 0\end{aligned}}\ . \tag{15.2.16}$$

The $L''_{\nu\mu}(0)$ parts drop out of the double sum. In the Markovian approximation, the time variation of the entropy is never negative. Relation (15.2.16), which exhibits no retardation in time, was an important starting point for the development of irreversible thermodynamics.

Exercise 15.2 Expand the accompanying entropy $S_{\{\mathscr{G}(h)\}}(t)$ over the relevant subset (15.1.16) in powers of $\langle \mathscr{G}_\mu(h)\rangle(t)-\langle \mathscr{G}_\nu\rangle_\beta$ and $h_\nu(t)$ up to and including the *2nd* order and compare the result with (15.2.7).

15.3 The Second Fluctuation-Dissipation Theorem

15.3.1 The Residual Force $f_\mu(t)$

The Mori equation for the dynamics of the expectation values $\langle \delta\mathscr{G}_\mu\rangle(t)$ at first contains the expectation value of the residual force, $\langle f_\mu(t)\rangle$, as seen from (11.4.21). This contribution drops out if one uses a linearized generalized canonical statistical operator (11.4.25) over the observation level $\{\mathscr{G}\}$. In the case that the observables $\mathscr{G}_\mu$ represent macroscopic quantities, the initial state $\rho(0)$ is based on knowledge of the macroscopic expectation values $\langle \mathscr{G}_\mu\rangle(0)$. One can therefore not expect that small, rapid variations of macroscopic quantities, as observed for example in the Brownian motion of a particle, which are the result of a detailed (microscopic) state, will be sufficiently described by the macroscopic relevant set of observables $\{\mathscr{G}\}$ for $\rho(0)$. Moreover, it is hardly possible to find a representative statistical operator for the initial state which allows the computation of the rapid variations as solutions to the Heisenberg equations or the Mori equations.

This problem does not occur, however, if one is interested in calculating the expectation values for products of the residual forces. They may be computed simply by using the equilibrium statistical operator $\mathscr{R}_\beta$. The goal of the following section is to demonstrate that the *equilibrium correlation functions of the spectral decomposition of the residual forces,*

$$f_\mu(\omega)=\frac{1}{\sqrt{2\pi\tau}}\int_0^\tau f_\mu(t)\mathrm{e}^{\mathrm{i}\omega t}\,dt=\frac{1}{\sqrt{2\pi\tau}}\,\frac{\exp[\mathrm{i}(\mathsf{QLQ}+\omega)\tau]-1}{\mathrm{i}(\mathsf{QLQ})+\omega)}\,f_\mu \tag{15.3.1}$$

for $\tau\to\infty$ can be expressed in terms of the components $L'_{\nu\mu}(\omega)$ of the Onsager–Casimir coefficients, which describe the entropy production (dissipation). This socalled *second fluctuation-dissipation theorem* is analogous to the first fluctuation-dissipation theorem (Sect. 9.8), which relates the equilibrium correlation functions of the spectral decomposition of the operators $\delta\mathscr{G}_\mu(\omega)$ to $\chi''_{\nu\mu}(\omega)$, i.e. to the energy absorbed.

15.3.2 Equilibrium Correlation Functions of $f_\mu(\omega)$

In order to relate the correlation functions $\langle [f_\mu(\omega)^\dagger, f_\nu(\omega)]_+\rangle_\beta$, to the Onsager coefficients, we need to recalculate them in terms of a Mori product. In the definition of the Mori product, there is an operator $(1-\exp(-\beta\hbar\mathsf{L})/\beta\hbar\mathsf{L}$; it is therefore expedient to rewrite the dynamics of $f_\mu(t)=\exp(\mathrm{i}\,\mathsf{QLQ}t)f_\mu$ in terms of dynamics with $\exp(\mathrm{i}\,\mathsf{L}t)$. For this purpose, we use the Mori equation. With the definition (15.3.1), we obtain after substituting for $t=t'+t''$ in the convolution integrals

$$f_\mu(\omega)=\frac{1}{\sqrt{2\pi\tau}}\int_0^\tau \dot{\mathscr{G}}_\mu(t)\mathrm{e}^{\mathrm{i}\omega t}\,dt-\sum_\varrho\frac{1}{\sqrt{2\pi\tau}}\int_0^\tau \mathscr{G}_\varrho(t)\mathrm{e}^{\mathrm{i}\omega t}\,dt\,\mathrm{i}\,\Omega_{\varrho\mu}$$

$$+\sum_\varrho\frac{1}{\sqrt{2\pi\tau}}\int_0^\tau \mathscr{G}_\varrho(t')\mathrm{e}^{\mathrm{i}\omega t'}\int_0^{\tau-t'}\gamma_{\varrho\mu}(t'')\mathrm{e}^{\mathrm{i}\omega t''}\,dt''\,dt'\;. \tag{15.3.2}$$

If we then compute the equilibrium correlation functions $\langle f_\mu(\omega)^\dagger f_\nu(\omega)\rangle_\beta$ using these expressions for $f_\mu(\omega)^\dagger$ and $f_\nu(\omega)$, we are able, provided the corresponding limits exist, to take first the thermodynamic limit and then the limit $\tau\to\infty$ in the factors $\int_0^{\tau-t'}\gamma_{\varrho\mu}(t'')\exp(\mathrm{i}\omega t'')dt''$. After this step, the operators are in the form (9.8.4). Therefore, as with (9.8.5), in the limit $\tau\to\infty$ we find the expression

$$\begin{aligned}&\langle f_\mu(\omega)^\dagger f_\nu(\omega)\rangle_\beta\\&=\left\langle\left\{\dot{\mathscr{G}}_\mu^\dagger-\sum_\varrho\mathscr{G}_\varrho^\dagger(-\mathrm{i}\,\Omega_{\varrho\mu}^*-\gamma_{\varrho\mu}^*(\omega))\right\}\delta(\mathsf{L}+\omega)\right.\\&\quad\left.\times\left\{\dot{\mathscr{G}}_\nu-\sum_\sigma\mathscr{G}_\sigma(\mathrm{i}\,\Omega_{\sigma\nu}-\gamma_{\sigma\nu}(\omega))\right\}\right\rangle_\beta\;.\end{aligned}\tag{15.3.3}$$

If we carry out the same calculation using the Mori product (8.2.14),

$$(f_\mu(\omega)|f_\nu(\omega))=\left\langle f_\mu(\omega)^\dagger\frac{1-\mathrm{e}^{-\beta\hbar\mathsf{L}}}{\beta\hbar\mathsf{L}}f_\nu(\omega)\right\rangle_\beta \tag{15.3.4}$$

we obtain an expression in which, instead of the δ-function as in (15.3.3), the operator

$$\frac{1-\mathrm{e}^{-\beta\hbar\mathsf{L}}}{\beta\hbar\mathsf{L}}\delta(\mathsf{L}+\omega)=-\frac{1-\mathrm{e}^{\beta\hbar\omega}}{\beta\hbar\omega}\delta(\mathsf{L}+\omega) \tag{15.3.5}$$

occurs. The comparison shows that the relation

$$\langle f_\mu(\omega)^\dagger f_\nu(\omega)\rangle_\beta=\frac{\beta\hbar\omega}{\mathrm{e}^{\beta\hbar\omega}-1}(f_\mu(\omega)|f_\nu(\omega)) \tag{15.3.6}$$

must hold. In order to compute the anticommutator of the residual forces, we require $\langle f_\nu(\omega)f_\mu(\omega)^\dagger\rangle_\beta$. To obtain it, we insert the definition (15.3.1) of $f_\mu(\omega)$

into (15.3.6),

$$\left\langle \frac{1}{\sqrt{2\pi\tau}} \int_0^\tau f_\mu(t)^\dagger e^{-i\omega t}\, dt \frac{1}{\sqrt{2\pi\tau}} \int_0^\tau f_\nu(t) e^{i\omega t}\, dt \right\rangle_\beta$$

$$= \frac{\beta\hbar\omega}{e^{\beta\hbar\omega}-1} \left(\frac{1}{\sqrt{2\pi\tau}} \int_0^\tau f_\mu(t) e^{i\omega t}\, dt \middle| \frac{1}{\sqrt{2\pi\tau}} \int_0^\tau f_\nu(t) e^{i\omega t}\, dt \right) . \qquad (15.3.7)$$

Making the replacements $f_\mu(t)^\dagger \leftrightarrow f_\nu(t)$ and $\omega \to -\omega$, we obtain with (8.2.10)

$$\langle f_\nu(\omega) f_\mu(\omega)^\dagger \rangle_\beta = -\frac{\beta\hbar\omega}{e^{-\beta\hbar\omega}-1} (f_\nu(\omega)|f_\mu(\omega))^*$$

$$= -\frac{\beta\hbar\omega}{e^{-\beta\hbar\omega}-1} (f_\mu(\omega)|f_\nu(\omega)) . \qquad (15.3.8)$$

With this result, using the definition (9.7.10) of $E(\beta,\omega)$, we obtain for the anticommutator

$$\tfrac{1}{2}\langle [f_\mu(\omega)^\dagger, f_\nu(\omega)]_+ \rangle_\beta = \beta E(\beta,\omega)(f_\mu(\omega)|f_\nu(\omega)) . \qquad (15.3.9)$$

On the other hand, we can relate $(f_\mu(\omega)|f_\nu(\omega))$ to the Onsager–Casimir coefficients via (15.3.1):

$$f_\mu(\omega) = \frac{1}{\sqrt{2\pi\tau}} \exp\left[\frac{i}{2}(\mathrm{QLQ}+\omega)\tau\right] \cdot 2 \frac{\sin(\mathrm{QLQ}+\omega)\tau/2}{\mathrm{QLQ}+\omega} f_\mu . \qquad (15.3.10)$$

From (15.3.10), we find using the Hermiticity of $\mathrm{QLQ}+\omega$ for $\tau\to\infty$ an expression in terms of a δ-function:

$$(f_\mu(\omega)|f_\nu(\omega)) = (f_\mu|\delta(\mathrm{QLQ}+\omega) f_\nu) . \qquad (15.3.11)$$

The right hand side is just the definition (11.3.18) of the dynamic Onsager–Casimir coefficients, so that we can combine (15.3.9 and 15.3.11) to give the final result,

$$\boxed{\frac{1}{2}\langle [f_\mu(\omega)^\dagger, f_\nu(\omega)]_+ \rangle_\beta = \frac{1}{\pi} \beta E(\beta,\omega) L'_{\mu\nu}(\omega)} . \qquad (15.3.12)$$

This *second fluctuation-dissipation theorem* thus connects the *equilibrium expectation values of the products of the spectrally decomposed residual forces* with the *dynamic Onsager–Casimir coefficients, which occur in the expression for the entropy production.*

16. Physical Examples

In this chapter, we shall apply the Mori theory to some particular physical systems which serve as examples. These have been chosen to be based on real physical problems, but are still sufficiently simple that they can serve to illustrate the basic features of the general theory. They thus do not provide an overview of possible applications, but rather are intended to help the reader to understand the theoretical formalism. For this reason, we begin with a model which can be solved exactly, and which allows us to carry out explicitly all the considerations of the preceding chapters.

16.1 A Heavy Particle in an Elastic Chain: A Model Which Can Be Solved Exactly – Rubin's Model

To illustrate the Mori theory, we consider a linear chain of particles which are joined by elastic bonds, having periodic boundary conditions and comprised of N particles of the same mass m and one heavy particle of mass $M \gg m$ at the position x_0 with momentum p_0 (see Fig. 16.1). We let the force constants of all the springs be the same, so that the Hamiltonian of the model is given by

$$\mathcal{H} = \sum_{\nu=1}^{N} \frac{p_\nu^2}{2m} + \frac{p_0^2}{2M} + \frac{k}{2} \sum_{\nu=0}^{N} (x_{\nu+1} - x_\nu)^2 \, , \quad x_{N+1} = x_0 \, . \tag{16.1.1}$$

We wish to study the correlation function

$$\Xi(t) = \frac{(p_0 | p_0(t))}{(p_0 | p_0)} \tag{16.1.2}$$

with the aid of the projection operator $\mathsf{P} = |p_0)\,(p_0(t)|/(p_0|p_0)$, i.e. with the integro-differential equation

$$\dot{\Xi}(t) = -\int_0^t \gamma(t')\,\Xi(t-t')\,dt' \, . \tag{16.1.3}$$

This specific system is intended to elucidate several different points. First of all, it demonstrates the action of the projector Q in the dynamics of the residual

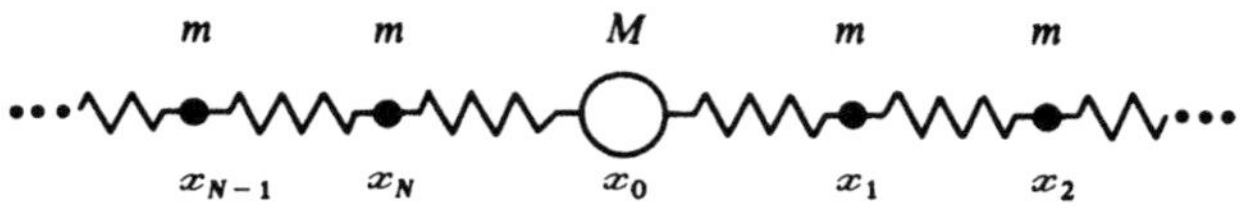

Fig. 16.1. A heavy particle of mass M in a chain of harmonic oscillators (model for Brownian motion)

force very clearly:

$$f(t)=e^{iQLt}QiL\,p_0 \ . \tag{16.1.4}$$

As we shall see, this is determined solely by the light particles of mass m. Furthermore, one can see explicitly how the thermodynamic limit leads to a relaxation behavior in the memory function. In considering approximations, it is important to recognize the different time scales of the memory function $\gamma(t)$ and the correlation function $\Xi(t)$. It turns out that the ratio of these time scales, τ_c/τ_R, is determined by the mass ratio m/M, so that small mass ratios permit the use of a Markovian approximation. Using the exact solution for $\Xi(t)$, we can see how it approaches an exponential function for small m/M. Physically, the chain may be regarded as a rough model for the Brownian motion of a heavy particle in a liquid.

16.1.1 Dynamics of the Residual Force

We begin with the calculation of the residual force (16.1.4). In order to understand the time evolution with $\exp(iQLt)$, we start from the following consideration: The Liouville operator which belongs to (16.1.1) does not lead out of a space of linear combinations of all the momenta and positions,

$$L\left\{\sum_{\nu=0}^{N}(a_\nu p_\nu+b_\nu x_\nu)\right\}=\sum_{\nu=0}^{N}(a'_\nu p_\nu+b'_\nu x_\nu) \ . \tag{16.1.5}$$

The same is true of the projection operator Q,

$$Q=1-\frac{|p_0)(p_0|}{(p_0|p_0)} \ , \tag{16.1.6}$$

which changes only the components in the direction of p_0. Therefore, all the powers of QL do not lead out of the subspace in question, and p_0 itself lies in the same subspace; so, to calculate $f(t)$, we need only to know the action of QL in this space. Using the relations

$$p_0=MiLx_0 \ , \tag{16.1.7}$$

$$(\mathcal{F}|L\mathcal{G})=\frac{1}{\hbar\beta}\langle[\mathcal{F}^\dagger,\mathcal{G}]\rangle \ , \tag{16.1.8}$$

$$(p_0|p_0)=M\frac{1}{\hbar\beta}i\langle[p_0,x_0]\rangle=\frac{M}{\beta} \ , \tag{16.1.9}$$

we find for the action of QL on an arbitrary element $\mathscr{F}$:

$$\mathrm{QL}\,\mathscr{F}=\mathrm{L}\,\mathscr{F}-\not{p}_0\frac{(\not{p}_0|\mathrm{L}\,\mathscr{F})}{(\not{p}_0|\not{p}_0)}$$

$$=\frac{1}{\hbar}[\mathscr{H},\mathscr{F}]-\frac{1}{\hbar}\frac{\not{p}_0}{M}\langle[\not{p}_0,\mathscr{F}]\rangle\;. \tag{16.1.10}$$

If, in particular, $\mathscr{F}$ is a linear combination of all the momenta and positions,

$$\mathscr{F}=\sum_{\nu=0}^{N}(a_\nu\not{p}_\nu+b_\nu x_\nu)\;, \tag{16.1.11}$$

then the commutator $[\not{p}_0,\mathscr{F}]$ gives a c-number, so that (16.1.10) may be written in the form

$$\mathrm{QL}\,\mathscr{F}=\frac{1}{\hbar}[\mathscr{H},\mathscr{F}]-\frac{1}{\hbar}\left[\frac{\not{p}_0^2}{2M},\mathscr{F}\right]\;, \tag{16.1.12}$$

i.e. as a commutator with a Hamiltonian

$$\hat{\mathscr{H}}=\sum_{\nu=1}^{N}\frac{\not{p}_\nu^2}{2m}+\frac{k}{2}\sum_{\nu=0}^{N}(x_{\nu+1}-x_\nu)^2\;, \tag{16.1.13}$$

giving finally, in compact form,

$$\mathrm{QL}\,\mathscr{F}=\frac{1}{\hbar}[\hat{\mathscr{H}},\mathscr{F}]=\hat{\mathrm{L}}\,\mathscr{F}\;. \tag{16.1.14}$$

With this, we find the residual force (16.1.4):

$$f(t)=\mathrm{e}^{\mathrm{iQL}t}\mathrm{QiL}\not{p}_0=\mathrm{e}^{\mathrm{i}\hat{\mathrm{L}}t}\mathrm{iL}\not{p}_0=\mathrm{e}^{\mathrm{i}\hat{\mathrm{L}}t}k\{x_N-2x_0+x_1\}\;. \tag{16.1.15}$$

It is thus determined by the dynamics of the particles of mass m when the particle of mass M is at rest. The dynamics of the heavy particle have been separated out. The explicit evaluation of expression (16.1.15) can be carried out using standard methods of lattice dynamics. The potential energy in $\hat{\mathscr{H}}$ can be expressed as a quadratic form in the N variables $x_\nu-x_0$, and this form is diagonalized by the matrix

$$c_{\nu\mu}=\sqrt{\frac{2}{N+1}}\sin\left(\pi\frac{\nu\mu}{N+1}\right)\;, \tag{16.1.16}$$

$$\sum_{\mu=1}^{N}c_{\nu\mu}c_{\mu\varrho}=\delta_{\nu\varrho} \tag{16.1.17}$$

therefore, $\hat{\mathscr{H}}$ may be represented as a superposition of noncoupled oscillators,

$$\hat{\mathscr{H}} = \sum_{\nu=1}^{N} \left(\frac{\hat{p}_\nu^2}{2m} + \frac{1}{2} m\omega_\nu^2 \hat{q}_\nu^2 \right) \tag{16.1.18}$$

using the normal coordinates

$$\begin{aligned} \hat{q}_\nu &= \sum_{\mu=1}^{N} c_{\nu\mu}(x_\mu - x_0) \ , \\ \hat{p}_\nu &= \sum_{\mu=1}^{N} c_{\nu\mu} p_\mu \ , \end{aligned} \tag{16.1.19}$$

with the eigenfrequencies

$$\omega_\nu = \sqrt{\frac{4k}{m}} \sin\left(\frac{\pi}{2} \frac{\nu}{N+1} \right) . \tag{16.1.20}$$

In order to calculate $f(t)$ from (16.1.15), one thus has only to decompose $x_N - x_0$ and $x_1 - x_0$ in normal coordinates according to (16.1.17 and 19) and to insert the time dependences $q_\nu(t)$ in the Heisenberg formulation. In this way, one obtains the result

$$f(t) = k \sum_{\nu,\mu=1}^{N} (c_{N\nu} + c_{1\nu}) \left\{ \cos\omega_\nu t \cdot c_{\nu\mu}(x_\mu - x_0) + \frac{\sin\omega_\nu t}{\omega_\nu} c_{\nu\mu} \frac{p_\mu}{m} \right\} . \tag{16.1.21}$$

The time scale of $f(t)$ is given by the frequencies ω_ν, i.e. by the frequency

$$\alpha = \sqrt{\frac{4k}{m}} . \tag{16.1.22}$$

16.1.2 The Memory Function

To calculate the memory function $\gamma(t)$ or the correlation function $(f(0)|f(t))$, we use the relation (16.1.8) with $\mathsf{QL}\, p_0 = \mathsf{L}\, p_0$. This leads to the commutator $[p_\nu, f(t)]$,

$$(f(0)|f(t)) = (\mathrm{i}\mathsf{L}\, p_0 | f(t)) = -\frac{\mathrm{i}}{\hbar\beta} \langle [p_0, f(t)] \rangle \ , \tag{16.1.23}$$

so that we find from (16.1.21) the following form for $(f(0)|f(t))$:

$$(f(0)|f(t)) = \frac{k}{\beta} \sum_{\nu,\mu=1}^{N} (c_{N\nu} + c_{1\nu}) \cos\omega_\nu t \cdot c_{\nu\mu} \ . \tag{16.1.24}$$

After carrying out the sum $\sum_{\mu=1}^{N} c_{\nu\mu}$, which enters only for odd values of ν, we may represent the memory function using (16.1.9, 20, 22, and 24) as

$$\gamma(t)=\frac{1}{(\not{p}_0|\not{p}_0)}(\not{f}(0)|\not{f}(t))$$

$$=\frac{m}{M}\alpha^2\frac{1}{N+1}\sum_{\nu=\pm 1,\pm 3\ldots}\cos^2\left(\frac{\pi}{2}\frac{\nu}{N+1}\right)\cos\left\{\alpha t\sin\left(\frac{\pi}{2}\frac{\nu}{N+1}\right)\right\}. \tag{16.1.25}$$

For finite N, this is an almost-periodic function (cf. Sect. 4.2). The thermodynamic limit $N\rightarrow\infty$, together with

$$\varphi_\nu=\frac{\pi}{2}\frac{\nu}{N+1}, \quad \nu=\pm 1, \pm 3, \ldots (\pm N) \tag{16.1.26}$$

and

$$\Delta\varphi=\pi\frac{1}{N+1} \quad (\Delta\nu=2) \tag{16.1.27}$$

leads to an integral representation of the Bessel functions $J_1(\alpha t)/\alpha t$,

$$\begin{aligned}\lim_{N\rightarrow\infty}\gamma(t)&=\frac{m}{M}\alpha^2\lim_{N\rightarrow\infty}\frac{1}{\pi}\sum_{\varphi_\nu}\cos^2\varphi_\nu\cdot\cos\{\alpha t\sin\varphi_\nu\}\Delta\varphi\\&=\frac{m}{M}\alpha^2\frac{1}{\pi}\int_{-\pi/2}^{\pi/2}\cos^2\varphi\cdot\cos\{\alpha t\sin\varphi\}\,d\varphi\\&=\frac{m}{M}\alpha^2\frac{1}{\pi}\int_{-1}^{1}\sqrt{1-u^2}\cos(\alpha t u)\,du=\frac{m}{M}\alpha^2\frac{J_1(\alpha t)}{\alpha t}\end{aligned}, \tag{16.1.28}$$

which goes to zero as $t\rightarrow\infty$ (see Fig. 16.2). It is useful to examine also the thermodynamic limit of the Fourier transform, $\int_0^\infty \gamma(t)\exp[(\mathrm{i}\omega-\eta)t]dt\ (\eta>0)$. For finite N from (16.1.28) with

$$z=\omega+\mathrm{i}\eta, \quad \eta>0 \tag{16.1.29}$$

we obtain

$$\gamma(z)=\int_0^\infty\gamma(t)\mathrm{e}^{\mathrm{i}zt}dt=\frac{m}{M}\alpha^2\frac{\mathrm{i}}{\pi}\sum_{\nu=\pm 1,\pm 3\ldots}\frac{\cos^2\varphi_\nu}{z+\alpha\sin\varphi_\nu}\Delta\varphi\ . \tag{16.1.30}$$

The integral over t defines a function which is holomorphic in the upper z-semiplane; its analytic continuation in the lower semiplane can be read from the right-hand side of (16.1.30), and it has poles on the real axis at the frequencies

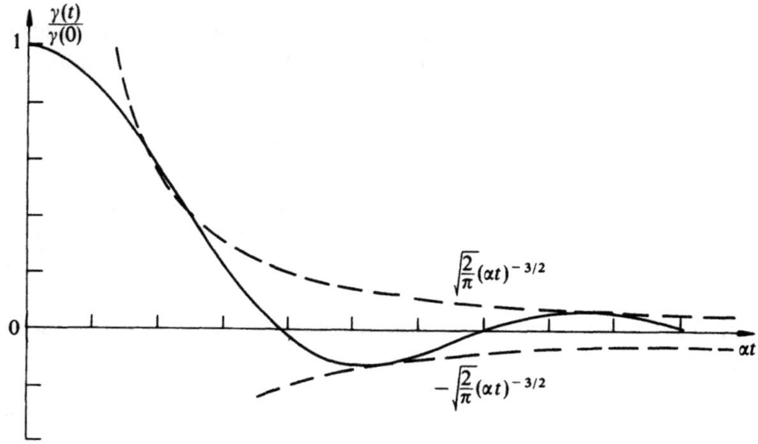

Fig. 16.2. The memory function $\gamma(t)/\gamma(0)=J_1(\alpha t)/\alpha t$ vs αt

$\pm\omega_\nu$ (for odd ν). In the limit $N\to\infty$, there are increasing numbers of poles in the interval $-\alpha<\omega<\alpha$, and the sum (16.1.30) becomes an integral, which for $\mathrm{Im}\,z>0$ yields

$$\lim_{N\to\infty}\gamma(z)=\frac{m}{M}\alpha\left\{\mathrm{i}\frac{z}{\alpha}+\sqrt{1-\frac{z^2}{\alpha^2}}\right\}$$

$$=\frac{m}{M}\alpha^2\int_0^\infty\frac{J_1(\alpha t)}{\alpha t}\mathrm{e}^{\mathrm{i}zt}dt\quad\left(-\frac{\pi}{2}\leq\mathrm{Arg}\sqrt{1-\frac{z^2}{\alpha^2}}<\frac{\pi}{2}\right);\qquad(16.1.31)$$

it represents the Laplace transform of $J_1(\alpha t)/\alpha t$. The function defined by (16.1.31), which is holomorphic in the upper z-half plane, may be continued in the lower half plane; it then gives one branch of the two-valued function

$$\frac{\mathrm{i}z}{\alpha}+\sqrt{1-\frac{z^2}{\alpha^2}},$$

whose branching line can be chosen arbitrarily between $z=\alpha$ and $z=-\alpha$ in the lower half plane, $\mathrm{Im}\,z\leqq 0$. After taking the limit $N\to\infty$ in (16.1.31), we can also take the limit $\eta=\mathrm{Im}\,z\to 0$.

16.1.3 Separation of the Time Scales

In this section, we wish to investigate the Markovian approximation for the correlation function (16.1.2). The time dependence of $\gamma(t)$ is shown in Fig. 16.2. For $t\to 0$, the function has a maximum which is apparent in the expansion of the

Bessel function

$$J_1(x)=\frac{x}{2}\left(1-\frac{1}{8}x^2+\dots\right) \tag{16.1.32}$$

for small x. At long times, in contrast, $\gamma(t)$ decreases only slowly, as $t^{-3/2}$, since $J_1(x)$ obeys the asymptotic expansion

$$J_1(x)=\sqrt{\frac{2}{\pi}}\frac{1}{x^{1/2}}\cos\left(x-\frac{3}{4}\pi\right)+R_1 , \tag{16.1.33}$$

$$|R_1|<\sqrt{\frac{2}{\pi}}\frac{3}{4}\frac{1}{x^{3/2}} . \tag{16.1.34}$$

Although $\gamma(t)$ depends on a smallness parameter m/M, it is thus not *a priori* clear whether a separation of the time scales of $\gamma(t)$ and $\Xi(t)$ is possible. To clarify these questions, we calculate the relaxation time τ_R and the correlation time τ_c from (13.1.6) and (13.1.27). For the relaxation time τ_R, using the integral

$$\int_0^\infty \frac{J_1(x)}{x}dx=1 \tag{16.1.35}$$

we find

$$\boxed{\frac{1}{\tau_R}=\Gamma=\int_0^\infty \gamma(t)\,dt=\frac{m}{M}\alpha} . \tag{16.1.36}$$

To determine τ_c from (13.1.27), we have to demonstrate the existence of the integral $\int_0^\infty |\psi(t)|dt$ in spite of the asymptotic behavior (16.1.33).

We find from (13.1.15), (16.1.28), and (16.1.35) for $\psi(t)/\psi(0)$:

$$\frac{\psi(t)}{\psi(0)}=\int_{\alpha t}^\infty \frac{J_1(x)}{x}dx . \tag{16.1.37}$$

Inserting (16.1.33) in this expression and integrating the angular function by parts, and estimating the residual terms with (16.1.34), we find a bound for the asymptotic behavior of $\psi(t)$:

$$\left|\frac{\psi(t)}{\psi(0)}\right|<ct^{-3/2} , \tag{16.1.38}$$

which guarantees convergence. The numerical value of τ_c is obtained according to (13.1.27):

$$\boxed{\tau_c = \int_0^\infty \frac{|\psi(t)|}{\psi(0)}\,dt = \frac{1}{\alpha}\times 1.8} \;, \tag{16.1.39}$$

where the integral

$$\int_0^\infty \left| \int_x^\infty \frac{J_1(x')}{x'}\,dx' \right| dx = 1.8 \tag{16.1.40}$$

has been evaluated numerically.

We have thus shown that the correlation function (16.1.2) is given by an exponential function using the results of Sect. 13.1.2 and (16.1.36 and 39):

$$\Xi(t) = \exp\left[-\frac{m}{M}\alpha t\right] + r(t) \;. \tag{16.1.41}$$

The error $r(t)$ cannot become larger than

$$\boxed{2\frac{\tau_c}{\tau_R} = 2\times 1.8\,\frac{m}{M}} \;. \tag{16.1.42}$$

With a mass ratio of $m/M < 10^{-2}$, $\Xi(t)$ can thus be distinguished from an exponential function only when the accuracy is extremely good.

16.1.4 Discussion of the Exact Solution $\Xi(t)$

Since we know the memory function $\gamma(t)$ in the above model, we can investigate the exact solution for $\Xi(t)$. Fourier (Laplace) transformation of the integro-differential equation (16.1.3) yields, with (11.4.8),

$$\Xi(t) = \frac{1}{2\pi}\int_{-\infty}^{\infty} \Xi(\omega)\mathrm{e}^{-\mathrm{i}\omega t}\,d\omega \;, \tag{16.1.43}$$

where

$$\Xi(\omega) = \frac{\mathrm{i}}{\omega + \mathrm{i}\gamma(\omega)} \;. \tag{16.1.44}$$

Here, $\gamma(\omega)$ is given by (16.1.31) with $\eta \to +0$. Depending on how one places the path of integration for the integral (16.1.43) in the lower ω-plane, one obtains

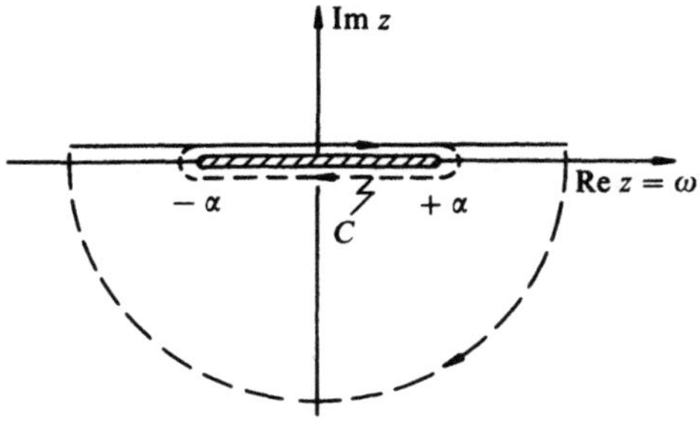

Fig. 16.3. The path of integration in (16.1.45)

various representations of the solution for $\Xi(t)$. It is expedient to continue $\gamma(z)$ in (16.1.31), and thus $\Xi(z)$, analytically from the upper half plane in such a way that $\Xi(z)$ has a cut on the real ω-axis between $z=\pm\alpha$. Because $\Xi(z)$ then shows no further singularities in the lower half plane, the path of integration in (16.1.43) can be closed with a semicircle in the lower half plane due to the factor $\exp(\mathrm{i}\omega t)$; it can then be contracted to a contour integral C around the cut on the real axis (see Fig. 16.3). Inserting $\gamma(z)$ from (16.1.31) into (16.1.43) and using the continuation in the lower half plane as described above gives the result for $\Xi(t)$:

$$\Xi(t)=\frac{1}{2\pi}\oint_{C}\frac{\left(1-\dfrac{m}{M}\right)\mathrm{i}z+\dfrac{m}{M}\sqrt{\alpha^2-z^2}}{\left(1-\dfrac{2m}{M}\right)z^2+\left(\dfrac{m}{M}\right)^2\alpha^2}\,\mathrm{e}^{-\mathrm{i}zt}\,dz$$

$$=\frac{1}{2\pi}\frac{2m}{M}\int_{-1}^{+1}\sqrt{1-u^2}\,\frac{1}{\left(1-\dfrac{2m}{M}\right)u^2+\left(\dfrac{m}{M}\right)^2}\,\mathrm{e}^{-\mathrm{i}u\alpha t}\,du\ . \qquad (16.1.45)$$

From (16.1.28), the factor $\sqrt{1-u^2}$ with $u^2\leqq 1$ is the Fourier transform of $J_1(\alpha t)/\alpha t$:

$$\int_{-1}^{1}\sqrt{1-u^2}\,\mathrm{e}^{-\mathrm{i}u\alpha t}\,du=\pi\frac{J_1(\alpha t)}{\alpha t}\ . \qquad (16.1.46)$$

The factor in (16.1.45),

$$\left(u^2+\left(\frac{m}{M}\right)^2\frac{1}{1-\dfrac{2m}{M}}\right)^{-1}\quad\text{with}\quad\frac{m}{M}<\frac{1}{2}\,,$$

can be regarded as the Fourier transform of

$$\exp\left(-\frac{m}{M}\frac{1}{\sqrt{1-\frac{2m}{M}}}|\alpha t|\right), \quad \int_{-\infty}^{\infty}\frac{1}{u^2+\left(\frac{m}{M}\right)^2\frac{1}{1-\frac{2m}{M}}}\mathrm{e}^{-iu\alpha t}du$$

$$=\pi\frac{M}{m}\sqrt{1-\frac{2m}{M}}\exp\left(-\frac{m}{M}\frac{1}{\sqrt{1-\frac{2m}{M}}}|\alpha t|\right). \tag{16.1.47}$$

Therefore, we may use the convolution theorem for Fourier transforms. Then, the correlation function $\Xi(t)$ (for $m/M<1/2$) takes on the form

$$\boxed{\Xi(t)=\frac{1}{2}\frac{1}{\sqrt{1-\frac{2m}{M}}}\int_{-\infty}^{\infty}\exp\left(-\frac{m}{M}\frac{1}{\sqrt{1-\frac{2m}{M}}}|\alpha t-\alpha t'|\right)\frac{J_1(\alpha t')}{\alpha t'}\alpha dt'}\ . \tag{16.1.48}$$

From this equation, we may read out the qualitative behavior of $\Xi(t)$ for $m/M \ll 1$. The exponential function of width $(M/m)\sqrt{1-2m/M}\,(1/\alpha)$ is shifted over the function $J_1(\alpha t')/(\alpha t')$ as t is increased. As long as αt is so small that the exponential function is located essentially within the principal maximum of $J_1(\alpha t')/(\alpha t')$ at $t'=0$, the main contribution to the integral comes from the neighborhood of $t'=0$ and an exponential behavior dominates $\Xi(t)$. However, when αt becomes sufficiently large that the exponential function overlaps practically only with the tail of $J_1(\alpha t')/(\alpha t')\sim(\alpha t')^{-3/2}$ at long times, the situation is quite different. Therefore, in the range $m/M\,[1/\sqrt{1-2m/M}]\alpha t \gg 1$, it is to be expected that we can approximate $\Xi(t)$ in (16.1.48) by using the asymptotic behavior for $J_1(\alpha t')/(\alpha t')$ from (16.1.33) and removing the factor $(\alpha t')^{-3/2}$, which varies slowly over the width $M/m\sqrt{1-2m/M}\,(1/\alpha)$, from the integral at $t'=t$:

$$\Xi(t)=\frac{1}{2}\frac{1}{\sqrt{1-\frac{2m}{M}}}\sqrt{\frac{2}{\pi}}\frac{1}{(\alpha t)^{3/2}}\int_{-\infty}^{\infty}\exp\left(-\frac{m}{M}\frac{1}{\sqrt{1-\frac{2m}{M}}}|\alpha t-\alpha t'|\right)$$

$$\times\cos\left(\alpha t'-\frac{3}{4}\pi\right)\alpha dt'+\ \dots\ . \tag{16.1.49}$$

In this way, we find for $\Xi(t)$:

$$\Xi(t)=\frac{m}{M}\frac{1}{\left(1-\frac{m}{M}\right)^2}\sqrt{\frac{2}{\pi}}\cos\left(\alpha t-\frac{3}{4}\pi\right)\frac{1}{(\alpha t)^{3/2}}+\dots \quad (16.1.50)$$

This result does, in fact, reproduce the exact asymptotic behavior of $\Xi(t)$ for $t\to\infty$, as one can show by an analysis of the integral representation (16.1.45) of $\Xi(t)$. The correlation function $\Xi(t)$ calculated from (16.1.48) for $m/M=0.2$ and 0.05 is shown in Fig. 16.4, together with the asymptotic solution (16.1.50) and the

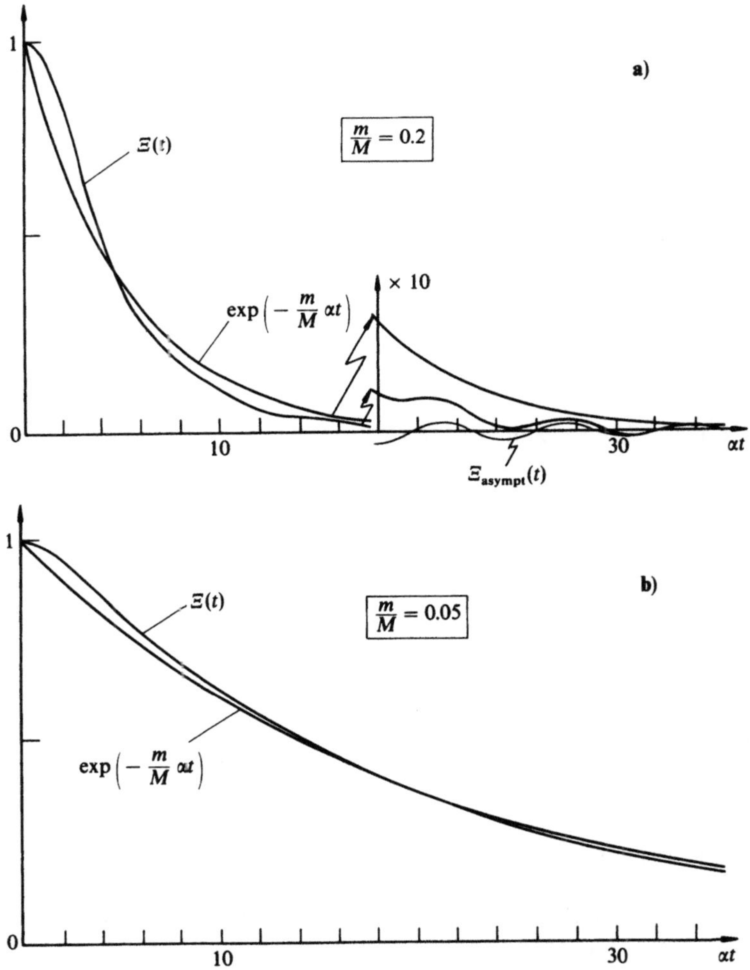

Fig. 16.4. The exact solution $\Xi(t)$ and the Markovian approximation $\exp[(-m/M)\alpha t]$ for a mass ratio **(a)** $m/M=0, 2$; **(b)** $m/M=0, 05$

Markovian approximation $\exp[(-m/M)\alpha t]$. One can readily see how $\Xi(t)$ rapidly approaches the exponential function for a decreasing mass ratio.

The asymptotic expansion of the Fourier integral: according to Erdélyi[1], the following relation holds for

$$f(x)=\int_{\alpha}^{\beta} e^{ixu}(u-\alpha)^{\lambda-1}(\beta-u)^{\mu-1}\,\Phi(u)\,du \ ,$$

when $\Phi(u)$ has N continuous derivatives and $0<\lambda\leqq 1$, $0<\mu\leqq 1$:

$$f(x)=B_N(x)-A_N(x)+O(x^{-N}) \ .$$

Here, A_N and B_N are given by

$$A_N(x)=\sum_{n=0}^{N-1}\frac{\Gamma(n+\lambda)}{n!}\exp\left[\frac{i\pi}{2}(n+\lambda-2)\right]x^{-n-\lambda}e^{ix\alpha}\frac{d^n}{d\alpha^n}\{(\beta-\alpha)^{\mu-1}\Phi(\alpha)\} \ ,$$

and

$$B_N(x)=\sum_{n=0}^{N-1}\frac{\Gamma(n+\mu)}{n!}\exp\left[\frac{i\pi}{2}(n-\mu)\right]x^{-n-\mu}e^{ix\beta}\frac{d^n}{d\beta^n}\{(\beta-\alpha)^{\lambda-1}\Phi(\beta)\} \ .$$

Application to (16.1.45) with $\lambda=\mu=1/2$ and

$$\Phi(u)=\frac{1-u^2}{\left(1-\frac{2m}{M}\right)u^2+\left(\frac{m}{M}\right)^2}$$

yields, for $N=2$, the result (16.1.50).

16.2 Spin-Bath Relaxation

In order to illustrate the theory for *one* Hermitian observable, we consider N independent spins $s=1/2$ in a magnetic field $\mathfrak{H}=H\mathbf{e}_z$, which are weakly coupled via an interaction $\mathscr{H}^{SB}$ to other degrees of freedom of the overall system, to a bath with the Hamiltonian $\mathscr{H}^{B}$. We wish to calculate the longitudinal relaxation function of the total spin $\mathscr{S}^z=\sum_{j=1}^{N}\sigma_j^z$,

$$\Phi(t)=\beta(\delta\mathscr{S}^z|\delta\mathscr{S}^z(t)) \ . \tag{16.2.1}$$

This function determines both the behavior of the expectation value $\langle\mathscr{S}^z\rangle(t)$ with time after switching off a magnetic-field perturbation, and the dynamic spin susceptibility $\chi_{zz}(\omega)$. Physically, the system termed the "bath" can be the phonons (one then refers to spin-lattice relaxation), or it can also be the

[1] A. Erdélyi: *Asymptotic Expansions*. Dover, 1956.

conduction electrons. The nature of the bath need not be precisely specified, as long as the unitary space for the total system is a product of subspaces for the spin system and the bath, and as long as the energy differences in the bath $\mathscr{H}^{B}$ become continuous in the thermodynamic limit [cf. Footnote 2 (p. 268)].

To treat this problem, we start with a Hamiltonian for the total system:

$$\mathscr{H}=\mathscr{H}^{S}+\mathscr{H}^{B}+\mathscr{H}^{SB} , \tag{16.2.2}$$

with

$$\mathscr{H}^{S}=-\gamma H\sum_{j=1}^{N}\sigma_j^z \tag{16.2.3}$$

and an interaction term

$$\mathscr{H}^{SB}=\sum_{j=1}^{N}(\mathscr{B}_j^+\sigma_j^- +\mathscr{B}_j^-\sigma_j^+ +\mathscr{B}_j^z\sigma_j^z) . \tag{16.2.4}$$

Here, the $\mathscr{B}_j^\alpha$ refer to bath operators which could be phonon operators in the case of a spin-lattice interaction, or spin density operators $\sigma^\alpha(\mathbf{r}_j)$ for an exchange interaction with the conduction electrons. For the bath equilibrium expectation values of these operators with $\exp(-\beta\mathscr{H}^{B})$, we assume

$$\langle\mathscr{B}_j^\pm\rangle_B=0 \tag{16.2.5}$$

and

$$\langle\mathscr{B}_j^z\rangle_B=\langle\mathscr{B}_k^z\rangle_B , \tag{16.2.6}$$

i.e. values which are independent of the position j.

To determine the relaxation function (16.2.1) for this system, we use the Mori theory with the projection operator onto $\delta\mathscr{S}^z=\sum_j(\sigma_j^z-\langle\sigma_j^z\rangle_\beta)$:

$$\mathsf{P}=\frac{|\delta\mathscr{S}^z)(\delta\mathscr{S}^z|}{(\delta\mathscr{S}^z|\delta\mathscr{S}^z)} . \tag{16.2.7}$$

The frequency Ω is zero in the case of one Hermitian observable; therefore, from (11.4.14) and (11.4.3) we find the exact relation

$$\frac{d}{dt}\Phi(t)=-\int_0^t\Phi(t-t')\gamma(t')\,dt' , \tag{16.2.8}$$

where the memory function $\gamma(t)$ is given from (11.2.23) by

$$\gamma(t)=\frac{1}{(\delta\mathscr{S}^z|\delta\mathscr{S}^z)}(\dot{\mathscr{S}}^z|e^{i\mathsf{QL}t}\mathsf{Q}\dot{\mathscr{S}}^z) . \tag{16.2.9}$$

We wish to base our calculations on the memory function in lowest order perturbation theory. The molecular field contributions of the bath operators $\langle\mathscr{B}_j^z\rangle_B$ are absorbed into $\mathscr{H}^0$. We thus divide the Hamiltonian $\mathscr{H}$ into the unperturbed term:

$$\mathscr{H}^0 = -\omega_0 \sum_j \sigma_j^z + \mathscr{H}^{\mathrm{B}} \tag{16.2.10}$$

with

$$\omega_0 = \gamma H - \frac{1}{N} \sum_j \langle \mathscr{B}_j^z \rangle_{\mathrm{B}} \tag{16.2.11}$$

and a perturbation term,

$$\mathscr{H}^1 = \sum_j (\mathscr{B}_j^+ \sigma_j^- + \mathscr{B}_j^- \sigma_j^+ + (\mathscr{B}_j^z - \langle \mathscr{B}_j^z \rangle_{\mathrm{B}}) \sigma_j^z) \ . \tag{16.2.12}$$

For this division, and for the observables $\mathscr{G} = \delta \sum_j \sigma_j^z$, the assumptions of Sect. 12.1 hold, so that the lowest order of the memory function is the second order in the perturbation.

If we take the square of a coupling constant as a smallness (=slowness) parameter in the sense of Sect. 13.1 for $\gamma(t)^{(2)}$, we can apply the Markovian approximation to (16.2.8) when the coupling is sufficiently weak[2]. From (13.1.5), in this case we find for the relaxation function:

$$\frac{d}{dt} \Phi(t) = -\Gamma \Phi(t) \ , \tag{16.2.13}$$

where the damping (13.1.7) has been specialized in perturbation theory (12.1.27) to

$$\Gamma = \int_0^\infty \gamma(t)^{(2)} dt = \frac{1}{(\delta \mathscr{S}^z | \delta \mathscr{S}^z)^{(0)}} L'(\omega = 0)^{(2)} \ . \tag{16.2.14}$$

In this expression, we shall separate out the contributions which are due to the bath.

The denominator in the damping (16.2.14) is, aside from a factor, just the isothermal static susceptibility of non-interacting spins, since the bath contributions drop out. The value of χ_{zz}^{T} is found to be

$$\chi_{zz}^{\mathrm{T}} = \beta (\delta \mathscr{S}^z | \delta \mathscr{S}^z)^{(0)} = \beta N \frac{\hbar^2}{4} \left(\cosh \frac{\beta \hbar \omega_0}{2} \right)^{-2} \ . \tag{16.2.15}$$

---

[2] One can see from the perturbation-theory formula for $\gamma(t)^{(2)}$ that its correlation time is determined by the dynamic bath correlation function $\langle \mathscr{B}_k^{-\alpha} \mathscr{B}_j^{\alpha}(t) \rangle_{\mathrm{B}}$. For a particular case, with a given bath, τ_c can thus be specified and one can discuss the validity of the Markovian approximation more precisely.

According to (12.1.27), the Onsager–Casimir coefficient $L'(\omega=0)$ in second order is given by

$$L'(0)^{(2)} = \mathrm{Re}\int_0^\infty (\dot{\mathscr{S}}^z|\mathrm{e}^{\mathrm{i}\mathsf{L}^0 t}\mathsf{Q}^0\dot{\mathscr{S}}^z)^{(0)}\mathrm{e}^{-\eta t}dt$$

$$= \pi(\dot{\mathscr{S}}^z|\delta(\mathsf{L}^0)\dot{\mathscr{S}}^z)^{(0)} = \pi\langle\dot{\mathscr{S}}^z\delta(\mathsf{L}^0)\dot{\mathscr{S}}^z\rangle_\beta^{(0)} \ . \tag{16.2.16}$$

The operator Q^0 drops out, since $\mathscr{S}^z$ commutes with $\mathscr{H}^0$ and thus

$$(\delta\mathscr{S}^z|\dot{\mathscr{S}}^z)^{(0)} = \left\langle \delta\mathscr{S}^z\frac{\mathrm{i}}{\hbar}[\mathscr{H}^1, \mathscr{S}^z]\right\rangle_\beta^{(0)} = 0 \tag{16.2.17}$$

holds. The last form in (16.2.16) is found when one takes (9.7.7) into account for $\omega=0$. For further evaluation of the Onsager–Casimir coefficient, we use

$$\dot{\mathscr{S}}^z = \frac{\mathrm{i}}{\hbar}[\mathscr{H}^1, \mathscr{S}^z] = \mathrm{i}\sum_j(\mathscr{B}_j^+ s_j^- - \mathscr{B}_j^- s_j^+) \tag{16.2.18}$$

and

$$\mathrm{e}^{\mathrm{i}\mathsf{L}^0 t}\mathscr{B}_j^\pm s_j^\mp = (\mathrm{e}^{\mathrm{i}\mathsf{L}_\mathrm{B} t}\mathscr{B}_j^\pm)\mathrm{e}^{\pm\mathrm{i}\omega_0 t}s_j^\mp . \tag{16.2.19}$$

The expectation values of products of spin and bath operators taken with $\exp(-\beta\mathscr{H}^0)$ can be factored; we therefore obtain

$$L'(0)^{(2)} = \pi\sum_j\{\langle s_j^- s_j^+\rangle_\mathrm{s}\langle\mathscr{B}_j^+\delta(\mathsf{L}^\mathrm{B}-\omega_0)\mathscr{B}_j^-\rangle_\mathrm{B}$$

$$+\langle s_j^+ s_j^-\rangle_\mathrm{s}\langle\mathscr{B}_j^-\delta(\mathsf{L}^\mathrm{B}+\omega_0)\mathscr{B}_j^+\rangle_\mathrm{B}\} \ , \tag{16.2.20}$$

where $\langle\ \rangle_\mathrm{s}$ refers to spin expectation values with $\exp(\beta\omega_0\mathscr{S}^z)$. This expression already reflects the essential result. The coefficient $L'(0)$ and thus the damping Γ are determined by the Fourier transforms of the *bath* correlation functions $\langle(\mathscr{B}_j^\alpha)^\dagger\mathscr{B}_j^\alpha(t)\rangle_\mathrm{B}$. The final result for the damping Γ may be written in various forms. The physical mechanism is most clearly expressed by introducing the dynamic *bath* susceptibilities

$$\chi_{kj}^{\alpha\alpha}(\omega) = \int_0^\infty -\frac{\mathrm{i}}{\hbar}\langle[(\mathscr{B}_k^\alpha)^\dagger, \mathscr{B}_j^\alpha(t)\rangle_\mathrm{B}\mathrm{e}^{\mathrm{i}\omega t}dt \ , \quad \alpha=\pm \ . \tag{16.2.21}$$

Using (9.1.27) for $(\mathscr{B}_j^\alpha)^\dagger = \mathscr{B}_j^{-\alpha}$; on decomposition according to (9.4.2), these are found to have the absorptive parts (9.4.11)

$$\chi_{kj}^{\alpha\alpha}(\omega)'' = \frac{\pi}{\hbar}(\mathrm{e}^{\beta\hbar\omega}-1)\langle\mathscr{B}_k^{-\alpha}\delta(\mathsf{L}^\mathrm{B}+\omega)\mathscr{B}_j^\alpha\rangle_\mathrm{B}$$

$$= -\frac{\pi}{\hbar}(\mathrm{e}^{-\beta\hbar\omega}-1)\langle\mathscr{B}_j^\alpha\delta(\mathsf{L}^\mathrm{B}-\omega)\mathscr{B}_k^{-\alpha}\rangle_\mathrm{B} \ . \tag{16.2.22}$$

After computing the expectation values $\langle \sigma_j^{\mp} \sigma_j^{\pm} \rangle_S$, we can summarize the damping (16.2.14) using (16.2.15, 20, and 22) in the form:

$$\boxed{\Gamma = \hbar^2 \frac{1}{\langle \mathscr{S}^z \rangle_S} \sum_j \chi_{jj}^{++}(\omega_0)'' \quad .} \tag{16.2.23}$$

Here, $\langle \mathscr{S}^z \rangle_S$ refers to the equilibrium expectation value

$$\langle \mathscr{S}^z \rangle_S = N \frac{\hbar}{2} \tanh \frac{\beta \hbar \omega_0}{2} \quad . \tag{16.2.24}$$

The bath susceptibilities $\chi_{jk}^{\alpha\gamma}(\omega)''$ describe, according to linear response theory, the energy absorbed per unit time by the bath when the latter is isolated from the spins and driven by *c*-number fields. Thus, result (16.2.23) can be interpreted as implying that the spins act on the bath with frequencies $\pm\omega_0$ and that the energy absorbed by the bath gives rise to the damping Γ of the spins $\langle \mathscr{S}^z \rangle(t)$.

For specific systems, e.g. the phonons or conduction electrons mentioned in the introduction, (16.2.23) can be carried further. One then arrives at the usual expressions for magnetic relaxation or the dynamic susceptibility, for which we find from (16.2.13) and (9.4.19)

$$\chi_{zz}(\omega) = \chi_{zz}^{T} \frac{\Gamma}{\Gamma - i\omega} \quad . \tag{16.2.25}$$

16.3 Magnetic Resonance

A simple example for the Markovian approximation in the case of *one non-Hermitian* observable is found in the study of magnetic resonance. We again start with the system described in Sect. 16.2, but we now inquire as to the transverse relaxation of the total spin $\mathscr{S}^x = \sum_j \sigma_j^x$,

$$\Phi_{xx}(t) = \beta(\delta\mathscr{S}^x | \delta\mathscr{S}^x(t)) \tag{16.3.1}$$

or, alternatively, the dynamic spin susceptibility $\chi_{xx}(\omega)$. Here, we are concerned mainly with the frequency shift of the resonance due to the coupling to the bath. To simplify the problem, we assume that the overall system is symmetric about the *z*-axis.

16.3.1 Reduction to a Single Equation for $\Phi_{++}(t)$

In order to obtain reasonable perturbation-theory expressions for a memory matrix, one needs, as discussed in Sect. 12.1, a set of observables $\mathscr{G}_\mu$ which form a subspace invariant with respect to L^0. We begin here with the observable $\mathscr{S}^x$:

according to (16.2.10), $\mathsf{L}^0\mathcal{S}^x$ leads us to $\mathcal{S}^y$. If we now include $\mathcal{S}^y$ in the space of the observables, we have defined a subspace which is invariant with respect to L^0. It is therefore reasonable to start from a projection operator P which projects onto the space spanned by $\mathcal{S}^x$ and $\mathcal{S}^y$ or, equivalently, by $\mathcal{S}^+$ and $\mathcal{S}^-$. For rotational symmetry about the z-axis, we have (cf. Appendix B1)

$$(\mathcal{S}^\pm|\mathcal{S}^\mp)=0 \quad \text{and} \quad \langle\mathcal{S}^\pm\rangle_\beta=0 \ , \tag{16.3.2}$$

so that the matrix $\boldsymbol{g}$ becomes diagonal:

$$\boldsymbol{g}=\begin{pmatrix} \dfrac{1}{(\mathcal{S}^+|\mathcal{S}^+)} & 0 \\ 0 & \dfrac{1}{(\mathcal{S}^-|\mathcal{S}^-)} \end{pmatrix} . \tag{16.3.3}$$

If we now write the system of coupled integro-differential equations for the relaxation functions $\beta(\delta\mathcal{S}^\alpha|\delta\mathcal{S}^\gamma(t))=\beta(\mathcal{S}^\alpha|\mathcal{S}^\gamma(t))$, with $\alpha,\ \gamma=\pm$, it can be decomposed into the two equations

$$\frac{d}{dt}\Phi_{++}(t)=\Phi_{++}(t)\mathrm{i}\Omega^{++}-\int_0^t\Phi_{++}(t-t')\gamma^{++}(t')\,dt' \ , \tag{16.3.4}$$

$$\frac{d}{dt}\Phi_{--}(t)=\Phi_{--}(t)\mathrm{i}\Omega^{--}-\int_0^t\Phi_{--}(t-t')\gamma^{--}(t')\,dt' \ , \tag{16.3.5}$$

owing to the rotational symmetry about the z-axis, which requires

$$\Phi_{\pm\mp}(t)=\beta(\mathcal{S}^\pm|\mathcal{S}^\mp(t))=0 \ . \tag{16.3.6}$$

The matrix elements $\Omega^{++}$and $\Omega^{--}$are given by

$$\begin{aligned}\Omega^{++}&=\frac{1}{(\mathcal{S}^+|\mathcal{S}^+)}(\mathcal{S}^+|\mathsf{L}\mathcal{S}^+) \ , \\ \Omega^{--}&=\frac{1}{(\mathcal{S}^-|\mathcal{S}^-)}(\mathcal{S}^-|\mathsf{L}\mathcal{S}^-)=-\Omega^{++} \ .\end{aligned} \tag{16.3.7}$$

The original definition of $\gamma^{++}(t)$ and $\gamma^{--}(t)$ contains the projector onto a two dimensional space. On the other hand, with a projector onto $\mathcal{S}^+$ *alone*, one also obtains an equation with the structure (16.3.4). In this equation, the frequency Ω is equal to Ω^{++}, and it, like (16.3.4), has the solution $\Phi_{++}(t)$; therefore, its integral kernel $\gamma(t)$ must be the same as $\gamma^{++}(t)$. This can be shown directly using the identities (3.4.6) and property (16.3.6). Equations (16.3.4 and 5) may therefore also be obtained from the Mori theory by applying the projector to $\mathcal{S}^+$ or $\mathcal{S}^-$. We can limit ourselves to one equation, however, since owing to the property (8.2.10) of the Mori product,

$$(\mathcal{S}^+|\mathcal{S}^+(t))=(\mathcal{S}^-|\mathcal{S}^-(t))^* \tag{16.3.8}$$

holds and thus (16.3.5) is the complex conjugate of (16.3.4). For the same reason, the quantity $\Phi_{xx}(t)$ which we are seeking can also be expressed in terms of $\Phi_{++}(t)$:

$$\Phi_{xx}(t)=\beta\left(\frac{\mathscr{S}^{+}+\mathscr{S}^{-}}{2}\middle|\frac{\mathscr{S}^{+}(t)+\mathscr{S}^{-}(t)}{2}\right)=\frac{1}{2}\operatorname{Re}\Phi_{++}(t)\ . \tag{16.3.9}$$

Therefore, to compute $\Phi_{xx}(t)$, we can employ (16.3.9), using the memory function $\gamma(t)$ defined by the projector on $\mathscr{S}^{+}$ and (16.3.4) for $\Phi_{++}(t)$ with the definition (16.3.7) of Ω.

16.3.2 Perturbation Theory and the Markovian Approximation

To treat (16.3.4), we start with the perturbation-theory expressions for Ω and $\gamma(t)$ up to second order, with the perturbation $\mathscr{H}^1$ again given by (16.2.12). We first consider the expansion of the frequency (12.1.18 and 19). From (16.2.10), we have

$$\mathsf{L}^0\mathscr{S}^{+}=-\omega_0\mathscr{S}^{+} \tag{16.3.10}$$

and since

$$(\mathscr{S}^{+}|\mathsf{L}^1\mathscr{S}^{+})^{(0)}=\sum_k\{(\mathscr{S}^{+}|\sigma_k^{+})_s\langle\mathscr{B}_k^z-\langle\mathscr{B}_k^z\rangle_B\rangle\rangle_B$$
$$-2(\mathscr{S}^{+}|\sigma_k^z)_s\langle\mathscr{B}_k^{+}\rangle_B\}=0 \tag{16.3.11}$$

vanishes, we obtain for Ω up to second order

$$\Omega=\Omega^{(0)}+\Omega^{(2)}=-\omega_0+\frac{1}{(\mathscr{S}^{+}|\mathscr{S}^{+})^{(0)}}(\mathscr{S}^{+}|\mathsf{L}^1\mathscr{S}^{+})^{(1)}\ . \tag{16.3.12}$$

We shall for the time being let the expression $(\mathscr{S}^{+}|\mathsf{L}^1\mathscr{S}^{+})^{(1)}$ stand as it is, without further evaluation.

For the expansion of the memory function, we use the form (12.1.29),

$$\gamma(t)=\frac{1}{(\mathscr{S}^{+}|\mathscr{S}^{+})^{(0)}}\{(\mathscr{S}^{+}|\mathsf{L}^1 e^{i\mathsf{L}^0 t}\mathsf{L}^1\mathscr{S}^{+})^{(0)}$$
$$+(\mathscr{S}^{+}|\mathsf{L}^0 e^{i\mathsf{L}^0 t}\mathsf{L}^1\mathscr{S}^{+})^{(1)}-(\mathsf{L}^0\mathscr{S}^{+}|e^{i\mathsf{L}^0 t}\mathsf{L}^1\mathscr{S}^{+})^{(1)}\}\ . \tag{16.3.13}$$

Here, we have taken into account the fact that the projector Q^0 which was originally present has dropped out due to (16.3.11).

If the coupling in $\mathscr{H}^1$ is sufficiently weak, we can apply the Markovian approximation to the integro-differential equation (16.3.4) for the relaxation function $\Phi_{++}(t)$, using the method outlined in Sect. 13.3.1. There, we split off the factor $\exp(i\Omega t)$ from the relaxation function and assumed $\exp(-i\Omega t)\Phi(t)$ to be slowly varying. In the present example, we shall separate the factor $\exp(i\Omega^{(0)}t)=\exp(-i\omega_0 t)$ from $\Phi_{++}(t)$, since the remainder frequency $\Omega^{(2)}$ for the variation $d/dt\,\{\exp(i\omega_0 t)\,\Phi_{++}(t)\}$, like $\gamma(t)$, yields only a contribution of second order. We can then transform (16.3.4), as we did with (13.3.7), to a differential equation for $\Phi_{++}(t)$:

$$\boxed{\frac{d}{dt}\Phi_{++}(t)=\mathrm{i}\Lambda\Phi_{++}(t)} \tag{16.3.14}$$

with

$$\mathrm{i}\Lambda=-\mathrm{i}\omega_0+\mathrm{i}\Omega^{(2)}-\int_0^\infty \mathrm{e}^{\mathrm{i}\omega_0 t}\gamma(t)^{(2)}\,dt\ . \tag{16.3.15}$$

We require that the coupling to be so weak that

$$\left|\mathrm{i}\Omega^{(2)}-\int_0^\infty \mathrm{e}^{\mathrm{i}\omega_0 t}\gamma(t)^{(2)}\,dt\right|\tau_c\ll 1 \tag{16.3.16}$$

holds. We obtain the quantum-statistical expression for $\mathrm{i}\Lambda$ from (16.3.12 and 13). Making use of the fact that from (16.3.10), $\mathscr{S}^+$ is an eigenvector of L^0, we can write $\mathrm{i}\Lambda$ in the following form:

$$\mathrm{i}\Lambda=-\mathrm{i}\omega_0+\frac{1}{(\mathscr{S}^+|\mathscr{S}^+)^0}\left\{\mathrm{i}(\mathscr{S}^+|\mathsf{L}^1\mathscr{S}^+)^{(1)}+\mathrm{i}\int_0^\infty \mathrm{e}^{-\eta t}\frac{d}{dt}(\mathrm{e}^{\mathrm{i}\mathsf{L}^0 t}\mathscr{S}^+|\mathrm{e}^{\mathrm{i}\mathsf{L}^0 t}\mathsf{L}^1\mathscr{S}^+)^{(1)}dt\right.$$

$$\left.-\int_0^\infty (\mathrm{e}^{\mathrm{i}\mathsf{L}^0 t}\mathscr{S}^+|\mathsf{L}^1\mathrm{e}^{\mathrm{i}\mathsf{L}^0 t}\mathsf{L}^1\mathscr{S}^+)^{(0)}\mathrm{e}^{-\eta t}\,dt\right\}\Bigg|_{\eta\to+0}\ . \tag{16.3.17}$$

The first integral may be evaluated, so that the terms with scalar products in first order just compensate each other[3]. In the last term, one can make use of the Hermiticity of L^0. Then we obtain the result for Λ,

$$\boxed{\mathrm{i}\Lambda=-\mathrm{i}\omega_0-\frac{1}{(\mathscr{S}^+|\mathscr{S}^+)^0}\int_0^\infty(\mathscr{S}^+|\mathsf{L}^1(-t)\mathsf{L}^1\mathscr{S}^+)^{(0)}\mathrm{e}^{-\eta t}\,dt\Bigg|_{\eta\to+0}}\ , \tag{16.3.18}$$

where

$$\mathsf{L}^1(-t)\equiv\mathrm{e}^{-\mathrm{i}\mathsf{L}^0 t}\mathsf{L}^1\mathrm{e}^{\mathrm{i}\mathsf{L}^0 t}. \tag{16.3.19}$$

[3] Here, we assume that the limit

$$\lim_{\eta\to 0}\eta\int_0^\infty(\mathrm{e}^{\mathrm{i}\mathsf{L}^0 t}\mathscr{S}^+|\mathrm{e}^{\mathrm{i}\mathsf{L}^0 t}\mathsf{L}^1\mathscr{S}^+)^{(1)}\,\mathrm{e}^{-\eta t}\,dt$$

is zero or negligibly small.

acts as a commutator, $\mathsf{L}^1(-t)\mathscr{X}=1/\hbar\,[\mathscr{H}^1(-t),\mathscr{X}]$, with

$$\mathscr{H}^1(-t)=\mathrm{e}^{-\mathrm{i}\mathsf{L}^0 t}\mathscr{H}^1 \;. \tag{16.3.20}$$

This method of eliminating $\Omega^{(2)}$ also works in the case of several observables, if they form a subspace which is invariant with respect to L^0.

16.3.3 Reduction to Bath Correlation Functions

After considering the general aspects of the problem, we now wish to decompose $\mathrm{i}\Lambda$ into the factors which are related to the spin and bath subsystems. For this purpose, we make use of the fact that $\mathscr{S}^+$ is an eigenvector of L^0 and that therefore, due to the definition of the Mori product (8.2.14), the relation

$$(\mathscr{S}^+|\mathscr{X})^{(0)}=\frac{\exp(\beta\hbar\omega_0)-1}{\beta\hbar\omega_0}\langle\mathscr{S}^-\mathscr{X}\rangle_\beta^{(0)} \tag{16.3.21}$$

is valid. In this way, we can relate the numerator and the denominator in (16.3.18) to expectation values which can be directly evaluated,

$$\mathrm{i}\Lambda=-\mathrm{i}\omega_0-\frac{1}{\langle\mathscr{S}^-\mathscr{S}^+\rangle^0}\int\limits_0^\infty\left\langle\mathscr{S}^-\frac{1}{\hbar}\left[\mathscr{H}^1(-t),\frac{1}{\hbar}[\mathscr{H}^1,\mathscr{S}^+]\right]\right\rangle_\beta^{(0)}\mathrm{e}^{-\eta t}dt\Bigg|_{\eta\to+0}\;. \tag{16.3.22}$$

It is useful here to apply the commutator rule:

$$[\mathscr{s}\mathscr{B},\mathscr{s}'\mathscr{B}']_-=\tfrac{1}{2}\{[\mathscr{s},\mathscr{s}']_-[\mathscr{B},\mathscr{B}']_++[\mathscr{s},\mathscr{s}']_+[\mathscr{B},\mathscr{B}']_-\} \tag{16.3.23}$$

since it will lead us to a form in which the linear response functions of the *bath* system appear. With the definitions

$$\mathrm{e}^{\mathrm{i}\mathsf{L}^{\mathrm{B}}t}\mathscr{B}_k^\alpha=\mathscr{B}_k^\alpha(t)\;, \tag{16.3.24}$$

$$\mathscr{B}_k^z-\langle\mathscr{B}_k^z\rangle_\mathrm{B}=\delta\mathscr{B}_k^z \tag{16.3.25}$$

we find, in fact,

$$\begin{aligned}&\frac{1}{\hbar^2}\langle\mathscr{S}^-[\mathscr{H}^1(-t),[\mathscr{H}^1,\mathscr{S}^+]]\rangle^{(0)}\\&\quad=\frac{1}{2}\sum_j\langle\mathscr{s}_j^-\mathscr{s}_j^+\rangle_\mathrm{s}\{\langle[\delta\mathscr{B}_j^z,\delta\mathscr{B}_j^z(t)]_+\rangle_\mathrm{B}+2\langle[\mathscr{B}_j^-,\mathscr{B}_j^+(t)]_+\rangle_\mathrm{B}\,\mathrm{e}^{\mathrm{i}\omega_0 t}\}\\&\qquad+\frac{1}{\hbar}\sum_{j\neq k}\langle\mathscr{s}_j^z\rangle_\mathrm{s}\langle\mathscr{s}_k^-\mathscr{s}_k^+\rangle_\mathrm{s}\{\langle[\mathscr{B}_j^z,\mathscr{B}_k^z(t)]_-\rangle_\mathrm{B}\\&\qquad-2\langle[\mathscr{B}_j^-,\mathscr{B}_k^+(t)]_-\rangle_\mathrm{B}\,\mathrm{e}^{\mathrm{i}\omega_0 t}\}\;.\end{aligned} \tag{16.3.26}$$

Thus, with the aid of the bath susceptibilities (16.2.21) for $\alpha = +, -, z$ and $j \neq k$, we can write $\mathrm{i}\Lambda$ in (16.3.22) in the form:

$$\begin{aligned}\mathrm{i}\Lambda = &-\mathrm{i}\omega_0 - \frac{1}{2}\frac{1}{N}\sum_j \int_0^\infty \mathrm{e}^{-\eta t}\{\langle[\delta\mathscr{B}_j^z, \delta\mathscr{B}_j^z(t)]_+\rangle_\mathrm{B} \\ &+ 2\langle[\mathscr{B}_j^-, \mathscr{B}_j^+(t)]_+\rangle_\mathrm{B}\mathrm{e}^{\mathrm{i}\omega_0 t}\}\,dt|_{\eta\to+0} \\ &-\mathrm{i}\frac{\langle\mathscr{S}^z\rangle_\mathrm{s}}{N}\frac{1}{N}\sum_{j\neq k}\{\chi_{jk}^{zz}(\omega=0) - 2\chi_{jk}^{++}(\omega_0)\} = \mathrm{i}(\Omega+\Delta\Omega) - \Gamma\ .\end{aligned} \tag{16.3.27}$$

In discussing $\mathrm{i}\Lambda$, we begin with the damping Γ. The real part of the anticommutators can, using (9.7.12), be expressed in terms of susceptibilities:

$$\int_0^\infty \langle[\delta\mathscr{B}_j^z, \delta\mathscr{B}_j^z(t)]_+\rangle_\mathrm{B}\mathrm{e}^{-\eta t}\,dt = \frac{2}{\beta}\lim_{\omega\to 0}\frac{1}{\omega}\chi_{jj}^{zz}(\omega)''\ , \tag{16.3.28}$$

$$\mathrm{Re}\int_0^\infty \langle[\mathscr{B}_j^-, \mathscr{B}_j^+(t)]_+\rangle_\mathrm{B}\mathrm{e}^{\mathrm{i}\omega_0 t - \eta t}dt = \frac{\hbar^2}{2}\frac{N}{\langle\mathscr{S}^z\rangle_\mathrm{s}}\chi_{jj}^{++}(\omega_0)''\ . \tag{16.3.29}$$

Therefore, using the decomposition (9.4.2) for the bath susceptibilities, we find for Γ

$$\begin{aligned}\Gamma = &\frac{1}{\beta N}\sum_j \lim_{\omega\to 0}\frac{1}{\omega}\chi_{jj}^{zz}(\omega)'' + \frac{1}{2}\frac{\hbar^2}{\langle\mathscr{S}^z\rangle_\mathrm{s}}\sum_j \chi_{jj}^{++}(\omega_0)'' \\ &+ 2\frac{\langle\mathscr{S}^z\rangle_\mathrm{s}}{N}\frac{1}{N}\sum_{j\neq k}\chi_{jk}^{++}(\omega_0)''\end{aligned} \tag{16.3.30}$$

The term $\chi_{jk}^{zz}(\omega=0)''$ in (16.3.27) makes no contribution to the double sum due to the symmetry relation (9.4.13). Comparison with the damping in the case of longitudinal relaxation (16.2.23) shows that now, terms for $j \neq k$ also appear. They indicate a *collective* damping of the spins. The *individual* spin damping terms, i.e. the terms with bath correlations at the same location, also show differences. The contribution due to the frequencies $\pm\omega_0$ is half as large as in the longitudinal case. Furthermore, an additional term $(1/\omega)\chi_{jj}^{zz}(\omega)$ appears in Γ for $\omega\to 0$; it originates from the coupling to $\mathscr{B}_j^z s_j^z$.

The frequency $\Omega+\Delta\Omega$ in (16.3.27) can, with the aid of the decomposition (9.4.2) and the definition (16.2.11) of ω_0, be expressed in the form

$$\begin{aligned}\Omega+\Delta\Omega = &-\left(\gamma H - \frac{1}{N}\sum_j\langle\mathscr{B}_j^z\rangle_\mathrm{B}\right) \\ &-\frac{1}{N}\sum_j \mathrm{Im}\int_0^\infty \langle[\mathscr{B}_j^-, \mathscr{B}_j^+(t)]_+\rangle_\mathrm{B}\mathrm{e}^{\mathrm{i}\omega_0 t - \eta t}dt \\ &-\frac{\langle\mathscr{S}^z\rangle_\mathrm{s}}{N}\frac{1}{N}\sum_{j\neq k}\{\chi_{jk}^{zz}(\omega=0)' - 2\chi_{jk}^{++}(\omega_0)'\}\ .\end{aligned} \tag{16.3.31}$$

Here, the second term in (16.3.27) makes no contribution, since it is real. This result (16.3.31) demonstrates the frequency shift relative to the resonance frequency γH of free spins. The shift $1/N\sum_j \langle \mathscr{B}_j^z \rangle_{\mathrm{B}}$, may be interpreted as a mean local magnetic field which the surroundings produce at the spin sites. The second term is caused by the effect of the bath on the dynamics of individual spins. It may also be written in terms of a susceptibility of the overall system[4]. The two last terms can be interpreted as molecular-field contributions of an effective spin-spin interaction produced by the bath.

For the relaxation function $\Phi_{++}(t)$, using $\Phi_{++}(0)=2\chi_{xx}^{\mathrm{T}}$ from (16.3.14), we find the result:

$$\Phi_{++}(t)=2\chi_{xx}^{\mathrm{T}}\mathrm{e}^{\mathrm{i}[(\Omega+\Delta\Omega)-\Gamma]t} \ . \qquad (16.3.32)$$

From this, the transverse dynamical spin susceptibility from (16.3.9) is found using (9.4.19):

$$\boxed{\chi_{xx}(\omega)=\chi_{xx}^{\mathrm{T}}\frac{1}{2}\left\{\frac{\Omega+\Delta\Omega+\mathrm{i}\Gamma}{\omega+\Omega+\Delta\Omega+\mathrm{i}\Gamma}+\frac{-\Omega-\Delta\Omega+\mathrm{i}\Gamma}{\omega-\Omega-\Delta\Omega+\mathrm{i}\Gamma}\right\}} \ . \qquad (16.3.33)$$

The expressions (16.3.31) for $\Omega+\Delta\Omega$ and (16.3.30) for Γ thus determine the resonance frequencies and linewidths of the susceptibility. As we already pointed out in Sect. 16.2, the bath susceptibilities may be evaluated for particular bath systems. They are of practical importance in the interpretation of nuclear and electron spin resonance experiments.

16.4 A Local Conservation Law

The preceding examples of the Mori theory all referred to observables which could be considered as slow owing to an internal interaction. A further group of slow observables are the spatial densities $\mathscr{u}(\mathfrak{r})$, whose integrals over the system volume V are constants of the motion:

$$\left[\int_V \mathscr{u}(\mathfrak{r})d^3x, \mathscr{H}\right]=\mathscr{O} \ . \qquad (16.4.1)$$

In such cases, the spatial Fourier components $\mathscr{u}_{\mathfrak{k}}$ are slow for small $\mathfrak{k}$, $\lim_{\mathfrak{k}\to 0}\dot{\mathscr{u}}_{\mathfrak{k}}=0$.

These densities may, physically, represent energy densities, particle densities, spin densities, etc. In the majority of cases, such as in hydrodynamics, several densities must be treated simultaneously. In order to make the principle clear, we

[4] It may be expressed in terms of $\beta(\mathscr{B}_j^+ \mathscr{s}_j^- | \mathscr{B}_j^+ \mathscr{s}_j^-)^{(0)}$ and $\chi_{jj}^{++}(\omega_0)'$.

shall discuss here the case of only a *single* density and shall assume that the coupling to other slow observables may be neglected. Possible examples are the spin density or the energy density in a solid. We wish to show how the Mori theory for the expectation value of such a density can lead us to an equation of the diffusion or heat-conduction type, and to the quantum-statistical definition of a diffusion constant.

In the following considerations, we take the operator $\mathscr{u}(\mathfrak{r})$ to represent a *macroscopic* density, which is obtained by spatial averaging over a microscopic density (cf. Sect. 4.1). The spatial Fourier series for $\mathscr{u}(\mathfrak{r})$ thus contains contributions from $\mathfrak{k}$-values for $|\mathfrak{k}| \leqq k_c$ only. As a result of (16.4.1), this density in the Heisenberg formulation obeys an equation of continuity:

$$\frac{\partial}{\partial t}\mathscr{u}(\mathfrak{r},t)+\operatorname{div}\vec{\mathscr{j}}(\mathfrak{r},t)=\mathcal{O} \ . \tag{16.4.2}$$

In Fourier-transformed space, the following results are obtained: the series for the operators are given by

$$\mathscr{u}(\mathfrak{r},\, t)=\sum_{|\mathfrak{k}|\leqq k_c} \mathscr{u}_{\mathfrak{k}}(t)e^{i\mathfrak{k}\mathfrak{r}} \ , \tag{16.4.3}$$

$$\vec{\mathscr{j}}(\mathfrak{r},\, t)=\sum_{|\mathfrak{k}|\leqq k_c} \vec{\mathscr{j}}_{\mathfrak{k}}(t)\, e^{i\mathfrak{k}\mathfrak{r}} \ , \tag{16.4.4}$$

with periodic boundary conditions. The operators

$$\mathscr{u}_{\mathfrak{k}}(t)=e^{i\mathsf{L}t}\mathscr{u}_{\mathfrak{k}} \tag{16.4.5}$$

and

$$\vec{\mathscr{j}}_{\mathfrak{k}}(t)=e^{i\mathsf{L}t}\vec{\mathscr{j}}_{\mathfrak{k}} \tag{16.4.6}$$

are connected by the following relation, due to the equation of continuity (16.4.2):

$$\mathsf{L}\mathscr{u}_{\mathfrak{k}}+\mathfrak{k}\vec{\mathscr{j}}_{\mathfrak{k}}=\mathcal{O} \ . \tag{16.4.7}$$

For $\mathfrak{k}\to 0$, this is just the conservation law (16.4.1). The currents for $\mathfrak{k}\to 0$ are assumed not to be constants of the motion,

$$\mathsf{L}\vec{\mathscr{j}}_{\mathfrak{k}=0}\neq\mathcal{O} \ . \tag{16.4.8}$$

16.4.1 Decoupling of the Fourier Components

In applying the Mori theory, we must first establish the nature of the set of observables $\mathscr{G}_\mu$. Here, the $\mathscr{G}_\mu$ are originally given by the $\mathscr{u}(\mathfrak{r})$ or by all $\mathscr{u}_{\mathfrak{k}}$ with $|\mathfrak{k}|\leqq k_c$. We would thus have to begin with a projector onto the set of *all* $\mathscr{u}_{\mathfrak{k}}$ with $|\mathfrak{k}|\leqq k_c$. However, in the case of translational invariance of $\mathscr{H}$, the treatment can be considerably simplified. We shall therefore assume that property here.

Then for all scalar products with different $\mathfrak{k}$, the following holds (Appendix B1):

$$(u_{\mathfrak{k}'}|u_{\mathfrak{k}}(t))=\delta_{\mathfrak{k}'\mathfrak{k}}(u_{\mathfrak{k}}|u_{\mathfrak{k}}(t)) \ , \quad \langle u_{\mathfrak{k}}\rangle_\beta=0 \quad \text{for} \quad \mathfrak{k}\neq 0 \ . \tag{16.4.9}$$

The Mori equations with a projector onto the space of the $u_{\mathfrak{k}}$ then give a system of non-coupled equations for the correlation functions $(u_{\mathfrak{k}}|u_{\mathfrak{k}}(t))$. Similarly to the argumentation in Sect. 16.3, it then follows that the integro-differential equation for a *fixed* $\mathfrak{k}$ is identical to the Mori equation which one obtains with the projector on the *single* non-Hermitian observable $u_{\mathfrak{k}}$,

$$\mathsf{P}=\frac{|u_{\mathfrak{k}})(u_{\mathfrak{k}}|}{(u_{\mathfrak{k}}|u_{\mathfrak{k}})} \ . \tag{16.4.10}$$

We may therefore use the Mori equation

$$\frac{\partial}{\partial t}(u_{\mathfrak{k}}|u_{\mathfrak{k}}(t))=\mathrm{i}\Omega(u_{\mathfrak{k}}|u_{\mathfrak{k}}(t))-\int_0^t(u_{\mathfrak{k}}|u_{\mathfrak{k}}(t-t'))\gamma(t')dt' \tag{16.4.11}$$

with the definitions

$$\Omega=\frac{1}{(u_{\mathfrak{k}}|u_{\mathfrak{k}})}(u_{\mathfrak{k}}|\mathsf{L}u_{\mathfrak{k}}) \tag{16.4.12}$$

and

$$\gamma(t)=\frac{1}{(u_{\mathfrak{k}}|u_{\mathfrak{k}})}(\dot{u}_{\mathfrak{k}}|\mathrm{e}^{\mathrm{i}\mathsf{QL}t}\mathsf{Q}\dot{u}_{\mathfrak{k}}) \ . \tag{16.4.13}$$

When the statistical operator $\rho(0)$ is defined by the linearized generalized statistical operator over the observation level $\mathcal{H}$ and all $u_{\mathfrak{k}}(|\mathfrak{k}|\leqq k_c)$, (16.4.11) is correspondingly valid for the expectation values also, since from (11.4.31) and (16.4.9), the relation

$$\langle u_{\mathfrak{k}}\rangle(t)=-\sum_{\mathfrak{k}'}\lambda_{\mathfrak{k}}(u_{\mathfrak{k}'}|u_{\mathfrak{k}}(t))=-\lambda_{\mathfrak{k}}(u_{\mathfrak{k}}|u_{\mathfrak{k}}(t)) \tag{16.4.14}$$

is satisfied.

The frequency Ω is zero when the densities $u(\mathrm{r})$ (as e.g. in the case of particle densities) at different locations commute. Then, from (11.2.15), taking account of the back transformation of $u_{\mathfrak{k}}$,

$$\begin{aligned}\Omega&=\frac{1}{(u_{\mathfrak{k}}|u_{\mathfrak{k}})}(u_{\mathfrak{k}}|\mathsf{L}u_{\mathfrak{k}})\\&=\frac{1}{(u_{\mathfrak{k}}|u_{\mathfrak{k}})}\frac{1}{\hbar\beta V^2}\int\langle[u(\mathrm{r}'),\ u(\mathrm{r})]\rangle_\beta \mathrm{e}^{\mathrm{i}\mathfrak{k}(\mathrm{r}'-\mathrm{r})}d^3x\,d^3x'=0 \ .\end{aligned} \tag{16.4.15}$$

The frequency Ω however also vanishes when $\mathcal{H}$ and $u(\mathrm{r})$ are invariant with respect to time reversal (cf. Appendix B2). We shall therefore take $\Omega=0$ in the following discussion.

The memory function can be simplified by using (16.4.7). From (16.4.15), the projector Q on $\mathsf{L}u_{\mathfrak{k}}$ drops out, so that after inserting (16.4.7), we have

$$\gamma(t)=\frac{1}{(u_{\mathfrak{k}}|u_{\mathfrak{k}})}\sum_{\alpha,\gamma}k_\alpha(\dot{j}^{\alpha}_{\mathfrak{k}}|\mathrm{e}^{\mathrm{iQL}t}\dot{j}^{\gamma}_{\mathfrak{k}})k_\gamma \ . \tag{16.4.16}$$

We thus obtain for the expectation values $\langle u_{\mathfrak{k}}\rangle(t)$ the equation

$$\frac{\partial}{\partial t}\langle u_{\mathfrak{k}}\rangle(t)=-\int_0^t\langle u_{\mathfrak{k}}\rangle(t-t')\gamma(t')\,dt' \ . \tag{16.4.17}$$

16.4.2 The Wavenumber as a Slowness Parameter

In the equation of motion (16.4.17), we are interested in small values of $\mathfrak{k}$. Therefore, we expand the memory function in $\mathfrak{k}\to 0$ and retain only the first nonvanishing term. In this way, we obtain the approximation

$$\gamma(t)=\frac{1}{(u_0|u_0)}\sum_{\alpha,\gamma}k_\alpha(\dot{j}^{\alpha}_{0}|\mathrm{e}^{\mathrm{i}\mathrm{L}t}\dot{j}^{\gamma}_{0})k_\gamma \ . \tag{16.4.18}$$

The projector Q can be left out for $\mathfrak{k}\to 0$, since from (16.4.7), we have

$$\lim_{\mathfrak{k}\to 0}\mathsf{QL} = \mathsf{L}+\lim_{\mathfrak{k}\to 0}\sum_{\alpha}\frac{|u_{\mathfrak{k}})(\dot{j}^{\alpha}_{\mathfrak{k}}|}{(u_{\mathfrak{k}}|u_{\mathfrak{k}})}k_\alpha=\mathsf{L} \ . \tag{16.4.19}$$

In (16.4.18) we have assumed that $\lim_{\mathfrak{k}\to 0}(u_{\mathfrak{k}}|u_{\mathfrak{k}})\equiv(u_0|u_0)$ is not zero[5] and is independent of direction.

For the further investigation of (16.4.17), we employ the approximation (16.4.18); then k^2 is a slowness parameter in the sense of Sect. 13.1. We can thus carry out a Markovian approximation for small $\mathfrak{k}$. We also assume rotational invariance, so that

$$(\dot{j}^{\alpha}_{0}|\mathrm{e}^{\mathrm{i}\mathrm{L}t}\dot{j}^{\gamma}_{0})=\delta_{\alpha\gamma}\tfrac{1}{3}(\dot{\vec{j}}_0|\mathrm{e}^{\mathrm{i}\mathrm{L}t}\dot{\vec{j}}_0) \tag{16.4.20}$$

holds. Then from (16.4.17 and 18), we obtain a simple result for $\langle u_{\mathfrak{k}}\rangle t)$:

$$\frac{\partial}{\partial t}\langle u_{\mathfrak{k}}\rangle(t)=-Dk^2\langle u_{\mathfrak{k}}\rangle(t) \ , \tag{16.4.21}$$

[5] In systems with a Coulomb interaction, the correlation function of the charge densities vanishes for $\mathfrak{k}\to 0$.

where D is defined by [6,7]

$$\boxed{D=\frac{1}{(u_0|u_0)}\frac{1}{3}\int_0^{\infty}(\vec{j}_0|\vec{j}_0(t))\,dt} \quad . \tag{16.4.22}$$

The back transformation of (16.4.21) into configuration space yields, according to (16.4.3), an equation of the *diffusion* or *heat conduction* type,

$$\boxed{\frac{\partial}{\partial t}\langle u(\mathrm{r})\rangle(t)=D\,\nabla^2\langle u(\mathrm{r})\rangle\,(t)} \quad . \tag{16.4.23}$$

The validity of (16.4.23) depends above all on whether the $\mathfrak{k}$ which occur in (16.4.3) are sufficiently small that the expansion (16.4.18) of $\gamma(t)$ in k_α is still reasonable. If we let ξ^{-1} be a limiting $\mathfrak{k}$-value, up to which the correlation function obeys the relation

$$\frac{1}{(u_{\mathfrak{k}}|u_{\mathfrak{k}})}(\dot{j}^{\alpha}_{\mathfrak{k}}|e^{iQLt}\dot{j}^{\gamma}_{\mathfrak{k}})=\frac{1}{(u_0|u_0)}(\dot{j}^{\alpha}_0|e^{iLt}\dot{j}^{\gamma}_0) \tag{16.4.24}$$

then it must fulfill

$$\xi k_c \leqq 1 \ . \tag{16.4.25}$$

In the transition to a differential equation, we must require that $\langle u_{\mathfrak{k}}\rangle(t)$ be practically constant within the correlation time of $(\vec{j}_0|\exp(iLt)\vec{j}_0)$, i.e. that

$$Dk_c^2\tau_c \ll 1 \tag{16.4.26}$$

is obeyed. If a ξ exists at all in (16.4.25), then the two conditions (16.4.25 and 26) can be fulfilled when k_c is sufficiently small, i.e. when the spatial averaging in the definition of the macroscopic densities is taken over a sufficiently large volume element ΔV.

In the given derivation of the diffusion equation (16.4.23), we assumed a *single* conservation law. In the case of *several* conservation laws, one can, in principle, treat the problem in just the same way. The Mori theory for several observables, but for the same wavenumber, is then applied; for small $\mathfrak{k}$, one obtains a system of coupled differential equations. In this way, for example, one can derive the linearized equations of hydrodynamics and the definitions of the corresponding material constants.

---

[6] In (16.4.22), for the transition to $\exp(iLt)$, one must first take $\mathfrak{k}\to 0$ and then integrate. Otherwise, one finds

$$\frac{1}{k^2}\int_0^{\infty}(\mathfrak{k}\vec{j}_{\mathfrak{k}})|e^{iLt}\,\mathfrak{k}\vec{j}_{\mathfrak{k}})\,dt=\frac{1}{k^2}\int_0^{\infty}(Lu_{\mathfrak{k}}|e^{iLt}Lu_{\mathfrak{k}})\,dt=0 \ .$$

[7] If the Mori theory in Markovian approximation is applied to $\vec{j}_0(t)$, one obtains from (16.4.22) a relation between the damping and the diffusion constant.

Part 4

Equations of Motion of the Relevant Parts of the Statistical Operator

In Part 3, we treated small deviations from equilibrium using the Mori–Langevin theory. We required that the statistical operators at the beginning of a process be expressable in terms of generalized canonical operators in a linear approximation. In this concluding part of the book, we consider more general initial situations, and discuss the basic aspects of some suitable methods for treating them. In principle, one could extend the Mori theory to include additional projectors and thus require the expectation value of the residual force to vanish even for situations further removed from equilibrium. It is, however, more convenient to construct a theory for such cases in the Schrödinger formulation and to carry out exact rearrangements of the von Neumann equation. In this way, one obtains an equation for a part of the statistical operator, which has – in particular in the case of numerous observables – a clearer structure than a set of multiple coupled Langevin equations. This is especially obvious when one wishes to describe portions of a system microscopically, since in that case, a very large set of (no longer macroscopic) observables has to be considered. The physical results are of course independent of whether one begins with the Schrödinger formulation or the Heisenberg formulation.

17. Mappings of the Statistical Operator onto a Relevant Part

17.1 The Concept of the Relevant Part, $\rho_{\text{rel}}(t)$

In general in a physical system, only the expectation values of a certain set of observables $\{\mathscr{F}\}$ are of interest[1]. These operators $\mathscr{F}_\nu$ do not generate a complete basis in the Liouville space of the system; thus, when one calculates the time dependence of the expectation value $\langle\mathscr{F}_\nu\rangle(t) = \text{Tr}(\mathscr{F}_\nu\rho(t))$, there are parts of the statistical operator $\rho(t)$ which do not contribute to the values $\langle\mathscr{F}_\nu\rangle(t)$. It is therefore expedient to decompose the statistical operator with respect to the observables of interest into a *relevant part*, $\rho_{\text{rel}}(t)$, and an irrelevant part, $\rho_{\text{irr}}(t)$, so that in taking the expectation values $\text{Tr}(\mathscr{F}_\nu\rho(t))$, only the relevant part contributes. For this decomposition

$$\rho(t) = \rho_{\text{rel}}(t) + \rho_{\text{irr}}(t) \tag{17.1.1}$$

we thus have

$$\text{Tr}(\mathscr{F}_\nu\rho(t)) = \text{Tr}(\mathscr{F}_\nu\rho_{\text{rel}}(t)) \tag{17.1.2}$$

and

$$\text{Tr}(\mathscr{F}_\nu\rho_{\text{irr}}(t)) = 0 \ . \tag{17.1.3}$$

This separation of the statistical operator $\rho(t)$ therefore depends on the set of observables $\{\mathscr{F}\}$, but is not uniquely determined by it alone. A relevant part which satisfies (17.1.1–3) may be specified in various ways, but it is in every case a mapping (even when explicitly time-dependent) of the statistical operator

$$\boxed{\rho_{\text{rel}}(t) = \mathscr{f}[t, \rho(t)]} \tag{17.1.4}$$

(compare Fig. 17.1).

In the simplest case, the relevant part, $\rho_{\text{rel}}(t)$ can depend *linearly* on $\rho(t)$, i.e. it can be constructed from $\rho(t)$ using a linear operator P in the Liouville space

---

[1] We take the observables $\mathscr{F}_\nu$ to be explicitly time *in*dependent.

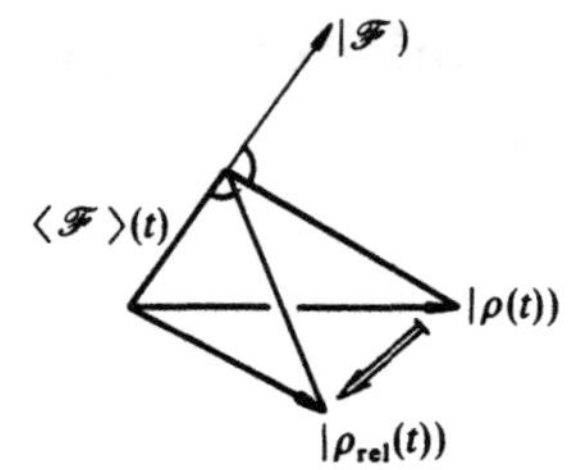

Fig. 17.1. Schematic representation of the mapping $\rho(t) \mapsto \rho_{\text{rel}}(t)$ in Liouville space (17.1.4). If one reads the traces in (17.1.2) as scalar products, then $\rho(t)$ and $\rho_{\text{rel}}(t)$ have the same components along the $\mathscr{F}$-axis, i.e. the same expectation values $\langle\mathscr{F}\rangle(t)$

(superoperator, cf. Sect. 3.1),

$$\rho_{\text{rel}}(t) = \mathsf{P}\rho(t) \ . \tag{17.1.5}$$

However, one can also apply a *nonlinear* construction rule for $\rho_{\text{rel}}(t) = \not{f}[\rho(t)]$ and, in particular, the case can be considered in which the relevant part $\rho_{\text{rel}}(t)$ is given by the accompanying generalized canonical statistical operator $\mathscr{R}_{\{\mathscr{G}\}}(t)$ over an observation level $\{\mathscr{G}\}$.

It is the goal of this chapter to eliminate the irrelevant part of the statistical operator from the von Neumann equation for $\rho(t)$. As we shall see, this is possible under certain conditions and leads to a complicated but exact integro-differential equation for the relevant part, $\rho_{\text{rel}}(t)$. This equation can then serve as the starting point for approximations and for the derivation of equations in closed form for the expectation values of interest, $\langle\mathscr{F}_\nu\rangle(t)$. In making the transition from the von Neumann equation to an equation for the relevant statistical operator $\rho_{\text{rel}}(t)$, one employs the essential property that the initial states of physical interest can also be characterized only by a certain set of statistical operators $\{\rho(t_0)\}$, e.g. by generalized canonical statistical operators $\mathscr{R}_{\{\mathscr{G}\}}$. The resulting equations for the relevant statistical operator therefore are valid only for a certain set of observables, $\{\mathscr{F}\}$, and for a certain set of statistical operators, $\{\rho(t_0)\}$, which describe the initial states under consideration[2].

17.2 Linear Relation Between $\rho_{\text{rel}}(t)$ and $\rho(t)$

First, we shall consider the case that the relevant part of the statistical operator with respect to a set of observables $\mathscr{F}_\nu$ can be obtained by a *linear* mapping,

[2] If one assumes the process to begin at a time $t_1 > t_0$ with the same set of initial states $\{\rho(t_0)\}$, then, in general, the initial states of the inverse process, $\{\rho^{\text{i}}(t_1)\}$, which lead to $\rho(t_2) = \mathsf{U}(t_2, t_1)\rho^{\text{i}}(t_1) = \rho(t_0)$ at a later time $t_2 > t_1$, no longer belong to the set $\{\rho(t_0)\}$. For this reason, an equation for $\rho_{\text{rel}}(t)$ is usually *not* valid for the formally possible inverse process.

$$\boxed{\rho_{rel}(t) = \not{p}[\rho(t)] = \mathsf{P}\rho(t)} \; . \tag{17.2.1}$$

The operator P which defines the mapping is thus independent of $\rho(t)$.[3]

17.2.1 Properties of the Operator P

For the determination of the relevant part according to (17.2.1), not just any linear operator P may be used. In the following, we list the requirements for P.

We imagine that a relevant part $\mathsf{P}\rho(t)$ has been generated from $\rho(t)$. On the other hand, if such a $\rho_{rel}(t)$ is already known in the form of a statistical operator, then the relevant part constructed from (17.2.1) is required to be identical with this given operator:

$$\rho_{rel}(t) = \mathsf{P}\rho_{rel}(t) \; . \tag{17.2.2}$$

According to the definition (17.2.1), this means that

$$\mathsf{P}\rho(t) = \mathsf{P}(\mathsf{P}\rho(t)) \; . \tag{17.2.3}$$

We therefore require in general that P be idempotent,

$$\boxed{\mathsf{P} = \mathsf{P}^2} \; . \tag{17.2.4}$$

For the given set of observables $\mathscr{F}_\nu$, the relevant part $\rho_{rel}(t)$ must yield the same expectation values as the total statistical operator $\rho(t)$. This means, according to (17.1.2) and (17.2.1), that

$$\mathrm{Tr}(\mathscr{F}_\nu \rho_{rel}(t)) = \mathrm{Tr}(\mathscr{F}_\nu \mathsf{P}\rho(t)) = \mathrm{Tr}(\mathscr{F}_\nu \rho(t)) = F_\nu(t) \; . \tag{17.2.5}$$

Thus, P must have the property

$$\boxed{\mathrm{Tr}(\mathscr{F}_\nu \mathsf{P}\mathscr{X}) = \mathrm{Tr}(\mathscr{F}_\nu \mathscr{X})} \; , \tag{17.2.6}$$

where $\mathscr{X}$ denotes an element in the Liouville space, for which the traces exist.

It is reasonable to require further of the definition of a relevant part that the initial states under consideration, $\rho(t_0)$, have no irrelevant part,

$$\rho(t_0) = \rho_{rel}(t_0) \; . \tag{17.2.7}$$

For a given set of $\{\rho(t_0)\}$, this requirement leads with (17.2.1) to the condition

$$\boxed{\mathsf{P}\rho(t_0) = \rho(t_0)} \tag{17.2.8}$$

[3] In Part 4, the symbol P denotes mappings onto $\rho_{rel}(t)$, whereas in all of Parts 2 and 3, it was used for the Mori projector.

for P. In Sect. 17.2.3, we shall see that this property of P plays an important role in obtaining a dynamical equation for $\rho_{rel}(t)$ in closed form.

17.2.2 Explicit Expressions for P

In order to make the linear mapping (17.2.1) clearer, we discuss some explicit expressions for P. We wish to assume that $\rho_{rel}(t)$ may be expanded in a time-independent basis ρ_μ[4] of a subspace $\mathfrak{L}_{\rho_{rel}}$ of $\mathfrak{L}$, of which $\rho_{rel}(t)$ is also always an element,

$$\rho_{rel}(t) = \sum_\mu \rho_\mu c_\mu(t) \in \mathfrak{L}_{\rho_{rel}} \ . \tag{17.2.9}$$

The coefficients $c_\mu(t)$ become, due to the requirement (17.2.5), functions of the expectation values $F_\nu(t)$[5] of the observables $\mathscr{F}_\nu$, since it follows from (17.2.5) that:

$$F_\nu(t) = \sum_\mu \mathrm{Tr}(\mathscr{F}_\nu \rho_\mu) c_\mu(t) \ . \tag{17.2.10}$$

We assume that the observables $\mathscr{F}_\nu$ lie in a subspace $\mathfrak{L}_{\mathscr{F}}$ of $\mathfrak{L}$ with the *same* dimensionality as $\mathfrak{L}_{\rho_{rel}}$,

$$\dim \mathfrak{L}_{\rho_{rel}} = \dim \mathfrak{L}_{\mathscr{F}} \tag{17.2.11}$$

and that the matrix

$$\mathrm{Tr}(\mathscr{F}_\nu \rho_\mu) \tag{17.2.12}$$

which is constant in time, has an inverse:

$$\sum_\mu \mathrm{Tr}(\mathscr{F}_\nu \rho_\mu) v_{\mu\lambda} = \delta_{\nu\lambda} \ . \tag{17.2.13}$$

Solution of (17.2.10) gives the result

$$c_\lambda(t) = \sum_\mu v_{\lambda\nu} F_\nu(t) \ . \tag{17.2.14}$$

Substituting in (17.2.9) yields:

$$\rho_{rel}(t) = \sum_{\mu,\nu} \rho_\mu v_{\mu\nu} F_\nu(t) = \sum_{\mu,\nu} \rho_\mu v_{\mu\nu} \mathrm{Tr}(\mathscr{F}_\nu \rho(t)) \tag{17.2.15}$$

i.e. $\rho_{rel}(t)$ *is a linear function of the given expectation values* $F_\nu(t)$ and therefore has

---

[4] The ρ_μ are at first taken to be some set of independent operators. It is not necessary that they have properties of a statistical operator.

[5] The unit operator 1 is taken to belong to the set of the $\mathscr{F}_\nu$ or to be a linear combination of them.

the form (17.2.1). For the superoperator P, we obtain the expression

$$\mathsf{P}\mathscr{X} = \sum_{\mu,\nu} \rho_\mu v_{\mu\nu} \operatorname{Tr}(\mathscr{F}_\nu \mathscr{X}) \tag{17.2.16}$$

i.e. a P which is independent of $\mathscr{X}$. It may be readily verified that this expression satisfies the conditions (17.2.4) and (17.2.6); (17.2.8) is also fulfilled, provided that the physical state of the system at the time t_0, $\rho(t_0)$, has the form (17.2.9), and thus lies in $\mathfrak{L}_{\rho_{rel}}$.

The mappings of $\mathfrak{L}$ onto $\mathfrak{L}_{\rho_{rel}}$ given by the form (17.2.9) or (17.2.15) or (17.2.16), and depending on the choice of the operators $\mathscr{F}_\nu$ and ρ_μ, comprise a whole series of different linear mappings. We note those cases which make use of the accompanying generalized canonical statistical operators (6.4.2) as the relevant statistical operator,

$$\rho_{rel}(t) = \mathscr{R}_{\{\mathscr{G}\}}(t) = \frac{1}{Z_{\{\mathscr{G}\}}(t)} \exp\left(-\sum_\nu \lambda_\nu(t)\mathscr{G}_\nu\right) \tag{17.2.17}$$

which, with a suitable choice of the set of observables $\mathscr{G}_\nu$, or with approximations, becomes a linear combination (17.2.9) of the operators ρ_ν (compare Exercise 17.1 and Sect. 18.1).

17.2.3 The Nakajima–Zwanzig Equation

With the requirements for P given in Sect. 17.2.1, we shall now proceed to derive an exact integro-differential equation for $\rho_{rel}(t)$ from the von Neumann equation. In order to make the principle clearer, we first consider the case that the Liouville operator is not time-dependent. For the rearrangements, it is expedient to define an operator Q by

$$\mathsf{P} + \mathsf{Q} = 1 \tag{17.2.18}$$

which, due to the idempotence of P, has the following properties:

$$\begin{aligned} \mathsf{Q}^2 &= \mathsf{Q} \,, \\ \mathsf{PQ} &= 0 \,. \end{aligned} \tag{17.2.19}$$

We start from the expression

$$\mathsf{P}\rho(t) = \mathsf{P}e^{-i\mathsf{L}(t-t_0)}\rho(t_0) \tag{17.2.20}$$

whose time derivative can be written in the form

$$\frac{d}{dt}\mathsf{P}\rho(t) = -i\mathsf{PL}\rho(t) = -i\mathsf{PLP}\rho(t) - i\mathsf{PLQ}\rho(t) \tag{17.2.21}$$

using the definition (17.2.18). The first term in (17.2.21) contains $\rho_{rel}(t)$. The second can be rewritten with the identity (3.4.7b), by making the following substitutions:

$$\mathsf{L}_1 = \mathsf{L}\mathsf{Q} , \tag{17.2.22}$$

$$\mathsf{L}_2 = \mathsf{L}\mathsf{P} \tag{17.2.23}$$

and applying it to $\rho(t_0)$. After multiplication by Q, we have

$$\begin{aligned} \mathsf{Q}\rho(t) &= \mathsf{Q}\,\mathrm{e}^{-\mathrm{i}\mathsf{L}(t-t_0)}\rho(t_0) \\ &= \mathsf{Q}\mathrm{e}^{-\mathrm{i}\mathsf{L}\mathsf{Q}(t-t_0)}\rho(t_0) - \mathrm{i}\int_{t_0}^{t} \mathsf{Q}\mathrm{e}^{-\mathrm{i}\mathsf{L}\mathsf{Q}(t-t')}\mathsf{L}\mathsf{P}\rho(t')\,dt' . \end{aligned} \tag{17.2.24}$$

On substituting (17.2.24) in (17.2.21), we obtain the Nakajima–Zwanzig equation

$$\boxed{\begin{aligned} \frac{d}{dt}\mathsf{P}\rho(t) = &-\mathrm{i}\mathsf{P}\mathsf{L}\mathsf{P}\rho(t) - \int_{t_0}^{t} \mathsf{P}\mathsf{L}\mathsf{Q}\mathrm{e}^{-\mathrm{i}\mathsf{L}\mathsf{Q}(t-t')}\mathsf{L}\mathsf{P}\rho(t')\,dt' \\ &- \mathrm{i}\mathsf{P}\mathsf{L}\mathsf{Q}\mathrm{e}^{-\mathrm{i}\mathsf{L}\mathsf{Q}(t-t_0)}\rho(t_0) . \end{aligned}} \tag{17.2.25}$$

It represents an *exact integro-differential equation* for $\rho_{\mathrm{rel}}(t) = \mathsf{P}\rho(t)$ with the inhomogeneous term $-\mathrm{i}\mathsf{P}\mathsf{L}\mathsf{Q}\exp[-\mathrm{i}\mathsf{L}\mathsf{Q}(t-t_0)]\rho(t_0)$. If one now requires property (17.2.8) of P, this term vanishes, since the idempotence of Q and the Taylor expansion of the exponential function lead to the relation (which is of practical use in some other problems, also):

$$\mathsf{Q}\mathrm{e}^{-\mathrm{i}\mathsf{L}\mathsf{Q}(t-t_0)} = \mathrm{e}^{-\mathrm{i}\mathsf{Q}\mathsf{L}(t-t_0)}\mathsf{Q} = \mathrm{e}^{-\mathrm{i}\mathsf{Q}\mathsf{L}\mathsf{Q}(t-t_0)}\mathsf{Q} . \tag{17.2.26}$$

Therefore, the inhomogeneous term in (17.2.25) can be reduced to the irrelevant part of the statistical operator $\rho(t_0)$,

$$\rho_{\mathrm{irr}}(t_0) = \rho(t_0) - \rho_{\mathrm{rel}}(t_0) = \mathsf{Q}\rho(t_0) \tag{17.2.27}$$

which, however, is zero according to requirement (17.2.8)[6].

In summary, we thus obtain from (17.2.25), using, $\rho_{\mathrm{rel}}(t) = \mathsf{P}\rho(t)$, the closed dynamical equation for $\rho_{\mathrm{rel}}(t)$:

$$\boxed{\frac{d}{dt}\rho_{\mathrm{rel}}(t) = -\mathrm{i}\mathsf{P}\mathsf{L}\mathsf{P}\rho_{\mathrm{rel}}(t) - \int_{t_0}^{t} \mathsf{P}\mathsf{L}\mathsf{Q}\mathrm{e}^{-\mathrm{i}\mathsf{L}\mathsf{Q}(t-t')}\mathsf{L}\mathsf{P}\rho_{\mathrm{rel}}(t')\,dt'} \quad . \tag{17.2.28}$$

[6] One can also drop this requirement, but then the operators under consideration, $\rho(t_0)$, must be carefully examined to verify whether the inhomogeneous term in (17.2.25) is negligible.

It has the form

$$\frac{d}{dt}\rho_{rel}(t) = -\mathrm{i}\mathsf{PLP}\rho_{rel}(t) - \int_{t_0}^{t} \mathsf{K}_{rel}(t-t')\rho_{rel}(t')\,dt' \tag{17.2.29}$$

and is valid for all $\rho(t_0)$ for which the relation $\mathsf{P}\rho(t_0) = \rho(t_0)$ exists. The operator

$$\mathsf{L}_{rel} = \mathsf{PLP} \tag{17.2.30}$$

and the operator $\mathsf{K}_{rel}(t)$, which can be written differently following (17.2.26):

$$\begin{aligned}\mathsf{K}_{rel}(t) &= \mathsf{PLQ}\,\mathrm{e}^{-\mathrm{i}\mathsf{LQ}t}\mathsf{LP} = \mathsf{PL}\,\mathrm{e}^{-\mathrm{i}\mathsf{QL}t}\mathsf{QLP}\\ &= \mathsf{PLQ}\mathrm{e}^{-\mathrm{i}\mathsf{QLQ}t}\mathsf{QLP}\ ,\end{aligned} \tag{17.2.31}$$

determine the dynamics of $\rho_{rel}(t)$ within the region of validity of (17.2.29) and thus play a similar role to that of the Liouville operator L in the dynamics of the total statistical operator. In this sense, (17.2.29), which is also called the *generalized master equation*[7], is completely general. Approximations can more readily be made from this equation than from the original von Neumann equation. This is a result of the fact that the unneeded degrees of freedom are no longer explicitly present in (17.2.29). They are hidden in the dynamics of $\mathsf{K}_{rel}(t)$, which is formally expressed by the time evolution with QLQ. The time dependence of $\mathsf{K}_{rel}(t-t')$ determines at which times $t' \leqq t$ in the past the relevant part $\rho_{rel}(t')$ gives a significant contribution to the integral in (17.2.29). Therefore, one also says that the time evolution of $\rho_{rel}(t)$ is determined by a memory equation.

If P has the form (17.2.16), then (17.2.28) becomes a closed system of linear integro-differential equations for the expectation values $F_\nu(t) = \mathrm{Tr}[\mathscr{F}_\nu\rho(t)]$. (An example is given in Exercise 17.1.) These equations also include the Mori equations (11.4.30) as a special case (cf. Exercise 18.3 and Appendix A).

17.2.4 Example: $\rho_{rel}(t)$ of a Subsystem

In order to clarify the formalism described above, we shall now consider the construction of $\rho_{rel}(t)$ and the corresponding dynamical equation in a concrete case, without, however, evaluating them further. To this end, we take up once again the example of Sect. 3.3, the dynamics of a subsystem. The Hamiltonian of the total system Σ is composed of the Hamiltonians: $\mathscr{H}^1$ and $\mathscr{H}^2$ of two subsystems Σ^1 and Σ^2, which are coupled by an interaction $\mathscr{H}^{12}$

$$\mathscr{H} = \mathscr{H}^1 + \mathscr{H}^2 + \mathscr{H}^{12}\ . \tag{17.2.32}$$

The unitary space of the total system Σ is required to be a *product space* of the unitary spaces of the two subsystems.

---

[7] This name comes from the special case of the Pauli master equation, in which $\rho_{rel}(t)$ is determined by the diagonal matrix elements of $\rho(t)$ in a $\mathscr{H}_0$ representation (cf. Exercise 17.1).

We are interested in the expectation values of the variables of system Σ^1. Furthermore, we assume that system Σ^2 is not strongly influenced by the coupling to system Σ^1, i.e. Σ^2 acts as a *thermal reservoir* (bath). We allow as a possible initial condition for the total system the situation where the subsystem Σ^2 remains in thermal equilibrium at a *fixed* temperature, while Σ^1 can be in an arbitrary initial state,

$$\rho(t_0) = \frac{\exp(-\beta\mathscr{H}^2)}{\mathrm{Tr}_2\exp(-\beta\mathscr{H}^2)}\rho^1(t_0) = \mathscr{R}^2\rho^1(t_0) \ . \tag{17.2.33}$$

The observables $\{\mathscr{F}\}$ in the sense of Sect. 17.2.1 are now *all* the variables of subsystem Σ^1. The relevant part $\rho_{\mathrm{rel}}(t)$ for the total system must therefore contain the exact statistical operator $\rho^1 = \mathrm{Tr}_2\rho$, reduced over the subsystem Σ^2. Since the latter is supposed to remain near an equilibrium state, we define as the relevant part of the total statistical operator

$$\rho_{\mathrm{rel}}(t) = \mathscr{R}^2\mathrm{Tr}_2\rho(t) = \mathsf{P}\rho(t) \ ,^8 \tag{17.2.34}$$

where we have introduced the operator P for an element $\mathscr{X}$ in the Liouville space of the total system, given by the relation

$$\mathsf{P}\mathscr{X} = \mathscr{R}^2\mathrm{Tr}_2\mathscr{X} \ . \tag{17.2.35}$$

One can readily show that the operator P defined in (17.2.35) has all the properties which we required in Sect. 17.2.1. From the calculational rules for the trace of a product space and the definition (17.2.35), we obtain directly the result that P is linear and obeys the relations

$$\mathsf{P} = \mathsf{P}\mathsf{P} \ , \tag{17.2.36}$$

$$\mathrm{Tr}\,\mathscr{F}_\nu\mathsf{P}\mathscr{X} = \mathrm{Tr}\,\mathscr{F}_\nu\mathscr{X} \ , \tag{17.2.37}$$

and

$$\mathsf{P}\rho(t_0) = \rho(t_0) \ . \tag{17.2.38}$$

The idempotence of P is demonstrated using definition (17.2.35), since $\mathrm{Tr}_2\mathscr{R}^2 = 1$ and $\mathrm{Tr}_2\mathscr{X}$ depends only on the variables of subsystem Σ^1,

$$\mathsf{P}(\mathsf{P}\mathscr{X}) = \mathscr{R}^2\,\mathrm{Tr}_2(\mathscr{R}^2\,\mathrm{Tr}_2\mathscr{X}) = \mathscr{R}^2\,\mathrm{Tr}_2\mathscr{X} = \mathsf{P}\mathscr{X} \ . \tag{17.2.39}$$

Condition (17.2.37) is fulfilled, because the observables $\mathscr{F}_\nu$, which we consider, belong to subsystem Σ^1 and thus act as a factor in taking the trace over

[8] $\rho_{\mathrm{rel}}(t)$ is also a direct product, but this is not a precondition for $\rho(t)$.

system Σ^2,

$$\mathrm{Tr}(\mathscr{F}_v \mathsf{P}\mathscr{X}) = \mathrm{Tr}(\mathscr{F}_v \mathscr{R}^2 \mathrm{Tr}_2 \mathscr{X}) = \mathrm{Tr}_1(\mathscr{F}_v \mathrm{Tr}_2(\mathscr{R}^2 \mathrm{Tr}_2 \mathscr{X}))$$
$$= \mathrm{Tr}_1(\mathscr{F}_v \mathrm{Tr}_2 \mathscr{X}) = \mathrm{Tr}(\mathscr{F}_v \mathscr{X}) \ . \qquad (17.2.40)$$

The validity of the third relation, (17.2.38), finally follows from the initial conditions (17.2.33) and the definition (17.2.35) of P:

$$\mathsf{P}\rho(t_0) = \mathscr{R}^2 \mathrm{Tr}_2 \rho(t_0) = \mathscr{R}^2 \mathrm{Tr}_2(\mathscr{R}^2 \rho^1(t_0)) = \mathscr{R}^2 \rho^1(t_0) = \rho(t_0) \ . \qquad (17.2.41)$$

It may also be shown that P (17.2.35) is contained as a special case in the system (17.2.16). From this point of view, the properties (17.2.36–38), which we have just demonstrated, are self evident.

The relevant part $\rho_{rel}(t)$ chosen in the example is thus generated by an operator P, which fulfills all the preconditions of the general theory. Therefore, this $\rho_{rel}(t)$ obeys the dynamical equation (17.2.28). Before we specify it for this particular example, we make some simplifications which are permitted by the definition of P. It turns out that in the products of P and Q with L, not always the whole Liouville operator is required. We decompose the Liouville operator corresponding to $\mathscr{H}$:

$$\mathsf{L} = \mathsf{L}^1 + \mathsf{L}^2 + \mathsf{L}^{12} \ , \qquad (17.2.42)$$

where the partial operators L^1, L^2, and L^{12} are defined by commutation relations with $\mathscr{H}^1$, $\mathscr{H}^2$, and $\mathscr{H}^{12}$. Then, using definition (17.2.35), we obtain

$$\mathsf{PLP} = \mathsf{L}^1\mathsf{P} + \mathsf{PL}^{12}\mathsf{P} \ , \qquad (17.2.43)$$

$$\mathsf{QLP} = \mathsf{QL}^{12}\mathsf{P} \ , \qquad (17.2.44)$$

and

$$\mathsf{PLQ} = \mathsf{PL}^{12}\mathsf{Q} \ . \qquad (17.2.45)$$

For the *proof* of (17.2.43–45), one starts by showing that L^1 commutes with P. By application to an element $\mathscr{X}$, we obtain with (17.2.35) and the properties of a product space:

$$\mathsf{PL}^1\mathscr{X} = \mathscr{R}^2 \mathrm{Tr}_2 \mathsf{L}^1 \mathscr{X} = \mathscr{R}^2 \mathsf{L}^1 \mathrm{Tr}_2 \mathscr{X} = \mathsf{L}^1 \mathscr{R}^2 \mathrm{Tr}_2 \mathscr{X} = \mathsf{L}^1 \mathsf{P} \mathscr{X} \ . \qquad (17.2.46)$$

Similarly, due to the invariance to cyclic permutations in taking the trace over the subsystem Σ^2, and to the equilibrium property $\mathsf{L}^2\mathscr{R}^2 = 0$, we find that

$$\mathsf{PL}^2\mathscr{X} = \frac{1}{\hbar} \mathscr{R}^2 \mathrm{Tr}_2[\mathscr{H}^2, \mathscr{X}] = 0 \qquad (17.2.47)$$

and

$$\mathsf{L}^2\mathsf{P}\mathscr{X} = \mathsf{L}^2 \mathscr{R}^2 \mathrm{Tr}_2 \mathscr{X} = (\mathsf{L}^2\mathscr{R}^2)\mathrm{Tr}_2 \mathscr{X} = 0 \ . \qquad (17.2.48)$$

With $\mathsf{PP} = \mathsf{P}$ and (17.2.46) and (17.2.47), one then obtains relation (17.2.43). Expressions (17.2.44) and (17.2.45) follow from the fact that P commutes with L^1 and L^2 (17.2.46, 47, and 48) taking into account $\mathsf{PQ} = 0$.

With these simplifications, we can now write down the dynamical equation (17.2.28) for $\rho_{\text{rel}}(t)$. After inserting expressions (17.2.43–45) and using the property $\mathsf{P}\rho_{\text{rel}}(t) = \rho_{\text{rel}}(t)$, it takes the form

$$\frac{d}{dt}\rho_{\text{rel}}(t) = -\,\mathrm{i}(\mathsf{L}^1 + \mathsf{PL}^{12})\rho_{\text{rel}}(t) - \int_{t_0}^{t} \mathsf{PL}^{12}\mathsf{Q}\,\mathrm{e}^{-\mathrm{i}\mathsf{LQ}(t-t')}\mathsf{QL}^{12}\rho_{\text{rel}}(t')dt' \;, \tag{17.2.49}$$

(where QLQ or QL instead of LQ could also occur in the exponent of the integral kernel operator).

The operators $\mathsf{PL}^{12}\mathsf{P}$, $\mathsf{PL}^{12}\mathsf{Q}$, and $\mathsf{QL}^{12}\mathsf{P}$ in (17.2.49) can also be expressed in terms of commutation relations, according to their definitions (cf. Exercise 17.2):

$$\mathsf{PL}^{12}\mathsf{P}\mathscr{X} = \mathsf{P}\frac{1}{\hbar}[\langle\mathscr{H}^{12}\rangle_2, \mathscr{X}] = \frac{1}{\hbar}[\langle\mathscr{H}^{12}\rangle_2, \mathsf{P}\mathscr{X}] \;, \tag{17.2.50}$$

$$\mathsf{PL}^{12}\mathsf{Q}\mathscr{X} = \mathsf{P}\frac{1}{\hbar}[\mathscr{H}^{12} - \langle\mathscr{H}^{12}\rangle_2, \mathscr{X}] \;, \tag{17.2.51}$$

and

$$\mathsf{QL}^{12}\mathsf{P}\mathscr{X} = \frac{1}{\hbar}[\mathscr{H}^{12} - \langle\mathscr{H}^{12}\rangle_2, \mathsf{P}\mathscr{X}] \;. \tag{17.2.52}$$

The quantity $\langle\mathscr{H}^{12}\rangle_2$ is the interaction averaged over Σ^2

$$\langle\mathscr{H}^{12}\rangle_2 = \mathrm{Tr}_2\mathscr{R}^2\mathscr{H}^{12} = \mathrm{Tr}_2\mathscr{H}^{12}\mathscr{R}^2 \tag{17.2.53}$$

and depends only upon the variables of system Σ^1.

The relevant part

$$\rho_{\text{rel}}(t) = \mathscr{R}^2\mathrm{Tr}_2\rho(t) \tag{17.2.54}$$

contains the reduced statistical operator $\rho^1 = \mathrm{Tr}_2\rho$ of the system Σ^1 as a factor; it alone is required to calculate the expectation values of observables $\mathscr{F}_\nu$ of the system Σ^1

$$\mathrm{Tr}(\mathscr{F}_\nu\rho_{\text{rel}}(t)) = \mathrm{Tr}_1(\mathscr{F}_\nu\rho^1(t)) \;. \tag{17.2.55}$$

It is therefore expedient to rewrite the dynamical equation for $\rho_{\text{rel}}(t)$ as a dynamical equation for $\rho^1(t)$. If we take the trace over subsystem Σ^2 with (17.2.49) after substituting the commutation relations (17.2.50–52), we obtain with

$\mathrm{Tr}_2 \mathsf{P}\mathscr{X} = \mathrm{Tr}_2\mathscr{X}$ the equation for $\rho^1(t) = \mathrm{Tr}_2\rho_{\mathrm{rel}}(t)$,

$$\boxed{\begin{aligned}\frac{d\rho^1}{dt} = &-\frac{\mathrm{i}}{\hbar}[\mathscr{H}^1 + \langle\mathscr{H}^{12}\rangle_2, \rho^1] \\ &-\frac{1}{\hbar^2}\int_{t_0}^{t} \mathrm{Tr}_2[\mathscr{H}^{12} - \langle\mathscr{H}^{12}\rangle_2, \mathrm{e}^{-\mathrm{i}\mathsf{LQ}(t-t')} \\ &\times [\mathscr{H}^{12} - \langle\mathscr{H}^{12}\rangle_2, \mathscr{R}^2\rho^1(t')]]\,dt'\end{aligned}} \quad . \tag{17.2.56}$$

We can now compare this equation of motion for the non-isolated system Σ^1 with the Liouville equation of the isolated system Σ^1:

$$\frac{d}{dt}\rho^1 = -\frac{\mathrm{i}}{\hbar}[\mathscr{H}^1, \rho^1] \ . \tag{17.2.57}$$

We see immediately that the operators P and Q have produced a separation of the interaction $\mathscr{H}^{12}$ into an averaged interaction $\langle\mathscr{H}^{12}\rangle_2$ and a "fluctuation" $\mathscr{H}^{12} - \langle\mathscr{H}^{12}\rangle_2$. The averaged portion $\langle\mathscr{H}^{12}\rangle_2$ acts like a supplemental energy in an external field. If one expands $\mathscr{H}^{12}$ in products of the observables of systems Σ^1 and Σ^2 (cf. Sect. 3.3), the observables from Σ^2 enter only as equilibrium expectation values, i.e. as c-numbers. The fluctuation $\mathscr{H}^{12} - \langle\mathscr{H}^{12}\rangle_2$, on the other hand, is responsible for the transport of heat from system Σ^2 to system Σ^1. Thus, if we interpret $\mathscr{H}^1 + \langle\mathscr{H}^{12}\rangle_2$ as the renormalized energy of system Σ^1, we obtain from (17.2.56) the following expression for the differential heat transfer (cf. (3.3.29)):

$$\begin{aligned}\mathrm{Tr}_1\left\{(\mathscr{H}^1 + \langle\mathscr{H}^{12}\rangle_2)\frac{d\rho^1}{dt}\right\} = \mathrm{Tr}\frac{1}{\hbar^2}\int_{t_0}^{t}&[\mathscr{H}^{12} - \langle\mathscr{H}^{12}\rangle_2, \mathscr{H}^1 + \langle\mathscr{H}^{12}\rangle_2] \\ &\times \mathrm{e}^{-\mathrm{i}\mathsf{LQ}(t-t')}[\mathscr{H}^{12} - \langle\mathscr{H}^{12}\rangle_2, \\ &\mathscr{R}^2\rho^1(t')]\,dt' \ .\end{aligned} \tag{17.2.58}$$

The dynamical equation (17.2.56) for $\rho^1(t)$ can be used as the starting point for the explanation of dissipative processes.

17.2.5 The Explicit Time Dependence of the Operators P and L

The theory of a linear mapping of $\rho(t)$ onto $\rho_{\mathrm{rel}}(t)$, described in Sects. 17.2.1 and 3, can be extended without difficulties to the case of an explicit time dependence of the mapping. Instead of definition (17.2.1), we now start with the general recipe for the relevant part:

$$\boxed{\rho_{\mathrm{rel}}(t) = \not{f}[t, \rho(t)] = \mathsf{P}(t)\rho(t)} \quad . \tag{17.2.59}$$

Here, $\mathsf{P}(t)$ is again a linear operator in Liouville space. Its time dependence can be given arbitrarily, depending on the problem at hand, and has in principle nothing to do with the explicit time dependence of the Liouville operator.

The requirements of Sect. 17.2.1 for the mapping (17.2.1) may be immediately adopted for the time-dependent case (17.2.59). Conditions (17.2.4 and 6) for P remain unchanged

$$\mathsf{P}(t) = \mathsf{P}(t)^2 \ , \tag{17.2.60}$$

$$\mathrm{Tr}\mathscr{F}_\nu \mathsf{P}(t)\mathscr{X} = \mathrm{Tr}\mathscr{F}_\nu\mathscr{X} \ , \tag{17.2.61}$$

while the condition (17.2.8) becomes

$$\mathsf{P}(t_0)\rho(t_0) = \rho(t_0) \ . \tag{17.2.62}$$

The dynamical equation for $\rho_{\mathrm{rel}}(t)$ can again be obtained with the aid of the identity (3.4.4b). After some rearrangements, we find (cf. Exercise 17.3)

$$\begin{aligned} \frac{d}{dt}\rho_{\mathrm{rel}}(t) = & -\mathrm{i}\{\mathsf{P}(t)\mathsf{L}(t)\mathsf{P}(t) + \mathrm{i}\dot{\mathsf{P}}(t)\mathsf{P}(t)\}\rho_{\mathrm{rel}}(t) \\ & - \int_{t_0}^{t} \mathsf{P}(t)\{\mathsf{L}(t) + \mathrm{i}\dot{\mathsf{P}}(t)\}\mathsf{Q}(t)\mathsf{V}(t, t')\mathsf{Q}(t') \\ & \times \{\mathsf{L}(t') - \mathrm{i}\dot{\mathsf{P}}(t')\}\mathsf{P}(t')\rho_{\mathrm{rel}}(t')\,dt' \ . \end{aligned} \tag{17.2.63}$$

In this equation, $\mathsf{V}(t, t')$ is defined by the differential equation

$$\frac{\partial}{\partial t}\mathsf{V}(t, t') = -\mathrm{i}\{\mathsf{Q}(t)\mathsf{L}(t) - \mathrm{i}\dot{\mathsf{P}}(t)\}\mathsf{V}(t, t') \tag{17.2.64}$$

and the initial condition $\mathsf{V}(t, t) = \mathsf{1}$. The operator $\mathsf{Q}(t)$ is given, in analogy to (17.2.18), by

$$\mathsf{P}(t) + \mathsf{Q}(t) = \mathsf{1} \ . \tag{17.2.65}$$

Exercise 17.1 Let the observables of interest be projectors $\mathscr{P}_\nu$ onto orthogonal spaces with $\sum_\nu \mathscr{P}_\nu = 1$. Choose a linear combination of all the $\mathscr{P}_\nu$ to be ρ_{rel}. What are the equations for the probabilities $w_\nu(t) = \langle\mathscr{P}_\nu\rangle(t)$?

Exercise 17.2 Verify equations (17.2.50 through 52).

Exercise 17.3 Derive the dynamical equation for $\rho_{\mathrm{rel}}(t)$, (17.2.63).

Start from the identity (3.4.4b), with $\mathsf{L}_1 = \mathsf{Q}(t)\mathsf{L}(t) - \mathrm{i}\dot{\mathsf{P}}(t)$ and carry out a partial integration of the integral which occurs over t'. (For the differentiation of $\mathsf{V}(t, t')$ w.r.t. t', see Exercise 3.5.)

17.3 Nonlinear Relation Between $\rho_{rel}(t)$ and $\rho(t)$

The preceding considerations of Sect. 17.2, dealing with methods of deriving a dynamical equation for $\rho_{rel}(t)$ from the von Neumann equation for $\rho(t)$, may also be extended to the case of a *nonlinear* relationship between $\rho_{rel}(t)$ and $\rho(t)$. In this sense, Sect. 17.2 represents a special case of the following theory. We assume that $\rho_{rel}(t)$ with respect to a set of observables is given by the general mapping

$$\boxed{\rho_{rel}(t) = \mathscr{f}[\rho(t)]} \ . \tag{17.3.1}$$

An explicit time dependence of the mapping will not be considered here.

17.3.1 Properties of the Mapping

The relevant part, $\rho_{rel}(t)$, of a statistical operator must have certain properties. These lead to restrictions on the possible mappings (17.3.1). In the following, we shall again list some conditions for $\mathscr{f}[\mathscr{X}]$, which we will use later. The symbol $\mathscr{X}$ always refers to the elements of the Liouville space for which the mapping (17.3.1) can be defined.

In the linear case, Sect. 17.2.1, we started with certain concepts about $\rho_{rel}(t)$ which led to the requirements (17.2.4), (17.2.6), and (17.2.8). Figuratively, we apply these considerations directly to the general case (17.3.1). We then arrive at the requirements for the mapping $\mathscr{f}[\mathscr{X}]$

$$\mathscr{f}[\mathscr{X}] = \mathscr{f}[\mathscr{f}[\mathscr{X}]] \ , \tag{17.3.2}$$

$$\mathrm{Tr}\,\mathscr{F}_v \mathscr{f}[\mathscr{X}] = \mathrm{Tr}\,\mathscr{F}_v \mathscr{X} \ , \tag{17.3.3}$$

$$\mathscr{f}[\rho(t_0)] = \rho(t_0) \ . \tag{17.3.4}$$

Equation (17.3.2) means that a relevant part which is given as a statistical operator is identical with its relevant part, while (17.3.3) requires that the relevant part of a statistical operator yield the same expectation values as the operator itself. The third condition, (17.3.4), requires that $\mathscr{f}[\mathscr{X}]$ be so chosen that the given initial states $\rho(t_0)$ have only a relevant part.

For a differential variation of $\mathscr{X}$, the variation of $\mathscr{f}[\mathscr{X}]$ is given by

$$d\mathscr{f}[\mathscr{X}] = \mathsf{P}[\mathscr{X}]\, d\mathscr{X} \ , \tag{17.3.5}$$

where, for a given $\mathscr{X}$, $\mathsf{P}[\mathscr{X}]$ is a linear operator in the Liouville space $\mathfrak{L}$ (a superoperator). It corresponds to the tangent of an ordinary function and can be

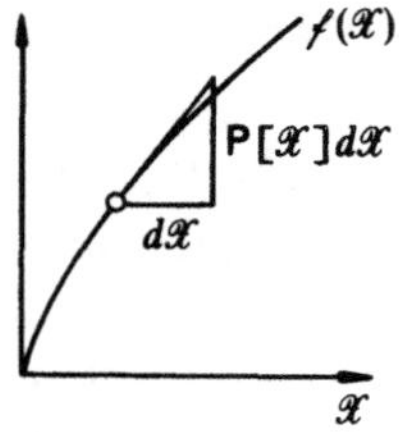

Fig. 17.2. Symbolic representation of a nonlinear mapping $\mathscr{X} \mapsto \not f[\mathscr{X}]$ and the definition of $\mathsf{P}[\mathscr{X}]$. In the particular case of the Nakajima–Zwanzig theory, $\not f[\mathscr{X}]$ is linear in $\mathscr{X}$ and $\not f[\mathscr{X}] = \mathsf{P}\mathscr{X}$, i.e. P is independent of $\mathscr{X}$

calculated from $\not f[\mathscr{X}]$.[9] This relationship is represented symbolically in Fig. 17.2.

We now require of the mapping $\not f[\mathscr{X}]$ that it lead to an operator $\mathsf{P}[\mathscr{X}]$ which possesses the property

$$\boxed{\mathsf{P}[\mathscr{X}] = \mathsf{P}[\not f[\mathscr{X}]]} \ . \tag{17.3.6}$$

This requirement is necessary because the operator $\mathsf{P}[\rho(t)]$ must be determined by $\rho_{\text{rel}}(t)$ in the derivation of a dynamical equation for $\rho_{\text{rel}}(t)$. It means that $\mathsf{P}[\rho(t)] = \mathsf{P}[\rho_{\text{rel}}(t) + \rho_{\text{irr}}(t)]$ should depend only on the relevant part $\rho_{\text{rel}}(t)$.

The differential variation of $\not f[\mathscr{X}]$ is obtained from (17.3.2) by using (17.3.5),

$$\begin{aligned} d\not f[\mathscr{X}] &= \mathsf{P}[\mathscr{X}]\,d\mathscr{X} = \mathsf{P}[\not f[\mathscr{X}]]\,d\not f[\mathscr{X}] \\ &= \mathsf{P}[\not f[\mathscr{X}]]\,\mathsf{P}[\mathscr{X}]\,d\mathscr{X} \ . \end{aligned} \tag{17.3.7}$$

If there are no limitations on the variation $d\mathscr{X}$, we can derive the relation

$$\mathsf{P}[\mathscr{X}] = \mathsf{P}[\not f[\mathscr{X}]]\cdot\mathsf{P}[\mathscr{X}] \tag{17.3.8}$$

and, using (17.3.6), the idempotence of $\mathsf{P}[\mathscr{X}]$,

$$\mathsf{P}[\mathscr{X}] = \mathsf{P}[\mathscr{X}]\cdot\mathsf{P}[\mathscr{X}] \ . \tag{17.3.9}$$

A further property of $\mathsf{P}[\mathscr{X}]$ is found from (17.3.3). If one applies this equation to a differential variation $d\mathscr{X}$, one obtains (with the definition (17.3.5)):

$$\operatorname{Tr}\mathscr{F}_\nu d\not f[\mathscr{X}] = \operatorname{Tr}\mathscr{F}_\nu \mathsf{P}[\mathscr{X}]\,d\mathscr{X} = \operatorname{Tr}\mathscr{F}_\nu\,d\mathscr{X} \ . \tag{17.3.10}$$

Choosing the variation $d\mathscr{X}$ to be

$$d\mathscr{X} = \mathscr{Y}\,d\alpha \ , \tag{17.3.11}$$

---

[9] If, in a representation of $d|\mathscr{X})$ in a basis $d|\mathscr{X}) = \sum|\mathscr{V}_\alpha)dc_\alpha$, all the dc_α except one can be set equal to zero, then the operator $\mathsf{P}[\mathscr{X}]$ is completely defined by (17.3.5). The theory given in this section can also be carried out in the space with the condition $\operatorname{Tr}\mathscr{X} = 1$. In this case, however, one must take into account the region in which $\mathsf{P}[\mathscr{X}]$ is defined, and we shall therefore not consider this route.

where α is a c-number, we find for the elements $\mathscr{Y}$ in Liouville space

$$\mathrm{Tr}\,\mathscr{F}_v \mathsf{P}[\mathscr{X}]\mathscr{Y} = \mathrm{Tr}\,\mathscr{F}_v \mathscr{Y} \ . \tag{17.3.12}$$

An important example of $\not{f}[\mathscr{X}]$, for which all of the requirements (17.3.2–4) and (17.3.6) are fulfilled, will be treated in Sect. 18.2.

One can best understand the meaning of the operator $\mathsf{P}[\mathscr{X}]$ by expanding the elements $\mathscr{Y}$ and $\mathscr{X}$ in $\mathscr{Y} = \not{f}[\mathscr{X}]$ in a basis $\mathscr{V}_\alpha$ in $\mathscr{L}$:

$$|\mathscr{X}) = \sum_\alpha \xi_\alpha |\mathscr{V}_\alpha) \ , \tag{17.3.13}$$

$$|\mathscr{Y}) = \sum_\alpha \eta_\alpha |\mathscr{V}_\alpha) \ . \tag{17.3.14}$$

Then the components η_α become ordinary functions of the components ξ_β,

$$\eta_\alpha = \eta_\alpha(\xi_1, \xi_2, \xi_3, \ldots) \ , \tag{17.3.15}$$

and one obtains the mapping of the components $d\xi_\beta$ as a matrix, as in ordinary vector analysis:

$$d\eta_\alpha = \sum_\beta \frac{\partial \eta_\alpha}{\partial \xi_\beta} d\xi_\beta = \sum_\beta \mathsf{P}_{\alpha\beta}[\mathscr{X}]\, d\xi_\beta \ . \tag{17.3.16}$$

Using a scalar product with $(\mathscr{V}_\alpha|\mathscr{V}_\alpha) = \delta_{\alpha\beta}$, this may also be written in the form

$$d\sum_\alpha \eta_\alpha|\mathscr{V}_\alpha) = \left\{\sum_{\alpha\beta} |\mathscr{V}_\alpha) \frac{\partial \eta_\alpha}{\partial \xi_\beta} (\mathscr{V}_\beta| \right\} d\sum_\sigma \xi_\sigma|\mathscr{V}_\sigma) = \mathsf{P}[\mathscr{X}]\, d\mathscr{X} \ . \tag{17.3.17}$$

17.3.2 Nonlinear Dynamical Equation for $\rho_{rel}(t)$

In Sect. 17.3.1, we collected the conditions which we require of a nonlinear mapping of $\rho(t)$ onto a relevant part $\rho_{rel}(t)$. Just as in the linear case, we now derive an equation for $\rho_{rel}(t)$ starting with the von Neumann equation. We first consider the case that L is not explicitly time-dependent. Formally, we proceed similarly to Sect. 17.2.3. The first step is to express the time derivative of $\rho_{rel}(t)$ using (17.3.1) and (17.3.5) as

$$\frac{d}{dt}\rho_{rel}(t) = \frac{d}{dt}\not{f}[\rho(t)] = \mathsf{P}[\rho(t)]\frac{d\rho}{dt} \ . \tag{17.3.18}$$

On the other hand, considering condition (17.3.6), we have

$$\mathsf{P}[\rho(t)] = \mathsf{P}[\rho_{rel}(t)] \ , \tag{17.3.19}$$

so that we can also write the time derivative of $\rho_{rel}(t)$, combining (17.3.18) and

(17.3.19) and using the von Neumann equation, as

$$\frac{d}{dt}\rho_{\text{rel}}(t) = -\mathrm{i}\mathsf{P}[\rho_{\text{rel}}(t)]\mathsf{L}\rho(t) \ . \tag{17.3.20}$$

We now relate the statistical operator $\rho(t)$ in (17.3.20) to $\rho_{\text{rel}}(t')$. To this end, we define $\mathsf{Q}[\rho_{\text{rel}}(t)]$ by

$$\mathsf{Q}[\rho_{\text{rel}}(t)] + \mathsf{P}[\rho_{\text{rel}}(t)] = 1 \ , \tag{17.3.21}$$

employ the identity (3.4.4b) with the decompositions of L,

$$\mathsf{L}_1(t) = \mathsf{Q}[\rho_{\text{rel}}(t)]\mathsf{L} \ , \tag{17.3.22}$$

$$\mathsf{L}_2(t) = \mathsf{P}[\rho_{\text{rel}}(t)]\mathsf{L} \tag{17.3.23}$$

and allow this identity to act upon $\rho(t_0)$. With (17.3.20), this leads to the relation

$$\rho(t) = \mathsf{T}(t, t_0)\rho(t_0) + \int_{t_0}^{t} \mathsf{T}(t, t')\frac{d}{dt'}\rho_{\text{rel}}(t')\,dt' \ . \tag{17.3.24}$$

The operator $\mathsf{T}(t, t')$ used here obeys the differential equation (cf. Exercise 3.5)

$$\frac{\partial}{\partial t}\mathsf{T}(t, t') = -\mathrm{i}\mathsf{Q}[\rho_{\text{rel}}(t)]\mathsf{L}\mathsf{T}(t, t') \ , \tag{17.3.25}$$

or

$$\frac{\partial}{\partial t'}\mathsf{T}(t, t') = \mathsf{T}(t, t')\mathrm{i}\mathsf{Q}[\rho_{\text{rel}}(t')]\mathsf{L} \tag{17.3.26}$$

and has the initial value

$$\mathsf{T}(t, t) = 1 \ . \tag{17.3.27}$$

By partial integration of (17.3.24) and with the differential equation (17.3.26), one thus obtains a relation between $\rho(t)$ and $\rho_{\text{rel}}(t)$:

$$\rho(t) = \rho_{\text{rel}}(t) - \mathrm{i}\int_{t_0}^{t} \mathsf{T}(t, t')\mathsf{Q}[\rho_{\text{rel}}(t')]\mathsf{L}\rho_{\text{rel}}(t')dt' + \mathsf{T}(t, t_0)\{\rho(t_0) - \rho_{\text{rel}}(t_0)\} \ . \tag{17.3.28}$$

The last term in (17.3.28) vanishes according to our requirement (17.3.4). Thus, the statistical operator $\rho(t)$ which occurs in the time derivative (17.3.20) of $\rho_{\text{rel}}(t)$ can be expressed in terms of $\rho_{\text{rel}}(t')$. Combining (17.3.20) and (17.3.28) leads to the dynamical equation for $\rho_{rel}(t)$:

$$\boxed{\begin{aligned}\frac{d}{dt}\rho_{rel}(t) = &-i\mathsf{P}[\rho_{rel}(t)]\mathsf{L}\rho_{rel}(t)\\ &-\int_{t_0}^{t}\mathsf{P}[\rho_{rel}(t)]\mathsf{L}\mathsf{T}(t,t')\mathsf{Q}[\rho_{rel}(t')]\mathsf{L}\rho_{rel}(t')\,dt'\end{aligned}}\ . \tag{17.3.29}$$

This equation, in contrast to (17.2.28), is *nonlinear*, since it depends on the quantity we wish to calculate, $\rho_{rel}(t)$, in a rather complicated way through $\mathsf{P}[\rho_{rel}(t)]$ as well as through the integral kernel $\mathsf{T}(t, t')$ [cf. (17.3.25)]. At first glance, it therefore seems to make little sense to derive such equations. However, we shall see in (18.2.23) that (17.3.29) provides a *good starting point for deriving nonlinear equations for the dynamics of expectation values.* Equation (17.3.29) of course reverts to the special case of (17.2.28) when the mapping is linear.

The *generalization* of (17.3.29) to the case of an *explicit time-dependence* of $\mathsf{L}(t)$ is relatively simple. In carrying out the rearrangements, one only need keep in mind that $\exp[-i\mathsf{L}(t-t_0)]$ is to be replaced by the time evolution $\mathsf{U}(t, t_0)$ according to $\mathsf{L}(t)$. Then, in analogy to (17.3.29), one obtains

$$\boxed{\begin{aligned}\frac{d}{dt}\rho_{rel}(t) = &-i\mathsf{P}[\rho_{rel}(t)]\mathsf{L}(t)\rho_{rel}(t)\\ &-\int_{t_0}^{t}\mathsf{P}[\rho_{rel}(t)]\mathsf{L}(t)\mathsf{T}(t,t')\mathsf{Q}[\rho_{rel}(t')]\mathsf{L}(t')\rho_{rel}(t')\,dt'\end{aligned}}\ , \tag{17.3.30}$$

where $\mathsf{T}(t, t')$ is given by (17.3.25) but with $\mathsf{L}(t)$ or by (17.3.26) but with $\mathsf{L}(t')$.

18. The Generalized Canonical Statistical Operator $\mathscr{R}(t)$ as $\rho_{\rm rel}(t)$

In the preceding Sects. 17.2 and 3, we showed dynamical equations for $\rho_{\rm rel}(t)$ but left the choice of an explicit expression for $\rho_{\rm rel}(t)$ open. As we demonstrated in Chap. 5, finding the extreme value of the uncertainty measure with given expectation values

$$\mathrm{Tr}(\mathscr{G}_\nu \rho) = G_\nu(t) = \mathrm{Tr}(\mathscr{G}_\nu \mathscr{R}_{\{\mathscr{G}\}}) \tag{18.1.1}$$

leads to the generalized canonical statistical operator, $\mathscr{R}_{\{\mathscr{G}\}}$. We now wish to consider the case that the time-accompanying operator (cf. Sect. 6.4) $\mathscr{R}_{\{\mathscr{G}\}}(t)$ is chosen as the relevant part of $\rho(t)$ with respect to the observables $\mathscr{G}_\nu$,

$$\rho_{\rm rel}(t) = \mathscr{R}_{\{\mathscr{G}\}}(t) \ . \tag{18.1.2}$$

The requirement $\rho(t_0) = \rho_{\rm rel}(t_0)$ means that an $\mathscr{R}_{\{\mathscr{G}\}}$ can be used as initial ρ.

The accompanying generalized canonical statistical operator $\mathscr{R}_{\{\mathscr{G}\}}$ over an observation level $\{\mathscr{G}\}$ is, by construction, a mapping of the statistical operator ρ at a particular fixed time,

$$\mathscr{R}_{\{\mathscr{G}\}} \equiv \frac{1}{Z_{\{\mathscr{G}\}}} \exp\left[-\sum_\nu \lambda_\nu(G_1[\rho], \ldots)\mathscr{G}_\nu \right] = \not{f}[\rho] \ . \tag{18.1.3}$$

If we wish to use $\mathscr{R}_{\{\mathscr{G}\}}(t)$ in the sense of the preceding paragraphs as $\rho_{\rm rel}(t)$, we must demonstrate that the mapping $\rho(t)$ onto $\mathscr{R}_{\{\mathscr{G}\}}(t)$ has the required properties.

18.1 The Linear Case

We first consider the simple case that $\mathscr{R}_{\{\mathscr{G}\}}$ can be expressed as a linear combination of constant basis elements ρ_ν:

$$\exp\left[-\ln Z \mathit{1} - \sum_\nu \lambda_\nu(t)\mathscr{G}_\nu \right] = \sum_\nu c_\nu(G_1(t), \ldots)\rho_\nu \ , \tag{18.1.4}$$

where the dimensionality of the space $\mathfrak{L}_{\rho_{\rm rel}}$ spanned by the ρ_ν is equal to the number of observables in the observation level[1] (c.f. 17.2.11).

---

[1] $\mathscr{G}_0 = \mathit{1}$ must also be counted, unless the unit operator is already included as a linear combination of the $\mathscr{G}_\nu$.

This property can, on the one hand, be a result of the particular algebraic structure of the $\mathcal{G}_\nu$'s, e.g. the algebra of spin operators or of projection operators. The latter case occurs for $\mathcal{G}_\nu \to \mathcal{P}_F$, whereby the generalized canonical statistical operator becomes a sum of microcanonical operators (Sect. 5.3.4):

$$\rho_{\text{rel}}(t) = \mathcal{R}_{\{\mathcal{P}_F\}}(t) = \sum_F \frac{w_F(t)}{d_F} \mathcal{P}_F \ . \tag{18.1.5}$$

This equation obviously is of the form (18.1.4), since here the transition

$$\mathcal{G}_\nu = \rho_\nu \to \mathcal{P}_F \tag{18.1.6}$$

holds.

On the other hand, if $\mathcal{R}_{\{\mathcal{G}\}}(t)$ in (18.1.3) is expanded around the point of thermal equilibrium,

$$\mathcal{R}_{\{\mathcal{G}\}}(t) = \mathcal{R}_\beta + \sum_{\nu\mu} \left.\frac{\partial \mathcal{R}}{\partial \lambda_\nu}\right|_{\text{eq}} \left.\frac{\partial \lambda_\nu}{\partial G_\mu}\right|_{\text{eq}} (G_\mu(t) - \langle \mathcal{G}_\mu \rangle_\beta) \ , \tag{18.1.7}$$

and only the linear terms are kept, we again obtain an equation of the form (18.1.4), if we set

$$\mathcal{G}_0 = 1, \mathcal{G}_1, \dots \quad \text{and} \quad \rho_\nu \to \mathcal{R}_\beta, \left.\frac{\partial \mathcal{R}}{\partial G_1}\right|_{\text{eq}}, \dots \ . \tag{18.1.8}$$

In these special cases (18.1.4) of (18.1.3), a linear mapping thus occurs; we have already analyzed it in Sect. 17.2.2,

$$\rho_{\text{rel}}(t) = \mathsf{P}\rho(t) \tag{18.1.9}$$

with a fixed projector P given by (17.2.16). Here, the existence of $\mathcal{R}_{\{\mathcal{G}\}}$ implies the existence of the matrix $v_{\nu\mu}$. The dynamics can be treated using the Nakajima–Zwanzig equation discussed in Sect. 17.2.3.[2] In the special case (18.1.7), the corresponding mapping (18.1.9) becomes equivalent to the Mori theory which we treated in Part 3 (compare Exercise 18.3).

18.2 The Robertson Equation

If the mapping (18.1.3) cannot be expressed in a basis of dimensionality $\dim \mathcal{L}_{\rho_{\text{rel}}} = \dim \mathcal{L}_{\mathcal{G}}$, we must return to the general equation (17.3.30) in Sect. 17.3.2 in order to write down the equation of motion for $\rho_{\text{rel}}(t) = \mathcal{R}_{\{\mathcal{G}\}}(t)$; we thus

[2] This may be accomplished by extending the set $\{\mathcal{G}\}$ to a larger set $\{\bar{\mathcal{G}}\}$ (cf. here also Sect. 12.2).

obtain the Robertson equation for $\mathscr{R}_{\{\mathscr{G}\}}(t)$ [3]. Of course, it contains the special cases which we have discussed above.

In order to apply (17.3.30) to $\mathscr{R}_{\{\mathscr{G}\}}(t)$, we must first demonstrate that the generally nonlinear mapping

$$\mathscr{R}_{\{\mathscr{G}\}}(t) = \not{f}[\rho(t)] = \rho_{\mathrm{rel}}(t) \tag{18.2.1}$$

has the properties (17.3.2–4) and (17.3.6). In the process, we shall calculate the operator $\mathsf{P}[\mathscr{R}_{\{\mathscr{G}\}}]$ which follows from (18.2.1) explicitly (18.2.19).

We employ the condition $\mathrm{Tr}\,\mathscr{R} = g_0$ in the procedure for constructing $\mathscr{R}$ and adopt the simplifying notation

$$\mathscr{G}_0 = \mathit{1} \ . \tag{18.2.2}$$

Then we may define the extension of the mapping (18.1.3) on $\mathscr{X}$ by

$$\not{f}[\mathscr{X}] = \exp\left(- \sum_{\nu=0}^{n} \lambda_\nu[\mathscr{X}]\mathscr{G}_\nu \right) , \tag{18.2.3}$$

where the c-number parameters $\lambda_\nu = \lambda_\nu[\mathscr{X}]$ are functionals of $\mathscr{X}$, determined by

$$\mathrm{Tr}\,\mathscr{G}_\nu \not{f}[\mathscr{X}] = \mathrm{Tr}\,\mathscr{G}_\nu \mathscr{X} \quad (\nu = 0, 1, \ldots, n) \ . \tag{18.2.4}$$

These parameters thus only depend on the values of the traces

$$\mathrm{Tr}\,\mathscr{G}_\nu \mathscr{X} = g_\nu[\mathscr{X}] \tag{18.2.5}$$

and therefore, the mapping (18.2.3) may also be written as a function of the c-number variables $g_\nu[\mathscr{X}]$ and of the operators $\mathscr{G}_\nu$:

$$\not{f}[\mathscr{X}] = \not{f}(g_\nu[\mathscr{X}], \mathscr{G}_\nu) \ . \tag{18.2.6}$$

The properties of $\not{f}[\mathscr{X}]$ required in Sect. 17.3.1 result directly from the definition. The first requirement (17.3.2) is met, since according to (18.2.4 and 5), we have

$$g_\nu[\mathscr{X}] = g_\nu[\not{f}[\mathscr{X}]] \tag{18.2.7}$$

and it therefore follows[4] from (18.2.6) that

$$\not{f}[\mathscr{X}] = \not{f}[\not{f}[\mathscr{X}]] \ . \tag{18.2.8}$$

The next condition (17.3.3) on the mapping is fulfilled automatically as a result of the construction procedure (18.2.4). Finally, if we choose a generalized

[3] If one only wishes to derive this equation for $\mathscr{R}_{\{\mathscr{G}\}}(t)$, the procedure can be simplified from the beginning. However, we wished to emphasize its place in the general theory of the relevant part of the statistical operator.

[4] This result is evident, since a generalized canonical operator $\mathscr{R}_{\{\mathscr{G}\}}$ constructed from the expectation values $\mathrm{Tr}\,\mathscr{G}_\nu \mathscr{R}_{\{\mathscr{G}\}}$ is identical to the given $\mathscr{R}_{\{\mathscr{G}\}}$.

canonical statistical operator over the observation level $\{\mathcal{G}\}$ as initial-state operator

$$\rho(t_0) = \mathcal{R}_{\{\mathcal{G}\}}(t_0) \ , \tag{18.2.9}$$

then, according to (18.2.8),

$$\not{f}[\mathcal{R}_{\{\mathcal{G}\}}(t_0)] = \mathcal{R}_{\{\mathcal{G}\}}(t_0) \tag{18.2.10}$$

i.e. the condition (17.3.4), that $\rho(t_0)$ have only a relevant part, is likewise fulfilled.

In order to apply the theory of Sect. 17.3.2, we must next calculate the operator $\mathsf{P}[\mathcal{X}]$ as defined in (17.3.5) from $\not{f}[\mathcal{X}]$. For this purpose, we use the representation (18.2.6) of $\not{f}[\mathcal{X}]$ as a function of the variables $g_\nu[\mathcal{X}]$ and thus obtain for the variation of the mapping due to a variation $d\mathcal{X}$ the following result:

$$d\not{f}[\mathcal{X}] = \sum_{\nu=0}^{n} \frac{\partial \not{f}(g[\mathcal{X}])}{\partial g_\nu} dg_\nu[\mathcal{X}] = \sum_{\nu=0}^{n} \frac{\partial \not{f}(g[\mathcal{X}])}{\partial g_\nu} \mathrm{Tr}\, \mathcal{G}_\nu d\mathcal{X} \ . \tag{18.2.11}$$

Comparing (18.2.11) and the definition (17.3.5), one then obtains an expression for the operator $\mathsf{P}[\mathcal{X}]$

$$\mathsf{P}[\mathcal{X}]\mathcal{Y} = \sum_{\nu=0}^{n} \frac{\partial \not{f}(g[\mathcal{X}])}{\partial g_\nu} \mathrm{Tr}\, \mathcal{G}_\nu \mathcal{Y} \ . \tag{18.2.12}$$

Here, $\mathcal{Y}$ indicates an arbitrary element of the Liouville space, for which the trace exists. The representation (18.2.12) of $\mathsf{P}[\mathcal{X}]$ may be rearranged somewhat. Considering $\not{f}[\mathcal{X}]$ as a function (18.2.6) of the $g_\nu[\mathcal{X}]$, one obtains from (18.2.4) (after multiplication by a number α)

$$\alpha\, \mathrm{Tr}\, \mathcal{G}_\nu \not{f}(g) = \alpha g_\nu \ , \tag{18.2.13}$$

$$\mathrm{Tr}\, \mathcal{G}_\nu \not{f}(\alpha g) = \alpha g_\nu \ . \tag{18.2.14}$$

From this, due to the same structure (18.2.3) of $\alpha\not{f}(g)$ and $\not{f}(\alpha g)$, the homogeneity property

$$\alpha \not{f}(g) = \not{f}(\alpha g) \tag{18.2.15}$$

follows.

With the aid of the associated Euler equation

$$\not{f}(g) = \sum_{\nu=0}^{n} \frac{\partial \not{f}}{\partial g_\nu} g_\nu \tag{18.2.16}$$

the derivative $\partial \not{f}/\partial g_0$ may be eliminated from (18.2.12), so that we can write $\mathsf{P}[\mathscr{X}]$ using (18.2.12) and (18.2.16) in the form

$$\mathsf{P}[\mathscr{X}]\mathscr{Y} = \frac{1}{g_0}\left(\not{f} - \sum_{\nu=1}^{n} \frac{\partial \not{f}}{\partial g_\nu} g_\nu\right) \mathrm{Tr}\, \mathscr{Y} + \sum_{\alpha=1}^{n} \frac{\partial \not{f}}{\partial g_\nu} \mathrm{Tr}\, \mathscr{G}_\nu \mathscr{Y} \; . \tag{18.2.17}$$

We still need to verify that the operator $\mathsf{P}[\mathscr{X}]$ calculated from $\not{f}[\mathscr{X}]$ meets the condition (17.3.6). The validity of the latter can be immediately seen in expression (18.2.17), since the derivatives $\partial \not{f}/\partial g_\nu$ are functions of the variables $g_\nu[\mathscr{X}]$, for which, however, (18.2.7) holds. Therefore, the property expressed in (18.2.7) is transferred to $\mathsf{P}[\mathscr{X}]$,

$$\mathsf{P}[\mathscr{X}] = \mathsf{P}[\not{f}[\mathscr{X}]] \; . \tag{18.2.18}$$

As a final step, we choose the particular value $\mathscr{X} = \rho(t)$, and thus obtain from (18.2.17), (18.2.18), and (18.2.1) the Kawasaki–Gunton operator:[5]

$$\boxed{\begin{aligned} \mathsf{P}[\mathscr{R}(t)]\mathscr{Y} &= \mathsf{P}[\rho(t)]\mathscr{Y} \\ &= \left\{\mathscr{R}(t) - \sum_{\nu=1}^{n} \frac{\partial \mathscr{R}(t)}{\partial \langle \mathscr{G}_\nu \rangle (t)} \langle \mathscr{G}_\nu \rangle (t)\right\} \mathrm{Tr}\, \mathscr{Y} \\ &\quad + \sum_{\nu=1}^{n} \frac{\partial \mathscr{R}(t)}{\partial \langle \mathscr{G}_\nu \rangle (t)} \mathrm{Tr}\, \mathscr{G}_\nu \mathscr{Y} \end{aligned}} \; . \tag{18.2.19}$$

Here, we have returned to the expectation values $\langle \mathscr{G}_\nu \rangle (t)$ with $g_\nu[\rho(t)] = \mathrm{Tr}\, \mathscr{G}_\nu \rho(t)$ and have inserted $g_0 = \mathrm{Tr}\, \rho(t) = 1$. In particular, it follows that

$$\mathsf{P}[\mathscr{R}(t)]\rho(t) = \mathsf{P}[\rho(t)]\rho(t) = \mathscr{R}(t) \tag{18.2.20}$$

which is a consequence of the Euler equation.

From our general investigations (17.3.9), we already know that the operator $\mathsf{P}[\mathscr{R}(t)]$ is idempotent. Furthermore, it also has the property

$$\mathsf{P}[\mathscr{R}(t)] \cdot \mathsf{P}[\mathscr{R}(t')] = \mathsf{P}[\mathscr{R}(t)] \; , \tag{18.2.21}$$

as one can readily demonstrate by an explicit calculation. This property is useful for some applications (cf. Exercise 18.1).

In the preceding considerations, we have demonstrated that the generalized canonical statistical operator $\mathscr{R}(t)$ over an observation level $\{\mathscr{G}\}$ can be used as

[5] The index referring to the observation level is suppressed, since we are always dealing with the same observation level.

the relevant part of $\rho(t)$ in the spirit of Sects. 17.3.1 and 17.3.2. Therefore, (17.3.30), the Robertson equation, holds for the accompanying operator $\mathscr{R}(t)$:

$$\boxed{\begin{aligned}\frac{d}{dt}\mathscr{R}(t) = &-\mathrm{i}\mathsf{P}[\mathscr{R}(t)]\mathsf{L}(t)\mathscr{R}(t)\\ &-\int_{t_0}^{t}\mathsf{P}[\mathscr{R}(t)]\mathsf{L}(t)\mathsf{T}(t,t')\mathsf{Q}[\mathscr{R}(t')]\mathsf{L}(t')\mathscr{R}(t')dt'\end{aligned}}\ , \tag{18.2.22}$$

where the operator $\mathsf{P}[\mathscr{R}(t)]$ is defined by (18.2.19) and the time evolution operator $\mathsf{T}(t, t')$ by (17.3.25) with $\mathsf{L}(t)$. As we have already emphasized in the general section, (18.2.22) is a *nonlinear* equation for $\mathscr{R}(t)$, since the operator P depends on $\mathscr{R}(t)$.

Instead of considering the equation for $\mathscr{R}(t)$ itself, it is often expedient to deal with the expectation values. If one takes the expectation values of the observables $\mathscr{G}_\nu$ with the aid of (18.2.22), one obtains [using (17.3.12) and (18.2.4)]

$$\boxed{\begin{aligned}\frac{d}{dt}\langle\mathscr{G}_\nu\rangle(t) = &-\mathrm{i}\,\mathrm{Tr}(\mathscr{G}_\nu\mathsf{L}(t)\mathscr{R}(t))\\ &-\int_{t_0}^{t}\mathrm{Tr}(\mathscr{G}_\nu\mathsf{L}(t)\mathsf{T}(t,t')\mathsf{Q}[\mathscr{R}(t')]\mathsf{L}(t')\mathscr{R}(t'))dt'\end{aligned}}\ . \tag{18.2.23}$$

This system of equations has various interpretations. On the one hand, the expectation values $\langle\mathscr{G}_\nu\rangle(t)$ may be expressed in terms of the Lagrange multipliers $\lambda_\nu(t)$, so that (18.2.23) can be read as a closed, *nonlinear* set of equations for the Lagrange multipliers. This viewpoint is of importance when one wishes to derive e.g. heat conductance equations for time-dependent temperatures. In this interpretation, it is reasonable to rearrange the right-hand side of (18.2.23) slightly, in order to make the connection clearer. Inserting the relation (cf. Exercise 18.4)

$$\mathsf{L}\mathscr{R} = -\sum_\mu\int_0^1\mathscr{R}^x(\mathsf{L}\mathscr{G}_\mu)\mathscr{R}^{1-x}dx\,\lambda_\mu\ , \tag{18.2.24}$$

we find (18.2.23) takes on the form

$$\frac{d}{dt}\langle\mathscr{G}_\nu\rangle(t) = \sum_\mu \mathrm{i}\omega_{\nu\mu}(t)\,\lambda_\mu(t) + \sum_\mu\int_{t_0}^{t} l_{\nu\mu}(t,t')\,\lambda_\mu(t')\,dt'\ . \tag{18.2.25}$$

The coefficients are given by

$$\mathrm{i}\omega_{\nu\mu}(t,\lambda(t)) = \mathrm{i}\,\mathrm{Tr}\left\{\mathscr{G}_\nu(t)\int_0^1\mathscr{R}(\lambda(t))^x\mathsf{L}(t)\mathscr{G}_\mu(t)\,\mathscr{R}(\lambda(t))^{-x}dx\,\mathscr{R}(\lambda(t))\right\}dx\ , \tag{18.2.26}$$

$$l_{\nu\mu}(t, t'; \lambda(t''))=$$

$$-\operatorname{Tr}\Big\{\mathsf{L}(t)\mathscr{G}_\nu\cdot\mathsf{T}(t,t')\mathsf{Q}[\mathscr{R}(\lambda(t')]$$

$$\times\int_0^1 \mathscr{R}(\lambda(t'))^x\,\mathsf{L}(t')\mathscr{G}_\mu\cdot\mathscr{R}(\lambda(t'))^{-x}dx\cdot\mathscr{R}(\lambda(t'))\Big\}\ . \qquad (18.2.27)$$

The matrix $\omega_{\nu\mu}$ depends via $\mathscr{R}(\lambda(t))$ on the $\lambda_\nu(t)$'s, while the $\lambda_\nu(t'')$ for times $t' \leqq t'' \leqq t$ determine the $l_{\nu\mu}$. The form (18.2.25) makes it clear that the Lagrange multipliers can be interpreted as driving forces in the sense of irreversible thermodynamics (compare Sect. 15.1.3).

On the other hand, if $\mathscr{R}(t)$ is written as a function of the expectation values $\langle\mathscr{G}_\nu\rangle(t)$, then (18.2.23) becomes a *closed, nonlinear system of integro-differential equations for the expectation values* $\langle\mathscr{G}_\nu\rangle(t)$. In this view, the system of equations (18.2.23) represents a good starting point (c.f. Exercise 18.5) for the derivation of specialized macroscopic equations. Depending on the choice of the observables $\mathscr{G}_\nu$, from single-particle densities to the position-dependent variables of hydrodynamics or to spin densities, (18.2.23) opens up a multitude of possibilities.

Exercise 18.1 Prove (18.2.21) using the definition (18.2.19) of $\mathsf{P}[\mathscr{R}(t)]$.

Exercise 18.2 For an observable $\mathscr{F}$, which does *not* belong to the observation level $\{\mathscr{G}\}$, use (17.3.28) to find the difference in the expectation values $\operatorname{Tr}(\mathscr{F}\rho(t))$ and $\operatorname{Tr}(\mathscr{F}\mathscr{R}_{\{\mathscr{G}\}}(t))$.

Exercise 18.3 Linearize the expectation-value equations in the Robertson theory around the equilibrium expectation values $\langle\mathscr{G}_\nu\rangle_\beta$ and show that the resulting equations are identical to those of the Mori theory. Take the Hamiltonian to be explicitly time-independent and to be contained in the observation level $\mathscr{G}_\nu = \mathscr{G}_\nu^\dagger$.

Exercise 18.4 Prove equation (18.2.24).

Exercise 18.5 Let an (in general time-dependent) Liouville operator of the form $\mathsf{L}(t) = \mathsf{L}^0(t) + \mathsf{L}^1$ be given. The dynamics of $\mathsf{L}^0(t)$ are assumed to be closed in the $\mathscr{G}_\nu$:

$$\mathsf{L}^0(t)\mathscr{G}_\mu = \sum_\nu \mathscr{G}_\nu \Omega^0_{\nu\mu}(t) \quad [\text{cf. } 12.1.2)]\ .$$

Simplify the Robertson equation (18.2.23), so that a starting point for a perturbation-theory expansion in L^1 is obtained.

Exercise 18.6 Consider interacting fermions with the Hamiltonian

$$\mathscr{H} = \sum_1 \varepsilon_1 a_1^\dagger a_1 + \frac{1}{2}\sum_{11'22'} V_{121'2'}a_1^\dagger a_2^\dagger a_{2'}a_{1'}\ .$$

Choose as the set of observables $\mathcal{G}_\nu$ all the single-particle operators $a_1^\dagger a_2$. Show that using only the first term of the Robertson equation (18.2.23), i.e. completely neglecting the integral, one obtains the time-dependent Hartree–Fock equations for $\langle a_1^\dagger a_2 \rangle(t)$ (calculation in a space of variable particle-number).

Appendix

A. Equivalence of the Nakajima–Zwanzig Equation and the Generalized-Operator Langevin Equation

The calculation of the expectation values of observables $\mathscr{F}_\nu$ is carried out, depending on whether one begins in the Schrödinger or the Heisenberg formulation, using

$$\begin{aligned}\langle \mathscr{F}_\nu \rangle(t) &= \mathrm{Tr}(\mathscr{F}_\nu \rho(t)) = \mathrm{Tr}(\mathscr{F}_\nu \mathrm{e}^{-\mathrm{i}\mathsf{L}t}\rho(0)) \\ &= \mathrm{Tr}(\mathrm{e}^{\mathrm{i}\mathsf{L}t}\mathscr{F}_\nu \cdot \rho(0)) \qquad (t_0 = 0)\ . \end{aligned} \tag{A.1}$$

Depending on the formulation employed, particular expressions for $\exp(-\mathrm{i}\mathsf{L}t)$ or for $\exp(\mathrm{i}\mathsf{L}t)$ are thus required.

We shall begin with the concept introduced in Chap. 17 (17.2.2),

$$\rho_{\mathrm{rel}}(t) = \mathsf{P}\rho(t) \tag{A.2}$$

and the associated Nakajima–Zwanzig equation of motion for the relevant statistical operator, (17.2.25), and, employing the expression for $\exp(-\mathrm{i}\mathsf{L}t)$ used there, we shall construct the corresponding operator Langevin equation in the Heisenberg formulation ($t_0 = 0$).

Making use of $\rho(t) = \exp(-\mathrm{i}\mathsf{L}t)\,\rho(0)$ in (17.2.25), we find for any expectation value, owing to (17.2.6):

$$\begin{aligned}\frac{d}{dt}\langle \mathscr{F}_\nu \rangle(t) = &-\mathrm{i}\,\mathrm{Tr}(\mathscr{F}_\nu \cdot \mathsf{LP}\mathrm{e}^{-\mathrm{i}\mathsf{L}t}\rho(0)) \\ &-\int_0^t \mathrm{Tr}(\mathscr{F}_\nu \cdot \mathsf{LQ}\mathrm{e}^{-\mathrm{i}\mathsf{LQ}(t-t')}\mathsf{LP}\mathrm{e}^{-\mathrm{i}\mathsf{L}t'}\rho(0))\,dt' \\ &-\mathrm{Tr}(\mathscr{F}_\nu \cdot \mathrm{i}\mathsf{LQ}\mathrm{e}^{-\mathrm{i}\mathsf{LQ}t}\rho(0))\ . \end{aligned} \tag{A.3}$$

The operator $\rho(0)$ can be chosen at will in this equation, so that the last term does not drop out in general. For further manipulations, we define an operator $\tilde{\mathsf{P}}$ belonging to P by

$$\mathrm{Tr}(\mathscr{X} \cdot \mathsf{P}\mathscr{Y}) = \mathrm{Tr}(\tilde{\mathsf{P}}\mathscr{X} \cdot \mathscr{Y}) = \mathrm{Tr}(\mathscr{Y} \cdot \tilde{\mathsf{P}}\mathscr{X})\ ^{1} \tag{A.4}$$

---

[1] $\mathsf{P}^2 = \mathsf{P}$ implies $\tilde{\mathsf{P}}^2 = \tilde{\mathsf{P}}$

for arbitrary $\mathscr{X}$ and $\mathscr{Y}$. Using (3.1.16), we can then bring (A.3) into the form

$$\frac{d}{dt}\langle \mathscr{F}_\nu \rangle(t) = \operatorname{Tr} \rho(0) \{ \mathrm{e}^{\mathrm{iL}t} \tilde{\mathsf{P}} \mathrm{i} \mathsf{L} \mathscr{F}_\nu - \int_0^t \mathrm{e}^{\mathrm{iL}t'} \tilde{\mathsf{P}} \mathsf{L} \mathrm{e}^{\mathrm{i}\tilde{\mathsf{Q}}\mathsf{L}(t-t')} \tilde{\mathsf{Q}} \mathsf{L} \mathscr{F}_\nu dt' + \mathrm{i} \mathrm{e}^{\mathrm{i}\tilde{\mathsf{Q}}\mathsf{L}t} \tilde{\mathsf{Q}} \mathsf{L} \mathscr{F}_\nu \} \ . \tag{A.5}$$

Since $\rho(0)$ can be chosen freely for $\mathscr{F}_\nu(t) = \exp(\mathrm{iL}t)\mathscr{F}_\nu$ the *generalized-operator Langevin equation*

$$\frac{d}{dt}\mathscr{F}_\nu(t) = \mathrm{e}^{\mathrm{iL}t} \tilde{\mathsf{P}} \dot{\mathscr{F}}_\nu + \mathrm{i} \int_0^t \mathrm{e}^{\mathrm{iL}t'} \tilde{\mathsf{P}} \mathsf{L} \mathrm{e}^{\mathrm{i}\tilde{\mathsf{Q}}\mathsf{L}(t-t')} \tilde{\mathsf{Q}} \dot{\mathscr{F}}_\nu dt' + \mathrm{e}^{\mathrm{i}\tilde{\mathsf{Q}}\mathsf{L}t} \tilde{Q} \dot{\mathscr{F}}_\nu \tag{A.6}$$

follows [compare (11.2.13)].

If the operator P has the form (17.2.16), it leads to

$$\tilde{\mathsf{P}}\mathscr{X} = \sum_{\lambda\mu} \mathscr{F}_\lambda v_{\mu\lambda} \operatorname{Tr}(\rho_\mu \mathscr{X}) \tag{A.7}$$

and one obtains a structure for (A.6) which is formally the same as that of (11.2.14). The Mori projector used throughout Part 3, now denoted by $\mathsf{P}_{\mathrm{Mori}}$, is thus a special case of $\tilde{\mathsf{P}}$. If we start with a projector P which is formed from the particular operators ρ_ν of (18.1.8), we have

$$\tilde{\mathsf{P}} = \mathsf{P}_{\mathrm{Mori}} \tag{A.8}$$

and the Nakajima–Zwanzig theory in the Schrödinger formulation becomes the Mori theory in the Heisenberg formulation (cf. Exercise 18.3).

When the physical initial state $\rho(0)$ at time $t_0 = 0$ fulfills the relation

$$\rho(0) = \rho_{\mathrm{rel}}(0) = \mathsf{P}\rho(0) \ , \tag{A.9}$$

and thus

$$\mathsf{Q}\rho(0) = \mathscr{O} \tag{A.10}$$

then we have

$$\operatorname{Tr}(\rho(0) \mathrm{e}^{\mathrm{i}\tilde{\mathsf{Q}}\mathsf{L}t} \tilde{\mathsf{Q}} \dot{\mathscr{F}}_\nu) = 0 \ . \tag{A.11}$$

The expectation value of the inhomogeneous term, i.e. the *residual force*,

$$\langle f_\nu \rangle(t) = 0 \tag{A.12}$$

vanishes at all times t.

The result of these considerations demonstrates explicitly how the projection operators in a description using a Nakajima–Zwanzig equation (17.2.25) must be related to a generalized-operator Langevin equation (A.6) – namely through

(A.4) – in order to yield the same expectation value equations. Which description one chooses to employ in a specific case is thus *a priori* unimportant. In the case of a very large set of observables, especially when one also wishes to describe a part of the overall system microscopically, the *single* Nakajima–Zwanzig equation for $\rho_{rel}(t)$ is thus considerably clearer. However, if only a few macroscopic observables $\mathscr{F}_\nu$ are required, a treatment using the Langevin equations is more accessible to physical interpretation of the type which we carried out in connection with the detailed treatment of the Mori theory in Part 3.

B. Symmetries

In this Appendix, we indicate some of the consequences of particular symmetry properties in space and time for the dynamics of a quantum statistical system. We assume that the reader is familiar with the use of group-theoretical methods in quantum mechanics. In Appendix B.1, we treat those symmetry properties which can be described by linear unitary symmetry operators. In Appendix B.2, we deal with properties which follow from time-reversal invariance, for the description of which antilinear unitary operators are required.

B.1 Linear Symmetry Operators $\mathsf{D}(g)$ in $\mathfrak{L}$

B.1.1 Properties of $\mathsf{D}(g)$

The elements g of a transformation group $\mathfrak{G}(g)$ are, as is well known, comprised of linear unitary operators $\mathscr{D}(g)$, i.e.

$$\mathscr{D}^{-1}(g) = \mathscr{D}^{\dagger}(g) \tag{B.1.1}$$

which for each g are particular functions of the operators $\vec{x}_\nu$, $\vec{p}_\nu$, and $\vec{\sigma}_\nu$.[2] The operators $\mathscr{D}(g)$ form a representation[2] of the group $\mathfrak{G}$ in $\mathfrak{U}$. The observables $\mathscr{F}$ of the physical system under consideration are thus mapped on to new observables $\mathscr{F}'$ by every transformation $\mathscr{D}(g)$ acording to

$$\mathscr{F}' = \mathscr{D}(g)\mathscr{F}\mathscr{D}^{-1}(g) = \mathsf{D}(g)\mathscr{F} \tag{B.1.2}$$

by which the *linear superoperators* $\mathsf{D}(g)$ in $\mathfrak{L}$ [cf. (3.1.8)] are defined. They also form a representation of $\mathfrak{G}$, i.e. the following relations hold:

[2] One could also consider explicitly time-dependent transformations, as required e.g. in transformation to a moving frame of reference (for example a Galilei transformation or a transformation to a rotating frame). However, we shall not consider such transformations further in the present context.

$$\mathsf{D}(g_2 g_1) = \mathsf{D}(g_2)\mathsf{D}(g_1) \quad [3]$$

$$\mathsf{D}(e) = 1 \tag{B.1.3}$$

$$\mathsf{D}(g^{-1}) = \mathsf{D}^{-1}(g) \ . [4]$$

From (B.1.1) and (B.1.2), it follows for the operator $(\mathsf{D}\mathscr{F})^\dagger$ which is adjoint to $\mathsf{D}\mathscr{F}$ in $\mathfrak{U}$ that

$$(\mathsf{D}(g)\mathscr{F})^\dagger = \mathsf{D}(g)(\mathscr{F}^\dagger) \ , \tag{B.1.4}$$

i.e. the adjoint operation commutes with transformations.

In a linear transformation (B.1.2), relations between observables are transformed into the same relations between the transformed operators; for example, it follows from (B.1.2) that

$$(\mathscr{F}\mathscr{G})' = \mathsf{D}(\mathscr{F}\mathscr{G}) = \mathsf{D}\mathscr{F} \cdot \mathsf{D}\mathscr{G} = \mathscr{F}'\mathscr{G}' \ . \tag{B.1.5}$$

The trace of the product $\mathscr{F}'\mathscr{G}'$ is therefore equal to the trace of $\mathscr{F}\mathscr{G}$, that is

$$\boxed{\mathrm{Tr}(\mathsf{D}(g)\mathscr{F} \cdot \mathsf{D}(g)\mathscr{G}) = \mathrm{Tr}(\mathscr{F}\mathscr{G})} \ . \tag{B.1.6}$$

If we set $\mathscr{G} = \rho$, so that

$$\rho' = \mathsf{D}(g)\rho \tag{B.1.7}$$

is the transformed statistical operator of the system, then (B.1.6) shows that the expectation value of $\mathscr{F}'$ taken in the state ρ' is equal to the expectation value of $\mathscr{F}$ taken in ρ.

If we transform *only* the observables according to (B.1.2) and take the expectation value of $\mathscr{F}'$ in the *non-transformed* state ρ,

$$\mathrm{Tr}(\mathscr{F}'\rho) = \mathrm{Tr}(\mathsf{D}\mathscr{F} \cdot \rho) = \mathrm{Tr}(\mathscr{F} \cdot \mathsf{D}^{-1}\rho) = \mathrm{Tr}(\mathscr{F}\tilde{\rho}) \ , \tag{B.1.8}$$

then we will find it to be the same as the expectation value of $\mathscr{F}$ in the transformed state $\tilde{\rho} = \mathsf{D}^{-1}(g)\rho$.

---

[3] For the operators $\mathscr{D}(g)$ applied to states $|\varphi\rangle$, "ray representations", i.e. representations up to a factor

$$\mathscr{D}(g_2, g_1) = \eta(g_2, g_1)\, \mathscr{D}(g_2)\mathscr{D}(g_1)$$

with $|\eta| = 1$, are of importance, e.g. in spatial rotations of systems with half-integral spins or in translations in a homogeneous magnetic field. For the superoperators $\mathsf{D}(g)$, this factor drops out according to (B.1.2); they form an actual representation of $\mathfrak{G}$.

[4] The adjoint of a superoperator and the unitarity of $\mathsf{D}(g)$ in $\mathfrak{L}$ connected with it will be considered only after explicitly chosing a scalar product $(\ |\ )$ in $\mathfrak{L}$ [cf. e.g. (8.2.5)]; it has, for the time being, nothing to do with the unitarity (B.1.1) of $\mathscr{D}(g)$ in $\mathfrak{U}$.

In particular, if an observable $\mathscr{F}$ is invariant with respect to a transformation in $\mathfrak{G}$:

$$\mathsf{D}(g)\mathscr{F} = \mathscr{F} \ , \tag{B.1.9}$$

then $\mathfrak{G}$ is called the *symmetry group* of $\mathscr{F}$ and the expectation value $\langle \mathscr{F} \rangle$ remains unchanged on transformation for any ρ, $\langle \mathscr{F}' \rangle = \langle \mathscr{F} \rangle$. If, on the other hand,

$$\mathsf{D}(g)\rho = \rho \tag{B.1.10}$$

holds, then the state of the system is invariant with respect to transformations in $\mathfrak{G}$, and the expectation value $\langle \mathscr{F} \rangle$ of an arbitrary observable is unchanged by the transformations in $\mathfrak{G}$.

B.1.2 Selection Rules

B.1.2.1) Irreducible Operators of a Group $\mathfrak{G}$. It frequently occurs that observables under consideration are not invariant with respect to a group of transformations, but rather that a set of observables of interest, $\mathscr{F}^{\mu}$, are transformed linearly into one another by $\mathsf{D}(g)$:

$$\mathsf{D}(g)\mathscr{F}^{\mu}_{m} = \sum_{\mu'} \mathscr{F}^{\mu'}_{m} D^{\mu'\mu}_{m}(g) \qquad (m \text{ fixed}) \ . \tag{B.1.11}$$

We shall assume that the coefficients $D^{\mu'\mu}_{m}(g)$ in this equation form an *irreducible* unitary representation Γ_m of the group $\mathfrak{G}$. The operators $\mathscr{F}^{\mu}_{m}$ are then termed *irreducible* with respect to the group $\mathfrak{G}$.[5] The well-known orthogonality relations

$$\sum_{g} D^{\mu'\mu *}_{m}(g)\, D^{\nu'\nu}_{n}(g) = \frac{N}{d_m}\delta_{mn}\delta^{\mu'\nu'}\delta^{\mu\nu} \tag{B.1.12}$$

hold between the matrices of two such irreducible representations Γ_m and Γ_n of a group $\mathfrak{G}$, where the sum extends over all the elements g of the group. N is the number of elements in the group and d_m is the dimensionality of Γ_m.[6] In this

---

[5] By applying a theorem from the theory of representations to the Liouville space, we can show that the irreducible operators $\mathscr{F}^{\mu}_{m}$ can be obtained from an arbitrary operator $\mathscr{F}$ by means of superoperators

$$\mathscr{F}^{\mu}_{m} = \frac{d_m}{N}\sum_{g} D^{\mu\alpha *}_{m}(g)\,\mathsf{D}(g)\mathscr{F} \quad (\text{fixed } \alpha) \ . \tag{B.1.11a}$$

[6] The orthogonality relations (B.1.12) are *a priori* valid for finite groups (N finite). However, they may be generalized to continuous groups by introducing a measure $d\mu_g$ of the number of group elements in the neighborhood of a particular element g, i.e. a density function $d\mu_g/dg$. For compact Lie groups, all irreducible representations are of finite dimensionality. We refer the reader to books on group theory and the theory of representations, e.g. M. Hamermesh: *Group Theory and Its Applications to Physical Problems* (London, Paris 1962).

equation, $m \neq n$ means two inequivalent irreducible representations and $m = n$ refers to the same irreducible representation.

B.1.2.2) Selection Rules for the Traces of Products of Two Irreducible Operators. We shall use the orthogonality relations (B.1.12) to determine which of the traces

$$\mathrm{Tr}(\mathscr{F}_m^{\mu\dagger}\mathscr{G}_n^{\nu}) \tag{B.1.13}$$

vanishes from symmetry considerations. We assume that $\mathscr{F}_m^{\mu}$ and $\mathscr{G}_n^{\nu}$ transform according to (B.1.11) with *irreducible* representations of Γ_m and Γ_n,

$$\begin{aligned} \mathsf{D}(g)\mathscr{F}_m^{\mu} &= \sum_{\mu'} \mathscr{F}_m^{\mu'} D_m^{\mu'\mu}(g) \quad (\Gamma_m) \\ \mathsf{D}(g)\mathscr{G}_n^{\nu} &= \sum_{\nu'} \mathscr{G}_n^{\nu'} D_n^{\nu'\nu}(g) \ , \quad (\Gamma_n) \end{aligned} \tag{B.1.14}$$

so that, using (B.1.4), the following relation also holds:

$$(\mathsf{D}(g)\mathscr{F}_m^{\mu})^{\dagger} = \mathsf{D}(g)(\mathscr{F}_m^{\mu})^{\dagger} = \sum_{\mu'} \mathscr{F}_m^{\mu'\dagger} D_m^{\mu'\mu}(g)^* \ . \tag{B.1.15}$$

Applying this relation (B.1.15) to expression (B.1.13), we find

$$\begin{aligned} \mathrm{Tr}(\mathscr{F}_m^{\mu\dagger}\mathscr{G}_n^{\nu}) &= \mathrm{Tr}(\mathsf{D}(g)\mathscr{F}_m^{\mu\dagger}\cdot\mathsf{D}(g)\mathscr{G}_n^{\nu}) \\ &= \sum_{\mu'\nu'} \mathrm{Tr}(\mathscr{F}_m^{\mu'\dagger}\mathscr{G}_n^{\nu'}) D_m^{\mu'\mu *}(g) D_n^{\nu'\nu}(g) \ . \end{aligned} \tag{B.1.16}$$

Summing over all the elements g of the group, we obtain with the orthogonality relations (B.1.12) the *selection rule*

$$\boxed{\mathrm{Tr}(\mathscr{F}_m^{\mu\dagger}\mathscr{G}_n^{\nu}) = \delta_{mn}\delta^{\mu\nu}\frac{1}{d_m}\sum_{\mu'} \mathrm{Tr}(\mathscr{F}_m^{\mu'\dagger}\mathscr{G}_m^{\mu'})} \ . \tag{B.1.17}$$

Thus, when $\mathscr{F}_m^{\mu}$ and $\mathscr{G}_n^{\nu}$ belong to two *inequivalent* irreducible representations of $\mathfrak{G}$ ($\Gamma_m \neq \Gamma_n$), the traces $\mathrm{Tr}(\mathscr{F}_m^{\mu\dagger}\mathscr{G}_n^{\nu})$ vanish. Furthermore, in the case that both belong to the same irreducible representation Γ_m, the traces $\mathrm{Tr}(\mathscr{F}_m^{\mu\dagger}\mathscr{G}_m^{\nu}) = 0$ when $\mu \neq \nu$.

Examples: a) The *rotation group* $O(3)$. The irreducible representations of this group are enumerated by orbital angular momentum quantum numbers $l = 0, 1, \ldots$. For example, if $\mathscr{H}$ or $\mathscr{R}_\beta$ is rotationally invariant (i.e. a scalar, $l = 0$), then the thermal expectation value of every vector or irreducible tensor operator vanishes ($l = 1, 2, \ldots$).

b) The *translation group* $T(3)$ with periodic boundary conditions. The Fourier transforms $\mathscr{u}_{\mathfrak{k}}$ of spatial densities $\mathscr{u}(\mathfrak{r})$ belong to the one-dimensional irreducible representaion $D_{\mathfrak{k}}(t) = e^{-i\mathfrak{k}t}$. Equation (B.1.17) then yields

$$\mathrm{Tr}(u_{\mathfrak{k}}^{\dagger} u_{\mathfrak{k}'}) = 0 \quad \text{for} \quad \mathfrak{k} \neq \mathfrak{k}' \tag{B.1.18}$$

$$\langle u_{\mathfrak{k} \neq 0} \rangle = 0 \tag{B.1.19}$$

for a translationally-invariant statistical operator.

B.1.2.3) Selection Rules for Mori Products. We assume that the Hamiltonian $\mathscr{H}$ has the symmetry group $\mathfrak{G}$,

$$\mathsf{D}(g)\mathscr{H} = \mathscr{H} \tag{B.1.20}$$

and thus that

$$\mathsf{D}(g)\mathsf{L} = \mathsf{L}\mathsf{D}(g) \tag{B.1.21}$$

holds. Then $\mathsf{D}(g)$ commutes with every function $f(\mathsf{L})$, as can be seen by employing a power-series expansion.

For the Mori products [(8.2.14) or also (8.2.5)] of two operators $\mathscr{F}$ and $\mathscr{G}$, we thus obtain using (B.1.4) and (B.1.6)

$$(\mathscr{F}|\mathscr{G}) = (\mathsf{D}(g)\mathscr{F}\,|\,\mathsf{D}(g)\mathscr{G}) \ . \tag{B.1.22}$$

Carrying out formally the same calculation as in Sect. B.1.2.2, we find for the Mori products of operators $\mathscr{F}_m^{\mu}$ and $\mathscr{G}_n^{\nu}$ (B.1.14) in irreducible representations Γ_m and Γ_n the symmetry relations:

$$\boxed{\begin{aligned}(\mathscr{F}_m^{\mu}|\mathscr{G}_n^{\nu}) &= (\mathsf{D}(g)\mathscr{F}_m^{\mu}|\,\mathsf{D}(g)\mathscr{G}_n^{\nu})\\ &= \sum_{\mu'\nu'} (\mathscr{F}_m^{\mu'}|\mathscr{G}_n^{\nu'}) D_m^{\mu'\mu *}(g) D_n^{\nu'\nu}(g)\\ &= \delta_{mn}\delta^{\mu\nu}\frac{1}{d_m}\sum_{\mu'}(\mathscr{F}_m^{\mu'}|\mathscr{G}_m^{\mu'})\end{aligned}}\ .^{7} \tag{B.1.23}$$

In contrast to (B.1.17), however, it is now essential that $\mathfrak{G}$ be the symmetry group of $\mathscr{H}$ (B.1.20 or 21).

Equation (B.1.23) therefore contains the assertion that $(\mathscr{F}_m^{\mu}|\mathscr{G}_n^{\nu}) = 0$ for $m \neq n$ and also that $(\mathscr{F}_m^{\mu}|\mathscr{G}_m^{\nu}) = 0$ for $\mu \neq \nu$. These *selection rules* are also important e.g. for extensions of the set $\{\mathscr{G}\}$ in the Mori theory (Sects. 11.2.2 and 12.2), since one can use them to eliminate observables which are decoupled for symmetry reasons.

---

[7] In contrast to (B.1.17), the adjoint sign † does not enter formally into these equations, since it is already contained in the definition of the scalar product, (8.2.14). The symbol Γ_m for the representation thus refers, corresponding to (B.1.14), to $\mathscr{F}_m^{\mu}$ and not to $\mathscr{F}_m^{\mu\dagger}$.

In linear response theory, the isothermal static susceptibility (Sect. 8.3.3) has the form (B.1.23). The dynamical quantities, such as the *response functions*, the *relaxation functions*, and the *dynamic susceptibilities* contain an additional function $f(\mathsf{L}, t)$ or $f(\mathsf{L}, \omega)$ within the scalar product. Without having to specify this function more precisely, we can conclude from the fact that $\mathsf{D}(g)$ and L commute that

$$\begin{aligned}(\mathscr{F}^{\mu}_{m} | f(\mathsf{L})\mathscr{G}^{\nu}_{n}) &= (\mathsf{D}(g)\mathscr{F}^{\mu}_{m} | \mathsf{D}(g) f(\mathsf{L})\mathscr{G}^{\nu}_{n}) \\ &= (\mathsf{D}(g)\mathscr{F}^{\mu}_{m} | f(\mathsf{L})\mathsf{D}(g)\mathscr{G}^{\nu}_{n}) \ . \end{aligned} \tag{B.1.24}$$

We can proceed as in (B.1.23) and conclude that for *all dynamic linear response functions* – formed from operators $\mathscr{G}^{\mu}_{m}$ irreducible with respect to the symmetry group $\mathfrak{G}$ of $\mathscr{H}$ – the general selection rule holds:

$$\boxed{(\mathscr{F}^{\mu}_{m} | f(\mathsf{L})\mathscr{G}^{\nu}_{n}) = \delta_{mn}\delta^{\mu\nu}\frac{1}{d_m}\sum_{\mu'}(\mathscr{F}^{\mu'}_{m} | f(\mathsf{L})\mathscr{G}^{\mu'}_{m})} \ . \tag{B.1.25}$$

Examples: a) If, in particular, the *rotation group* $O(3)$ is a symmetry group of $\mathscr{H}$, then all the Mori products of irreducible tensors of differing rank l are zero.

Thus, if $\mathscr{F}$ is a scalar ($l = 0$) and the $\mathscr{A}^i$ are the components of a vector $\vec{\mathscr{A}}(l = 1)$ in configuration space, then the products

$$(\mathscr{F} | f(\mathsf{L})\mathscr{A}^i) = 0 \quad (i = x, y, z \quad \text{or} \quad +, -, z) \tag{B.1.26}$$

vanish.

For the components of any two vector operators $\vec{\mathscr{A}}$ and $\vec{\mathscr{B}}$, as a result of $\delta^{\mu\nu}$ in (B.1.25), we find for the tensor

$$(\mathscr{A}^i | f(\mathsf{L})\mathscr{B}^k) = \begin{pmatrix} s & 0 & 0 \\ 0 & s & 0 \\ 0 & 0 & s \end{pmatrix} \quad (i, k = x, y, z \quad \text{or} \quad +, -, z) \tag{B.1.27}$$

with

$$s = \tfrac{1}{3}(\vec{\mathscr{A}} | f(\mathsf{L})\vec{\mathscr{B}})) \ .$$

The static and dynamic susceptibilities χ^{ik} and $\chi^{ik}(\omega)$, as well as the relaxation tensor $\varphi^{ik}(t)$, therefore have no off-diagonal elements when $\mathscr{H}$ has $O(3)$-symmetry. The expression for the diffusion tensor (16.4.20) is a further example of (B.1.27).

b) In the case of *rotational symmetry about the z-axis only*, $O(2)$, for which just one-dimensional irreducible representations exist, the δ_{mn} in (B.1.25) results in

$$(\mathscr{A}_i | f(\mathsf{L})\mathscr{B}_k) = 0 \quad \text{for} \quad i \neq k \quad \text{with} \quad i, k = +, -, z \tag{B.1.28}$$

for the off-diagonal elements; but here, in contrast to (B.1.27), the diagonal elements are not necessarily all equal [cf. as examples (16.3.2, 6, and 8)].

c) If the *translation group* $T(3)$ is a symmetry group of $\mathscr{H}$, then from (B.1.25) we find e.g. for the Fourier transforms $u_{\mathfrak{k}}$ of the densities $u(\mathfrak{r})$ the following selection rule:

$$(u_{\mathfrak{k}}|u_{\mathfrak{k}'}) = 0 \quad \text{for} \quad \mathfrak{k} \neq \mathfrak{k}' , \tag{B.1.29}$$

as was used for example in (16.4.9).

We now turn to the application of spatial symmetries in the Mori theory (Sect. 11.2). For the *special case* that the set $\{\mathscr{G}\}$ consists of irreducible operators $\mathscr{G}_\nu \to \mathscr{G}^\mu_m$ ($\mu = 1, \ldots d_m$; $m = 1, \ldots$)[8] with respect to a symmetry group of the Hamiltonian $\mathscr{H}$, the metric $g^{\mu\nu}_{mn}$ is determined according to (11.2.3) by

$$\sum_{n\nu} g^{\mu\nu}_{mn}(\mathscr{G}^\nu_n|\mathscr{G}^\lambda_l) = \delta_{ml}\delta^{\mu\lambda} . \tag{B.1.30}$$

From the property (B.1.23) of $(\mathscr{G}^\nu_n|\mathscr{G}^\lambda_l)$, we obtain:

$$g^{\mu\nu}_{mn} = \delta_{mn}\delta^{\mu\nu}\frac{d_m}{\sum_{\mu'}(\mathscr{G}^{\mu'}_m|\mathscr{G}^{\mu'}_m)} . \tag{B.1.31}$$

The projector P onto the set $\mathscr{G}^\mu_m$,

$$\mathsf{P} = \sum_{\substack{\mu\nu\\ mn}} |\mathscr{G}^\mu_m) g^{\mu\nu}_{mn} (\mathscr{G}^\nu_n| \tag{B.1.32}$$

can be simplified using (B.1.31) to give

$$\mathsf{P} = \sum_{m\mu} |\mathscr{G}^\mu_m)\frac{d_m}{\sum_{\mu'}(\mathscr{G}^{\mu'}_m|\mathscr{G}^{\mu'}_m)}(\mathscr{G}^\mu_m| . \tag{B.1.33}$$

With regard to the Mori formulas (11.2.17, 23), (11.3.15, 18, 19), we examine the effect of the symmetry operators $\mathsf{D}(g)$ of $\mathscr{H}$ on P. Using (B.1.22), we obtain

$$\mathsf{D}(g)\mathsf{P}|\mathscr{X}) = \sum_{m\mu} |\mathsf{D}(g)\mathscr{G}^\mu_m)\frac{d_m}{\sum_{\mu'}(\mathscr{G}^{\mu'}_m|\mathscr{G}^{\mu'}_m)}(\mathsf{D}(g)\mathscr{G}^\mu_m|\mathsf{D}(g)\mathscr{X}) \tag{B.1.34}$$

and insert the transformation behavior (B.1.14) of the $\mathscr{G}^\mu_m$. From the unitarity of the $D^{\mu'\mu}_m(g)$-matrices we find

$$\sum_\mu D^{\mu'\mu}_m(g) D^{\mu''\mu *}_m(g) = \delta^{\mu'\mu''} \tag{B.1.35}$$

[8] The $\mathscr{G}^\mu_m$ are, for example, multipole operators of a spin system.

and thus

$$\mathsf{D}(g)\mathsf{P} = \mathsf{P}\mathsf{D}(g) \tag{B.1.36}$$

and therefore also

$$\boxed{\mathsf{D}(g)\mathsf{Q} = \mathsf{Q}\mathsf{D}(g)} \; . \tag{B.1.37}$$

The projectors P or Q formed from the set of the irreducible operators $\mathscr{G}^{\mu}_{m}$ thus commute with all the $\mathsf{D}(g)$.

Furthermore, since according to the assumption (B.1.20) we also have $\mathsf{DL} = \mathsf{LD}$, it then follows e.g. for the residual force in the Mori theory (11.2.17):

$$\begin{aligned} \mathsf{D}(g)f^{\mu}_{m}(t) &= \mathsf{D}(g)\mathrm{i}e^{\mathrm{i}\mathsf{QL}t}\mathsf{QL}\mathscr{G}^{\mu}_{m} = \mathrm{i}e^{\mathrm{i}\mathsf{QL}t}\mathsf{QLD}(g)\mathscr{G}^{\mu}_{m} \\ &= \sum_{\mu'} f^{\mu'}_{m}(t)D^{\mu'\mu}_{m}(g) \; . \end{aligned} \tag{B.1.38}$$

For the dynamic Onsager–Casimir coefficients (11.3.15), generated using irreducible operators $\mathscr{G}^{\mu}_{m}$ with respect to the symmetry group $\mathfrak{G}$ of $\mathscr{H}$, (B.1.38) yields the selection rules:

$$L^{\mu\nu}_{mn}(\omega) = \delta_{mn}\delta^{\mu\nu}\frac{l}{d_m}\sum_{\mu'}\mathsf{L}^{\mu'\mu'}_{mm}(\omega) \; . \tag{B.1.39}$$

Thus, the off-diagonal elements vanish, just as we found for the dynamic susceptibility. The same results therefore hold as those shown in the examples (B.1.26) through (B.1.28). The selection rules (B.1.39) are valid in addition to the symmetry relations (11.3.20) which followed from the Hermiticity of QLQ. The same is true of the reciprocity relations which follow from time-reversal invariance. These will be treated in Appendix B.2.

B.2 The Antilinear Time-Reversal Operator Θ in $\mathfrak{L}$

B.2.1 Properties of Θ

The term *time reversal* refers to the transformation:

$$\mathscr{F} \to \mathscr{F}' = \Theta\mathscr{F} = \theta\mathscr{F}\theta^{-1} \tag{B.2.1}$$

which has the following physical properties:

$$\begin{aligned} \Theta\vec{x}_{\nu} &= \theta\vec{x}_{\nu}\theta^{-1} = x_{\nu} \\ \Theta\vec{p}_{\nu} &= \theta\vec{p}_{\nu}\theta^{-1} = -\vec{p}_{\nu} \\ \Theta\vec{s}_{\nu} &= \theta\vec{s}_{\nu}\theta^{-1} = -\vec{s}_{\nu} \; . \end{aligned} \tag{B.2.2}$$

These are taken into account by an operator θ which is *anti*linear in $\mathfrak{U}$ and has the mathematical properties:

$$\theta c = c^*\theta \quad \text{(commutation of } \theta \text{ with a complex number } c) \tag{B.2.3}$$

$$\langle\varphi|\theta\psi\rangle = \langle\theta^\dagger\varphi|\psi\rangle^* = \langle\psi|\theta^\dagger\varphi\rangle \quad \text{(definition of the adjoint } \theta^\dagger) \tag{B.2.4}$$

and

$$\theta^\dagger = \theta^{-1} \quad \text{(unitarity of } \theta \text{ in } \mathfrak{U}) \ .^9 \tag{B.2.5}$$

For the superoperator Θ, we thus have the relations:

$$\Theta c = c^*\Theta \tag{B.2.6}$$

$$(\Theta\mathscr{F})^\dagger = \Theta\mathscr{F}^\dagger \tag{B.2.7}$$

$$\Theta(\mathscr{F}\mathscr{G}) = \Theta\mathscr{F}\cdot\Theta\mathscr{G} \tag{B.2.8}$$

$$\Theta^2 = 1 \ . \tag{B.2.9}$$

Equations (B.2.7) and (B.2.8) are formally the same as in the case of linear superoperators [cf. (B.1.4) and (B.1.5)]. The group property (B.2.9) of Θ is a result of (B.2.2), *in*dependent of the number of fermions, i.e. independently of the statistics.

The antilinear operator Θ has the following important property:

$$\mathrm{Tr}(\Theta\mathscr{F}) = \mathrm{Tr}\,\mathscr{F}^\dagger = (\mathrm{Tr}\,\mathscr{F})^* \ , \tag{B.2.10}$$

since, in a complete basis $|v_m\rangle$ in $\mathfrak{U}$, taking (B.2.4) into account, we have

$$\mathrm{Tr}\,(\theta\mathscr{F}\theta^{-1}) = \sum_n \langle v_n|\theta\mathscr{F}\theta^{-1}v_n\rangle = \sum_n \langle\theta^\dagger v_n|\mathscr{F}\theta^\dagger v_n\rangle^* = (\mathrm{Tr}\,\mathscr{F})^* \ .$$

Together with (B.2.8), it follows from this that

$$\boxed{\mathrm{Tr}(\Theta\mathscr{F}\cdot\Theta\mathscr{G}) = \mathrm{Tr}(\mathscr{F}^\dagger\mathscr{G}^\dagger) = \mathrm{Tr}(\mathscr{F}\mathscr{G})^*} \tag{B.2.11}$$

and also

$$\mathrm{Tr}\,(\Theta\mathscr{F}\cdot\mathscr{G}) = \mathrm{Tr}\,(\mathscr{F}\cdot\Theta\mathscr{G})^* \ . \tag{B.2.12}$$

It must be taken into consideration that in (B.2.11), for the antilinear superoperator Θ, in contrast to the linear superoperators (B.1.6), the complex conjugate of $\mathrm{Tr}\,(\mathscr{F}\mathscr{G})$ appears.

---

[9] Further, it also holds that $\theta^2 = (-1)^N$, when N is the number of particles with half-integral spin.

The antilinear property (B.2.3) or (B.2.6), which provides that on commutation of θ or of Θ with a complex number c, the complex conjugate c^* results, means physically that the fundamental commutation relations (2.1.1) and (2.1.2) are *form invariant* with respect to the transformations (B.2.2). The same is true of the equations of motion (3.2.1) and (3.2.2), both in the Neumann formulation (3.2.5) and in the Heisenberg formulation (3.2.10), when these are *not only* transformed according to Θ, *but also* the (real c-number) time t is substituted by $t \to t' = -t$. This is the reason that the operators θ and Θ are termed *time-reversal operators*, although they by themselves do *not* change the time t.

B.2.2 Symmetry Properties Resulting from Time-Reversal Invariance

B.2.2.1) Even and Odd Operators with Respect to Time Reversal. When the Hamiltonian of a physical system exhibits a microscopic invariance property with respect to time reversal owing to the fact that only the electromagnetic interaction is considered, one can derive general symmetry properties for the macroscopic, dynamical quantities discussed in this book, such as susceptibilities and Onsager–Casimir coefficients. To this end, it is necessary that the macroscopic observables $\mathscr{F}$ in question have a certain simple behavior with respect to time reversal: we wish to consider those operators $\mathscr{F}$ which have the property

$$\boxed{\Theta\mathscr{F} = \varepsilon_{\mathscr{F}}\mathscr{F}_-^\dagger} \ . \tag{B.2.13}$$

Here, $\varepsilon_{\mathscr{F}}$ can be either $+1$ or -1 for each $\mathscr{F}$,

$$\varepsilon_{\mathscr{F}} = \pm 1 \ . \tag{B.2.14}$$

If $\mathscr{F}$ depends on a (c-number) magnetic field $\mathfrak{H}$, $\mathscr{F} = \mathscr{F}(\mathfrak{H})$, then we assume the definition

$$\mathscr{F}_- = \mathscr{F}(-\mathfrak{H}) \ . \tag{B.2.15}$$

Thus, in (B.2.13), $\mathscr{F}_-^\dagger$ refers to the adjoint operator $\mathscr{F}^\dagger(-\mathfrak{H})$. For Hermitian operators (observables), the † sign is unnecessary. If $\mathscr{F}$ is *in*dependent of magnetic fields, then the index $_-$ does not occur in (B.2.13).

Examples: a) $\mathfrak{H}$-independent observables:

$$\vec{x}, \vec{p}^{\,2}, \frac{\vec{p}^{\,2}}{2m} + V(\vec{x}), \vec{\ell}\vec{s}: \quad \varepsilon = +1 \tag{B.2.16}$$

$$\vec{p}, \vec{\ell}, \vec{s}: \quad \varepsilon = -1 \ ; \tag{B.2.17}$$

b) Non-Hermitian, $\mathfrak{H}$-independent operators[10] [cf. (4.1.2)]:

$$n_{\mathfrak{k}} = \int n(\mathfrak{r})\,\mathrm{e}^{\mathrm{i}\mathfrak{k}\mathfrak{r}}\,d^3x: \quad \varepsilon = +1 \tag{B.2.18}$$

$$\vec{\sigma}_{\mathfrak{k}} = \int \vec{\sigma}(\mathfrak{r})\,\mathrm{e}^{\mathrm{i}\mathfrak{k}\mathfrak{r}}\,d^3x: \quad \varepsilon = -1 \tag{B.2.19}$$

$$\sigma_{\pm} = \sigma_x \pm \mathrm{i}\sigma_y: \quad \varepsilon = -1\ . \tag{B.2.20}$$

c) Hermitian $\mathfrak{H}$-dependent observables:

$$\mathscr{H} = \frac{1}{2m}\left(\vec{p} - \frac{q}{c}\mathfrak{A}(\vec{x})\right)^2 + q\varphi(\vec{x}): \quad \varepsilon = +1 \tag{B.2.21}$$

$$\boxed{\Theta\mathscr{H} = \mathscr{H}_-}\ ^{11,12}: \quad \varepsilon = +1 \tag{B.2.22}$$

$$\left.\begin{aligned}&\overset{\circ}{\vec{x}} = \vec{v} = \frac{1}{m}\left(\vec{p} - \frac{q}{c}\mathfrak{A}(\vec{x})\right)\\ &\Theta\vec{v} = -\vec{v}_-\end{aligned}\right\}: \quad \varepsilon = -1\ . \tag{B.2.23}$$

From the symmetry of $\mathscr{H}$ with respect to time reversal, (B.2.22), it follows immediately that

$$\Theta\mathrm{e}^{-\beta\mathscr{H}} = \mathrm{e}^{-\beta\mathscr{H}_-}\ , \tag{B.2.24}$$

from which we find, using (B.2.10), that the partition function must be an *even* function of the magnetic field,

$$Z(\beta, \mathfrak{H}) = Z(\beta, -\mathfrak{H})\ . \tag{B.2.25}$$

If, in general, two operators $\mathscr{F}$ and $\mathscr{G}$ are considered, both having the property (B.2.13) with respect to time reversal, then it follows from (B.2.11) for the trace of the product:

$$\mathrm{Tr}(\mathscr{F}\mathscr{G}) = \varepsilon_{\mathscr{F}}\varepsilon_{\mathscr{G}}\,\mathrm{Tr}(\mathscr{F}\mathscr{G})_-\ , \tag{B.2.26}$$

from which one can obtain useful symmetry relations concerning the reversal of the sign of magnetic fields: the trace is an even function of the magnetic field $\mathfrak{H}$ when $\varepsilon_{\mathscr{F}}\varepsilon_{\mathscr{G}} = 1$, and an odd function in $\mathfrak{H}$ when $\varepsilon_{\mathscr{F}}\varepsilon_{\mathscr{G}} = -1$. For example, with

---

[10] The non-Hermitian combination $b = \alpha x + \mathrm{i}\beta p$ (α, β real) does *not* belong in this category, since it obeys $\Theta b = b$. However, using the results that follow, it is not difficult to extend these considerations to systems in which such oscillator normal coordinates occur. In order to keep the treatment reasonably compact, we shall not do this here.

[11] This relation also holds for spin system with Zeeman and exchange energies,

$$\mathscr{H} = -\gamma\mathfrak{H}\sum_{\nu\mu}\vec{\sigma}_\nu + \sum_{\nu\mu} J_{\nu\mu}\vec{\sigma}_\nu\vec{\sigma}_\mu\ .$$

[12] Reversing the sign of the magnetic field $\mathfrak{H}$ or of the vector potential $\mathfrak{A}$ means physically that the currents producing the field $\mathfrak{H}$ must also be reversed by a time-reversal transformation.

(B.2.25) we find the following relation for the thermal expectation value of the magnetization:

$$\langle \vec{m} \rangle_\beta(\mathfrak{H}) = -\langle \vec{m} \rangle_\beta(-\mathfrak{H})\ .^{13} \tag{B.2.27}$$

B.2.2.2) Time-Reversal Invariance in Mori Products. We now assume *in general* that the Hamiltonian obeys relation (B.2.22) with respect to time reversal,

$$\boxed{\Theta \mathscr{H} = \mathscr{H}_-}\ , \tag{B.2.28}$$

so that we also have

$$\boxed{\Theta \mathsf{L} = \mathsf{L}_- \Theta}\ . \tag{B.2.29}$$

For the Mori product [cf. (8.2.14)], taken with operators $\mathscr{F}$ and $\mathscr{G}$ but with $\mathscr{H}_-$, we write:

$$(\mathscr{F} \,\underline{|}\, \mathscr{G}) = \mathrm{Tr}\left(\mathscr{F}^\dagger \frac{1 - \exp(-\beta\hbar \mathsf{L}_-)}{\beta\hbar \mathsf{L}_-} \mathscr{G} \mathscr{R}_{\beta_-} \right)\ , \tag{B.2.30}$$

so that (B.2.11), (B.2.24), and (B.2.25) yield

$$\boxed{(\mathscr{F} \,|\, \mathscr{G}) = (\Theta\mathscr{F} \,\underline{|}\, \Theta\mathscr{G})^*}\ . \tag{B.2.31}$$

With the projector (11.2.2),

$$\mathsf{P} = \sum_{\lambda\mu} |\mathscr{G}_\lambda) g_{\lambda\mu} (\mathscr{G}_\mu| \tag{B.2.32}$$

employing the transformation $\mathfrak{H} \to -\mathfrak{H}$, we associate another projector:

$$\mathsf{P}_- = \sum_{\lambda\mu} |\mathscr{G}_{\lambda_-}) g_{\lambda\mu_-} (\mathscr{G}_{\mu_-} \underline{|}\ . \tag{B.2.33}$$

If we *assume* that $\mathscr{G}_\mu^\dagger$ as well as $\mathscr{G}_\mu$ is a member of the set $\{\mathscr{G}\}$, then the corresponding is true also of $\mathscr{G}_{\mu_-}$ and of $\mathscr{G}_{\mu_-}^\dagger$, $\mathscr{G}_{\mu_-}^\dagger \in \{\mathscr{G}_-\}$. Instead of (B.2.33), we may then also write

$$\mathsf{P}_- = \sum_{\lambda\mu} |\mathscr{G}_{\lambda_-}^\dagger) g_{\lambda\mu_-}^* (\mathscr{G}_{\mu_-}^\dagger \underline{|}\ . \tag{B.2.34}$$

[13] However, one cannot conclude from this that $\langle \vec{m} \rangle_\beta(0)$ must be equal to zero. The left- and right-hand limits $\mathfrak{H} \to \pm 0$ can namely differ, as is the case e.g. for ferromagnets due to symmetry-breaking; a discontinuity in $\langle \vec{m} \rangle_\beta$ at $\mathfrak{H} = 0$ is completely compatible with (B.2.27).

Equations (B.2.31 and 33) will be applied in the following section to the expressions occurring in the Kubo and Mori theories. Here, we *assume* that the operators $\mathscr{G}_\nu$ of the observation level $\{\mathscr{G}\}$ possess the time-reversal property

$$\boxed{\Theta\mathscr{G}_\nu = \varepsilon_\nu \mathscr{G}^\dagger_{\nu_-}} \qquad \varepsilon_\nu = \pm 1 \ . \tag{B.2.35}$$

Then we have from (B.2.31):

$$(\mathscr{G}_\mu|\mathscr{G}_\nu) = \varepsilon_\mu\varepsilon_\nu(\mathscr{G}_\mu|\mathscr{G}_\nu)_- \ , \tag{B.2.36}$$

i.e. the general isothermal susceptibility has the properties

$$\chi_{\mu\nu} = \chi^*_{\nu\mu} = \varepsilon_\mu\varepsilon_\nu\chi_{\mu\nu_-} \ , \tag{B.2.37}$$

where the first equation expresses the Hermiticity of $\chi_{\mu\nu}$ [for fixed $\mathfrak{H}$; compare (8.3.16)], while the second equation follows from the assumptions (B.2.28) and (B.2.35) concerning the behavior of the observables with respect to time reversal. For Hermitian observables $\mathscr{G}_\nu$, $\chi_{\mu\nu}$ is real and symmetric, and is also even or odd in $\mathfrak{H}$, depending on whether $\varepsilon_\mu\varepsilon_\nu = +1$ or -1.

From (B.2.29) for the dynamics we find

$$\Theta e^{iLt} = e^{-iL_-t}\Theta \tag{B.2.38}$$

so that we obtain from (B.2.30 and 31) the following symmetry relations for the relaxation functions (9.3.10 and 11):

$$\Phi_{\mu\nu}(t) = \Phi^*_{\nu\mu}(-t) = \varepsilon_\mu\varepsilon_\nu\Phi_{\mu\nu}(-t)_- \ . \tag{B.2.39}$$

With regard to the treatment of the memory function and the Onsager–Casimir coefficients, we require the effect of Θ on the projection operator P (11.2.2) of the Mori theory. To this end, we start by investigating the time-symmetry behavior of the Hermitian metric matrix $g_{\mu\nu}(\mathfrak{H})$ (11.2.3), which is defined as the inverse of the static susceptibility $\chi_{\mu\nu}(\mathfrak{H})$. In analogy to (B.2.37) we have

$$g_{\mu\nu} = g^*_{\nu\mu} = \varepsilon_\mu\varepsilon_\nu g_{\mu\nu_-} \ ; \tag{B.2.40}$$

from (11.4.13), it namely follows that $\varepsilon_\mu\varepsilon_\nu g_{\mu\nu_-}$ is the inverse of $\chi_{\nu\lambda}$:

$$\begin{aligned}\sum_\nu \varepsilon_\mu\varepsilon_\nu g_{\mu\nu_-}\chi_{\nu\lambda} &= \varepsilon_\mu\varepsilon_\lambda\sum_\nu \varepsilon^2_\nu g_{\mu\nu_-}\chi_{\nu\lambda_-} \\ &= \varepsilon_\mu\varepsilon_\lambda\beta\delta_{\mu\lambda} = \beta\delta_{\mu\lambda} \ .\end{aligned} \tag{B.2.41}$$

With this, we obtain the following expression for the application of Θ to P, taking account of (B.2.6), (B.2.31), and (B.2.35):

$$\begin{aligned}\Theta|\mathsf{P}\mathscr{X}) &= \sum_{\lambda\mu} |\Theta\mathscr{G}_\lambda)\cdot g^*_{\lambda\mu}(\mathscr{G}_\mu|\mathscr{X})^* \\ &= \sum_{\lambda\mu} |\Theta\mathscr{G}_\lambda) g^*_{\lambda\mu}(\Theta\mathscr{G}_\mu \,\underline{|}\, \Theta\mathscr{X}) \\ &= \sum_{\lambda\mu} |\mathscr{G}^\dagger_{\lambda_-})\varepsilon_\lambda g^*_{\lambda\mu}\varepsilon_\mu(\mathscr{G}^\dagger_{\mu_-} \,\underline{|}\, \Theta\mathscr{X}) \\ &= \sum_{\lambda\mu} |\mathscr{G}^\dagger_{\lambda_-}) g^*_{\lambda\mu_-}(\mathscr{G}^\dagger_{\mu_-} \,\underline{|}\, \Theta\mathscr{X}) \ .\end{aligned} \tag{B.2.42}$$

Therefore, from the definition (B.2.34), we have

$$\boxed{\Theta\mathsf{P} = \mathsf{P}_-\Theta} \tag{B.2.43}$$

and, correspondingly,

$$\Theta\mathsf{Q} = \mathsf{Q}_-\Theta \ . \tag{B.2.44}$$

For the *transformed* residual force [cf. (11.2.21)], we thus find

$$\Theta\mathscr{f}_\mu = -\mathrm{i}\,\Theta\mathsf{QL}\mathscr{G}_\mu = -\mathrm{i}\mathsf{Q}_-\mathsf{L}_-\Theta\mathscr{G}_\mu = -\varepsilon_\mu\mathrm{i}\mathsf{Q}_-\mathsf{L}_-\mathscr{G}^\dagger_{\mu_-} \ ,$$

and therefore

$$\Theta\mathscr{f}_\mu = -\varepsilon_\mu\mathscr{f}^\dagger_{\mu_-} \tag{B.2.45}$$

furthermore,

$$\Theta\mathscr{f}_\mu(t) = \Theta \mathrm{e}^{+\mathrm{i}\mathsf{QL}t}\mathscr{f}_\mu = \mathrm{e}^{-\mathrm{i}\mathsf{Q}_-\mathsf{L}_-t}\,\Theta\mathscr{f}_\mu$$

leading to (c.f. Exercise 11.2)

$$\Theta\mathscr{f}_\mu(t) = -\varepsilon_\mu\mathscr{f}^\dagger_\mu(-t)_- \ . \tag{B.2.46}$$

Its correlation function, the unnormalized *memory function* [cf. (11.2.23)],

$$\begin{aligned} l_{\mu\nu}(t) &= (\mathscr{f}_\mu|\mathrm{e}^{\mathrm{i}\mathsf{QL}t}\mathscr{f}_\nu) = (\Theta\mathscr{f}_\mu \,\underline{|}\, \Theta\mathrm{e}^{\mathrm{i}\mathsf{QL}t}\mathscr{f}_\nu)^* \\ &= \varepsilon_\mu\varepsilon_\nu(\mathscr{f}^\dagger_{\mu_-} \,\underline{|}\, \mathrm{e}^{-\mathrm{i}\mathsf{Q}_-\mathsf{L}_-t}\mathscr{f}^\dagger_{\nu_-})^* \end{aligned} \tag{B.2.47}$$

thus has the time-reversal symmetries:

$$l_{\mu\nu}(t) = l^*_{\nu\mu}(-t) = \varepsilon_\mu\varepsilon_\nu l_{\mu\nu}(-t)_- \ , \tag{B.2.48}$$

where the first equation results from relation (11.3.11), i.e. the Hermiticity of QLQ. The second equation expresses the fact that $l_{\mu\nu}(t, \mathfrak{H})$ is even or odd in t and

$\mathfrak{H}$, depending on whether $\varepsilon_\mu \varepsilon_\nu = +1$ or -1. For Hermitian observables $\mathscr{G}_\mu$, $l_{\mu\nu}(t, \mathfrak{H})$ is real.

In a similar manner, we may discuss the behavior of other quantities with respect to exchange of indices, time reversal, and reversal of the magnetic field. These *reciprocity relations* of the linear theory are collected in the following table (Sect. B.2.2.3).

Using the formulas collected in the preceding Appendices B.1 and B.2, symmetry relations for quantities arising in nonlinear theories may also be derived.

B.2.2.3) Table of Symmetry Relations of the Linear Response

Definition	L or QLQ Hermitian	$\Theta\mathscr{H} = \mathscr{H}_-$ $\Theta\mathscr{G}_\mu = \varepsilon_\mu \mathscr{G}^\dagger_{\mu_-}$; $\mathscr{G}_\nu, \mathscr{G}^\dagger_\nu \in \{\mathscr{G}\}$
Sect. 8.3.3	$\chi_{\mu\nu} = \chi^*_{\nu\mu}$	$= \varepsilon_\mu \varepsilon_\nu \chi_{\mu\nu_-}$
Sect. 9.3.1	$\varphi_{\mu\nu}(t) = -\varphi^*_{\nu\mu}(-t)$	$= -\varepsilon_\mu \varepsilon_\nu \varphi_{\mu\nu}(-t)_-$
Sect. 9.3.2	$\Phi_{\mu\nu}(t) = \Phi^*_{\nu\mu}(-t)$	$= \varepsilon_\mu \varepsilon_\nu \Phi_{\mu\nu}(-t)_-$
Sect. 9.4.1	$\chi'_{\mu\nu}(\omega) = \chi'_{\nu^\dagger\mu^\dagger}(-\omega)$	$= \varepsilon_\mu \varepsilon_\nu \chi'_{\mu\nu}(-\omega)_-$
Sect. 9.4.1	$\chi''_{\mu\nu}(\omega) = -\chi''_{\nu^\dagger\mu^\dagger}(-\omega)$	$= -\varepsilon_\mu \varepsilon_\nu \chi''_{\mu\nu}(-\omega)_-$
Sect. 11.2.1	$g_{\mu\nu} = g^*_{\nu\mu}$	$= \varepsilon_\mu \varepsilon_\nu g_{\mu\nu_-}$
Sect. 11.2.1	$\omega_{\mu\nu} = \omega^*_{\nu\mu}$	$= -\varepsilon_\mu \varepsilon_\nu \omega_{\mu\nu_-}$
Sect. 11.3.2	$l_{\mu\nu}(t) = l^*_{\nu\mu}(-t)$	$= \varepsilon_\mu \varepsilon_\nu l_{\mu\nu}(-t)_-$
Sect. 11.3.2	$L'_{\mu\nu}(\omega) = L'_{\nu^\dagger\mu^\dagger}(-\omega)$	$= \varepsilon_\mu \varepsilon_\nu L'_{\mu\nu}(-\omega)_-$
Sect. 11.3.2	$L''_{\mu\nu}(\omega) = -L''_{\nu^\dagger\mu^\dagger}(-\omega)$	$= -\varepsilon_\mu \varepsilon_\nu L''_{\mu\nu}(-\omega)_-$

For $\mathscr{G}_\mu = \mathscr{G}^\dagger_\mu$:

$$\chi_{\mu\nu},\ \varphi_{\mu\nu}(t),\ \Phi_{\mu\nu}(t),\ g_{\mu\nu},\ \mathrm{i}\omega_{\mu\nu},\ l_{\mu\nu}(t) = \text{real}$$

and

Sect. 9.4.1 $\quad \chi_{\mu\nu}(\omega) = \varepsilon_\mu \varepsilon_\nu \chi_{\nu\mu}(\omega)_-$

Sect. 11.3.2 $\quad L_{\mu\nu}(\omega) = \varepsilon_\mu \varepsilon_\nu L_{\nu\mu}(\omega)_-$

(Onsager–Casimir reciprocity relations)

The minus index means $F_-(\mathfrak{H}) = F(-\mathfrak{H})$

Solutions to the Exercises*

2.1(p. 11). From the eigenvalue equation of ρ

$$\rho|v_m\rangle = r_m|v_m\rangle \left(0 \leqq r_m \leqq 1, \sum_m r_m = 1\right)$$

(m denumerates the eigenstates $|v_m\rangle$), it follows that

$$\mathrm{Tr}\,\rho^2 = \sum_m r_m^2 \leqq \left(\sum_m r_m\right)^2 = (\mathrm{Tr}\,\rho)^2 = 1 \ .$$

The equals sign is valid only when one eigenvalue $r_1 = 1$ and therefore all the other eigenvalues vanish, i.e. when ρ represents a pure state $\mathscr{P}_{|\varphi\rangle}$.

3.1(p. 16). For a spin, $s = \frac{1}{2}$, the four dyads $||u_m\rangle\langle u_{m'}|)$ obtained from the two eigenstates $|u_m\rangle$ of σ_z form a complete basis set in the associated four-dimensional Liouville space. Instead of the dyads, one may also use the independent operators 1, σ_x, σ_y, and σ_z, as may best be seen from the representation of the Pauli spin matrices. An arbitrary operator, in particular the statistical operator ρ, may therefore be expressed as a linear combination

$$\rho = a_0 1 + \sum_i a_i \sigma_i \ .$$

From the well-known properties of the spin-$\frac{1}{2}$ operators

$$\mathrm{Tr}\, 1 = 2, \quad \mathrm{Tr}\,\sigma_i = 0, \quad \mathrm{Tr}\,\sigma_i\sigma_k = \frac{\hbar^2}{2}\delta_{ik}$$

it follows that

$$1 = \mathrm{Tr}\,\rho = 2a_0$$

$$\langle\sigma_i\rangle = \mathrm{Tr}\,\rho\sigma_i = a_i \mathrm{Tr}\,\sigma_i^2 = a_i\frac{\hbar^2}{2} \ ,$$

*The exercises marked with an asterisk are intended to complement and extend the theory presented in the text.

so that one obtains the form

$$\rho = \frac{1}{2} 1 + \frac{2}{\hbar^2} \sum_i \langle s_i \rangle s_i$$

of an arbitrary statistical operator ρ for a spin $s = \frac{1}{2}$.

3.2(p. 16). For $\mathcal{X}(c) = \exp[(c/\hbar)\mathcal{H}]\mathcal{F} \cdot \exp[-(c/\hbar)\mathcal{H}]$, differentiation with respect to c yields the following differential equation:

$$\frac{d}{dc}\mathcal{X}(c) = \frac{1}{\hbar}[\mathcal{H}, \mathcal{X}(c)] \ ,$$

which, using the definition of the Liouville operator (3.1.12), can also be written in the form

$$\frac{d}{dc}\mathcal{X}(c) = \mathsf{L}\mathcal{X}(c) \ .$$

Formal integration of this equation yields for $\mathcal{X}(c)$

$$\mathcal{X}(c) = e^{c\mathsf{L}}\mathcal{X}(0) = e^{c\mathsf{L}}\mathcal{F} \ .$$

On inserting the definition of $\mathcal{X}(c)$, one obtains from this the identity (3.1.14).

3.3(p. 20). The explicitly time-dependent Hamiltonian of a forced oscillator is given by

$$\mathcal{H}(t) = \frac{p^2}{2m} + \frac{m\omega^2}{2} x^2 - xK(t) \ .$$

The Heisenberg equations of motion,

$$\frac{dx^{\mathrm{H}}}{dt} = \frac{p^{\mathrm{H}}(t)}{m}, \qquad \frac{dp^{\mathrm{H}}(t)}{dt} = -m\omega^2 x^{\mathrm{H}}(t) + K(t)1 \ ,$$

have the same solutions as in classical mechanics in this simple case:

$$x^{\mathrm{H}}(t) = x(0)\cos\omega t + \frac{1}{m\omega} p(0)\sin\omega t + \frac{1}{m\omega}\int_0^t \sin\omega(t-t')K(t')dt' 1 \ ,$$

$$p^{\mathrm{H}}(t) = -m\omega x(0)\sin\omega t + p(0)\cos\omega t + \int_0^t \cos\omega(t-t')K(t')dt' 1 \ ,$$

as can be verified by differentiation. On inserting this solution, one obtains the following result for the Hamiltonian in the Heisenberg formulation:

$$\begin{aligned}\mathscr{H}^{\mathrm{H}}(t) &= \frac{m\omega^2}{2}\,x^2(0) + \frac{1}{2m}\,p^2(0) - x(0)\left\{\omega\int_0^t \sin\omega t'\, K(t')\,dt' + K(t)\cos\omega t\right\}\\ &\quad + \frac{p(0)}{m\omega}\left\{\omega\int_0^t \cos\omega t'\, K(t')\,dt' - K(t)\sin\omega t\right\}\\ &\quad + \frac{1}{2m}\iint_0^t \cos\omega(t''-t')K(t'')K(t')\,dt''\,dt'\,1\\ &\quad - \frac{K(t)}{m\omega}\int_0^t \sin\omega(t-t')K(t')\,dt'\,1 \ .\end{aligned}$$

This result differs considerably from the simple Hamiltonian in the Schrödinger formulation

$$\mathscr{H}^{\mathrm{S}}(t) = \frac{m\omega^2}{2}\,x^2(0) + \frac{1}{2m}\,p^2(0) - x(0)K(t) \ ,$$

although for both formulations, similar equations hold:

$$\frac{d\mathscr{H}^{\mathrm{H}}(t)}{dt} = \left(\frac{\partial\mathscr{H}^{\mathrm{H}}}{\partial t}\right)_{\mathrm{ex}} = -\,x^{\mathrm{H}}(t)\dot{K}(t) \ ,$$

$$\frac{d\mathscr{H}^{\mathrm{S}}(t)}{dt} = \left(\frac{\partial\mathscr{H}^{\mathrm{S}}}{\partial t}\right)_{\mathrm{ex}} = -\,x(0)\dot{K}(t) \ .$$

This difference between $\mathscr{H}^{\mathrm{S}}(t)$ and $\mathscr{H}^{\mathrm{H}}(t)$ is the reason that the Schrödinger formulation is preferred for treating explicitly time-dependent problems (cf. Part 2), although the two formulations are of course equivalent.

3.4(p. 26). We assume that the reduced statistical operator ρ^1 is a projection operator onto a pure state $|\varphi_0^1\rangle$. Then the following relation holds:

$$\rho^1 = \mathrm{Tr}_2\,\mathscr{P}_{|\Phi\rangle} = |\varphi_0^1\rangle\langle\varphi_0^1| \ .$$

If we introduce an orthogonal basis $|\varphi_\nu^1\rangle$ for Σ^1, which contains $|\varphi_0^1\rangle$, and use an orthogonal basis $|\varphi_\alpha^2\rangle$ for Σ^2, we find

$$\begin{aligned}\langle\varphi_\nu^1|\rho^1\varphi_\nu^1\rangle &= \sum_\alpha \langle\varphi_\nu^1\varphi_\alpha^2|\mathscr{P}_{|\Phi\rangle}|\varphi_\nu^1\varphi_\alpha^2\rangle\\ &= \sum_\alpha |\langle\varphi_\nu^1\varphi_\alpha^2|\Phi\rangle|^2 = \delta_{\nu 0} \ , \qquad \text{i.e.}\end{aligned}$$

$$\langle\varphi_\nu^1\varphi_\alpha^2|\Phi\rangle = 0$$

for all α, if $\nu \neq 0$. Since $|\varphi_\nu^1 \varphi_\alpha^2\rangle$ represents a basis for the total system, we thus find by writing the expansion for $|\Phi\rangle$

$$|\Phi\rangle = \sum_{\nu\alpha} |\varphi_\nu^1 \varphi_\alpha^2\rangle\langle\varphi_\nu^1 \varphi_\alpha^2|\Phi\rangle$$

$$= \sum_{\alpha} |\varphi_0^1 \varphi_\alpha^2\rangle\langle\varphi_0^1 \varphi_\alpha^2|\Phi\rangle = |\varphi_0^1\rangle \sum_{\alpha} |\varphi_\alpha^2\rangle\langle\varphi_0^1 \varphi_\alpha^2|\Phi\rangle$$

$$= |\varphi_0^1\rangle|\psi^2\rangle \ .$$

This means that the statistical operator ρ^1 obtained from $\mathscr{P}_{|\Phi\rangle}$ is only then a projector on a pure state, when $|\Phi\rangle$ is a product state. Therefore, in the general case, one requires a statistical operator ρ^1 to describe a subsystem, even when the total system is in a pure state. No system is completely isolated from its surroundings; thus, this statement can also be used to justify the introduction of statistical operators in general.

3.5(p. 29). If the differential equation for $\mathsf{U}(t, t_0)$

$$\frac{\partial}{\partial t} \mathsf{U}(t, t_0) = -\mathrm{i}\mathsf{L}(t)\mathsf{U}(t, t_0)$$

is differentiated with respect to t_0 and the order of the derivatives is interchanged, we obtain

$$\frac{\partial}{\partial t}\frac{\partial}{\partial t_0} \mathsf{U}(t, t_0) = -\mathrm{i}\mathsf{L}(t)\frac{\partial}{\partial t_0} \mathsf{U}(t, t_0) \ .$$

This differential equation for $\partial\mathsf{U}/\partial t_0$ can be integrated with the aid of $\mathsf{U}(t, t_0)$:

$$\frac{\partial}{\partial t_0} \mathsf{U}(t, t_0) = \mathsf{U}(t, t_0)\frac{\partial}{\partial t_0} \mathsf{U}(t, t_0)|_{t=t_0} \ .$$

The initial value, $\partial\mathsf{U}(t, t_0)/\partial t_0|_{t=t_0}$ is found from

$$\mathsf{U}(t, t_0) = 1 - \mathrm{i}\int_{t_0}^{t} \mathsf{L}(t')\mathsf{U}(t', t_0)\,dt'$$

by differentiating with respect to t_0 to be

$$\frac{\partial}{\partial t_0} \mathsf{U}(t, t_0)|_{t=t_0} = \mathrm{i}\mathsf{L}(t_0) \ .$$

With this, we obtain

$$\frac{\partial}{\partial t_0} \mathsf{U}(t, t_0) = \mathsf{U}(t, t_0)\mathrm{i}\mathsf{L}(t_0) \ .$$

4.1(p. 36). a) The spin density $\vec{s}(\mathrm{r})$ is defined by

$$\vec{s}(\mathrm{r}) = \sum_{\nu=1}^{N} \vec{s}_\nu \delta(\mathrm{r} - \vec{x}_\nu) .$$

Using the commutation relations of the spin operators,

$$[s_{\nu x}, s_{\mu y}] = -\frac{\hbar}{\mathrm{i}} \delta_{\nu\mu} s_{\nu z}, \ldots$$

and taking into account that

$$\delta(\mathrm{r} - \vec{x}_\nu)\delta(\mathrm{r}' - \vec{x}_\nu) = \delta(\mathrm{r} - \mathrm{r}')\delta(\mathrm{r} - \vec{x}_\nu)$$

we find after a simple calculation

$$[s_x(\mathrm{r}), s_y(\mathrm{r}')] = -\frac{\hbar}{\mathrm{i}} \delta(\mathrm{r} - \mathrm{r}') s_z(\mathrm{r}), \ldots .$$

b) For the components of the momentum current density,

$$j_\alpha(\mathrm{r}) = \frac{1}{2} \sum_{\nu=1}^{N} \{p_{\nu\alpha} \delta(\mathrm{r} - \vec{x}_\nu) + \text{c.c.}\}$$

we find, applying the commutation relation

$$[p_{\nu\alpha}, f(\mathrm{r} - \vec{x}_\mu)] = -\frac{\hbar}{\mathrm{i}} \delta_{\nu\mu} \frac{\partial}{\partial x_\alpha} f(\mathrm{r} - \vec{x}_\mu)$$

and using the property of the δ-function

$$\mathscr{F}(\vec{x}_\mu)\delta(\mathrm{r} - \vec{x}_\mu) = \mathscr{F}(\mathrm{r})\delta(\mathrm{r} - \vec{x}_\mu)$$

the following commutation relations:

$$[j_\alpha(\mathrm{r}), j_\beta(\mathrm{r}')] = \frac{\hbar}{\mathrm{i}} \left\{ \frac{\partial}{\partial x_\beta} \delta(\mathrm{r} - \mathrm{r}') j_\alpha(\mathrm{r}') - \frac{\partial}{\partial x'_\alpha} \delta(\mathrm{r} - \mathrm{r}') j_\beta(\mathrm{r}) \right\} .$$

Macroscopic Density: If we convert to macroscopic densities (by averaging with a function $f(\mathrm{r}_1)$, which essentially vanishes except within a volume element ΔV),

$$\mathscr{J}_\alpha(\mathrm{r}) = \int_{-\infty}^{\infty} f(\mathrm{r} - \mathrm{r}_1) j_\alpha(\mathrm{r}_1) d^3x_1 ,$$

then, by integration, we obtain the commutation relations

$$[\mathscr{I}_\alpha(\mathbf{r}), \mathscr{I}_\beta(\mathbf{r}')] = \frac{\hbar}{\mathrm{i}}\int d^3x_1 d^3x_2 f(\mathbf{r}-\mathbf{r}_1)f(\mathbf{r}'-\mathbf{r}_2)$$

$$\times\left\{\frac{\partial}{\partial x_{1\beta}}\delta(\mathbf{r}_1-\mathbf{r}_2)\mathscr{j}_\alpha(\mathbf{r}_2) - \frac{\partial}{\partial x_{2\alpha}}\delta(\mathbf{r}_1-\mathbf{r}_2)\mathscr{j}_\beta(\mathbf{r}_1)\right\}$$

$$= \frac{\hbar}{\mathrm{i}}\left\{\frac{\partial}{\partial x_\beta}\int d^3x_1 f(\mathbf{r}-\mathbf{r}_1)f(\mathbf{r}'-\mathbf{r}_1)\mathscr{j}_\alpha(\mathbf{r}_1)\right.$$

$$\left. - \frac{\partial}{\partial x'_\alpha}\int d^3x_1 f(\mathbf{r}-\mathbf{r}_1)f(\mathbf{r}'-\mathbf{r}_1)\mathscr{j}_\beta(\mathbf{r}_1)\right\} .$$

If the function $f(\mathbf{r}_1)$ is nonvanishing only within a cube $\Delta V = L^3$, then the functions $f(\mathbf{r}-\mathbf{r}_1)$ and $f(\mathbf{r}'-\mathbf{r}_1)$ overlap only when the distance $|\mathbf{r}-\mathbf{r}'|$ is smaller than the body diagonal, $\sqrt{3}L$. At larger distances, the macroscopic densities commute. Independent of the choice of the particular function $f(\mathbf{r}_1)$, we thus find approximately

$$[\mathscr{I}_\alpha(\mathbf{r}), \mathscr{I}_\beta(\mathbf{r}')] \approx 0 \quad \text{for} \quad |\mathbf{r}-\mathbf{r}'| > (\Delta V)^{1/3}.$$

***4.2**(p. 42). Due to the double sum over i and j in the Hamiltonian, we can assume without restriction that $J_{ij} = J(\mathbf{r}_i - \mathbf{r}_j)$ is symmetric, $J_{ij} = J_{ji}$.

In the Heisenberg formulation, employing $\sigma_{\pm j} = \sigma_{xj} \pm \mathrm{i}\sigma_{yj}$, we have

$$\left\langle \sum_{j=1}^{N} \sigma_{xj} \right\rangle(t) = \sum_{j=1}^{N} \mathrm{Tr}(\sigma_{xj}(t)\rho(0)) = \frac{1}{2}\sum_{j=1}^{N} \mathrm{Tr}(\sigma_{+j}(t)\rho(0)) + \text{h.c.} ,$$

where $\sigma_{+j}(t)$ and $\sigma_{zj}(t)$ are given by

$$\frac{d}{dt}\sigma_{+j}(t) = \frac{\mathrm{i}}{\hbar}[\mathscr{H}, \sigma_{+j}] = 2\mathrm{i}\sum_{l(\neq j)} J_{jl}\sigma_{zl}(t)\sigma_{+j}(t) \qquad \text{and}$$

$$\frac{d}{dt}\sigma_{zl}(t) = \frac{\mathrm{i}}{\hbar}[\mathscr{H}, \sigma_{zl}] = 0 .$$

From this, we find for $\sigma_{+j}(t)$

$$\sigma_{+j}(t) = \exp\left(2\mathrm{i}\sum_{l(\neq j)} J_{jl}\sigma_{zl}t\right)\sigma_{+j}(0)$$

and for the desired expectation value, $\langle \sigma_{+j} \rangle(t)$,

$$\begin{aligned}
\operatorname{Tr} \sigma_{+j}(t)\rho(0) &= 0 + \frac{\beta\gamma H_x}{\operatorname{Tr} 1} \sum_{i=1}^{N} \operatorname{Tr}\left\{\sigma_{+j}(t)\frac{\sigma_{+i}+\sigma_{-i}}{2}\right\} \\
&= \frac{\beta\gamma H_x}{2\operatorname{Tr} 1} \operatorname{Tr}\left\{\prod_{l(\neq j)} \mathrm{e}^{2\mathrm{i}J_{lj}\sigma_{zl}t}\sigma_{+j}\sigma_{-j}\right\} \\
&= \beta\gamma H_x \frac{\hbar^2}{4} \prod_{l(\neq j)} \cos(J_{lj}\hbar t) \ .
\end{aligned}$$

From this, we calculate the final result

$$\left\langle \sum_{j=1}^{N} \sigma_{xj} \right\rangle(t) = \beta\gamma H_x \frac{\hbar^2}{4} \sum_{j=1}^{N} \prod_{l(\neq j)} \cos(J_{lj}\hbar t) \ ,$$

and, in particular, for $J_{ij} = J_0(N)$,

$$\left\langle \sum_{j=1}^{N} \sigma_{xj} \right\rangle(t) = \beta\gamma H_x \frac{\hbar^2}{4} N(\cos J_0\hbar t)^{N-1} \ .$$

We find for the expectation value of $\mathscr{H}$

$$\begin{aligned}
\langle \mathscr{H} \rangle(t) = \langle \mathscr{H} \rangle(0) = \operatorname{Tr}(\mathscr{H}\rho(0)) &= 0 - \frac{\beta}{\operatorname{Tr} 1} \operatorname{Tr} \mathscr{H}^2 \\
&= -2\beta\left(\frac{\hbar^2}{4}\right)^2 \sum_{i\neq j} (J_{ij})^2 \ ,
\end{aligned}$$

so that in a model with $J_{ij} = J_0(N)$, we must set

$$J_0(N) = \frac{1}{\sqrt{N-1}} J$$

in order to guarantee that $\langle \mathscr{H} \rangle \sim N$. In this case, in the thermodynamic limit we thus have

$$\begin{aligned}
\lim_{N\to\infty} \frac{1}{N} \left\langle \sum_{j=1}^{N} \sigma_{xj} \right\rangle(t) &= \lim_{N\to\infty} \beta\gamma H_x \frac{\hbar^2}{4} \exp\left[(N-1)\ln\left(\cos\frac{J\hbar t}{\sqrt{N-1}}\right)\right] \\
&= \beta\gamma H_x \frac{\hbar^2}{4} \exp\left(-\frac{J^2\hbar^2}{2} t^2\right) \ .
\end{aligned}$$

This physical example thus clearly illustrates the behavior as a function of time which was described in general in Sect. 4.2.

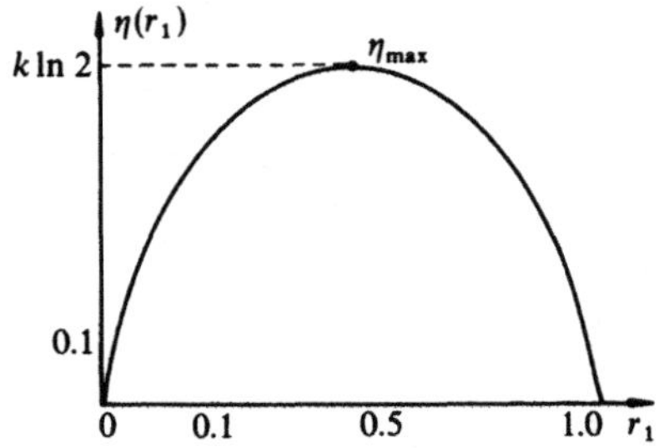

Fig. E.1. The uncertainty measure $\eta[\rho] = -k\sum_{m=1}^{2} r_m \ln r_m$ as a function of the eigenvalue r_1

5.1 (p. 50). In a space $\mathfrak{U}_2$, the uncertainty measure $\eta[\rho]$ of the statistical operator ρ is given as a function of its eigenvalues r_1 and r_2 by

$$\eta = -kr_1 \ln r_1 - kr_2 \ln r_2 \quad (r_1 + r_2 = 1; 0 \leqq r_m \leqq 1)$$

$$= -k[r_1 \ln r_1 + (1 - r_1) \ln(1 - r_1)] \ .$$

For $r_1 = 0$ or $r_2 = 0$, the uncertainty measure vanishes; we have a pure state. For $r_1 = r_2 = \frac{1}{2}$, we obtain the maximum value of the uncertainty measure, $\eta_{\max} = k \ln 2$ (cf. Fig. E.1).

In a space $\mathfrak{U}_3$, the uncertainty measure $\eta[\rho]$ is given as a function of its eigenvalues r_1, r_2, and r_3 by

$$\eta = -k\sum r_m \ln r_m \qquad (\sum r_m = 1; \quad 0 \leqq r_m \leqq 1) \ .$$

Figure E.2 shows some curves of constant uncertainty measure in a triangular coordinate system. The maximum uncertainty measure $\eta_{\max} = k \ln 3$, is at the

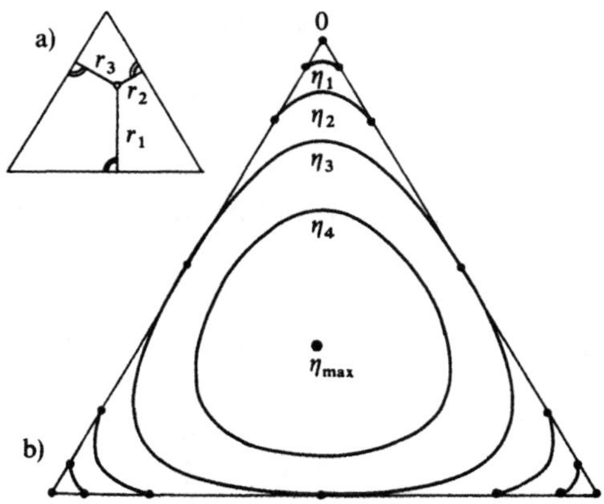

Fig. E.2. **(a)** Definition of the triangular coordinates r_1, r_2, r_3 $(0 \leqq r_m \leqq 1)$, $\sum_{m=1}^{3} r_m = 1$) **(b)** Curves of constant uncertainty measure $\eta[\rho] = -k\sum_{m=1}^{3} r_m \ln r_m$ in the triangular coordinate system: $\eta_1 = k(1/3)\ln 2$; $\eta_2 = k(2/3)\ln 2$; $\eta_3 = k \ln 2$, $\eta_4 = k(4/3) \ln 2$; $\eta_{\max} = k \ln 3$

midpoint ($r_1 = r_2 = r_3 = \frac{1}{3}$). The value $\eta = k \ln 2$ holds along the closed curve which just touches the limiting triangle. The curves for $\eta = \text{const.} < k \ln 2$ are curve segments in the corners of the triangle. The vertices of the triangle correspond to the pure states, $\eta = 0$.

5.2(p. 50). The observable $\mathscr{F}$ has simple eigenvalues,

$$\mathscr{F}|u_F\rangle = F|u_F\rangle \ ,$$

so that the probabilities w_F are given by

$$w_F = \langle u_F|\rho u_F\rangle \ .$$

Due to the assumption $[\mathscr{M}, \rho] = 0$, the probabilities w_M obey

$$w_M = \langle v_m|\rho v_m\rangle = r_m \ ,$$

where $|v_m\rangle$ are the common eigenvectors of $\mathscr{M}$ and ρ and r_m are the eigenvalues of ρ. For the difference $\eta_{\mathscr{F}} - \eta_{\mathscr{M}}$, we thus find, using

$$\sum_F |u_F\rangle\langle u_F| = 1 \quad \text{and}$$

$$\rho = \sum_m |v_m\rangle r_m \langle v_m| \ ,$$

the following result:

$$\begin{aligned}\eta_{\mathscr{F}} - \eta_{\mathscr{M}} &= k\left\{\sum_M w_M \ln w_M - \sum_F w_F \ln w_F\right\} \\ &= k \sum_{m,F} |\langle u_F|v_m\rangle|^2 r_m \{\ln r_m - \ln w_F\} \ .\end{aligned}$$

Using the inequality

$$\ln x \geqq 1 - \frac{1}{x}$$

and choosing $x = r_m/w_F$, we thus obtain

$$\eta_{\mathscr{F}} - \eta_{\mathscr{M}} \geqq k \sum_{mF} |\langle u_F|v_m\rangle|^2 (r_m - w_F) = k\left(\sum_m r_m - \sum_F w_F\right) = k(1-1) = 0 \ .$$

5.3(p. 50). a) The results R = rain, S = snow, and 0 = no precipitation have the following probabilities on June 1:

$$w_R = 0.4 \ , \qquad w_S = 0 \ , \qquad w_0 = 0.6 \ ;$$

and on January 1, the probabilities

$$w_R = 0.65 \ , \qquad w_S = 0.15 \ , \qquad w_0 = 0.2 \ .$$

The measure of the indeterminateness of the weather prediction (a) is therefore (k set equal to 1)

$$\eta^{(a)}_{\text{June}} = -0.4\ln 0.4 - 0 - 0.6\ln 0.6 = 0.6730$$

$$\eta^{(a)}_{\text{Jan}} = -0.65\ln 0.65 - 0.15\ln 0.15 - 0.2\ln 0.2 = 0.8865 \ .$$

It follows from $\eta^{(a)}_{\text{Jan}} > \eta^{(a)}_{\text{June}}$ that the weather prediction (a) is more uncertain in January than in June.

b) The events P = precipitation, 0 = no precipitation have on June 1 the probabilities

$$w_P = 0.4 \ , \qquad \text{and} \qquad w_0 = 0.6 \ ,$$

and on January 1, the probabilities

$$w_P = 0.8 \ , \qquad \text{and} \qquad w_0 = 0.2 \ .$$

The indeterminateness of prediction (b) is therefore

$$\eta^{(b)}_{\text{June}} = -0.4\ln 0.4 - 0.6\ln 0.6 = 0.6730$$

$$\eta^{(b)}_{\text{Jan}} = -0.8\ln 0.8 - 0.2\ln 0.2 = 0.5004 \qquad \text{i.e.}$$

$$\eta^{(b)}_{\text{Jan}} < \eta^{(b)}_{\text{June}} \ .$$

Weather prediction (b) is thus more uncertain in June than in January.

5.4(p. 61). Due to the possibility of carrying out linear transformations within an observation level, we may write

$$\mathcal{R}_{x,x^2,p,p^2} = \exp(-\lambda_0 - \lambda_1 x - \lambda_2 x^2 - \lambda_3 p - \lambda_4 p^2)$$

equally well in the form

$$\mathcal{R} = \exp[-\mu_0 - \mu_1(b^\dagger(a) - \alpha^*)(b(a) - \alpha)] \ .$$

The quantities μ_0, μ_1, α, α^*, and a then take on the role of the Lagrange multipliers which are calculated from

$$\operatorname{Tr}\mathcal{R} = 1 \ ; \qquad \operatorname{Tr}(\mathcal{R}x) = \langle x\rangle \ ; \qquad \operatorname{Tr}(\mathcal{R}p) = \langle p\rangle \ ;$$

$$\operatorname{Tr}(\mathcal{R}x^2) - (\operatorname{Tr}\mathcal{R}x)^2 = (\Delta x)^2 \ ; \qquad (\operatorname{Tr}\mathcal{R}p^2) - (\operatorname{Tr}\mathcal{R}p)^2 = (\Delta p)^2 \ .$$

If one takes into account the fact that the operators $\mathscr{b}^\dagger - \alpha^*$ and $\mathscr{b} - \alpha$ act formally like the creation and annihilation operators of a harmonic oscillator, the Lagrange multipliers are found as functions of the given expectation values to be

$$\mu_0 = \ln\left(\frac{\Delta x\,\Delta p}{\hbar} + \frac{1}{2}\right), \quad \mu_1 = \ln\left(1 + \frac{1}{\dfrac{\Delta x\,\Delta p}{\hbar} - \dfrac{1}{2}}\right),$$

$$\operatorname{Re}\alpha = \sqrt{\frac{\Delta p}{2\hbar\,\Delta x}}\langle x\rangle, \qquad \operatorname{Im}\alpha = \sqrt{\frac{\Delta x}{2\hbar\,\Delta p}}\langle p\rangle, \qquad a = \frac{\Delta p}{\Delta x}.$$

***5.5**(p. 61). The generalized canonical statistical operator with $\mathscr{G}_\nu \to \mathscr{N}_j = a_j{}^\dagger a_j$ is given according to (5.2.6) by

$$\mathscr{R}_{\{\mathscr{N}_j\}} = \frac{\exp\left(-\sum_j \lambda_j \mathscr{N}_j\right)}{\operatorname{Tr}\exp\left(-\sum_j \lambda_j \mathscr{N}_j\right)}.$$

The relation between the Lagrange multipliers λ_j and the expectation values $\langle \mathscr{N}_j\rangle$ can be calculated directly from

$$\langle \mathscr{N}_j\rangle = -\frac{\partial \ln Z}{\partial \lambda_j} \qquad \text{with}$$

$$Z = \operatorname{Tr}\exp\left(-\sum_j \lambda_j \mathscr{N}_j\right) = \sum_{n_1=0,1;\, n_2=0,1;\,\dots} \exp\left(-\sum_j \lambda_j n_j\right)$$
$$= \prod_j [1 + \exp(-\lambda_j)].$$

One obtains

$$\langle \mathscr{N}_j\rangle = \frac{1}{\exp(\lambda_j) + 1} \qquad \text{or} \qquad \lambda_j = \ln\frac{1 - \langle \mathscr{N}_j\rangle}{\langle \mathscr{N}_j\rangle}$$

and thus, after eliminating λ_j, for the entropy (5.2.19):

$$S_{\{\mathscr{N}_j\}} = k\left\{\sum_j \lambda_j \langle \mathscr{N}_j\rangle + \ln Z\right\} = k\sum_j\left\{\langle \mathscr{N}_j\rangle \ln\frac{1 - \langle \mathscr{N}_j\rangle}{\langle \mathscr{N}_j\rangle} + \ln\frac{1}{1 - \langle \mathscr{N}_j\rangle}\right\}$$
$$= -k\sum_j \{(1 - \langle \mathscr{N}_j\rangle)\ln(1 - \langle \mathscr{N}_j\rangle) + \langle \mathscr{N}_j\rangle \ln\langle \mathscr{N}_j\rangle\}.$$

***5.6**(p. 61). Only the unitary space with two modes is considered. The generalized canonical statistical operator over the relevant subset $b_i^\dagger b_j$, $i, j = 1, 2$, is

$$\mathscr{R} = \frac{\exp\left(-\sum_{lj} \lambda_{lj} b_j^\dagger b_l\right)}{\mathrm{Tr}\exp\left(-\sum_{lj} \lambda_{lj} b_j^\dagger b_l\right)} ,$$

where the matrix λ_{ij} is Hermitian and determined by $\langle b_i^\dagger b_j \rangle$. $\mathscr{R}$ can be put in the form

$$\mathscr{R} = \frac{\exp\left(-\sum_{k} \lambda_k a_k^\dagger a_k\right)}{\mathrm{Tr}\exp\left(-\sum_{k} \lambda_k a_k^\dagger a_k\right)} ,$$

by diagonalizing λ_{ij} using a unitary matrix φ_{kj},

$$\lambda_{lj} = \sum_{k=1,2} \varphi_{kl} \lambda_k \varphi_{kj}^*$$

and introducing new Bose operators

$$a_k = \sum_i \varphi_{ki} b_i .$$

With the above form, all of the expectation values can be readily calculated. One obtains

$$\mathrm{Tr}(a_k^\dagger a_l \mathscr{R}) = \delta_{kl} \frac{1}{\mathrm{e}^{\lambda_k} - 1} .$$

On the other hand, we have

$$\mathrm{Tr}(a_k^\dagger a_l \mathscr{R}) = \sum_{ij} \varphi_{ki}^* \varphi_{lj} \mathrm{Tr}(b_i^\dagger b_j \mathscr{R}) = \sum_{ij} \varphi_{ki}^* \varphi_{lj} \langle b_i^\dagger b_j \rangle ,$$

so that the matrix of the expectation values, $\langle b_i^\dagger b_j \rangle$, is also diagonalized by φ_{kj}, and the eigenvalues of $\langle b_i^* b_j \rangle$ just give $\langle a_k^\dagger a_k \rangle$:

$$\langle a_k^\dagger a_k \rangle = \frac{1}{\mathrm{e}^{\lambda_k} - 1} = \frac{1}{2}\{\langle \mathscr{N}_1 \rangle + \langle \mathscr{N}_2 \rangle \pm \sqrt{(\langle \mathscr{N}_1 \rangle - \langle \mathscr{N}_2 \rangle)^2 + 4j \langle \mathscr{N}_1 \rangle \langle \mathscr{N}_2 \rangle}\} .$$

Therefore, the entropy, which follows from (5.2.19), is

$$S = k \sum_{l} \left\{ \lambda_l \langle a_l^\dagger a_l \rangle + \ln \frac{1}{1 - \exp(-\lambda_l)} \right\}$$

$$= \sum_{l=1}^{2} L(\langle a_l^\dagger a_l \rangle) \ , \qquad \text{where}$$

$$L(x) = k\{(x+1)\ln(x+1) - x \ln x\} \ .$$

We have thus expressed the entropy through the expectation values $\langle \mathcal{b}_i^\dagger \mathcal{b}_j \rangle = \langle \mathcal{N}_i \rangle$ and the degree of coherence, j.

5.7(p. 70). The system Σ^1 is taken to have a nondecomposable observation level, with partial energies $\mathcal{H}_\nu^1$ ($\nu = 1, \ldots, n$); the same is true of system Σ^2, with partial energies $\mathcal{H}_\mu^2$ ($\mu = 1, \ldots, n$). If Σ^1 and Σ^2 are thermally isolated from one another, the set of partial energies $\mathcal{H}_\nu^1$, $\mathcal{H}_\mu^2$ (ν, $\mu = 1, \ldots, n$) forms an observation level for the total system, i.e. the total system is described by the generalized canonical statistical operator

$$\mathcal{R}(\beta_\nu^1, \beta_\mu^2) = \frac{\exp(-\sum \beta_\nu^1 \mathcal{H}_\nu^1)}{\operatorname{Tr} \exp(-\sum \beta_\nu^1 \mathcal{H}_\nu^1)} \frac{\exp(-\sum \beta_\mu^2 \mathcal{H}_\mu^2)}{\operatorname{Tr} \exp(-\sum \beta_\mu^2 \mathcal{H}_\mu^2)} \ .$$

If an interaction energy $\mathcal{H}^{12}$ between Σ^1 and Σ^2 is switched on during a time interval Δt,

$$\mathcal{H} = \sum_{\nu} \mathcal{H}_\nu^1 + \sum_{\mu} \mathcal{H}_\mu^2 + \mathcal{H}^{12} \ ,$$

then the partial energies change according to

$$\Delta \langle \mathcal{H}_\nu^1 \rangle = \langle \mathcal{H}_\nu^1 \rangle(t + \Delta t) - \langle \mathcal{H}_\nu^1 \rangle(t) = \operatorname{Tr} \mathcal{R}(e^{iL\Delta t} \mathcal{H}_\nu^1 - \mathcal{H}_\nu^1)$$

and correspondingly for the partial energies $\Delta \langle \mathcal{H}_\mu^2 \rangle$.

Choosing an interaction $\mathcal{H}^{12}$ which commutes with all the $\mathcal{H}_\nu^1$, and $\mathcal{H}_\mu^2$ ($\nu, \mu \neq 1$),

$$\mathsf{L}^{12} \mathcal{H}_\nu^1 = 0 \quad \text{and} \quad \mathsf{L}^{12} \mathcal{H}_\mu^2 = 0 \quad \text{for} \quad \nu, \mu \neq 1 \ ,$$

we find also

$$\mathsf{L} \mathcal{H}_\nu^1 = 0 \quad \text{and} \quad \mathsf{L} \mathcal{H}_\mu^2 = 0 \quad \text{for} \quad \nu, \mu \neq 1 \ ,$$

so that only $\Delta \langle \mathcal{H}_1^1 \rangle$ and $\Delta \langle \mathcal{H}_1^2 \rangle$ are nonvanishing. An exchange of energy occurs only between $\mathcal{H}_1^1$ and $\mathcal{H}_1^2$. All the other partial energies $\langle \mathcal{H}_\nu^1 \rangle$ and $\langle \mathcal{H}_\mu^2 \rangle$ ($\nu, \mu \neq 1$) remain constant.

To show that the Lagrange multipliers β_ν are legitimate *temperatures* in the sense of the Zeroth Law of thermodynamics, in the nondecomposable case also, we consider the composite system made up of Σ^1 and Σ^2, in which in the case mentioned above, energy exchange occurs only between $\mathscr{H}_1^1$ and $\mathscr{H}_1^2$. We consider the partial energies to be sufficient as an observation level for the description of the total system. However, due to the energetic contact, as in the case of an ordinary thermal contact, $\mathscr{H}_1^1 + \mathscr{H}_1^2$ must be considered as *one* partial energy, which is associated with *one* Lagrange multiplier β_1. Comparison of the statistical operators before and after the energy exchange thus shows that the result of the contact can be summarized by $\beta_1^1 = \beta_1^2 = \beta_1$. All other β's of Σ^1 and Σ^2 may remain different. Then β_1 is a contact variable for the chosen thermal equilibrium between $\mathscr{H}_1^1$ and $\mathscr{H}_1^2$, i.e. β_1 can be interpreted as an empirical temperature. The same arguments naturally are valid for every other β_ν within a nondecomposable observation level.

6.1(p. 74). a) We treat the time evolution in the Heisenberg formulation:

$$x^{\mathrm{H}}(t) = \mathrm{e}^{\mathrm{i}\mathrm{L}t} x = x + p\frac{t}{m} \; ; \qquad p^{\mathrm{H}}(t) = \mathrm{e}^{\mathrm{i}\mathrm{L}t} p = p \; ,$$

so that the expectation values at time t, calculated with $\rho(0) = \mathscr{R}(0)$, are given by

$$\langle p\rangle(t) = \langle p\rangle(0) \; , \qquad \langle x\rangle(t) = \langle x\rangle(0) + \langle p\rangle(0)\frac{t}{m} \; ,$$

$$(\Delta p)(t) = (\Delta p)(0) \; , \qquad (\Delta x)^2(t) = (\Delta x)^2(0) + (\Delta p)^2(0)\frac{t^2}{m^2} \; .$$

The generalized canonical statistical operator at time t constructed from them is

$$\mathscr{R}_{x,x^2,p,p^2}(t) = \exp[-\mu_0(t) - \mu_1(t)(b^\dagger - \alpha^*(t))(b - \alpha(t))]$$

and has the Lagrange multipliers (cf. Exercise 5.4)

$$\mu_0(t) = \ln\left(\frac{\Delta x(t)\,\Delta p(t)}{\hbar} + \frac{1}{2}\right) \; ; \qquad \mu_1(t) = \ln\left(1 + \frac{1}{\dfrac{\Delta x(t)\,\Delta p(t)}{\hbar} - \dfrac{1}{2}}\right) \; ;$$

$$\mathrm{Re}\,\alpha(t) = \sqrt{\frac{\Delta p(t)}{2\hbar\,\Delta x(t)}}\langle x\rangle(t) \; , \qquad \mathrm{Im}\,\alpha = \sqrt{\frac{\Delta x(t)}{2\hbar\,\Delta p(t)}}\langle p\rangle(t) \; ,$$

$$a(t) = \frac{\Delta p(t)}{\Delta x(t)} \; .$$

b) After the time evolution, we find

$$\mathrm{Tr}\{\rho(t_1)(px + xp)\} = \mathrm{Tr}\{\mathscr{R}(0)(p^{\mathrm{H}}(t_1)x^{\mathrm{H}}(t_1) + x^{\mathrm{H}}(t_1)p^{\mathrm{H}}(t_1))\}$$

$$= 2\langle x\rangle(0)\langle p\rangle(0) + 2\langle p^2\rangle(0)\frac{t_1}{m} \; .$$

On the other hand, the generalized canonical statistical operator $\mathscr{R}(t_1)$ constructed according to (a) for time t_1 gives the following result:

$$\mathrm{Tr}\{\mathscr{R}(t_1)(\not{p}\not{x} + \not{x}\not{p})\} = 2\langle\not{x}\rangle(0)\langle\not{p}\rangle(0) + 2\langle\not{p}\rangle^2(0)\frac{t_1}{m} .$$

The difference is thus

$$\frac{2}{m}(\Delta p)^2(0)t_1 .$$

c) With $\mathscr{R}(t_1)$ as the new initial state at time t_1, the von Neumann equation yields for $t_2 \geqq t_1$

$$\rho(t) = \mathrm{e}^{-\mathrm{i}L(t_2-t_1)}\mathscr{R}(t_1)$$

and therefore for the expectation values of $\mathscr{F} = \not{x}, \not{x}^2, \not{p}$, and $\not{p}^2$

$$\mathrm{Tr}(\mathscr{F}\rho(t)) = \mathrm{Tr}\{\mathscr{F}\cdot\mathrm{e}^{-\mathrm{i}L(t_2-t_1)}\mathscr{R}(t_1)\} = \mathrm{Tr}\{\mathrm{e}^{\mathrm{i}L(t_2-t_1)}\mathscr{F}\cdot\mathscr{R}(t_1)\} .$$

The time dependences can be evaluated, with the following result:

$$\mathrm{e}^{\mathrm{i}L(t_2-t_1)}\not{p} = \not{p} ;$$

$$\mathrm{e}^{\mathrm{i}L(t_2-t_1)}\not{p}^2 = \not{p}^2 ;$$

$$\mathrm{e}^{\mathrm{i}L(t_2-t_1)}\not{x} = \not{x} + \not{p}\frac{t_2-t_1}{m} ;$$

$$\mathrm{e}^{\mathrm{i}L(t_2-t_1)}\not{x}^2 = \not{x}^2 + \not{p}^2\frac{(t_2-t_1)^2}{m^2} + (\not{p}\not{x} + \not{x}\not{p})\frac{t_2-t_1}{m} .$$

The values of $\mathrm{Tr}\{\mathscr{F}(t_2)\rho(t_2)\}$, which are being sought, have thus been expressed in terms of $\mathrm{Tr}\{\mathscr{F}\mathscr{R}(t_1)\}$, which are given in parts (a) and (b). For $t_2 \geqq t_1$, we obtain

$$\langle\not{p}\rangle(t_2) = \mathrm{Tr}\{\not{p}\mathscr{R}(t_1)\} = \langle\not{p}\rangle(0) ;$$

$$\langle\not{p}^2\rangle(t_2) = \mathrm{Tr}\{\not{p}^2\mathscr{R}(t_1)\} = \langle\not{p}^2\rangle(0) ;$$

$$\langle\not{x}\rangle(t_2) = \mathrm{Tr}\{\not{x}\mathscr{R}(t_1)\} + \frac{t_2-t_1}{m}\mathrm{Tr}\{\not{p}\mathscr{R}(t_1)\} = \langle\not{x}\rangle(0) + \langle\not{p}\rangle(0)\frac{t_2}{m} ;$$

$$\begin{aligned}\langle\not{x}^2\rangle(t_2\rangle &= \mathrm{Tr}\{\not{x}^2\mathscr{R}(t_1)\} + \frac{(t_2-t_1)^2}{m^2}\mathrm{Tr}\{\not{p}^2\mathscr{R}(t_1)\} \\ &\quad + \frac{t_2-t_1}{m}\mathrm{Tr}\{(\not{p}\not{x} + \not{x}\not{p})\mathscr{R}(t_1)\} \\ &= \langle\not{x}^2\rangle(0) + 2\langle\not{x}\rangle(0)\langle\not{p}\rangle(0)\frac{t_2}{m} \\ &\quad + \langle\not{p}^2\rangle(0)\frac{t_2^2}{m^2} - (\Delta p)^2(0)\frac{2t_1}{m^2}(t_2-t_1) .\end{aligned}$$

The dynamics, however, which occur between $t = 0$ and $t = t_2$ following the von Neumann equation, have already been calculated in (a). Comparison shows that the transition from $\rho(t_1)$ to $\mathcal{R}(t_1)$ changes only the expectation value $\langle x^2\rangle(t_2)$, which is given by a calculation as in (a):

$$\langle x^2\rangle(t_2) = \mathrm{Tr}\{x^2 \mathrm{e}^{-\mathrm{i}\mathsf{L}t_2}\rho(0)\}$$

$$= \langle x^2\rangle(0) + 2\langle x\rangle(0)\langle p\rangle(0)\frac{t_2}{m} + \langle p^2\rangle(0)\frac{t_2^2}{m^2} \ .$$

7.1(p. 89). For the decomposition of $\mathsf{L}(t)$ into $\mathsf{L}_1(t) = -\mathsf{M}\hat{h}\theta(t)$ and $\mathsf{L}_2 = \mathsf{L}$, the identity (3.4.4b) takes on the form:

$$\mathsf{U}(t, t_0) = \mathsf{U}_1(t, t_0) - \mathrm{i}\int_{t_0}^{t} \mathsf{U}_1(t, t')\mathsf{L}\mathsf{U}(t', t_0)\,dt' \ .$$

The evolution operator $\mathsf{U}_1(t, t_0)$ in this expression is determined by

$$\frac{d}{dt}\mathsf{U}_1(t, t_0) = -\mathrm{i}\mathsf{L}_1(t)\mathsf{U}_1(t, t_0)$$

with $\mathsf{U}_1(t_0, t_0) = \mathbf{1}$. Using $[\mathsf{L}_1(t), \mathsf{L}_1(t')] = 0$, it may be written as

$$\mathsf{U}_1(t, t_0) = \exp\left(\mathrm{i}\mathsf{M}\hat{h}\int_{t_0}^{t}\theta(t')\,dt'\right) \ .$$

In the limit $t_0 = -0$, $t = +0$, it therefore follows that

$$\mathsf{U}(+0, -0) = \mathsf{U}_1(+0, -0) = \mathbf{1} \ .$$

The statistical operator thus remains unchanged immediately after the sudden switching on of a field:

$$\rho(+0) = \mathsf{U}(+0, -0)\rho(-0) = \rho(-0) \ .$$

This is approximately valid for very short switching times (sudden approximation).

***7.2**(p. 90). a) In analogy to the solution of Exercise 7.1, we find on decomposing $\mathsf{L}(t)$ into $\mathsf{L}_1(t) = -\mathsf{M}\hat{h}\delta(t)$ and $\mathsf{L}_2 = \mathsf{L}$ the evolution operator

$$\mathsf{U}_1(t, t_0) = \exp\left(\mathrm{i}\mathsf{M}\hat{h}\int_{t_0}^{t}\delta(t')\,dt'\right) ,$$

so that in the limit $t_0 = -0$, $t = +0$ from the integral equation for $\mathsf{U}(t, t_0)$, we obtain

$$\mathsf{U}(+0, -0) = \mathsf{U}_1(+0, -0) = \mathrm{e}^{\mathrm{i}\mathsf{M}\hat{h}} \quad \text{or}$$

$$\rho(+0) = \mathrm{e}^{\mathrm{i}\mathsf{M}\hat{h}}\rho(-0) \ .$$

b) For times t between the δ-function field spikes, $t_\nu < t < t_{\nu+1}$, the equation

$$\frac{\partial}{\partial t}\mathsf{U}(t, t_\nu + 0) = -\,\mathrm{i}\mathsf{L}\mathsf{U}(t, t_\nu + 0)$$

holds; then

$$\mathsf{U}(t, t_\nu + 0) = \mathrm{e}^{-\mathrm{i}\mathsf{L}(t - t_\nu)} .$$

Together with the result from (a) for $\mathsf{U}(t_\nu + 0, t_\nu - 0)$, it follows for $t > t_n$

$$\begin{aligned}\mathsf{U}(t, t_0) &= \mathsf{U}(t, t_n + 0)\mathsf{U}(t_n + 0, t_n - 0)\mathsf{U}(t_n - 0, t_{n-1} + 0)\dots\mathsf{U}(t_1 - 0, t_0)\\ &= \mathrm{e}^{-\mathrm{i}\mathsf{L}(t - t_n)}\mathrm{e}^{\mathrm{i}\mathsf{M}\hat{h}^{(n)}}\mathrm{e}^{-\mathrm{i}\mathsf{L}(t_n - t_{n-1})}\dots\mathrm{e}^{-\mathrm{i}\mathsf{L}(t_1 - t_0)} .\end{aligned}$$

The statistical operator $\rho(t)$ for $t > t_n$ and $\mathsf{L}\rho(t_0) = \mathsf{L}\mathscr{R}_\beta = \mathcal{O}$ is then

$$\rho = \mathrm{e}^{-\mathrm{i}\mathsf{L}(t - t_n)}\mathrm{e}^{\mathrm{i}\mathsf{M}\hat{h}^{(n)}}\mathrm{e}^{-\mathrm{i}\mathsf{L}(t_n - t_{n-1})}\mathrm{e}^{\mathrm{i}\mathsf{M}\hat{h}^{(n-1)}}\mathrm{e}^{-\mathrm{i}\mathsf{L}(t_{n-1} - t_{n-2})}\dots\mathrm{e}^{\mathrm{i}\mathsf{M}\hat{h}^{(1)}}\mathscr{R}_\beta .$$

If we take the expectation value of an observable $\mathscr{F}$ with this operator, we see how $\langle\mathscr{F}\rangle(t) = \mathrm{Tr}\{\mathscr{F}\rho(t)\}$ depends on all the $\hat{h}^{(\nu)}$ [compare (7.1.3)].

8.1(p. 95). The element $|\tilde{\mathscr{A}}_2)$ is a linear combination of $\mathscr{A}_1$ and $\mathscr{A}_2$. Therefore, we have

$$(\mathscr{A}_1|\tilde{\mathscr{A}}_2) = \frac{1}{\sqrt{\Delta_2\Delta_1}}\begin{vmatrix}(\mathscr{A}_1|\mathscr{A}_1) & (\mathscr{A}_1|\mathscr{A}_1)\\ (\mathscr{A}_1|\mathscr{A}_2) & (\mathscr{A}_1|\mathscr{A}_2)\end{vmatrix} = 0 .$$

In the same way, one can see that $\tilde{\mathscr{A}}_3$ is orthogonal to $\mathscr{A}_1$ and $\mathscr{A}_2$ and thus to $\tilde{\mathscr{A}}_2$.

Only the component with $\mathscr{A}_2$ contributes to the normalization of $\tilde{\mathscr{A}}_2$:

$$(\tilde{\mathscr{A}}_2|\tilde{\mathscr{A}}_2) = \frac{1}{\sqrt{\Delta_2\Delta_1}}(\mathscr{A}_1|\mathscr{A}_1)(\mathscr{A}_2|\tilde{\mathscr{A}}_2) = 1 .$$

The corresponding argument applies to the normalization of $|\tilde{\mathscr{A}}_3)$,

$$(\tilde{\mathscr{A}}_3|\tilde{\mathscr{A}}_3) = \frac{1}{\sqrt{\Delta_3\Delta_2}}\Delta_2(\mathscr{A}_3|\tilde{\mathscr{A}}_3) = 1 .$$

The determinants Δ_ν are determinants of a Hermitian matrix and therefore real. The given expressions for $|\tilde{\mathscr{A}}_\nu)$ may be obtained from (8.1.21 and 8.1.22). They are useful for dealing with the projection operator P onto the space $\mathfrak{L}_{\{\mathscr{A}\}}$.

8.2(p. 100) a) It follows from the definition (8.2.5) with $\mathscr{H}|n\rangle = E_n|n\rangle$ (n denotes the states) that

$$(\mathscr{F}|\mathscr{G}) = \frac{1}{\beta}\frac{1}{Z}\int_0^\beta \sum_{n,m} e^{-\beta E_n}\langle n|\mathscr{F}^\dagger m\rangle e^{-\alpha E_m}\langle m|\mathscr{G}n\rangle e^{\alpha E_n}\,d\alpha$$

$$= \frac{1}{Z}\sum_{n,m}\frac{\exp(-\beta E_m)-\exp(-\beta E_n)}{\beta(E_n-E_m)}\langle n|\mathscr{F}^\dagger m\rangle\langle m|\mathscr{G}n\rangle \ .$$

b) In the expression

$$(\mathscr{F}|\mathscr{F}) = \frac{1}{\beta}\frac{1}{Z}\int_0^\beta \sum_{n,m} e^{-\beta E_n}e^{\alpha(E_n-E_m)}|\langle m|\mathscr{F}n\rangle|^2\,d\alpha$$

with $\mathscr{G} = \mathscr{F}$, all the terms in the sum are $\geqq 0$. Therefore, $(\mathscr{F}|\mathscr{F}) \geqq 0$. $(\mathscr{F}|\mathscr{F})$ can be zero only when each of the terms in the sum is zero, i.e. when the matrix elements $\langle m|\mathscr{F}n\rangle$ are zero for all m and n. However, since the $|n\rangle$ represent a complete basis set, this is only the case when $\mathscr{F} = \mathscr{O}$.

8.3(p. 100). From the definition of the scalar product, (8.2.5), using $\langle\mathscr{A}\rangle_\beta^* = \langle\mathscr{A}^\dagger\rangle_\beta$, we obtain

$$(\mathscr{F}|\mathscr{G})^* = \frac{1}{\beta}\int_0^\beta \mathrm{Tr}(\mathscr{R}_\beta e^{\alpha\mathscr{H}}\mathscr{G}^\dagger e^{-\alpha\mathscr{H}}\mathscr{F})\,d\alpha \ ,$$

from which, with $[\mathscr{R}_\beta, e^{\alpha\mathscr{H}}] = \mathscr{O}$ and the invariance of the trace to cyclic permutations, we have

$$(\mathscr{F}|\mathscr{G})^* = (\mathscr{G}|\mathscr{F}) \ .$$

If the integration variable α in $(\mathscr{F}|\mathscr{G})^*$ is replaced by $\alpha' = \beta - \alpha$ we obtain the second relation contained in (8.2.10),

$$(\mathscr{F}|\mathscr{G})^* = \frac{1}{\beta}\int_0^\beta \mathrm{Tr}(e^{-\alpha'\mathscr{H}}\mathscr{G}^\dagger e^{\alpha'\mathscr{H}}\mathscr{R}_\beta\mathscr{F})\,d\alpha'$$

$$= \frac{1}{\beta}\int_0^\beta \mathrm{Tr}(\mathscr{R}_\beta\mathscr{F}e^{-\alpha'\mathscr{H}}\mathscr{G}^\dagger e^{\alpha'\mathscr{H}})\,d\alpha' = (\mathscr{F}^\dagger|\mathscr{G}^\dagger) \ .$$

8.4(p. 100). From the invariance of the trace to cyclic permutations and $[\mathscr{H}, \mathscr{R}_\beta] = \mathscr{O}$, it follows for arbitrary $\mathscr{F}$ and $\mathscr{G}$ that

$$(\mathscr{F}|\mathsf{L}\mathscr{G}) = \frac{1}{\beta}\int_0^\beta \mathrm{Tr}\left\{\mathscr{R}_\beta\mathscr{F}^\dagger e^{-\alpha\mathscr{H}}\frac{1}{\hbar}[\mathscr{H},\mathscr{G}]e^{\alpha\mathscr{H}}\right\}d\alpha$$

$$= -\frac{1}{\beta}\int_0^\beta \mathrm{Tr}\left\{\mathscr{R}_\beta\frac{1}{\hbar}[\mathscr{H},\mathscr{F}^\dagger]e^{-\alpha\mathscr{H}}\mathscr{G}e^{\alpha\mathscr{H}}\right\}d\alpha \ .$$

Using the relation for the observables,

$$-\frac{1}{\hbar}[\mathscr{H},\mathscr{F}^\dagger] = \frac{1}{\hbar}[\mathscr{H},\mathscr{F}]^\dagger$$

we can demonstrate the Hermiticity of L in the Mori product:

$$(\mathscr{F}|\mathsf{L}\mathscr{G}) = (\mathsf{L}\mathscr{F}|\mathscr{G}) .$$

8.5(p. 100). a) From the definition (8.2.5) of the scalar product, it follows that

$$\begin{aligned}(\mathscr{F}|\mathsf{L}\mathscr{G}) &= -\frac{1}{\beta\hbar}\int_0^\beta \mathrm{Tr}\left\{\mathscr{R}_\beta\mathscr{F}^\dagger\frac{d}{d\alpha}(\mathrm{e}^{-\alpha\mathscr{H}}\mathscr{G}\mathrm{e}^{\alpha\mathscr{H}})\right\}d\alpha \\ &= \frac{1}{\beta\hbar}\mathrm{Tr}\{\mathscr{R}_\beta\mathscr{F}^\dagger\mathscr{G} - \mathscr{F}^\dagger\mathscr{R}_\beta\mathscr{G}\} \\ &= \frac{1}{\beta\hbar}\langle[\mathscr{F}^\dagger, \mathscr{G}]\rangle_\beta .\end{aligned}$$

b) From the rule for the differentiation of exponential operators, (3.4.11), we find for the derivative of a generalized canonical statistical operator

$$\mathscr{R}(\beta, \lambda) = \frac{1}{Z(\beta, \lambda)}\exp\left(-\beta\mathscr{H} - \sum_\nu \lambda_\nu\mathscr{F}_\nu^\dagger\right)$$

with respect to λ_ν:

$$\frac{\partial}{\partial\lambda_\nu}\mathscr{R}(\beta, \lambda)|_{\lambda=0} = -\frac{1}{\beta}\int_0^\beta \mathscr{R}_\beta\mathrm{e}^{\alpha\mathscr{H}}(\mathscr{F}_\nu^\dagger - \langle\mathscr{F}_\nu^\dagger\rangle_\beta)\mathrm{e}^{-\alpha\mathscr{H}}\,d\alpha ,$$

so that a Mori product $(\delta\mathscr{F}_\nu|\mathscr{G})$ can be written in the form

$$(\delta\mathscr{F}_\nu|\mathscr{G}) = -\mathrm{Tr}\left\{\left.\frac{\partial\mathscr{R}(\beta, \lambda)}{\partial\lambda_\nu}\right|_{\lambda=0}\mathscr{G}\right\} .$$

In particular, for $(\delta\mathscr{F}_\nu|\mathsf{L}\mathscr{G}) = (\mathscr{F}_\nu|\mathsf{L}\mathscr{G})$, it follows that

$$(\mathscr{F}_\nu|\mathsf{L}\mathscr{G}) = -\mathrm{Tr}\left\{\left.\frac{\partial\mathscr{R}(\beta, \lambda)}{\partial\lambda_\nu}\right|_{\lambda=0}\mathsf{L}\mathscr{G}\right\} = \mathrm{Tr}\left\{\mathsf{L}\left.\frac{\partial\mathscr{R}(\beta, \lambda)}{\partial\lambda_\nu}\right|_{\lambda=0}\mathscr{G}\right\} .$$

On the other hand, by differentiating the expression

$$\left[\mathscr{H} + \sum_\nu \frac{\lambda_\nu}{\beta}\mathscr{F}_\nu^\dagger, \mathscr{R}(\beta, \lambda)\right] = \mathscr{O}$$

with respect to λ_ν, we obtain

$$\mathsf{L}\left.\frac{\partial\mathscr{R}(\beta, \lambda)}{\partial\lambda_\nu}\right|_{\lambda=0} + \frac{1}{\beta\hbar}[\mathscr{F}_\nu^\dagger, \mathscr{R}_\beta] = \mathscr{O} ,$$

from which we find

$$(\mathscr{F}_v|\mathsf{L}\mathscr{G}) = \frac{1}{\beta\hbar}\,\mathrm{Tr}\{[\mathscr{R}_\beta, \mathscr{F}_v^\dagger]\mathscr{G}\} = \frac{1}{\beta\hbar}\langle[\mathscr{F}_v^\dagger, \mathscr{G}]\rangle_\beta \,.$$

With this method, one can in a simple manner find relations to the *higher derivatives of* $\mathscr{R}(\beta, \lambda)$.

8.6(p. 106). For $\beta \to 0$, the Mori product (8.2.5) is given by the expression $(\mathscr{F}|\mathscr{G})_{\beta\to 0} = \mathrm{Tr}(\mathscr{F}^\dagger\mathscr{G})/\mathrm{Tr}\,1$. The expansion of the adiabatic susceptibility (8.3.30) in powers of β then yields for the first nonvanishing term:

$$\chi^{\mathrm{ad}} = \beta\left\{\frac{\mathrm{Tr}(\mathscr{M}^2)}{\mathrm{Tr}\,1} - \frac{1}{\mathrm{Tr}\,1}\frac{[\mathrm{Tr}(\mathscr{M}\mathscr{H})]^2}{\mathrm{Tr}(\mathscr{H}^2)}\right\} .$$

The terms in $(1|\mathscr{M})$ and $(1|\mathscr{H})$ make no contribution in zeroth order, owing to $\mathrm{Tr}\,\mathscr{M} = 0$, $\mathrm{Tr}\,\mathscr{H} = 0$. To evaluate the above expression, it is expedient to write $\mathscr{H} = -H\mathscr{M} + \mathscr{H}_{\mathrm{int}}$. As may be seen directly or shown from time-reversal invariance, $\mathrm{Tr}(\mathscr{M}\mathscr{H}_{\mathrm{int}})$ is zero. With the definition of the Curie constant, $C = (1/Nk)\,\mathrm{Tr}(\mathscr{M}^2)/\mathrm{Tr}\,1$, and the abbreviation $kb = (1/N)\,\mathrm{Tr}(\mathscr{H}_{\mathrm{int}})^2/\mathrm{Tr}\,1$, we thus obtain the result

$$\chi^{\mathrm{ad}} = \beta\frac{\mathrm{Tr}(\mathscr{M}^2)}{\mathrm{Tr}\,1}\left\{1 - \frac{H^2\,\mathrm{Tr}(\mathscr{M}^2)}{H^2\,\mathrm{Tr}(\mathscr{M}^2) + \mathrm{Tr}(\mathscr{H}_{\mathrm{int}})^2}\right\} = N\frac{C}{T}\frac{b}{CH^2 + b} .$$

For the values of the constants, we find

$$C = \frac{1}{N}\frac{1}{k}\gamma^2\sum_{ij}\frac{\mathrm{Tr}(\sigma_i^z\sigma_j^z)}{\mathrm{Tr}\,1} = \frac{1}{k}\gamma^2\frac{1}{3}\frac{\mathrm{Tr}(\vec{\sigma}_1^2)}{\mathrm{Tr}\,1} = \frac{1}{k}\gamma^2\hbar^2\frac{s(s+1)}{3}$$

and

$$kb = \frac{1}{N}\sum_{\substack{i\neq j\\ k\neq l}}\frac{1}{\hbar^4}J_{ij}J_{kl}\frac{\mathrm{Tr}(\vec{\sigma}_i\vec{\sigma}_j)(\vec{\sigma}_k\vec{\sigma}_l)}{\mathrm{Tr}\,1} = \frac{1}{N}\frac{2}{\hbar^4}\sum_{i\neq j}J_{ij}^2\frac{\mathrm{Tr}(\vec{\sigma}_i\vec{\sigma}_j)^2}{\mathrm{Tr}\,1'}$$

$$= \frac{1}{N}\frac{2}{\hbar^4}\sum_{i\neq j}J_{ij}^2\frac{\mathrm{Tr}\{(\sigma_i^x)^2(\sigma_j^x)^2 + (\sigma_i^y)^2(\sigma_j^y)^2 + (\sigma_i^z)^2(\sigma_j^z)^2\}}{\mathrm{Tr}\,1}$$

$$= 2\frac{[s(s+1)]^2}{3}\frac{1}{N}\sum_{i\neq j}J_{ij}^2 .$$

9.1(p. 114). Using the notation $\omega_0 = \gamma H$, we have for the observables $\mathscr{S}^\mu$ ($\mu = +, -, 0$):

$$\mathsf{L}\mathscr{S}^\mu = \frac{1}{\hbar}[\mathscr{H}, \mathscr{S}^\mu] = -\mu\omega_0\mathscr{S}^\mu .$$

From this, it follows for the response function (9.1.10):

$$\varphi_{\nu\mu}(t) = -\frac{\mathrm{i}}{\hbar}\langle[\mathscr{S}^{-\nu}, \mathrm{e}^{\mathrm{iL}t}\mathscr{S}^{\mu}]\rangle_{\beta} = -\frac{\mathrm{i}}{\hbar}\mathrm{e}^{-\mathrm{i}\mu\omega_0 t}\langle[\mathscr{S}^{-\nu}, \mathscr{S}^{\mu}]\rangle_{\beta} .$$

In this way we obtain

$$\varphi_{++}(t) = \varphi_{--}(t)^* = 2\mathrm{i}\langle\mathscr{S}^{z}\rangle_{\beta}\mathrm{e}^{-\mathrm{i}\omega_0 t} .$$

All the other $\varphi_{\nu\mu}(t)$ vanish, since only $\langle\mathscr{S}^{z}\rangle_{\beta} \neq 0$. The response functions for the Cartesian components $\mathscr{S}^{x}$, $\mathscr{S}^{y}$, and $\mathscr{S}^{z}$ may be obtained as linear combinations of the $\varphi_{\nu\mu}(t)$ for $\nu, \mu = +, -, 0$.

9.2(p. 114). Applying Schwarz's inequality (8.1.6) to the definition (9.1.12) of $\varphi_{MF}(t)$, we obtain with $\mathsf{L} = \mathsf{L}^{\dagger}$

$$|\varphi_{MF}(t)|^2 \leqq \beta^2(\mathscr{M}|\mathscr{M})(\mathrm{i}\mathsf{L}\mathrm{e}^{\mathrm{iL}t}\mathscr{F}|\mathrm{i}\mathsf{L}\mathrm{e}^{\mathrm{iL}t}\mathscr{F}) = \beta^2(\mathscr{M}|\mathscr{M})(\mathsf{L}\mathscr{F}|\mathsf{L}\mathscr{F})$$

or

$$|\varphi_{MF}(t)|^2 \leqq \beta^2(\mathsf{L}\mathscr{M}|\mathsf{L}\mathscr{M})(\mathrm{e}^{\mathrm{iL}t}\mathscr{F}|\mathrm{e}^{\mathrm{iL}t}\mathscr{F}) = \beta^2(\mathsf{L}\mathscr{M}|\mathsf{L}\mathscr{M})(\mathscr{F}|\mathscr{F}) .$$

9.3(p. 118). The response function $\varphi(t)$ is found from (9.1.17) to be given by

$$\varphi(t) = \Phi(0)\Gamma\mathrm{e}^{-\Gamma t}, \quad t > 0 .$$

With $\mathscr{M} = \mathscr{M}^{\dagger}$, we find from this the response (9.2.8),

$$\langle\delta\mathscr{M}\rangle(t) = \hbar\Phi(0)\Gamma\,\mathrm{Im}\int_0^t \mathrm{e}^{\mathrm{i}\omega_0(t-t_1)-\Gamma t_1}\,dt_1$$

$$= \hbar\Phi(0)\left\{\frac{\Gamma^2}{\Gamma^2+\omega_0^2}\sin\omega_0 t - \frac{\omega_0\Gamma}{\Gamma^2+\omega_0^2}(\cos\omega_0 t - \mathrm{e}^{-\Gamma t})\right\} .$$

After the transients have died out ($t \gg \Gamma^{-1}$), the result (9.2.9) applies, with the susceptibility $\chi(\omega) = \Phi(0)(\Gamma - \mathrm{i}\omega_0)^{-1}$ which follows from $\varphi(t)$.

9.4(p. 120). Denote the eigenvectors of $\mathscr{H}$ by $|n\rangle$ and the eigenvalues by E_n; then the eigenelements of L can be represented by $|n\rangle\langle m|$. Thus, we find

$$\mathscr{Y} = f(\mathsf{L})\mathscr{X} = f(\mathsf{L})\sum_{n,m}|n\rangle\langle n|\mathscr{X}m\rangle\langle m| = \sum_{n,m} f\left(\frac{E_n - E_m}{\hbar}\right)|n\rangle\langle n|\mathscr{X}m\rangle\langle m| .$$

From this, it follows that

$$\mathscr{Y}^{\dagger} = \sum_{n,m} f^*\left(\frac{E_n - E_m}{\hbar}\right)|m\rangle\langle n|\mathscr{X}m\rangle^*\langle n|$$

$$= f^*(-\mathsf{L})\sum_{n,m}|m\rangle\langle n|\mathscr{X}m\rangle^*\langle n| = f^*(-\mathsf{L})\mathscr{X}^{\dagger} .$$

Specializing $f(\mathsf{L})$ to $e^{i\mathsf{L}t}$ leads to the immediately apparent result

$$(e^{i\mathsf{L}t}\mathscr{X})^\dagger = (e^{(i/\hbar)\mathscr{H}t}\mathscr{X}e^{(-i/\hbar)\mathscr{H}t})^\dagger = e^{i\mathsf{L}t}\mathscr{X}^\dagger \ .$$

For the other two functions, we obtain

$$\left(\frac{1}{z+\mathsf{L}}\mathscr{X}\right)^\dagger = \frac{1}{z^*-\mathsf{L}}\mathscr{X}^\dagger \quad \text{and} \quad (\delta(\mathsf{L}+\omega)\mathscr{X})^\dagger = \delta(\mathsf{L}-\omega)\mathscr{X}^\dagger \ .$$

9.5(p. 129). The relaxation function $\Phi(t)$ can be written in the entire range $t \geqq 0$ as $\Phi(t) = \tilde{\Phi}(t) + \Phi_1 e^{-\Gamma t}$, where $\tilde{\Phi}(t)$ vanishes for $t > \tau_1$, and $\Phi(0) = \tilde{\Phi}(0) + \Phi_1 = \chi^{\mathrm{T}}$. Therefore, in the region $\omega\tau_1 \ll 1$, we obtain for the susceptibility:

$$\begin{aligned}\chi(\omega) &= -\int_0^\infty \dot{\Phi}(t)e^{i\omega t}\,dt = -\int_0^{\tau_1} \dot{\tilde{\Phi}}(t)\{1 + \ \ldots\}\,dt + \Phi_1\Gamma\int_0^\infty e^{(-\Gamma+i\omega)t}\,dt \\ &= \tilde{\Phi}(0) + \Phi_1\frac{\Gamma}{\Gamma - i\omega} = \chi^{\mathrm{T}} - \Phi_1 + \Phi_1\frac{\Gamma}{\Gamma - i\omega} \ ,\end{aligned}$$

i.e. for the real and imaginary parts,

$$\chi'(\omega) = \Phi_1\frac{\Gamma^2}{\Gamma^2+\omega^2} + \chi^{\mathrm{T}} - \Phi_1 \ , \quad \chi''(\omega) = \Phi_1\frac{\omega\Gamma}{\Gamma^2+\omega^2} \ .$$

The quantity Φ_1 is, from (9.4.11), necessarily positive. The real part $\chi'(\omega)$ decreases with increasing ω from χ^{T} at $\omega = 0$ to a level $\chi^{\mathrm{T}} - \Phi_1$ which is maintained for frequencies $\Gamma \ll \omega \ll \tau_1^{-1}$. The function $\tilde{\Phi}(t)$ determines the behavior of $\chi(\omega)$ at higher frequencies, $\omega\tau_1 > 1$.

9.6(p. 129). a) From (9.4.11), we find the general relation

$$i\int_{-\infty}^{\infty} \chi''(\omega)e^{-i\omega t}\,d\omega = -\pi\beta(\delta\mathscr{M}|i\mathsf{L}e^{i\mathsf{L}t}\delta\mathscr{M}) = \pi\varphi(t) \ , \quad t \gtrless 0 \ .$$

Evaluating the integral with $\chi''(\omega) = \chi^{\mathrm{T}}\Gamma\omega(\Gamma^2+\omega^2)^{-1}$ yields $\varphi(t) = \chi^{\mathrm{T}}\Gamma e^{-\Gamma t}$ for $t > 0$ and $\varphi(t) = -\chi^{\mathrm{T}}\Gamma e^{\Gamma t}$ for $t < 0$.[1]

For $\Phi(t)$ and $\chi(\omega)$, we find by inserting $\varphi(t)$ into (9.1.18) or (9.1.27), respectively,

$$\Phi(t) = \chi^{\mathrm{T}}e^{-\Gamma|t|} \quad \text{and} \quad \chi(\omega) = \chi^{\mathrm{T}}\frac{\Gamma}{\Gamma - i\omega} \ .$$

[1] The computation of $\chi'(\omega)$ using the Kramers–Kronig relation (9.4.28) and reverse transforming $\chi(\omega)$ as in (9.1.30) yields $\varphi(t)$ only for $t > 0$.

b) The expansion of $\chi(z)$ in z^{-1} is given by

$$\frac{\chi(z)}{\chi^{\mathrm{T}}} = \frac{\Gamma}{\mathrm{i}z}\frac{1}{\frac{\Gamma}{\mathrm{i}z}-1} = \sum_{n=1}^{\infty} -\left(\frac{\Gamma}{\mathrm{i}z}\right)^n .$$

The derivatives of $\varphi(t)$ at $t=+0$ are found from $\chi^{\mathrm{T}}\Gamma\mathrm{e}^{-\Gamma t}$ to be $\varphi^n(+0) = \chi^{\mathrm{T}}(-1)^n\Gamma^{n+1}$, so that the relation

$$\chi(z) = \sum_{n=1}^{\infty} (\mathrm{i})^n \varphi^{(n-1)}(+0)\frac{1}{z^n}$$

is obeyed. In the computation of $\langle\omega^n\rangle$, we first obtain (with the aid of the residual theorem)

$$\frac{1}{\pi}\int_{-\infty}^{\infty} \chi^{\mathrm{T}}\frac{\Gamma}{\omega^2+\Gamma^2}\left(\frac{\mathrm{e}^{-\mathrm{i}\omega\Delta}-1}{-\mathrm{i}\Delta}\right)^n d\omega = \chi^{\mathrm{T}}\left(\frac{\mathrm{e}^{-\Gamma\Delta}-1}{-\mathrm{i}\Delta}\right)^n ,$$

so that in the limit $\Delta \to +0$, we have

$$\langle\omega^n\rangle = (\mathrm{i})^n(-1)^n\Gamma^n\chi^{\mathrm{T}} = -(\mathrm{i})^n\varphi^{(n-1)}(+0) ,$$

i.e. the last line of (9.4.48) is verified.

9.7(p. 129). Evaluation of the commutator yields the result for $\mathsf{L}n_{\mathfrak{k}}$:

$$\mathsf{L}n_{\mathfrak{k}} = \frac{1}{\hbar}\left[\mathscr{H}, \sum_j \mathrm{e}^{-\mathrm{i}\mathfrak{k}\vec{x}_j}\right] = -\frac{1}{2m}\sum_j\{(\mathfrak{k}\,\vec{p}_j)\mathrm{e}^{-\mathrm{i}\mathfrak{k}\vec{x}_j} + \mathrm{e}^{-\mathrm{i}\mathfrak{k}\vec{x}_j}(\mathfrak{k}\,\vec{p}_j)\} ,$$

so that, using the Kubo identity (8.2.18) for the second moment, we find

$$\langle\omega^2\rangle = \beta(n_{\mathfrak{k}}|\mathsf{L}^2 n_{\mathfrak{k}}) = \frac{1}{\hbar}\langle[n_{-\mathfrak{k}}, \mathsf{L}n_{\mathfrak{k}}]\rangle_\beta = \frac{Nk^2}{m} .$$

***9.8**(p. 134). With $\mathscr{A}_1 = \mathit{1}$, $\mathscr{A}_2 = \mathscr{H}$, $\mathscr{A}_3 = \mathscr{H}^2$, and the orthonormalized elements $\tilde{\mathscr{A}}_\nu$ from Exercise 8.1, we find

$$\beta(\mathscr{M}|\mathsf{Q}_{\{\mathit{1},\mathscr{H},\mathscr{H}^2\}}\mathscr{F}) = \beta(\mathscr{M}|\mathscr{F}) - \beta\sum_{\nu=1}^{3}(\mathscr{M}|\tilde{\mathscr{A}}_\nu)(\tilde{\mathscr{A}}_\nu|\mathscr{F}) .$$

The determinants which enter this expression may be written in terms of derivatives w.r.t. β, using $D = -\partial/\partial\beta = kT^2\,\partial/\partial T$ and the relation $\langle\mathscr{G}\mathscr{H}^n\rangle_\beta = (D+\langle\mathscr{H}\rangle_\beta)^n\langle\mathscr{G}\rangle_\beta$:

$$\begin{vmatrix} 1 & \langle\mathscr{G}\rangle_\beta \\ \langle\mathscr{H}\rangle_\beta & \langle\mathscr{G}\mathscr{H}\rangle_\beta \end{vmatrix} = \begin{vmatrix} 1 & 0 \\ 0 & D\langle\mathscr{G}\rangle_\beta \end{vmatrix} ,$$

$$\begin{vmatrix} 1 & \langle\mathscr{H}\rangle_\beta & \langle\mathscr{G}\rangle_\beta \\ \langle\mathscr{H}\rangle_\beta & \langle\mathscr{H}^2\rangle_\beta & \langle\mathscr{G}\mathscr{H}\rangle_\beta \\ \langle\mathscr{H}^2\rangle_\beta & \langle\mathscr{H}^3\rangle_\beta & \langle\mathscr{G}\mathscr{H}^2\rangle_\beta \end{vmatrix} = \begin{vmatrix} 1 & 0 & 0 \\ 0 & D\langle\mathscr{H}\rangle_\beta & D\langle\mathscr{G}\rangle_\beta \\ 0 & D^2\langle\mathscr{H}\rangle_\beta & D^2\langle\mathscr{G}\rangle_\beta \end{vmatrix} .$$

In this way, specializing $\mathscr{G}$ to $\mathscr{M}^\dagger$, $\mathscr{F}$, $\mathscr{H}$, and $\mathscr{H}^2$ and taking note of (8.3.3), we obtain

$$\frac{1}{\langle\mathscr{F}\rangle_\beta}\beta(\mathscr{M}|\mathsf{Q}_{\{1,\mathscr{H},\mathscr{H}^2\}}\mathscr{F}) = \frac{1}{\langle\mathscr{F}\rangle_\beta}\frac{\partial}{\partial h}\langle\mathscr{F}\rangle_\beta|_{h=0} - \beta\frac{D\langle\mathscr{M}^\dagger\rangle_\beta}{D\langle\mathscr{H}\rangle_\beta}\frac{D\langle\mathscr{F}\rangle_\beta}{\langle\mathscr{F}\rangle_\beta}$$

$$-\beta\frac{(D^2\langle\mathscr{M}^\dagger\rangle_\beta)D\langle\mathscr{H}_\beta\rangle - (D\langle\mathscr{M}^\dagger\rangle_\beta)D^2\langle\mathscr{H}\rangle_\beta}{2(D\langle\mathscr{H}\rangle_\beta)^3 + (D^3\langle\mathscr{H}\rangle_\beta)D\langle\mathscr{H}\rangle_\beta - (D^2\langle\mathscr{H}\rangle_\beta)^2}$$

$$\times\left\{\frac{D^2\langle\mathscr{F}\rangle_\beta}{\langle\mathscr{F}\rangle_\beta} - \frac{D^2\langle\mathscr{H}\rangle_\beta}{D\langle\mathscr{H}\rangle_\beta}\frac{D\langle\mathscr{F}\rangle_\beta}{\langle\mathscr{F}\rangle_\beta}\right\}.$$

In the thermodynamic limit for intensive (i.e. not extensive) field parameters, the quantities $\langle\mathscr{M}\rangle_\beta$ and $\langle\mathscr{H}\rangle_\beta$, which are derivatives of the free energy, become proportional to the volume V of the system.

The first two terms of

$$\frac{1}{\langle\mathscr{F}\rangle_\beta}\beta(\mathscr{M}|\mathsf{Q}_{\{1,\mathscr{H},\mathscr{H}^2\}}\mathscr{F})$$

therefore are found for $V\to\infty$ to have finite limiting values; in contrast, the term obtained from $\tilde{\mathscr{A}}_3$ vanishes as V^{-1} owing to $(D\langle\mathscr{H}\rangle_\beta)^3 \propto V^3$,

$$\lim_{V\to\infty}\frac{1}{\langle\mathscr{F}\rangle_\beta}\beta(\mathscr{M}|\mathsf{Q}_{\{1,\mathscr{H},\mathscr{H}^2\}}\mathscr{F}) = \lim_{V\to\infty}\frac{1}{\langle\mathscr{F}\rangle_\beta}\beta(\mathscr{M}|\mathsf{Q}_{\{1,\mathscr{H}\}}\mathscr{F}).$$

These considerations may be extended to higher powers of $\mathscr{H}$. This result does not, however, mean that for every system $\chi^{\text{is}} = \chi^{\text{ad}}$, since in the case of degenerate eigenvalues of $\mathscr{H}$, not all the constants of the motion can be given in terms of powers of $\mathscr{H}$.

9.9(p. 134). Setting $\mathscr{A}_1 = 1$, $\mathscr{A}_2 = \delta\mathscr{H}_1$, and $\mathscr{A}_3 = \delta\mathscr{H}_2$, we obtain from (8.1.17) the projector onto 1, $\mathscr{H}_1$, and $\mathscr{H}_2$ in the form $\sum_{\nu\mu}|\mathscr{A}_\nu)g_{\nu\mu}(\mathscr{A}_\mu|$, where $\Delta = \det(\mathscr{A}_\nu|\mathscr{A}_\mu)$ represents the inverse matrix of $(\mathscr{A}_\nu|\mathscr{A}_\mu)$:

$$g_{\nu\mu} = \frac{1}{\Delta}\begin{pmatrix}\Delta & 0 & 0\\ 0 & (\delta\mathscr{H}_2|\delta\mathscr{H}_2) & -(\delta\mathscr{H}_1|\delta\mathscr{H}_2)\\ 0 & -(\delta\mathscr{H}_2|\delta\mathscr{H}_1) & (\delta\mathscr{H}_1|\delta\mathscr{H}_1)\end{pmatrix}.$$

Inserting $\mathscr{A}_\nu$ and $g_{\nu\mu}$ yields the explicit expression for

$$\beta(\mathscr{M}_\alpha|\mathsf{Q}_{\{1,\mathscr{H}_1,\mathscr{H}_2\}}\mathscr{F}) = \beta\left\{(\mathscr{M}_\alpha|\mathscr{F}) - \sum_{\nu\mu}(\mathscr{M}_\alpha|\mathscr{A}_\nu)g_{\nu\mu}(\mathscr{A}_\mu|\mathscr{F})\right\}.$$

For a generalized canonical statistical operator

$$\mathscr{R}(\beta_1, \beta_2, h_1, \ldots) = \frac{1}{Z(\beta_1, \beta_2, h_1 \ldots)} \times \exp[-\beta_1 \mathscr{H}_1(h_1 \ldots) - \beta_2 \mathscr{H}_2(h_1 \ldots)] ,$$

the heats are $DQ_1 = \mathrm{Tr}(d\mathscr{R}\mathscr{H}_1)$ and $DQ_2 = \mathrm{Tr}(d\mathscr{R}\mathscr{H}_2)$ [cf. (5.3.26)]. The variation of an expectation value $\langle \mathscr{F} \rangle$ is given by $d\langle \mathscr{F} \rangle = \mathrm{Tr}(d\mathscr{R}\mathscr{F})$. In particular, for $h = 0$ and $\beta_1 = \beta_2$, we find

$$d\mathscr{R} = \sum_\alpha \left.\frac{\partial \mathscr{R}}{\partial h_\alpha}\right|_{\substack{h=0 \\ \beta_1=\beta_2=\beta}} dh_\alpha + \left.\frac{\partial \mathscr{R}}{\partial \beta_1}\right|_{\substack{h=0 \\ \beta_1=\beta_2=\beta}} d\beta_1 + \left.\frac{\partial \mathscr{R}}{\partial \beta_2}\right|_{\substack{h=0 \\ \beta_1=\beta_2=\beta}} d\beta_2 ,$$

and, using the Mori product, we obtain for the adiabatic case:

$$0 = DQ_1 = \sum_\alpha \beta\, dh_\alpha(\mathscr{M}_\alpha|\delta\mathscr{H}_1) - d\beta_1(\delta\mathscr{H}_1|\delta\mathscr{H}_1) - d\beta_2(\delta\mathscr{H}_2|\delta\mathscr{H}_1) ,$$

$$0 = DQ_2 = \sum_\alpha \beta\, dh_\alpha(\mathscr{M}_\alpha|\delta\mathscr{H}_2) - d\beta_1(\delta\mathscr{H}_1|\delta\mathscr{H}_2) - d\beta_2(\delta\mathscr{H}_2|\delta\mathscr{H}_2) ,$$

where we have used

$$\mathscr{M}_\alpha^\dagger = - \left.\frac{\partial \mathscr{H}_1(h)}{\partial h_\alpha}\right|_{h=0} - \left.\frac{\partial \mathscr{H}_2(h)}{\partial h_\alpha}\right|_{h=0} .$$

Solving the equations for $d\beta_1$ and $d\beta_2$ and inserting in

$$d\langle \mathscr{F} \rangle = \sum_\alpha \beta\, dh_\alpha(\delta\mathscr{M}_\alpha|\mathscr{F}) - d\beta_1(\delta\mathscr{H}_1|\mathscr{F}) - d\beta_2(\delta\mathscr{H}_2|\mathscr{F})$$

leads to

$$\begin{aligned} \chi^{\mathrm{ad(1)ad(2)}}_{M_\alpha F} &= \beta(\mathscr{M}_\alpha|\mathscr{F}) - \beta\langle \mathscr{M}_\alpha^\dagger \rangle \langle \mathscr{F} \rangle \\ &\quad - \beta \frac{(\mathscr{M}_\alpha|\delta\mathscr{H}_1)(\delta\mathscr{H}_2|\delta\mathscr{H}_2) - (\mathscr{M}_\alpha|\delta\mathscr{H}_2)(\delta\mathscr{H}_2|\delta\mathscr{H}_1)}{(\delta\mathscr{H}_1|\delta\mathscr{H}_1)(\delta\mathscr{H}_2|\delta\mathscr{H}_2) - (\delta\mathscr{H}_1|\delta\mathscr{H}_2)(\delta\mathscr{H}_2|\delta\mathscr{H}_1)} (\delta\mathscr{H}_1|\mathscr{F}) \\ &\quad - \beta \frac{(\mathscr{M}_\alpha|\delta\mathscr{H}_2)(\delta\mathscr{H}_1|\delta\mathscr{H}_1) - (\mathscr{M}_\alpha|\delta\mathscr{H}_1)(\delta\mathscr{H}_1|\delta\mathscr{H}_2)}{(\delta\mathscr{H}_1|\delta\mathscr{H}_1)(\delta\mathscr{H}_2|\delta\mathscr{H}_2) - (\delta\mathscr{H}_1|\delta\mathscr{H}_2)(\delta\mathscr{H}_2|\delta\mathscr{H}_1)} (\delta\mathscr{H}_2|\mathscr{F}) , \end{aligned}$$

i.e. to the previously derived expression for $\beta(\mathscr{M}_\alpha|\mathsf{Q}_{\{I,\mathscr{H}_1,\mathscr{H}_2\}}\mathscr{F})$. When $\mathscr{H}_1$ and $\mathscr{H}_2$ are constants of the motion, i.e. when $[\mathscr{H}_1, \mathscr{H}_2] = \mathscr{O}$ holds, and only $\mathscr{H}_1$ and $\mathscr{H}_2$ couple in $\chi^{\mathrm{is}}_{M_\alpha F}$, then $\chi^{\mathrm{is}}_{M_\alpha F}$ may be represented as a double adiabatic susceptibility.

***9.10**(p. 134). In the energy representation (Exercise 8.2a), we find for

$$\chi^{\mathrm{is}}_{M_\alpha F} = \beta\left(\mathscr{M}_\alpha \middle| \frac{\mathsf{L}}{\mathsf{L} + i\eta} \mathscr{F} \right)$$

the relation

$$\chi^{\text{is}}_{M_\alpha F} = \frac{1}{Z}\sum_{m,n} \frac{\exp(-\beta E_m) - \exp(-\beta E_n)}{E_n - E_m} \langle n|\mathscr{M}^\dagger_\alpha m\rangle \frac{E_m - E_n}{E_m - E_n + \mathrm{i}\hbar\eta} \langle m|\mathscr{F} n\rangle$$

$$= \frac{1}{Z} \sum_{\substack{m,n \\ (E_n \neq E_m)}} \exp(-\beta E_m) \left\{ \langle m|\mathscr{F} n\rangle \frac{\langle n|\mathscr{M}^\dagger_\alpha m\rangle}{E_n - E_m} + \frac{\langle m|\mathscr{M}^\dagger_\alpha n\rangle}{E_n - E_m} \langle n|\mathscr{F} m\rangle \right\}.$$

Differentiating the eigenvalue equation $\mathscr{H}(h)|m(h)\rangle = E_m(h)|m(h)\rangle$ w.r.t. h_α or h^*_α yields, for $h = 0$ and $E_m \neq E_n$

$$\left\langle n \left| \frac{\partial}{\partial h_\alpha} m(h) \right\rangle \right|_{h=0} = \frac{\langle n|\mathscr{M}^\dagger_\alpha m\rangle}{E_n - E_m} \quad \text{or} \quad \frac{\partial}{\partial h_\alpha} \langle m(h)|n\rangle \Big|_{h=0} = \frac{\langle m|\mathscr{M}^\dagger_\alpha n\rangle}{E_n - E_m} .$$

With these expressions, we can eliminate the matrix elements of $\mathscr{M}^\dagger$ to obtain:

$$\chi^{\text{is}}_{M_\alpha F} = \frac{1}{Z}\sum_m \mathrm{e}^{-\beta E_m} \left\{ \sum_n \langle m|\mathscr{F} n\rangle \left\langle n \left| \frac{\partial}{\partial h_\alpha} m(h) \right\rangle \right|_{h=0} \right.$$

$$+ \sum_n \frac{\partial}{\partial h_\alpha} \langle m(h)|n\rangle \Big|_{h=0} \langle n|\mathscr{F} m\rangle$$

$$\left. - \sum_{\substack{n \\ (E_n = E_m)}} \langle m|\mathscr{F} n\rangle \left(\left\langle n \left| \frac{\partial}{\partial h_\alpha} m(h) \right\rangle \right|_{h=0} + \frac{\partial}{\partial h_\alpha} \langle n(h)|m\rangle \Big|_{h=0} \right) \right\} .$$

The first two sums over n may be carried out with the aid of the completeness relation. The last sum makes no contribution, owing to

$$\frac{\partial}{\partial h_\alpha} \langle n(h)|m(h)\rangle = \frac{\partial}{\partial h_\alpha} \delta_{mn} = 0 ,$$

so that χ^{is} can be written in the form:

$$\chi^{\text{is}}_{M_\alpha F} = \frac{1}{Z}\sum_m \mathrm{e}^{-\beta E_m} \frac{\partial}{\partial h_\alpha} \langle m(h)\,|\,\mathscr{F} m(h)\rangle \Big|_{h=0} .$$

In the limit of a slowly-varying field h_α, the expectation value of an arbitrary observable is given in response theory by

$$\langle \mathscr{F} \rangle(t) = \langle \mathscr{F} \rangle_\beta + \sum_\alpha \chi^{\text{is}}_{M_\alpha F} h_\alpha(t) .$$

Therefore, the representation of χ^{is} obtained above shows that the statistical operator in this case varies according to

$$\rho(t) = \frac{1}{Z}\sum_m \mathrm{e}^{-\beta E_m} \left\{ |m\rangle\langle m| + \sum_\alpha h_\alpha(t) \frac{\partial}{\partial h_\alpha} |m(h)\rangle\langle m(h)| \Big|_{h=0} \right\} ,$$

i.e. according to the linear approximation of

$$\frac{1}{Z}\sum_m \mathrm{e}^{-\beta E_m}\,|m(h(t))\rangle\langle m(h(t))|$$

("adiabatic theorem" of quantum mechanics).

From the eigenvalue equation of $\mathscr{H}(h)$, we obtain

$$\left\langle m(h)\left|\frac{\partial \mathscr{H}(h)}{\partial h_\gamma^*}\right. m(h)\right\rangle = \frac{\partial E_m(h)}{\partial h_\gamma^*} \ ,$$

so that, for $\mathscr{F} = \mathscr{M}_\gamma$, it follows that

$$\frac{\partial}{\partial h_\alpha}\langle m(h)|\,\mathscr{M}_\gamma m(h)\rangle\bigg|_{h=0} = -\left.\frac{\partial^2 E_m(h)}{\partial h_\alpha \partial h_\gamma^*}\right|_{h=0} + \left\langle m\left|\left.\frac{\partial^2 \mathscr{H}(h)}{\partial h_\alpha \partial h_\gamma^*}\right|_{h=0}\right. m\right\rangle \ .$$

Then $\chi_{\alpha\gamma}^{\mathrm{is}}$ takes on the form:

$$\chi_{\alpha\gamma}^{\mathrm{is}} = -\frac{1}{Z}\sum_m \left.\frac{\partial^2 E_m(h)}{\partial h_\alpha \partial h_\gamma^*}\right|_{h=0} \mathrm{e}^{-\beta E_m} + \left\langle \left.\frac{\partial^2 \mathscr{H}(h)}{\partial h_\alpha \partial h_\gamma^*}\right|_{h=0}\right\rangle_\beta \ .$$

This shows clearly that $\chi_{\alpha\gamma}^{\mathrm{is}}$ depends sensitively on small perturbations of the Hamiltonian, since they can strongly influence the curvature of the terms $E_m(h)$.

9.11(p. 134). For N independent ions, we obtain from the solution of Exercise 9.10

$$\chi_{MM}^{\mathrm{is}} = -N\frac{1}{Z_{\mathrm{ion}}}\sum_m \frac{\partial^2 \varepsilon_m(H)}{\partial H^2}\,\mathrm{e}^{-\beta\varepsilon_m(H)} \ ,$$

where $\varepsilon_m(H)$ denote the energy eigenvalues for one ion ($h = 0$). In the expansion in terms of β, the term

$$\sum_m \frac{\partial^2}{\partial H^2}\varepsilon_m(H) = \frac{\partial^2}{\partial H^2}\,\mathrm{Tr}_{\mathrm{ion}}(\mathscr{H}_{\mathrm{ion}}) = 0$$

makes no contribution, so that in the high-temperature limit, we obtain

$$\chi_{MM}^{\mathrm{is}} = N\beta\sum_m \frac{1}{3}\varepsilon_m(H)\frac{\partial^2}{\partial H^2}\varepsilon_m(H) \ .$$

For $\mathscr{H}_{\mathrm{ion}}$, we find the eigenvalues

$$\varepsilon_0(H) = -\frac{2}{3}D \ , \qquad \varepsilon_\pm(H) = \frac{D}{3} \pm \sqrt{E^2 + (\gamma\hbar H)^2} \ ,$$

so that, with the Curie constant from Exercise 8.6, $C = \frac{2}{3}k^{-1}\gamma^2\hbar^2$, the result can be cast in the form:

$$\chi_{MM}^{\mathrm{is}} = N\frac{C}{T}\frac{E^2}{E^2 + (\gamma\hbar H)^2} \ .$$

In a three dimensional unitary space with nondegenerate ε_m, there are only two constants of the motion in addition to the unit operator. Therefore, from Exercise 9.9 for the case under consideration, χ_{MM}^{is}, we can write

$$\chi_{MM}^{\mathrm{is}} = \beta(\mathcal{M}|\mathsf{Q}_{\{I,\mathcal{H}_1,\mathcal{H}_2\}}\mathcal{M}) = \chi^{\mathrm{ad}(1)\mathrm{ad}(2)}$$

if we decompose $\mathcal{H}$ into the two commuting partial Hamiltonians

$$\mathcal{H}_1 = \sum_j \left\{ -\gamma H \sigma_j^z + \frac{1}{\hbar^2} E\{(\sigma_j^x)^2 - (\sigma_j^y)^2\} \right\} \qquad \text{and}$$

$$\mathcal{H}_2 = \sum_j \frac{D}{\hbar^2}\left\{(\sigma_j^z)^2 - \frac{2}{3}\hbar^2\right\} .$$

The result for χ_{MM}^{is} can thus be obtained from the solution of Exercise 9.9, which can be specialized for $\beta \to 0$ using $\mathrm{Tr}(\mathcal{H}_1\mathcal{H}_2) = 0$ and $\mathrm{Tr}(\mathcal{H}_2\mathcal{M}) = 0$ to

$$\chi_{MM}^{\mathrm{is}} = \chi_{MM}^{\mathrm{ad}(1)\mathrm{ad}(2)} = \beta\left\{\frac{\mathrm{Tr}(\mathcal{M}^2)}{\mathrm{Tr}\, I} - \frac{1}{\mathrm{Tr}\, I}\frac{[\mathrm{Tr}(\mathcal{M}\mathcal{H}_1)]^2}{\mathrm{Tr}(\mathcal{H}_1)^2}\right\} .$$

In the high-temperature approximation, $\mathcal{H}_2$ does not contribute to χ_{MM}^{is}. By contrast, for χ_{MM}^{ad}, we obtain

$$\chi_{MM}^{\mathrm{ad}} = \beta\left\{\frac{\mathrm{Tr}(\mathcal{M}^2)}{\mathrm{Tr}\, I} - \frac{1}{\mathrm{Tr}\, I}\frac{[\mathrm{Tr}(\mathcal{M}\mathcal{H})]^2}{\mathrm{Tr}(\mathcal{H}^2)}\right\} = N\frac{C}{T}\frac{E^2 + \frac{1}{3}D^2}{E^2 + \frac{1}{3}D^2 + (\gamma\hbar H)^2} ,$$

where

$$\frac{1}{N}\frac{1}{k}\frac{\mathrm{Tr}(\mathcal{H}^2)}{\mathrm{Tr}\, I} = CH^2 + \frac{1}{k}\frac{2}{3}\left(E^2 + \frac{1}{3}D^2\right)$$

holds.

***9.12**(p.137). To calculate the work, one need only take the time interval between t_1 and t_0 into account, since $\dot{h}(t) = 0$ for $t > t_1$. From (9.6.2) we thus obtain with the aid of the response formula (9.1.9):

$$A = -\langle\mathcal{M}\rangle_\beta \hat{h} + \frac{1}{2}\left\langle\frac{\partial^2\mathcal{H}(h)}{\partial h^2}\bigg|_{h=0}\right\rangle_\beta \hat{h}^2$$

$$- \int_{t'=t_0}^{t_1} \dot{h}(t') \int_{t''=0}^{t'-t_0} h(t'-t'')\,\varphi_{MM}(t'')\,dt'\,dt''$$

$$= -\langle\mathcal{M}\rangle_\beta \hat{h} + \frac{1}{2}\left\langle\frac{\partial^2\mathcal{H}(h)}{\partial h^2}\bigg|_{h=0}\right\rangle_\beta \hat{h}^2$$

$$- \frac{1}{2}\frac{\hat{h}^2}{\Delta t^2}\int_0^{\Delta t} (\Delta t - t'')^2\,\varphi_{MM}(t'')dt'' .$$

a) When the integrals $\int_0^\infty \varphi(t'')t''dt''$ and $\int_0^\infty \varphi(t'')\, t''^2 dt''$ exist, we find in the limit $\Delta t \to \infty$:

$$A_{\Delta t\to\infty} = -\langle \mathcal{M}\rangle_\beta \hat{h} + \frac{1}{2}\left\{ -\chi^{\mathrm{ad}}_{MM} + \left\langle \frac{\partial^2 \mathcal{H}(h)}{\partial h^2}\bigg|_{h=0}\right\rangle_\beta \right\}\hat{h}^2 \ .$$

The expression in curly brackets is, from (8.3.27), just the adiabatic susceptibility w.r.t. $\partial\mathcal{H}(h)/\partial h$, because of $\partial\mathcal{H}(h)/\partial h|_{h=0} = -\mathcal{M}$:

$$-\chi^{\mathrm{ad}}_{MM} + \left\langle \frac{\partial^2 \mathcal{H}(h)}{\partial h^2}\bigg|_{h=0}\right\rangle_\beta = \frac{\partial}{\partial h}\left\langle \frac{\partial \mathcal{H}(h)}{\partial h}\right\rangle\bigg|^{\mathrm{ad}}_{h=0} \ .$$

Since the derivative of the internal energy $U = \mathrm{Tr}\{\mathcal{H}(h)\mathcal{R}_\beta(h)\}$ at constant entropy is given by:

$$\left(\frac{\partial U}{\partial h}\right)_S = \mathrm{Tr}\left\{\frac{\partial \mathcal{H}(h)}{\partial h}\mathcal{R}_\beta(h)\right\} \ ,$$

we find [cf. (5.3.6)] that the work performed may be written as a quasi-static adiabatic energy variation:

$$A_{\Delta t\to\infty} = \left(\frac{\partial U}{\partial h}\right)_S^{h=0} \hat{h} + \frac{1}{2}\left(\frac{\partial^2 U}{\partial h^2}\right)_S^{h=0} \hat{h}^2 \ .$$

b) For the case of a sudden change in the field, $\Delta t \to 0$, the integral over t'' makes no contribution,

$$A_{\Delta t\to 0} = -\langle \mathcal{M}\rangle_\beta \hat{h} + \frac{1}{2}\left\langle \frac{\partial^2 \mathcal{H}(h)}{\partial h^2}\bigg|_{h=0}\right\rangle_\beta \hat{h}^2$$

$$= \left\langle \frac{\partial \mathcal{H}(h)}{\partial h}\bigg|_{h=0}\right\rangle_\beta \hat{h} + \frac{1}{2}\left\langle \frac{\partial^2 \mathcal{H}(h)}{\partial h^2}\bigg|_{h=0}\right\rangle_\beta \hat{h}^2 \ .$$

The work performed, $A_{\Delta t\to 0}$, is thus given by terms in the expansion of $\langle \mathcal{H}(\hat{h})\rangle_\beta - \langle \mathcal{H}\rangle_\beta$. This may be understood in a straightforward way, since $\rho(+t_0) = \rho(-t_0) = \mathcal{R}_\beta(h=0)$ [cf. Exercise 7.1] and $\langle \mathcal{H}(\hat{h})\rangle(t)$ is constant for $t > t_1$ [cf. (3.2.23)].

Comparing the work performed in case (a) with that in case (b) for equal final amplitudes $\hat{h}$, we find

$$A_{\Delta t\to 0} - A_{\Delta t\to\infty} = \tfrac{1}{2}\chi^{\mathrm{ad}}_{MM}\hat{h}^2 = \tfrac{1}{2}\beta(\mathcal{M}|\mathrm{Q}_{\{1,\mathcal{H}\}}\mathcal{M})\hat{h}^2 \geqq 0 \ ,$$

as required by the Second Law.

9.13(p. 140). Using (9.7.5) with $\mathcal{F}(t) = \mathfrak{a}_{\mathfrak{k}}^-(t)$ and $\mathcal{M} = \mathfrak{a}^+_{-\mathfrak{k}}$, we find the relation:

$$\langle [\mathfrak{a}^+_{-\mathfrak{k}}, \mathfrak{a}^-_{\mathfrak{k}}]_+\rangle_\beta\, \delta(\omega + \omega_{\mathfrak{k}}) = -\coth\frac{\beta\hbar\omega}{2}\langle [\mathfrak{a}^+_{-\mathfrak{k}}, \mathfrak{a}^-_{\mathfrak{k}}]_-\rangle_\beta\, \delta(\omega + \omega_{\mathfrak{k}}) \ .$$

From

$$\sum_{\mathfrak{k}} [\sigma^{+}_{-\mathfrak{k}}, \sigma^{-}_{\mathfrak{k}}]_{+} = \sum_{l,j} \sum_{\mathfrak{k}} e^{i\mathfrak{k}(\mathfrak{R}_l - \mathfrak{R}_j)} [\sigma_l^{+}, \sigma_j^{-}]_{+} = N \sum_{j} [\sigma_j^{+}, \sigma_j^{-}]_{+} = N^2 \hbar^2 \cdot 1$$

and

$$[\sigma^{+}_{-\mathfrak{k}}, \sigma^{-}_{\mathfrak{k}}]_{-} = \sum_{l,j} e^{i\mathfrak{k}(\mathfrak{R}_l - \mathfrak{R}_j)} [\sigma_l^{+}, \sigma_j^{-}]_{-} = 2\hbar \sum_{j} \sigma_j^{z}$$

we obtain from the above relation by integrating over ω and summing over $\mathfrak{k}$:

$$\frac{1}{\frac{N\hbar}{2}} \left\langle \sum_j \sigma_j^z \right\rangle_\beta = \frac{1}{\frac{1}{N} \sum_{\mathfrak{k}} \coth \frac{\beta\hbar\omega_{\mathfrak{k}}}{2}} = \frac{1}{1 + \frac{2}{N} \sum_{\mathfrak{k}} \frac{1}{e^{\beta\hbar\omega_{\mathfrak{k}}} - 1}} .$$

For the special case of low temperatures, we have:

$$\frac{1}{\frac{N\hbar}{2}} \left\langle \sum_j \sigma_j^z \right\rangle_\beta = 1 - \frac{2}{N} \sum_{\mathfrak{k}} \frac{1}{e^{\beta\hbar\omega_{\mathfrak{k}}} - 1} + \dots .$$

This result may be interpreted in terms of quasiparticles, if we take $1/[\exp(\beta\hbar\omega_{\mathfrak{k}}) - 1]$ to be the occupation number $\langle \mathcal{N}_{\mathfrak{k}} \rangle_\beta$ for particles of energy $\hbar\omega_{\mathfrak{k}}$.

9.14(p. 144). a) With the aid of

$$\dot{x}(t) = \frac{p(t)}{m} = -x\omega_0 \sin \omega_0 t + \frac{p}{m} \cos \omega_0 t = \frac{1}{im} \sqrt{\frac{\hbar m \omega_0}{2}} (e^{-i\omega_0 t} b - e^{-i\omega_0 t} b^\dagger)$$

we find the spectral decomposition of $\delta\dot{x}(t) = \dot{x}(t)$:

$$\begin{aligned} \dot{x}(\omega) &= \frac{1}{\sqrt{2\pi\tau}} \int_0^\tau \dot{x}(t)\, e^{i\omega t}\, dt \\ &= \frac{1}{\sqrt{2\pi\tau}} \frac{1}{im} \sqrt{\frac{\hbar m\omega_0}{2}} \left\{ \frac{e^{i(\omega - \omega_0)\tau} - 1}{i(\omega - \omega_0)} b - \frac{e^{i(\omega + \omega_0)\tau} - 1}{i(\omega + \omega_0)} b^\dagger \right\} , \end{aligned}$$

from which it follows that

$$\begin{aligned} \frac{1}{2} \langle \dot{x}(\omega)^\dagger \dot{x}(\omega) + \dot{x}(\omega)\dot{x}(\omega)^\dagger \rangle_\beta = \frac{1}{2m} \langle \mathcal{H} \rangle_\beta &\left\{ \frac{1}{\pi} \frac{\tau}{2} \left(\frac{\sin(\omega - \omega_0)\tau/2}{(\omega - \omega_0)\tau/2} \right)^2 \right. \\ &\left. + \frac{1}{\pi} \frac{\tau}{2} \left(\frac{\sin(\omega + \omega_0)\tau/2}{(\omega + \omega_0)\tau/2} \right)^2 \right\} . \end{aligned}$$

Here, $b^\dagger$ and b are the usual creation and annihilation operators,

$$b^\dagger = \frac{1}{\sqrt{2\hbar}}\left(\sqrt{m\omega_0}x - \frac{\mathrm{i}}{\sqrt{m\omega_0}}p\right)$$

with $[b, b^\dagger] = 1$. In the limit $\tau \to \infty$, we find

$$\frac{1}{2}\langle \dot{x}(\omega)^\dagger \dot{x}(\omega) + \dot{x}(\omega)\dot{x}(\omega)^\dagger \rangle_\beta = \frac{1}{2m}\langle \mathcal{H} \rangle_\beta (\delta(\omega - \omega_0) + \delta(\omega + \omega_0)) \ .$$

b) From the response function

$$\varphi_{x\dot{x}}(t) = -\frac{\mathrm{i}}{\hbar}\langle [x, \dot{x}(t)] \rangle_\beta = -\frac{\mathrm{i}}{\hbar}\left\langle \left[x, \frac{p}{m}\cos\omega_0 t\right]\right\rangle_\beta = \frac{1}{m}\cos\omega_0 t$$

using (9.4.8), we obtain

$$\chi''_{x\dot{x}}(\omega) = \frac{1}{2\mathrm{i}}\int_{-\infty}^{\infty} \varphi_{x\dot{x}}(t)\, \mathrm{e}^{\mathrm{i}\omega t - \eta|t|}\, dt = \frac{\pi}{2\mathrm{i}m}(\delta(\omega - \omega_0) + \delta(\omega + \omega_0)) \ ,$$

so that

$$\frac{\mathrm{i}}{\pi}E(\beta, \omega)\,\chi''_{x\dot{x}}(\omega) = \frac{1}{2m}(E(\beta, \omega_0)\,\delta(\omega - \omega_0) + E(\beta, -\omega_0)\,\delta(\omega + \omega_0))$$

$$= \frac{1}{2m}\langle \mathcal{H} \rangle_\beta (\delta(\omega - \omega_0) + \delta(\omega + \omega_0))$$

holds. For the fluctuations, from (9.8.11) we thus again obtain the result (a).

***9.15**(p. 146). The Hamiltonian under the influence of the external potentials $\mathfrak{A}(\mathrm{r}, t)$ and $\phi(\mathrm{r}, t)$ is given by

$$\mathcal{H}(t) = \sum_i \frac{1}{2m}\left(\vec{p}_i - \frac{q}{c}\mathfrak{A}(x_i, t)\right)^2 + q\sum_i \phi(\vec{x}_i, t) + V(\vec{x}_1, \ldots, \vec{x}_N)$$

$$= \mathcal{H} - \frac{q}{mc}\int \vec{p}(\mathrm{r})\,\mathfrak{A}(\mathrm{r}, t)\, d^3x + q\int n(\mathrm{r})\phi(\mathrm{r}, t)\, d^3x$$

$$+ \frac{q^2}{2mc^2}\int n(\mathrm{r})\mathfrak{A}^2(\mathrm{r}, t)\, d^3x \ .$$

Here,

$$\vec{p}(\mathrm{r}) = \frac{1}{2}\sum_i (\vec{p}_i\,\delta(\mathrm{r} - \vec{x}_i) + \delta(\mathrm{r} - \vec{x}_i)\,\vec{p}_i)$$

is the canonical momentum-current density, and

$$n(\mathrm{r}) = \sum_i \delta(\mathrm{r} - \vec{x}_i)$$

is the particle density. The *linearized* external perturbation is therefore given by

$$\sum_\alpha \mathcal{M}_\alpha h_\alpha(t) \to \frac{q}{mc}\int \vec{p}(\mathrm{r})\,\mathfrak{A}(\mathrm{r},t)\,d^3x - q\int n(\mathrm{r})\,\phi(\mathrm{r},t)\,d^3x \ ,$$

i.e. the index α denotes the components of the potentials and the position r. The observable we are seeking,

$$\vec{j}(\mathrm{r}) = \frac{q}{2}\sum_i \left\{\frac{1}{m}\left(\vec{p}_i - \frac{q}{c}\mathfrak{A}(\vec{x}_i,t)\right)\delta(\mathrm{r}-\vec{x}_i) + \delta(\mathrm{r}-\vec{x}_i)\frac{1}{m}\left(\vec{p}_i - \frac{q}{c}\mathfrak{A}(\vec{x}_i,t)\right)\right\}$$

$$= \frac{q}{m}\,\vec{p}(\mathrm{r}) - \frac{q^2}{mc}\,n(\mathrm{r})\,\mathfrak{A}(\mathrm{r},t)$$

depends on the vector potential, i.e. it corresponds to $j_\nu(\mathrm{r},\mathfrak{A}) \to \mathcal{F}(h)$. From (9.9.13), with $\langle \vec{p}(\mathrm{r})\rangle_\beta = 0$, we thus find

$$\langle \vec{j}(\mathrm{r})\rangle(t) = -\,\mathfrak{A}(\mathrm{r},t)\frac{q^2}{mc}\langle n(\mathrm{r})\rangle_\beta$$

$$-\frac{1}{c}\left(\frac{q}{m}\right)^2 \beta\int_0^t \sum_{\nu=1}^3 \int A_\nu(\mathrm{r}',t-t')\,(p_\nu(\mathrm{r}')|\mathrm{iL}\,\mathrm{e}^{\mathrm{iL}t'}\,\vec{p}(\mathrm{r}))\,d^3x'\,dt'$$

$$+\frac{q^2}{m}\beta\int_0^t\int \phi(\mathrm{r}',t-t')\,(n(\mathrm{r}')|\mathrm{iL}\mathrm{e}^{\mathrm{iL}t'}\vec{p}(\mathrm{r}))\,d^3x'\,dt' \ .$$

If we wish to apply response theory to gauge-transformed potentials $\mathfrak{A} \to \mathfrak{A} + \mathrm{grad}\,\Lambda$ and $\phi \to \phi - (1/c)(\partial/\partial t)\Lambda$, they must likewise vanish for $t = 0$, i.e. $\Lambda(\mathrm{r}, t = 0)$ must be a constant. The response to the gauge-transformed potentials yields, in addition to the response to $\mathfrak{A}$ and ϕ,

$$\vec{\delta} = -\frac{q^2}{mc}\langle n(\mathrm{r})\rangle_\beta\ \mathrm{grad}\,\Lambda(\mathrm{r},t)$$

$$-\frac{1}{c}\left(\frac{q}{m}\right)^2 \beta\int_0^t \sum_{\nu=1}^3 \int \frac{\partial}{\partial x'_\nu}\Lambda(\mathrm{r}',t-t')\,(p_\nu(\mathrm{r}')|\mathrm{iL}\mathrm{e}^{\mathrm{iL}t'}\vec{p}(\mathrm{r}))\,d^3x'\,dt'$$

$$+\frac{q^2}{mc}\beta\int\int_0^t \frac{\partial}{\partial t'}\Lambda(\mathrm{r}',t-t')(n(\mathrm{r}')|\mathrm{iL}\,\mathrm{e}^{\mathrm{iL}t'}\,\vec{p}(\mathrm{r}))\,d^3x'\,dt' \ .$$

Integration by parts of the third term over t' cancels out the first term, due to

$$(n(\mathrm{r}')|\mathrm{iL}\,p_\nu(\mathrm{r})) = \frac{\mathrm{i}}{\beta\hbar}\langle[n(\mathrm{r}'), p_\nu(\mathrm{r})]\rangle_\beta = \frac{1}{\beta}\langle n(\mathrm{r})\rangle_\beta\,\frac{\partial}{\partial x'_\nu}\,\delta(\mathrm{r}-\mathrm{r}') \ .$$

The second term may be combined with the remaining part of the third term to give a divergence, because of $\mathrm{iL}n(\mathbf{r}') + (1/m)\,\mathrm{div}\,\vec{p}(\mathbf{r}') = \mathcal{O}$:

$$\vec{\delta} = -\frac{1}{c}\left(\frac{q}{m}\right)^2 \beta\Bigg\{\int_0^t \sum_{\nu=1}^3 \int \frac{\partial}{\partial x'_\nu} A(\mathbf{r}', t-t')\,(p_\nu(\mathbf{r}')|\mathrm{iL}e^{\mathrm{iL}t'}\vec{p}(\mathbf{r}))\, d^3x'\, dt'$$

$$+\int_0^t \int A(\mathbf{r}', t-t')\, m(-\mathrm{iL}n(\mathbf{r}')|\mathrm{iL}e^{\mathrm{iL}t'}\vec{p}(\mathbf{r}))\, d^3x'\, dt'\Bigg\}$$

$$= -\frac{1}{c}\left(\frac{q}{m}\right)^2 \beta \int_0^t \int \sum_{\nu=1}^3 \frac{\partial}{\partial x'_\nu}\{A(\mathbf{r}', t-t')\,(p_\nu(\mathbf{r}')|\mathrm{iL}e^{\mathrm{iL}t'}\vec{p}(\mathbf{r}))\}\, d^3x'\, dt' \;.$$

Therefore, applying Gauss' theorem, we find $\vec{\delta} = 0$.

10.1(p. 156). As in Exercise 9.1, using the notation $\omega_0 = \gamma H$, we have $\mathsf{L}\mathcal{S}^\mu = -\mu\omega_0\mathcal{S}^\mu$ ($\mu = +, -, 0$), i.e. $e^{\mathrm{iL}t}\mathcal{S}^\mu = \exp(-\mathrm{i}\mu\omega_0 t)\mathcal{S}^\mu$. We therefore find the response functions from (10.1.8) to be:

$$\varphi_{+0,+}(t_1, t_2) = \varphi^*_{-0,-}(t_1, t_2) = -\frac{1}{\hbar^2}\langle[\mathcal{S}^-, [\mathcal{S}^z, \mathcal{S}^+]]\rangle_\beta\, e^{-\mathrm{i}\omega_0(t_1+t_2)}$$

$$= 2\langle\mathcal{S}^z\rangle_\beta\, e^{-\mathrm{i}\omega_0(t_1+t_2)}$$

$$\varphi_{+-,0}(t_1, t_2) = \varphi^*_{-+,0}(t_1, t_2) = -\frac{1}{\hbar^2}\langle[\mathcal{S}^-, [\mathcal{S}^+, \mathcal{S}^z]]\rangle_\beta\, e^{-\mathrm{i}\omega_0 t_1}$$

$$= -2\langle\mathcal{S}^z\rangle_\beta\, e^{-\mathrm{i}\omega_0 t_1} \;.$$

All the other $\varphi_{\nu\mu,\lambda}(t_1, t_2)$ vanish, since of the $\langle\mathcal{S}^\mu\rangle_\beta$, only $\langle\mathcal{S}^z\rangle_\beta$ is nonzero.

***10.2**(p. 161). Using $\varphi_{M\dot{M}}(t_1) = \dot{\varphi}_{MM}(t_1)$ and $\varphi_{MM}(0) = 0$, we find for $\langle\mathcal{F}_\parallel\rangle_2\,(t_{\mathrm{I}})$ from (10.2.23) and (10.1.11) the following expression:

$$\langle\mathcal{F}_\parallel\rangle_2\,(t_{\mathrm{I}}) = \int_{t_2=0}^{\Delta t}\int_{t_1=0}^{\Delta t - t_2} h(t_{\mathrm{I}} - t_1 - t_2)\, h(t_{\mathrm{I}} - t_2)\, \varphi_{MMF_\parallel}(t_1)\, dt_1\, dt_2$$

$$= \frac{\langle\delta\mathcal{H}\,\mathcal{F}\rangle}{\langle\delta\mathcal{H}\,\delta\mathcal{H}\rangle}\frac{\hbar^2}{(\Delta t)^2}\int_{t_2=0}^{\Delta t}(\Delta t - t_2)\int_{t_1=0}^{\Delta t - t_2}(\Delta t - t_2 - t_1)$$

$$\times\, \dot{\varphi}_{MM}(t_1)\, dt_1\, dt_2$$

$$= \frac{\langle\delta\mathcal{H}\,\mathcal{F}\rangle}{\langle\delta\mathcal{H}\,\delta\mathcal{H}\rangle}\frac{\hbar^2}{(\Delta t)^2}\int_0^{\Delta t}\frac{(\Delta t)^2 - t''^2}{2}\,\varphi_{MM}(t'')\, dt'' \;.$$

In the limit $\Delta t \to \infty$, with $\int_0^\infty \varphi_{MM}(t)dt = \chi^{\mathrm{is}}_{MM} = \chi^{\mathrm{ad}}_{MM}$, we obtain

$$\langle\mathcal{F}_\parallel\rangle^{(2)}(t_{\mathrm{I}}) = \frac{\langle\delta\mathcal{H}\,\mathcal{F}\rangle}{\langle\delta\mathcal{H}\,\delta\mathcal{H}\rangle}\,\chi^{\mathrm{ad}}_{MM}\,\frac{\hbar^2}{2} \;.$$

The susceptibility $\chi_{MMF_\parallel}(\omega_1, \omega_2)$ in (10.2.25) can, using (9.4.16) and $\chi^{\mathrm{T}}_{M\dot{M}} = \beta(\mathcal{M}|\mathrm{i}\mathrm{L}\mathcal{M}) = 0$, be put into the form

$$\chi_{MMF_\parallel}(\omega_1, \omega_2) = \frac{\mathrm{i}}{\omega_2 + \mathrm{i}\eta_2}(-\mathrm{i}\omega_1)\chi_{MM}(\omega_1)\frac{\langle\delta\mathcal{H}\,\mathcal{F}\rangle}{\langle\delta\mathcal{H}\,\delta\mathcal{H}\rangle},$$

so that the variation of $\mathcal{F}_\parallel$ is found in terms of the limit of the susceptibility

$$\langle\mathcal{F}_\parallel\rangle_2(t_1) = \lim_{\omega_1\to 0} \chi_{MMF_\parallel}(\omega_1, \omega_1)\frac{\hat{h}^2}{2} = \lim_{\omega_1\to 0} \tilde{\chi}_{MMF_\parallel}(\omega_1, \omega_1)\hat{h}^2 .$$

The part $\langle\mathcal{F}_\perp\rangle^{(2)}(t_1)$ can be most readily obtained from (10.1.20),

$$\begin{aligned}\langle\mathcal{F}_\perp\rangle_2(t_1) &= \int_0^{\Delta t}\int_0^{\Delta t} h(t_1 - t')\,h(t_1 - t'')\,\tilde{\varphi}_{MMF_\parallel}(t', t'')dt'dt'' \\ &\quad + \frac{1}{2}\int_0^{\Delta t}\{h(t_1 - t')\}^2\,\varphi_{M_{\alpha\alpha}F}(t')\,dt' \\ &= \hat{h}^2\int_0^{\Delta t}\int_0^{\Delta t}\frac{(\Delta t - t')(\Delta t - t'')}{(\Delta t)^2}\,\tilde{\varphi}_{MMF_\perp}(t', t'')\,dt'\,dt'' \\ &\quad + \frac{1}{2}\hat{h}^2\int_0^{\Delta t}\frac{(\Delta t - t')^2}{(\Delta t)^2}\,\varphi_{M_{\alpha\alpha}F}(t')dt' .\end{aligned}$$

When the integrals

$$\iint_0^\infty t't''\,\tilde{\varphi}_{MMF_\perp}(t', t'')\,dt'dt'' \quad \text{and} \quad \int_0^\infty t'^2\varphi_{M_{\alpha\alpha}F}(t')dt'$$

exist, the limit $\Delta t \to \infty$ leads to

$$\langle\mathcal{F}_\perp\rangle_2(t_1) = \tilde{\chi}_{MMF_\perp}(0, 0)\hat{h}^2 + \tfrac{1}{2}\chi_{M_{\alpha\alpha}F}(0)\,\hat{h}^2.$$

The total response, $\langle\mathcal{F}\rangle_2(t_1)$, for $\Delta t \to \infty$ thus may be expressed as a limiting value of the total susceptibility,

$$\langle\mathcal{F}\rangle_2(t_1) = \lim_{\omega\to 0}\{\tilde{\chi}_{MMF}(\omega, \omega) + \tfrac{1}{2}\chi_{M_{\alpha\alpha}F}(\omega)\}\hat{h}^2 .$$

***10.3**(p. 166). The response $\langle\mathcal{F}_\parallel\rangle_2(t)$ can be found from (10.3.18) by taking the limit $\omega_2 \to \omega_1$, or it can be calculated from (10.1.11) with $h(t) = \hat{h}\,\mathrm{Re}\{\mathrm{i}\exp(-\mathrm{i}\omega_0 t)\}$. In the latter case, one obtains

$$\langle\mathcal{F}_\parallel\rangle_2(t) = \int_{t_1=0}^{t}\int_{t_2=0}^{t-t_1} h(t - t_1 - t_2)\,h(t - t_2)\,\varphi_{MMF_\parallel}(t_1)\,dt_1\,dt_2$$

$$= \frac{\hbar^2}{2} \operatorname{Re} \int_{t_1=0}^{t} \int_{t_2=0}^{t-t_1} e^{-i\omega_0(t-t_1-t_2)}$$

$$\times \{e^{i\omega_0(t-t_2)} - e^{-i\omega_0(t-t_2)}\} \varphi_{MMF_\parallel}(t_1)\, dt_1\, dt_2$$

$$= \frac{\hbar^2}{2} \operatorname{Re} \int_{t_1=0}^{t} \left\{(t-t_1)\, e^{i\omega_0 t_1} - \frac{1}{2i\omega_0} e^{i\omega_0 t_1}(e^{-2i\omega_0 t_1} - e^{-2i\omega_0 t})\right\}$$

$$\times \varphi_{MMF_\parallel}(t_1) dt_1 \ .$$

In the steady state, $t \to \infty$, one is thus led with the aid of (10.2.23) to

$$\langle \mathscr{F}_\parallel \rangle_2(t) = \frac{\langle \delta\mathscr{H}\, \mathscr{F} \rangle}{\langle \delta\mathscr{H}\, \delta\mathscr{H} \rangle} \frac{\hbar^2}{2} \operatorname{Re} \left\{ t\chi_{M\dot{M}}(\omega_0) + i \frac{\partial}{\partial \omega_0} \chi_{M\dot{M}}(\omega_0) \right.$$

$$\left. - \frac{1}{2i\omega_0} \chi_{M\dot{M}}(-\omega_0) + \frac{1}{2i\omega_0} \chi_{M\dot{M}}(\omega_0) e^{-2i\omega_0 t} \right\} .$$

Using $\chi_{M\dot{M}}(\omega) = i\omega\chi_{MM}(\omega) = -i\omega\{\chi'_{MM}(\omega) + i\chi''_{MM}(\omega)\}$, this result may be summarized in the form

$$\langle \mathscr{F}_\parallel \rangle_2(t) = \frac{\langle \delta\mathscr{H}\, \mathscr{F} \rangle}{\langle \delta\mathscr{H}\, \delta\mathscr{H} \rangle} \frac{\hbar^2}{2} \left\{ \omega_0 \frac{\partial}{\partial \omega_0} \chi'_{MM}(\omega_0) + \chi'_{MM}(\omega_0) \sin^2 \omega_0 t \right.$$

$$\left. + \chi''_{MM}(\omega_0) \left(\omega_0 t - \frac{1}{2} \sin 2\omega_0 t \right) \right\} .$$

A term which increases linearly with t is found, which describes the energy absorption. The response $\langle \mathscr{F}_\perp \rangle_2(t)$ is found by taking the limit $\omega_2 \to \omega_1 = \omega_0$ from (10.3.5) and (10.3.14); it is

$$\langle \mathscr{F}_\perp \rangle_2(t) = \frac{\hbar^2}{2} \operatorname{Re} \left\{ \tilde{\chi}_{MMF_\perp}(\omega_0, -\omega_0) - \tilde{\chi}_{MMF_\perp}(\omega_0, \omega_0)\, e^{-2i\omega_0 t} \right.$$

$$\left. + \frac{1}{2} \chi_{M_{\alpha\alpha}F}(0) - \frac{1}{2} \chi_{M_{\alpha\alpha}F}(2\omega_0)\, e^{-2i\omega_0 t} \right\} .$$

The total response, $\langle \mathscr{F} \rangle_2(t) = \langle \mathscr{F}_\parallel \rangle_2\, (t) + \langle \mathscr{F}_\perp \rangle_2(t)$, thus consists of a constant term, an oscillation with frequency $2\omega_0$, and a term which increases linearly with t.

10.4(p. 166). The exact solution for $t > t^{(2)}$ from Exercise 7.2,

$$\langle \mathscr{F} \rangle(t) = \operatorname{Tr} \left\{ \exp[-i\mathsf{L}(t - t^{(2)})] \exp\left[i \sum_\gamma \mathsf{M}_\gamma^{(\dagger)} \hat{h}_\gamma^{(2)} \right] \right.$$

$$\left. \times \exp[-i\mathsf{L}(t^{(2)} - t^{(1)})] \exp\left[i \sum_\alpha \mathsf{M}_\alpha^{(\dagger)} \hat{h}_\alpha^{(1)} \right] \mathscr{R}_\beta \right\} ,$$

has the following terms which are of second order in the amplitudes:

$$\langle \mathscr{F} \rangle_2(t) = \frac{1}{2} \sum_{\alpha\alpha'} \hat{h}_\alpha^{(1)} \hat{h}_{\alpha'}^{(1)} (\mathrm{i})^2 \, \mathrm{Tr} \{ \mathscr{F} \mathrm{e}^{-\mathrm{iL}(t-t^{(1)})} \mathsf{M}_\alpha^{(\dagger)} \mathsf{M}_{\alpha'}^{(\dagger)} \mathscr{R}_\beta \}$$

$$+ \frac{1}{2} \sum_{\gamma\gamma'} \hat{h}_\gamma^{(2)} \hat{h}_{\gamma'}^{(2)} (\mathrm{i})^2 \, \mathrm{Tr} \{ \mathscr{F} \mathrm{e}^{-\mathrm{iL}(t-t^{(2)})} \mathsf{M}_\gamma^{(\dagger)} \mathsf{M}_{\gamma'}^{(\dagger)} \mathscr{R}_\beta \}$$

$$+ \sum_{\alpha\gamma} \hat{h}_\alpha^{(1)} \hat{h}_\gamma^{(2)} (\mathrm{i})^2 \, \mathrm{Tr} \{ \mathscr{F} \mathrm{e}^{-\mathrm{iL}(t-t^{(2)})} \mathsf{M}_\gamma^{(\dagger)} \mathrm{e}^{-\mathrm{iL}(t^{(2)}-t^{(1)})} \mathsf{M}_\alpha^{(\dagger)} \mathscr{R}_\beta \} \, .$$

The first two terms each represent the response to a single δ-function pulse. The interference term is, from (10.1.8), given by

$$\sum_{\alpha\gamma} \hat{h}_\alpha^{(1)} \hat{h}_\gamma^{(2)} \varphi_{M_\alpha M_\gamma F}(t^{(2)} - t^{(1)}, \; t - t^{(2)})$$

[cf. (10.3.9)].

11.1(p. 186). From the symmetry of the Hamiltonian with respect to rotations about the z-axis and to translations, it follows for the matrix

$$(\mathscr{S}_{\mathfrak{k}'}^{\mu'} | \mathscr{S}_{\mathfrak{k}}^{\mu}) = \delta_{\mathfrak{k}'\mathfrak{k}} \delta_{\mu'\mu} (\mathscr{S}_{\mathfrak{k}}^{\mu} | \mathscr{S}_{\mathfrak{k}}^{\mu}) \, .$$

For the same reasons, we find

$$(\mathscr{S}_{\mathfrak{k}'}^{\mu'} | \mathsf{L} \mathscr{S}_{\mathfrak{k}}^{\mu}) = \delta_{\mathfrak{k}'\mathfrak{k}} \delta_{\mu'\mu} \frac{1}{\beta\hbar} \langle [\mathscr{S}_{-\mathfrak{k}}^{-\mu}, \mathscr{S}_{\mathfrak{k}}^{\mu}] \rangle = \delta_{\mathfrak{k}'\mathfrak{k}} \delta_{\mu'\mu} (-\mu) \frac{2}{\beta} \left\langle \sum_j s_z^j \right\rangle \, .$$

The frequency matrix (11.2.15) is thus given by

$$\Omega_{\mathfrak{k}'\mathfrak{k}}^{\mu'\mu} = -\mu \delta_{\mathfrak{k}'\mathfrak{k}} \delta_{\mu'\mu} \Omega_{\mathfrak{k}}^{\mu} \, ,$$

where $\Omega_{\mathfrak{k}}^{\pm}$ denotes $2\langle \mathscr{S}^z \rangle / [\beta (\mathscr{S}_{\mathfrak{k}}^{\pm} | \mathscr{S}_{\mathfrak{k}}^{\pm})]$ and, from (8.2.10), we find $\Omega_{\mathfrak{k}}^{+} = \Omega_{-\mathfrak{k}}^{-}$. The temperature dependence of the spin wave frequency is therefore determined by that of the permanent magnetization and of the transverse susceptibility.

If we take only the frequency matrix into account in the dynamics of the observables $\mathscr{S}_{\mathfrak{k}}^{\pm}$, then the relation

$$\mathscr{S}_{\mathfrak{k}}^{\pm}(t) = \mathrm{e}^{\mp \mathrm{i} \Omega_{\mathfrak{k}}^{\pm} t} \mathscr{S}_{\mathfrak{k}}^{\pm}$$

holds (spin wave approximation). In the limit $T \to 0$, the frequency can be expressed explicitly in terms of the parameters of $\mathscr{H}$. For $\gamma H > 0$, the ground state is given by the product state $|0\rangle = |s^{(1)}\rangle |s^{(2)}\rangle \ldots |s^{(N)}\rangle$, so that the low-temperature limit of $\langle \mathscr{S}^z \rangle$ is given by $\lim_{T \to 0} \langle \mathscr{S}^z \rangle = N\hbar s$. For a static susceptibility, the limit $T \to 0$ for a nondegenerate ground state gives

$$\lim_{T\to 0} \beta(\mathscr{F}\,|\,\delta\mathscr{G}) = \int_0^\infty \langle 0|\mathscr{F}^\dagger\, e^{-\alpha(\mathscr{H}-E_0)}\delta\mathscr{G}\,|0\rangle\, d\alpha$$

$$+ \int_0^\infty \langle 0|\delta\mathscr{G}\, e^{-\alpha(\mathscr{H}-E_0)}\,\mathscr{F}^\dagger\,|0\rangle\, d\alpha \ ,$$

where $\delta\mathscr{G}$ denotes $\mathscr{G} - \langle 0|\mathscr{G}|0\rangle$ [cf. Exercise 8.2].

Using the exact relations

$$(\mathscr{H} - E_0)\mathscr{S}_{\mathfrak{k}}^-\,|0\rangle = \{\gamma\hbar H + 2s[J(0) - J(\mathfrak{k})]\}\ \mathscr{S}_{\mathfrak{k}}^-\,|0\rangle \ ,$$

$$J(\mathfrak{k}) = \sum_j e^{-i\mathfrak{k}(\mathfrak{R}_j - \mathfrak{R}_i)} J(\mathfrak{R}_j - \mathfrak{R}_i)$$

we therefore find

$$\lim_{T\to 0} \beta(\mathscr{S}_{\mathfrak{k}}^-\,|\,\mathscr{S}_{\mathfrak{k}}^-) = \frac{1}{\gamma\hbar H + 2s[J(0) - J(\mathfrak{k})]}\langle 0|\,\mathscr{S}_{-\mathfrak{k}}^+\mathscr{S}_{\mathfrak{k}}^-\,|0\rangle$$

$$= \frac{N\cdot 2\hbar^2 s}{\gamma\hbar H + 2s[J(0) - J(\mathfrak{k})]}$$

and thus for the frequency:

$$\lim_{T\to 0} \hbar\Omega_{\mathfrak{k}}^- = \gamma\hbar H + 2s[J(0) - J(\mathfrak{k})] \ .$$

This result for $T\to 0$ agrees with that of the Holstein–Primakoff theory.

11.2(p. 187). Differentiation of

$$\mathscr{f}_\mu(t) = e^{i\mathsf{QL}t}\mathscr{f}_\mu$$

yields

$$\dot{\mathscr{f}}_\mu(t) = i\mathsf{QL}\mathscr{f}_\mu(t) \ .$$

We form the Hermitian conjugate equation, noting that conjugate formation commutes with differentiation,

$$\dot{\mathscr{f}}_\mu^\dagger(t) = -\,i(\mathsf{QL}\mathscr{f}_\mu(t))^\dagger \ .$$

Due to the Hermiticity of $\mathscr{H}$, we have $(\mathsf{L}\mathscr{X})^\dagger = -\,\mathsf{L}\mathscr{X}^\dagger$. On the other hand, we have the relation $(\mathsf{P}\mathscr{X})^\dagger = \mathsf{P}\mathscr{X}^\dagger$. This latter relation has *nothing* to do with the fact that P is a Hermitian operator in $\mathfrak{L}$; rather, it follows from (8.2.10), with the important precondition that for each $\mathscr{G}_\mu$ in the set $\{\mathscr{G}\}$, there is a $\mathscr{G}_\mu^\dagger$ which also belongs to the set. With this condition, we find from (11.2.3 and 11.2.4) the relation $g^*_{\mu\nu} = g_{\mu^\dagger\nu^\dagger}$, which is also required for the proof. With these results, we obtain the differential equation for $f^\dagger_\mu(t)$:

$$\dot{f}^\dagger_\mu(t) = \mathrm{iQL} f^\dagger_\mu(t)$$

with the solution

$$f^\dagger_\mu(t) = \mathrm{e}^{\mathrm{iQL}t} f^\dagger_\mu = \mathrm{e}^{\mathrm{iQLQ}t} f^\dagger_\mu \ .$$

11.3(p. 192). The integro-differential equation for $\Xi(t)$ (13.1.1):

$$\dot{\Xi}(t) = -\int_0^t \gamma(t-t')\,\Xi(t')dt' \ ,$$

yields, after differentiation w.r.t. t and using $\gamma(t) = \gamma_0 \mathrm{e}^{-t/\tau_c}$, the differential equation

$$\ddot{\Xi}(t) + \frac{1}{\tau_c}\dot{\Xi}(t) + \gamma_0 \Xi(t) = 0$$

with initial conditions $\Xi(0) = 1$ and $\dot{\Xi}(0) = 0$.

For $4\gamma_0\tau_c^2 < 1$, the damped solution is obtained:

$$\Xi(t) = \frac{1}{2}\left\{\left(1 - \frac{1}{\sqrt{1-4\gamma_0\tau_c^2}}\right)\exp\left[-\frac{t}{2\tau_c}\left(1+\sqrt{1-4\gamma_0\tau_c^2}\right)\right]\right.$$
$$\left. + \left(1 + \frac{1}{\sqrt{1-4\gamma_0\tau_c^2}}\right)\exp\left[-\frac{t}{2\tau_c}\left(1-\sqrt{1-4\gamma_0\tau_c^2}\right)\right]\right\} ,$$

which for the case that $\gamma_0\tau_c^2 \ll 1$ leads to the approximation

$$\Xi(t) \approx \mathrm{e}^{-\gamma_0\tau_c t} - \gamma_0\tau_c^2\mathrm{e}^{-t/\tau_c} \ .$$

For times $t \gg \tau_c$, or from $\gamma_0\tau_c^2 \ll 1$, we can neglect the second term, obtaining

$$\Xi(t) \approx \mathrm{e}^{-\Gamma t}$$

where the integral over the memory function

$$\Gamma = \int_0^\infty \gamma(t)dt = \gamma_0\tau_c = \frac{1}{\tau_R}$$

manifests itself as the reciprocal of the relaxation time τ_R (Markovian approximation [cf. Sect. (13.1)]). This limiting case of the solution corresponds to neglecting the term $\ddot{\Xi}$ in the differential equation for $\Xi(t)$.

In the case that $4\gamma_0\tau_c^2 > 1$, the solution exhibits damped oscillations:

$$\Xi(t) = e^{-t/2\tau_c}\left\{\cos\left(\frac{t}{2\tau_c}\sqrt{4\gamma_0\tau_c^2 - 1}\right) + \frac{1}{\sqrt{4\gamma_0\tau_c^2 - 1}}\sin\left(\frac{t}{2\tau_c}\sqrt{4\gamma_0\tau_c^2 - 1}\right)\right\} .$$

In the limit $4\gamma_0\tau_c^2 \gg 1$, the oscillation frequency is $\sqrt{\gamma_0}$.

11.4(p. 192). Equation (11.4.9) for the resolvent $\mathsf{G}(z)$ may be written using $z = \omega + i\eta$ as

$$(\mathsf{L} + z)\,\mathsf{G}(z) = \mathrm{i}\cdot\mathbf{1} \ .$$

We insert $\mathbf{1} = \mathsf{P} + \mathsf{Q}$ (11.2.5) and multiply from the right by P:

$$(\mathsf{L} + z)\cdot\mathsf{PG}(z)\mathsf{P} + (\mathsf{L} + z)\cdot\mathsf{QG}(z)\mathsf{P} = \mathrm{i}\mathsf{P} \ .$$

Multiplying from the left by P or Q and using the projector properties (11.2.7 and 9) yields the two equations

$$\mathsf{P}(\mathsf{L} + z)\mathsf{P}\cdot\mathsf{PG}(z)\mathsf{P} + \mathsf{PLQ}\cdot\mathsf{QG}(z)\mathsf{P} = \mathrm{i}\mathsf{P} \qquad \text{and}$$

$$\mathsf{QLP}\cdot\mathsf{PG}(z)\mathsf{P} + (\mathsf{QLQ} + z)\cdot\mathsf{QG}(z)\mathsf{P} = 0 \ .$$

Solving the second equation for $\mathsf{QG}(z)\,\mathsf{P}$,

$$\mathsf{QG}(z)\,\mathsf{P} = -\,(\mathsf{QLQ} + z)^{-1}\cdot\mathsf{QLP}\cdot\mathsf{PG}(z)\,\mathsf{P} \ ,$$

and substituting into the first one yields the relation

$$\{z + \mathsf{PLP} - \mathsf{PLQ}(\mathsf{QLQ} + z)^{-1}\cdot\mathsf{QLP}\}\cdot\mathsf{PG}(z)\,\mathsf{P} = \mathrm{i}\mathsf{P} \ ,$$

i.e. (11.4.11) for $\mathsf{PG}(z)\mathsf{P}$.

***12.1**(p. 197). From the definition of the scalar product:

$$(\mathscr{F}|\mathscr{G}) = \frac{1}{\beta}\frac{1}{Z}\int_0^\beta \mathrm{Tr}\{\mathscr{F}^\dagger\, e^{-\alpha\mathscr{H}}\mathscr{G}\, e^{(\alpha-\beta)\mathscr{H}}\}\, d\alpha$$

and the expansions which follow from (3.4.11),

$$(e^{-\alpha\mathscr{H}})^1 = -\int_0^\alpha e^{-\alpha'\mathscr{H}^0}\mathscr{H}^1\, e^{(\alpha'-\alpha)\mathscr{H}^0}\, d\alpha' \ ,$$

$$(e^{(\alpha-\beta)\mathscr{H}})^1 = -\int_0^{\beta-\alpha} e^{-\alpha'\mathscr{H}^0}\mathscr{H}^1\, e^{(\alpha'-\beta+\alpha)\mathscr{H}^0}\, d\alpha' \ ,$$

$$\left(\frac{1}{Z}\right)^1 = \frac{1}{Z^0}\beta\langle\mathscr{H}^1\rangle^0 \ ,$$

we find:

$$(\mathscr{F}|\mathscr{G})^1 = -\frac{1}{\beta}\int\limits_{\alpha=0}^{\beta}\int\limits_{\alpha'=0}^{\alpha} \langle \mathscr{F}^\dagger \, e^{-\alpha'\mathscr{H}^0} \mathscr{H}^1 \, e^{(\alpha'-\alpha)\mathscr{H}^0} \mathscr{G} \, e^{\alpha\mathscr{H}^0} \rangle^0 \, d\alpha \, d\alpha'$$

$$-\frac{1}{\beta}\int\limits_{\alpha=0}^{\beta}\int\limits_{\alpha'=0}^{\beta-\alpha} \langle \mathscr{F}^\dagger \, e^{-\alpha\mathscr{H}^0} \mathscr{G} \, e^{-\alpha'\mathscr{H}^0} \mathscr{H}^1 \, e^{(\alpha'+\alpha)\mathscr{H}^0} \rangle^0 \, d\alpha \, d\alpha'$$

$$+ \beta\langle \mathscr{H}^1 \rangle^0 (\mathscr{F}|\mathscr{G})^0 \; .$$

Making the substitution $\alpha' = \alpha'' - \alpha$ in the second term and then renaming the variables α, α'' allows us to summarize the result in the following form:

$$(\mathscr{F}|\mathscr{G})^1 = -\frac{1}{\beta}\int\limits_{\alpha=0}^{\beta}\int\limits_{\alpha'=0}^{\alpha} \{\langle \mathscr{F}^\dagger (e^{-\alpha' h\mathsf{L}^0} \mathscr{H}^1)(e^{-\alpha h\mathsf{L}^0}\mathscr{G})\rangle^0$$

$$+ \langle \mathscr{F}^\dagger (e^{-\alpha' h\mathsf{L}^0} \mathscr{G})(e^{-\alpha h\mathsf{L}^0}\mathscr{H}^1\rangle^0\} \, d\alpha \, d\alpha' + \beta\langle \mathscr{H}^1\rangle^0 (\mathscr{F}|\mathscr{G})^0 \; .$$

The expressions containing three observables which occur are invariant w.r.t. cyclic permutation. From

$$\mathscr{R}^0_\beta \mathscr{A} = (e^{-\beta h\mathsf{L}^0}\mathscr{A})\mathscr{R}^0_\beta$$

we derive

$$\langle \mathscr{A}(e^{-\alpha' h\mathsf{L}^0}\mathscr{B})(e^{-\alpha h\mathsf{L}^0}\mathscr{C})\rangle^0 = \langle (e^{-\alpha' h\mathsf{L}^0}\mathscr{B})(e^{-\alpha h\mathsf{L}^0}\mathscr{C})(e^{-\beta h\mathsf{L}^0}\mathscr{A})\rangle^0$$

$$= \langle \mathscr{B}(e^{(\alpha'-\alpha)h\mathsf{L}^0}\mathscr{C})(e^{(\alpha'-\beta)h\mathsf{L}^0}\mathscr{A})\rangle^0 \; ,$$

from which, with a corresponding substitution of the variables of integration, we find

$$\int\limits_{\alpha=0}^{\beta}\int\limits_{\alpha'=0}^{\alpha} \langle \mathscr{A}(e^{-\alpha' h\mathsf{L}^0}\mathscr{B})(e^{-\alpha h\mathsf{L}^0}\mathscr{C})\rangle^0 \, d\alpha \, d\alpha'$$

$$= \int\limits_{\alpha=0}^{\beta}\int\limits_{\alpha'=0}^{\alpha} \langle \mathscr{B}(e^{-\alpha' h\mathsf{L}^0}\mathscr{C})(e^{-\alpha h\mathsf{L}^0}\mathscr{A})\rangle^0 \, d\alpha \, d\alpha' \; .$$

Furthermore, we have

$$\frac{1}{\beta}\int\limits_{\alpha=0}^{\beta}\int\limits_{\alpha'=0}^{\alpha} \{\langle \mathscr{A}(e^{-\alpha' h\mathsf{L}^0}\cdot \mathit{1})(e^{-\alpha h\mathsf{L}^0}\mathscr{C})\rangle^0$$

$$+ \langle \mathscr{A}(e^{-\alpha' h\mathsf{L}^0}\mathscr{C})(e^{-\alpha h\mathsf{L}^0}\cdot \mathit{1})\rangle^0\} \, d\alpha \, d\alpha' = \beta(\mathscr{A}^\dagger|\mathscr{C})^0 \; ,$$

so that, in the expression for $(\mathscr{F}|\mathscr{G})^1$, the last term $\beta\langle\mathscr{H}^1\rangle^0(\mathscr{F}|\mathscr{G})^0$ can be replaced by making the substitution $\mathscr{H}^1 \to \mathscr{H}^1 - \langle \mathscr{H}^1\rangle^0$ in the first two terms. In the case that $\mathsf{L}^0\mathscr{A}_\nu = \Omega^0_\nu \mathscr{A}_\nu$, i.e. $\mathsf{L}^0\mathscr{A}^\dagger_\nu = -\,\Omega^0_\nu\,\mathscr{A}^\dagger_\nu$, we obtain with the aid of a cyclic permutation

$$(\mathscr{A}_\nu|\mathscr{A}_\mu)^1 = -\frac{1}{\beta}\int_{\alpha=0}^{\beta}\int_{\alpha'=0}^{\alpha}\{\langle(\mathscr{H}^1 - \langle\mathscr{H}^1\rangle^0(\mathrm{e}^{-\alpha' h \mathsf{L}^0}\mathscr{A}_\mu)\,\mathrm{e}^{-\alpha h \mathsf{L}^0}\mathscr{A}_\nu^\dagger\rangle^0$$

$$+ \langle(\mathscr{H}^1 - \langle\mathscr{H}^1\rangle^0)\,(\mathrm{e}^{-\alpha' h \mathsf{L}^0}\mathscr{A}_\nu^\dagger)\,\mathrm{e}^{-\alpha h \mathsf{L}^0}\mathscr{A}_\mu\rangle^0\}\,d\alpha\,d\alpha'$$

$$= -\frac{1}{\beta}\int_{\alpha=0}^{\beta}\int_{\alpha'=0}^{\alpha}\{\mathrm{e}^{-\alpha' h\Omega_\mu^0 + \alpha h\Omega_\nu^0}\langle(\mathscr{H}^1 - \langle\mathscr{H}^1\rangle^0)\mathscr{A}_\mu\mathscr{A}_\nu^\dagger\rangle^0$$

$$+ \mathrm{e}^{\alpha' h\Omega_\nu^0 - \alpha h\Omega_\mu^0}\langle(\mathscr{H}^1 - \langle\mathscr{H}^1\rangle^0)\,\mathscr{A}_\nu^\dagger\,\mathscr{A}_\mu\rangle^0\}\,d\alpha\,d\alpha' \;.$$

13.1(p. 209). From (11.4.8), it follows that

$$\Xi_{\mathrm{exact}}(\omega) - \Xi_{\mathrm{Mark}}(\omega) = \frac{1}{\omega + \mathrm{i}\gamma(\omega)}\,\frac{1}{\omega + \mathrm{i}\Gamma}(\gamma(\omega) - \Gamma) \;.$$

The difference $\gamma(\omega) - \Gamma$ can be estimated as:

$$|\gamma(\omega) - \Gamma| = |\omega|\left|\int_0^\infty \psi(t)\,\mathrm{e}^{\mathrm{i}\omega t}\,dt\right| \leqq |\omega|\int_0^\infty |\psi(t)|\,dt = |\omega|\,\tau_{\mathrm{c}}\,\Gamma$$

by using

$$\gamma(\omega) = -\int_0^\infty \dot{\psi}(t)\,\mathrm{e}^{\mathrm{i}\omega t}dt = \Gamma + \mathrm{i}\omega\int_0^\infty \psi(t)\,\mathrm{e}^{\mathrm{i}\omega t}dt \;.$$

It therefore follows that

$$\left|\frac{\Xi_{\mathrm{exact}}(\omega) - \Xi_{\mathrm{Mark}}(\omega)}{\Xi_{\mathrm{exact}}(\omega)}\right| \leqq \frac{|\omega|}{\sqrt{\omega^2 + \Gamma^2}}\,\Gamma\tau_{\mathrm{c}} \leqq \Gamma\tau_{\mathrm{c}} = \frac{\tau_{\mathrm{c}}}{\tau_{\mathrm{R}}} \;.$$

***13.2**(p. 215). Choosing $\delta\mathscr{G}_\nu = \delta\mathscr{H}_\nu$ and $\nu = 1, 2, 3$, one obtains the desired susceptibility in perturbation theory and in the Markovian approximation (13.2.21):

$$\chi_{MM}(\omega) = \frac{1}{H^2}\left(\chi^{\mathrm{T}}\frac{\Lambda}{\omega\cdot\mathbf{1} + \Lambda}\right)_{11} \;.$$

Here, in lowest order, $\chi^{\mathrm{T}}_{\nu\mu}$ is given by

$$\chi^{\mathrm{T}}_{\nu\mu} = \beta(\delta\mathscr{H}_\nu|\delta\mathscr{H}_\mu)^0 = \beta\langle\delta\mathscr{H}_\nu\delta\mathscr{H}_\mu\rangle^0 = \beta(g^{-1})_{\nu\mu}$$

and $\Lambda_{\nu\mu}$ from (12.1.27) and (13.2.7) by

$$\mathrm{i}\Lambda_{\nu\mu} = -\sum_\varrho g_{\nu\varrho}\left(\mathsf{L}^1\mathscr{H}_\varrho\left|\frac{\mathrm{i}}{\mathsf{L}^0 + \mathrm{i}\eta}\mathsf{L}^1\mathscr{H}_\mu\right.\right)^0 = -\sum_\varrho g_{\nu\varrho}L_{\varrho\mu}(0) \;.$$

The matrix $\Omega_{\nu\mu}$ and the projector Q^0 in $L_{\varrho\mu}$ drop out, due to $[\mathscr{H}_\nu, \mathscr{H}_\mu] = \mathcal{O}$. Decomposing $L_{\varrho\mu}(0)$ from (11.3.17) into

$$L'_{\varrho\mu}(0) = \pi(\mathsf{L}^1\mathscr{H}_\varrho|\delta(\mathsf{L}^0)\mathsf{L}^1\mathscr{H}_\mu)^0 = \pi\langle(\mathsf{L}^1\mathscr{H}_\varrho)^\dagger\delta(\mathsf{L}^0)\mathsf{L}^1\mathscr{H}_\mu\rangle^0_\beta$$

$$= \frac{1}{2}\int_{-\infty}^{\infty}\left\langle\frac{\mathrm{i}}{\hbar}[\mathscr{H}^1,\mathscr{H}_\varrho]\mathrm{e}^{\mathrm{i}\mathsf{L}^0 t}\frac{\mathrm{i}}{\hbar}[\mathscr{H}^1,\mathscr{H}_\mu]\right\rangle^0_\beta dt$$

and $L''_{\varrho\mu}(0)$, we see that, $L''_{\varrho\mu}(0)$ makes no contribution, since

$$L''_{\varrho\mu}(0) = \left(\mathsf{L}^1\mathscr{H}_\varrho\left|\frac{\mathsf{L}^0}{\mathsf{L}^{02}+\eta^2}\mathsf{L}^1\mathscr{H}_\mu\right.\right)^0$$

$$= \frac{1}{\beta\hbar^3}\left\{\left\langle[\mathscr{H}_\varrho,\mathscr{H}^1]\frac{1}{\mathsf{L}^{02}+\eta^2}[\mathscr{H}^1,\mathscr{H}_\mu]\right\rangle^0\right.$$

$$\left.-\left\langle\frac{1}{\mathsf{L}^{02}+\eta^2}[\mathscr{H}^1,\mathscr{H}_\mu]\cdot[\mathscr{H}_\varrho,\mathscr{H}^1]\right\rangle^0_\beta\right\}$$

vanishes due to $[\mathscr{H}_\nu,\mathscr{H}_\mu]=0$ and $\mathscr{H}^0=\Sigma\mathscr{H}_\nu$. According to (13.2.7), the matrix $\mathrm{i}\Lambda_{\nu\mu}$ consists only of the damping matrix $\Gamma_{\nu\mu}$, which has real, non-negative eigenvalues [cf. (13.2.16)]:

$$\mathrm{i}\Lambda_{\nu\mu} = -\Gamma_{\nu\mu} = -\sum_\varrho g_{\nu\varrho}L'_{\varrho\mu}(0)\ .$$

For the discussion of $\chi_{MM}(\omega)$, one can invert the matrix $\omega\cdot\boldsymbol{1}+\boldsymbol{\Lambda}$. For a general investigation, however, it is more expedient to start with the eigenvalue problem

$$\boldsymbol{\Gamma c}^\sigma = \boldsymbol{g L}'(0)\boldsymbol{c}^\sigma = \Gamma_\sigma\boldsymbol{c}^\sigma\ .$$

From the equation

$$\boldsymbol{g}^{1/2}\boldsymbol{L}'(0)\boldsymbol{g}^{1/2}\boldsymbol{g}^{-1/2}\boldsymbol{c}^\sigma = \Gamma_\sigma\boldsymbol{g}^{-1/2}\boldsymbol{c}^\sigma$$

it may be seen that the $\boldsymbol{g}^{-1/2}\boldsymbol{c}^\sigma$ form an orthogonal and complete basis, being the eigenvectors of a real symmetric matrix. Therefore, the matrix $\Gamma_{\nu\mu}$ has the decomposition

$$\Gamma_{\nu\mu} = \sum_\sigma c^\sigma_\nu\Gamma_\sigma\frac{\sum_\varrho c^\sigma_\varrho\chi^{\mathrm{T}}_{\varrho\mu}}{\sum_{\alpha\gamma}c^\sigma_\alpha\chi^{\mathrm{T}}_{\alpha\gamma}c^\sigma_\gamma}\ ,$$

which, with the definition $\tilde{\mathscr{H}}_\sigma = \sum_\nu c^\sigma_\nu\mathscr{H}_\nu$, leads to the representation:

$$\chi_{MM}(\omega) = \sum_\sigma\beta(\delta\mathscr{M}|\delta\tilde{\mathscr{H}}_\sigma)^0\frac{\Gamma_\sigma}{\Gamma_\sigma-\mathrm{i}\omega}\frac{1}{(\delta\tilde{\mathscr{H}}_\sigma|\delta\tilde{\mathscr{H}}_\sigma)^0}(\delta\tilde{\mathscr{H}}_\sigma|\delta\mathscr{M})^0\ .$$

Due to

$$\delta(\mathsf{L}^0)\mathsf{L}^1\left(\sum_\nu \mathscr{H}_\nu\right) = -\,\delta(\mathsf{L}^0)\mathsf{L}^0\mathscr{H}^1 = \mathcal{O}\,,$$

$\Gamma_{\nu\mu}$ always has the eigenvector $c_\varrho^1 = 1$ corresponding to the eigenvalue $\Gamma_1 = 0$ with $\delta\mathscr{H}_1 = \delta\mathscr{H}^0$, so that the sum over σ in $\chi_{MM}(\omega)$ runs only over $\sigma = 2, 3$. The frequency dependence of $\chi_{MM}(\omega)$ is thus determined by a superposition of two Lorenz curves [see Fig. 9.8 (p. 132)].

Particular frequency ranges: a) For $\omega = 0$, we find

$$\begin{aligned}\chi_{MM}(0) &= \sum_{\sigma=2}^{3} \beta(\delta\mathscr{M}\,|\,\delta\tilde{\mathscr{H}}_\sigma)^0 \frac{1}{(\delta\tilde{\mathscr{H}}_\sigma|\delta\mathscr{H}_\sigma)^0}(\delta\tilde{\mathscr{H}}_\sigma|\delta\mathscr{M})^0 \\ &= \beta\left(\delta\mathscr{M}\,\middle|\left\{1 - \frac{|\delta\mathscr{H}^0)(\delta\mathscr{H}^0|}{(\delta\mathscr{H}^0|\delta\mathscr{H}^0)^0}\right\}\delta\mathscr{M}\right)^0 = \chi_{MM}^{\mathrm{ad}}\,,\end{aligned}$$

as a result of the validity of the relations

$$(\delta\tilde{\mathscr{H}}_\sigma|\delta\tilde{\mathscr{H}}_\lambda)^0 = \delta_{\sigma\lambda}(\delta\tilde{\mathscr{H}}_\sigma|\delta\tilde{\mathscr{H}}_\sigma)^0 \qquad \text{and}$$

$$\delta\mathscr{M} = \sum_{\sigma=1}^{3} |\delta\tilde{\mathscr{H}}_\sigma)\frac{1}{(\delta\tilde{\mathscr{H}}_\sigma|\delta\tilde{\mathscr{H}}_\sigma)^0}(\delta\tilde{\mathscr{H}}_\sigma|\delta\mathscr{M})^0$$

which can be used to obtain $\chi(0)$.

b) In the frequency range $\Gamma_2 \ll \omega \ll \Gamma_3$, $\Gamma_2/(\Gamma_2 - \mathrm{i}\omega)$ can be neglected, and $\Gamma_3/(\Gamma_3 - \mathrm{i}\omega)$ can be set equal to 1, so that in this range, it follows that

$$\begin{aligned}\chi_{MM}(\omega) &= \beta(\delta\mathscr{M}\,|\,\delta\tilde{\mathscr{H}}_3)^0 \frac{1}{(\delta\tilde{\mathscr{H}}_3|\delta\tilde{\mathscr{H}}_3)^0}(\delta\tilde{\mathscr{H}}_3|\delta\mathscr{M})^0 \\ &= \beta\left(\delta\mathscr{M}\,\middle|\left\{1 - \sum_{\sigma=1}^{2}\frac{|\delta\tilde{\mathscr{H}}_\sigma)(\delta\tilde{\mathscr{H}}_\sigma|}{(\delta\tilde{\mathscr{H}}_\sigma|\delta\tilde{\mathscr{H}}_\sigma)^0}\right\}\delta\mathscr{M}\right)^0 \\ &= \beta(\delta\mathscr{M}\,|\,\mathsf{Q}_{\{1,\tilde{\mathscr{H}}_1,\tilde{\mathscr{H}}_2\}}\delta\mathscr{M})^0\,.\end{aligned}$$

Since $\tilde{\mathscr{H}}_1 = \mathscr{H}^0$, the operators 1, $\mathscr{H}^0 - \tilde{\mathscr{H}}_2$ and $\tilde{\mathscr{H}}_2$ span the same space as 1, $\tilde{\mathscr{H}}_1$, and $\tilde{\mathscr{H}}_2$. Therefore, with the result of Exercise 9.9, $\chi_{MM}(\omega)$ can be interpreted in the region $\Gamma_2 \ll \omega \ll \Gamma_3$ as the double-adiabatic susceptibility of the partial Hamiltonians $\mathscr{H}^0 - \tilde{\mathscr{H}}_2$ and $\tilde{\mathscr{H}}_2$.

14.1(p. 239). The continued fraction we are seeking can be obtained by calculation of the orthogonal $\hat{\mathscr{G}}_\nu$. Using (14.1.14 and 12), the equations of motion:

$$m\mathrm{i}\mathsf{L}x_\nu = p_\nu\,, \qquad \mathrm{i}\mathsf{L}p_\nu = k(x_{\nu+1} - 2x_\nu + x_{\nu-1})$$

and the Kubo identity, $\hat{\mathscr{G}}_2$ and $\hat{\mathscr{G}}_3$ may be immediately written down. For the determination of the $\hat{\mathscr{G}}_\nu$ ($\nu = 4, 5, \ldots$), we write the result for $\hat{\mathscr{G}}_2$ and $\hat{\mathscr{G}}_3$ with $\mu = 1$ in the expedient form

$$\hat{\mathscr{G}}_{2\mu} = \mathrm{i}m\left(-\frac{k}{m}\right)^{\mu}(x_{\lambda+\mu} - x_{\lambda+\mu-1} - x_{\lambda-\mu+1} + x_{\lambda-\mu}) ,$$

$$\hat{\mathscr{G}}_{2\mu+1} \sim \left(-\frac{k}{m}\right)^{\mu}(p_{\lambda+\mu} + p_{\lambda-\mu}) ,$$

where the expressions

$$(\mathscr{G}_1|\mathscr{G}_1) = \frac{m}{\beta} , \quad (\hat{\mathscr{G}}_{2\mu}|\hat{\mathscr{G}}_{2\mu}) = \left(\frac{k}{m}\right)^{2\mu-1}\frac{2m}{\beta} , \quad (\hat{\mathscr{G}}_{2\mu+1}|\hat{\mathscr{G}}_{2\mu+1}) = \left(\frac{k}{m}\right)^{2\mu}\frac{2m}{\beta}$$

hold. The next step leads to

$$\begin{aligned}\hat{\mathscr{G}}_{2\mu+2} &= \mathsf{L}\hat{\mathscr{G}}_{2\mu+1} - \frac{k}{m}\hat{\mathscr{G}}_{2\mu} \\ &= -\mathrm{i}\left(-\frac{k}{m}\right)^{\mu}(\dot{p}_{\lambda+\mu} + \dot{p}_{\lambda-\mu}) \\ &\quad + \mathrm{i}m\left(-\frac{k}{m}\right)^{\mu+1}(x_{\lambda+\mu} - x_{\lambda+\mu-1} - x_{\lambda-\mu+1} + x_{\lambda-\mu}) , \\ &= \mathrm{i}m\left(-\frac{k}{m}\right)^{\mu+1}(x_{\lambda+\mu+1} - x_{\lambda+\mu} - x_{\lambda-\mu} + x_{\lambda-\mu-1})\end{aligned}$$

with the normalization

$$(\hat{\mathscr{G}}_{2\mu+2}|\hat{\mathscr{G}}_{2\mu+2}) = \frac{1}{\beta\hbar}\langle[\hat{\mathscr{G}}^{\dagger}_{2\mu+1}, \hat{\mathscr{G}}_{2\mu+2}]\rangle = \left(\frac{k}{m}\right)^{2\mu+1}\frac{2m}{\beta} .$$

Correspondingly, we find

$$\begin{aligned}\hat{\mathscr{G}}_{2\mu+3} &= \mathsf{L}\hat{\mathscr{G}}_{2\mu+2} - \frac{m}{k}\hat{\mathscr{G}}_{2\mu+1} \\ &= \left(-\frac{k}{m}\right)^{\mu+1}(p_{\lambda+\mu+1} - p_{\lambda+\mu} - p_{\lambda-\mu} + p_{\lambda-\mu-1}) \\ &\quad + \left(-\frac{k}{m}\right)^{\mu+1}(p_{\lambda+\mu} + p_{\lambda-\mu}) \\ &= \left(-\frac{k}{m}\right)^{\mu+1}(p_{\lambda+\mu+1} + p_{\lambda-\mu-1}) \qquad \text{with}\end{aligned}$$

$$(\hat{\mathscr{G}}_{2\mu+3}|\hat{\mathscr{G}}_{2\mu+3}) = \frac{1}{\beta\hbar}\langle[\hat{\mathscr{G}}^{\dagger}_{2\mu+2}, \hat{\mathscr{G}}_{2\mu+3}]\rangle = \left(\frac{k}{m}\right)^{2\mu+2}\frac{2m}{\beta} .$$

The results for $\hat{\mathcal{G}}_{2(\mu+1)}$ and $\hat{\mathcal{G}}_{2(\mu+1)+1}$ are thus obtained from the starting formulas for $\hat{\mathcal{G}}_{2\mu}$ and $\hat{\mathcal{G}}_{2\mu+1}$ by replacing μ in them with $\mu+1$. Therefore, the formulas hold generally for $\mu = 1, \ldots (N-1)/2$. The continued-fraction coefficients are then given by

$$\frac{(\hat{\mathcal{G}}_2|\hat{\mathcal{G}}_2)}{(\mathcal{G}_1|\mathcal{G}_1)} = 2\frac{k}{m}\,, \qquad \frac{(\hat{\mathcal{G}}_{\nu+1}|\hat{\mathcal{G}}_{\nu+1})}{(\hat{\mathcal{G}}_\nu|\hat{\mathcal{G}}_\nu)} = \frac{k}{m} \quad (\nu = 2, \ldots, N-1)$$

and

$$(\hat{\mathcal{G}}_\nu|\mathsf{L}\hat{\mathcal{G}}_\nu) = \frac{1}{\beta\hbar}\langle[\hat{\mathcal{G}}_\nu^\dagger, \hat{\mathcal{G}}_\nu]\rangle = 0.$$

The memory function $\gamma_{nn}(z)$ vanishes for $n = N$, since then $\mathsf{L}\hat{\mathcal{G}}_N = k/m\hat{\mathcal{G}}_{N-1}$ and therefore $f_n = \mathsf{Q}\mathrm{i}\mathsf{L}\hat{\mathcal{G}}_n = \mathcal{O}$. For the continued fraction (14.2.6) of

$$\Xi(z) = \frac{1}{(p_\lambda|p_\lambda)}\left(p_\lambda\left|\frac{\mathrm{i}}{z+\mathsf{L}}p_\lambda\right.\right),$$

we then find, choosing $n = N$(odd N):

$$\Xi(z) = \cfrac{\mathrm{i}}{z - 2\frac{k}{m}\cfrac{1}{z - \frac{k}{m}\cfrac{1}{z - \frac{k}{m}\cfrac{1}{z \ldots - \frac{k}{m}\cfrac{1}{z - \frac{k}{m}\frac{1}{z}}}}}}\,.$$

For finite N, it represents a rational function of z. In the limit $N \to \infty$, this is however no longer true. For $\gamma(z)$ in

$$\Xi(z) = \frac{\mathrm{i}}{z + \mathrm{i}\gamma(z)}$$

we then find the relation

$$\mathrm{i}\gamma(z) = -2\frac{k}{m}\frac{1}{z + \frac{1}{2}\mathrm{i}\gamma(z)}\,,$$

which leads to

$$\gamma(z) = \mathrm{i}z + \sqrt{\frac{4k}{m} - z^2}\,, \quad \text{or to} \quad \Xi(z) = \frac{1}{\sqrt{\frac{4k}{m} - z^2}}\,.$$

The continued fractions of $\gamma(z)$ or $\Xi(z)$ for finite N and $\operatorname{Im} z > 0$ have the property $\operatorname{Re}\gamma(z) > 0$, $\operatorname{Re}\Xi(z) > 0$; thus, the root which occurs in the limit $N \to \infty$ is to be extracted with

$$-\frac{\pi}{2} < \operatorname{Arg}\sqrt{\frac{4k}{m} - z^2} < \frac{\pi}{2} , \qquad \operatorname{Im} z > 0 .$$

15.1 (p. 248). For (15.1.23), it follows from $\mathsf{L}\mathscr{H}_2 = -\mathsf{L}\mathscr{H}_1$ and the Kubo identity (8.2.18), using $\mathscr{G}_\mu \to \mathscr{H}_\mu$ and $\lambda_\mu \to \beta_\mu$ ($\mu = 1, 2$), that $(\mathscr{H}_\nu|\mathrm{i}\mathsf{L}\mathscr{H}_\mu) = 0$ and

$$(\mathscr{f}_\nu|\mathscr{f}_\mu(t)) = (\mathrm{i}\mathsf{L}\mathscr{H}_\nu|e^{\mathrm{i}\mathsf{QL}t}\mathsf{Q}\mathrm{i}\mathsf{L}\mathscr{H}_\mu) = \begin{pmatrix} +(\mathscr{f}_1|\mathscr{f}_1(t)) & -(\mathscr{f}_1|\mathscr{f}_1(t)) \\ -(\mathscr{f}_1|\mathscr{f}_1(t)) & +(\mathscr{f}_1|\mathscr{f}_1(t)) \end{pmatrix} .$$

Equations (15.1.23) then assume the form

$$\frac{d\langle\mathscr{H}_1\rangle(t)}{dt} = \int_0^t \{\beta_1(t-t') - \beta_2(t-t')\}(\mathscr{f}_1|\mathscr{f}_1(t'))dt' ,$$

$$\frac{d\langle\mathscr{H}_2\rangle(t)}{dt} = \int_0^t \{-\beta_1(t-t') + \beta_2(t-t')\}(\mathscr{f}_1|\mathscr{f}_1(t'))dt' ,$$

or, as one-sided Fourier (Laplace) transforms as in (15.1.24),

$$\frac{d\langle\mathscr{H}_1\rangle}{dt}(\omega) = \{\beta_1(\omega) - \beta_2(\omega)\}L_{11}(\omega) ,$$

$$\frac{d\langle\mathscr{H}_2\rangle}{dt}(\omega) = \{-\beta_1(\omega) + \beta_2(\omega)\}L_{11}(\omega) .$$

One recognizes the temperature difference as an internal driving force. The total energy remains constant, $d\langle\mathscr{H}\rangle/dt = 0$, because no work is performed on the adiabatically isolated system Σ. In the Markovian approximation (15.1.23a), we find in particular

$$\frac{d\langle\mathscr{H}_1\rangle(t)}{dt} = \{\beta_1(t) - \beta_2(t)\}L_{11}(0) ,$$

$$\frac{d\langle\mathscr{H}_2\rangle(t)}{dt} = \{-\beta_1(t) + \beta_2(t)\}L_{11}(0) .$$

15.2 (p. 252). The accompanying $S_{\{\mathscr{G}(h)\}}(t)$ is, by construction, a function of the variables $G_\nu = \langle\mathscr{G}_\nu(h)\rangle(t)$ and $h_\nu(t)$ [cf. (p. 245) and (p. 248)]. For the first derivatives of S, we find from the total differential (15.2.2):

$$\left(\frac{\partial S}{\partial G_\mu}\right)_h = k\lambda_\mu^* \quad \text{and} \quad \left(\frac{\partial S}{\partial h_\nu}\right)_G = -k\sum_\mu \lambda_\mu^* \operatorname{Tr}\left\{\mathscr{R}\frac{\partial\mathscr{G}_\mu(h)}{\partial h_\nu}\right\} .$$

The functions $\lambda_\mu(G, h)$ have the property

$$\lambda_\mu(\langle \mathscr{G} \rangle_\beta, h = 0) = \beta\delta_{\mu 0} \ .$$

The first derivatives in the Taylor expansion are therefore

$$\left(\frac{\partial S}{\partial G_\mu}\right)_h \bigg|_{h=0, \langle \mathscr{G} \rangle_\beta} = k\beta\delta_{\mu 0} \ , \qquad \left(\frac{\partial S}{\partial h_\nu}\right)_G \bigg|_{h=0, \langle \mathscr{G} \rangle_\beta} = - k\beta \operatorname{Tr}\left(\frac{\partial \mathscr{H}}{\partial h_\nu} \mathscr{R}_\beta\right) .$$

The second derivatives at $\langle \mathscr{G}_\nu \rangle_\beta$ and $h = 0$ are

$$\frac{\partial^2 S}{\partial G_\mu \partial G_\nu^*}\bigg|_{h=0, \langle \mathscr{G} \rangle_\beta} = k\left(\frac{\partial \lambda_\mu}{\partial G_\nu}\right)_h^* \bigg|_{h=0, \langle \mathscr{G} \rangle_\beta} ,$$

$$\begin{aligned}
\frac{\partial^2 S}{\partial h_\nu^* \partial G_\mu}\bigg|_{h=0, \langle \mathscr{G} \rangle_\beta} &= \frac{\partial^2 S}{\partial G_\mu \partial h_\nu^*}\bigg|_{h=0, \langle \mathscr{G} \rangle_\beta} \\
&= k\left(\frac{\partial \lambda_\mu}{\partial h_\nu}\right)_G^* \bigg|_{h=0, \langle \mathscr{G} \rangle_\beta} \\
&= - k\sum_\varrho \left(\frac{\partial \lambda_\varrho}{\partial G_\mu}\right)_h \bigg|_{h=0, \langle \mathscr{G} \rangle_\beta} \left\langle \frac{\partial \mathscr{G}_\varrho}{\partial h_\nu}\right\rangle_\beta^* + k\beta\left(\frac{\partial M_\nu}{\partial G_\mu}\right)_h ,
\end{aligned}$$

$$\frac{\partial^2 S}{\partial h_\mu \partial h_\nu^*}\bigg|_{h=0, \langle \mathscr{G} \rangle_\beta} = - k\sum_\varrho \left(\frac{\partial \lambda_\varrho}{\partial h_\nu}\right)_G^* \bigg|_{h=0, \langle \mathscr{G} \rangle_\beta} \left\langle \frac{\partial \mathscr{G}_\varrho}{\partial h_\mu}\right\rangle_\beta .$$

Here, $M_\nu = \operatorname{Tr}(\mathscr{R}\mathscr{M}_\nu(h))$ denotes a special variable G_ν, such that

$$\left(\frac{\partial M_\nu}{\partial G_\mu}\right)_h = \delta_{M_\nu G_\mu}$$

holds. For the same reason, the term

$$- k\beta\left(\frac{\partial}{\partial h_\nu^*} \operatorname{Tr}\left\{\mathscr{R}\frac{\partial \mathscr{H}(h)}{\partial h_\mu}\right\}\right)_G = k\beta\left(\frac{\partial M_\mu}{\partial h_\nu^*}\right)_G \quad \text{in} \quad \frac{\partial^2 S}{\partial h_\mu \partial h_\nu^*}\bigg|_{h=0, \langle \mathscr{G} \rangle_\beta}$$

makes no contribution.

The second derivatives have thus all been expressed in terms of

$$\left(\frac{\partial \lambda_\mu}{\partial G_\nu}\right)_h \bigg|_{h=0, \langle \mathscr{G} \rangle_\beta} .$$

This matrix is found from the relation

$$\sum_\sigma \left(\frac{\partial \lambda_\mu}{\partial G_\sigma}\right)_h \left(\frac{\partial G_\sigma}{\partial \lambda_\nu}\right)_h = \delta_{\nu\mu}$$

as the inverse matrix of

$$\left(\frac{\partial G_\sigma}{\partial \lambda_\varrho}\right)_h\Bigg|_{h=0,\,\langle\mathscr{G}\rangle_\beta} = -(\delta\mathscr{G}_\varrho|\delta\mathscr{G}_\sigma)$$

and is given by

$$\left(\frac{\partial \lambda_\mu}{\partial G_\nu}\right)_h\Bigg|_{h=0,\,\langle\mathscr{G}\rangle_\beta} = -g_{\nu\mu}\ .$$

The result for the expansion of $S_{\{\mathscr{G}(h)\}}(t) = S(G,h)$ is thus found as

$$\begin{aligned} S_{\{\mathscr{G}(h)\}}(t) = S_{\{\mathscr{H}\}} &- \frac{k}{2}\sum_{\nu\mu}\left(G_\mu - \langle\mathscr{G}_\mu\rangle_\beta - \sum_\varrho\left\langle\frac{\partial\mathscr{G}_\mu}{\partial h_\varrho}\right\rangle_\beta h_\varrho\right) \\ &\times g_{\mu\nu}\left(G_\nu - \langle\mathscr{G}_\nu\rangle_\beta - \sum_\sigma\left\langle\frac{\partial\mathscr{G}_\nu}{\partial h_\sigma}\right\rangle_\beta h_\sigma\right)^* \\ &+ k\beta\left\{\langle\mathscr{H}(h)\rangle(t) - \langle\mathscr{H}\rangle_\beta + \sum_\nu G_\nu h_\nu^* - \frac{1}{2}\sum_{\nu\mu} h_\mu\left\langle\frac{\partial\mathscr{G}_\nu}{\partial h_\mu}\right\rangle_\beta h_\nu^*\right\}\ . \end{aligned}$$

In this form, only the instantaneous values of the observables $\langle\mathscr{G}_\nu(h)\rangle(t) = G_\nu$ enter. No restrictions on the dynamics of the $\langle\mathscr{G}_\nu(h)\rangle(t)$ were required.

If, however, we use the condition of adiabatic isolation,

$$\begin{aligned} \frac{d}{dt}\langle\mathscr{H}(h)\rangle(t) &= \left\langle\left(\frac{\partial\mathscr{H}(h)}{\partial t}\right)_{\mathrm{ex}}\right\rangle(t) = \sum_\nu\left\langle\frac{\partial\mathscr{H}(h)}{\partial h_\nu^*}\right\rangle(t)\dot{h}_\nu^*(t) \\ &= -\sum_\nu\langle\mathscr{G}_\nu(h)\rangle(t)\dot{h}_\nu^*(t)\ , \end{aligned}$$

then the following relation holds:

$$\begin{aligned} \langle\mathscr{H}(h)\rangle(t) &= \langle\mathscr{H}\rangle(0) - \int_0^t\sum_\nu\langle\mathscr{G}_\nu(h)\rangle(t')\dot{h}_\nu^*(t')dt' \\ &= \langle\mathscr{H}\rangle(0) - \sum_\nu\langle\mathscr{G}_\nu(h)\rangle(t)h_\nu^*(t) + \frac{1}{2}\sum_{\mu\nu}h_\mu(t)\left\langle\frac{\partial\mathscr{G}_\nu}{\partial h_\mu}\right\rangle_\beta h_\nu^*(t) \\ &\quad + \int_0^t\sum_\nu h_\nu^*(t')\frac{d}{dt'}\left\{\langle\mathscr{G}_\nu(h)\rangle(t') - \sum_\mu\left\langle\frac{\partial\mathscr{G}_\nu}{\partial h_\mu}\right\rangle_\beta h_\mu(t')\right\}dt'\ . \end{aligned}$$

Setting this expression into $S_{\{\mathscr{G}(h)\}}(t)$ and employing the linear Kubo formula for all the $\langle\mathscr{G}_\nu(h)\rangle(t)$ which occur, we obtain the functional expansion of $S_{\{\mathscr{G}(h)\}}(t)$ in terms of the fields up to second order (15.2.7). Here, we have employed the convention $h_\mu(t) = 0$ for $\mathscr{G}_\mu \neq \mathscr{M}_\nu$, as in Chap. 15; i.e. only those fields were considered which enter $\mathscr{H}(h)$.

***17.1**(p. 294). We make use of the Nakajima–Zwanzig equation for $\mathcal{F}_\nu = \mathcal{P}_\nu$ and $\rho_\mu = \mathcal{P}_\mu$ with the projector (17.2.16). The latter is found with the help of the matrix $v_{\nu\mu} = \delta_{\nu\mu} d_\mu$ which is the inverse of

$$\mathrm{Tr}(\mathcal{F}_\nu \rho_\mu) = \mathrm{Tr}(\mathcal{P}_\nu \mathcal{P}_\mu) = \delta_{\nu\mu}\mathrm{Tr}(\mathcal{P}_\mu) = \delta_{\nu\mu} d_\mu$$

and is given by

$$\mathcal{P}\mathcal{X} = \sum_\mu \mathcal{P}_\mu \frac{1}{d_\mu}\mathrm{Tr}(\mathcal{P}_\mu \mathcal{X}) \ .$$

On inserting in (17.2.28) and taking the expectation values $\langle \mathcal{P}_\nu \rangle = w_\nu$ we obtain

$$\frac{d}{dt} w_\nu(t) = -\int_{t_0}^{t} \sum_\mu \mathrm{Tr}(\mathcal{P}_\nu \mathsf{L}\mathsf{Q}e^{-i\mathsf{L}\mathsf{Q}(t-t')}\mathsf{L}\mathcal{P}_\mu)\frac{1}{d_\mu} w_\mu(t')dt' \ .$$

The term PLP makes no contribution, since the $\mathcal{P}_\nu$ commute:

$$\mathrm{Tr}(\mathcal{P}_\nu \mathsf{L}\mathcal{P}_\mu) = \frac{1}{\hbar}\mathrm{Tr}([\mathcal{P}_\mu, \mathcal{P}_\nu]\mathcal{H}) = 0 \ .$$

With the abbreviation

$$-\mathrm{Tr}(\mathcal{P}_\nu \mathsf{L}\mathsf{Q}e^{-i\mathsf{L}\mathsf{Q}t}\mathsf{L}\mathcal{P}_\mu)\frac{1}{d_\mu} = p_{\nu\mu}(t)$$

we may write the result in the form

$$\frac{dw_\nu(t)}{dt} = \int_{t_0}^{t} \sum_\mu p_{\nu\mu}(t-t') w_\mu(t')dt' \ ,$$

where the coefficients $p_{\nu\nu}(t-t')$ can also be eliminated by using $\sum_\mu p_{\mu\nu}(t-t') = 0$; we thus finally obtain the form of rate equations or balance equations:

$$\frac{dw_\nu(t)}{dt} = \int_{t_0}^{t} \sum_\mu \{p_{\nu\mu}(t-t')w_\mu(t') - p_{\mu\nu}(t-t')w_\nu(t')\}\, dt' \ .$$

17.2(p. 294). With the definition (17.2.35) of P and the definition of L^{12} as a commutation relation, one obtains for $\mathsf{PL}^{12}\mathsf{P}$.

$$\mathsf{PL}^{12}\mathsf{P}\mathcal{X} = \mathcal{R}^2\mathrm{Tr}_2\left\{\frac{1}{\hbar}[\mathcal{H}^{12}, \mathcal{R}^2\,\mathrm{Tr}_2\mathcal{X}]\right\} = \mathcal{R}^2\mathrm{Tr}_2\left\{\frac{1}{\hbar}\mathcal{H}^{12}\mathcal{R}^2(\mathrm{Tr}_2\mathcal{X})\right\}$$

$$- \mathcal{R}_2\mathrm{Tr}_2\left\{\mathcal{R}^2(\mathrm{Tr}_2\mathcal{X})\frac{1}{\hbar}\mathcal{H}^{12}\right\} \ .$$

Taking into account that $\mathrm{Tr}_2\mathcal{X}$ depends only on the variables of system 1 and that therefore $\mathscr{R}^2(\mathrm{Tr}_2\mathscr{X}) = (\mathrm{Tr}_2\mathscr{X})\mathscr{R}^2$, we find due to the cyclic invariance of the trace in Σ^2 and using (17.2.53),

$$\mathsf{PL}^{12}\mathsf{P}\mathscr{X} = \mathscr{R}^2\frac{1}{\hbar}\langle\mathscr{H}^{12}\rangle_2\mathrm{Tr}_2\mathscr{X} - \mathscr{R}^2(\mathrm{Tr}_2\mathscr{X})\frac{1}{\hbar}\langle\mathscr{H}^{12}\rangle_2 \ .$$

For $\langle\mathscr{H}^{12}\rangle_2$, which contains no variables of Σ^2, we can write either

$$\mathsf{PL}^{12}\mathsf{P}\mathscr{X} = \mathscr{R}^2\mathrm{Tr}_2\left\{\frac{1}{\hbar}\langle\mathscr{H}^{12}\rangle_2\mathscr{X} - \mathscr{X}\frac{1}{\hbar}\langle\mathscr{X}^{12}\rangle_2\right\} = \mathsf{P}\left[\frac{1}{\hbar}\langle\mathscr{H}^{12}\rangle_2, \mathscr{X}\right]$$

or

$$\mathsf{PL}^{12}\mathsf{P}\mathscr{X} = \frac{1}{\hbar}\langle\mathscr{H}^{12}\rangle_2\mathscr{R}^2\mathrm{Tr}_2\mathscr{X} - \mathscr{R}^2(\mathrm{Tr}_2\mathscr{X})\frac{1}{\hbar}\langle\mathscr{H}^{12}\rangle_2$$

$$= \frac{1}{\hbar}[\langle\mathscr{H}^{12}\rangle_2, \mathsf{P}\mathscr{X}] \ .$$

Using these results, we can obtain the expressions (17.2.51) and (17.2.52) for $\mathsf{PL}^{12}\mathsf{Q}\mathscr{X}$ and $\mathsf{QL}^{12}\mathsf{P}\mathscr{X}$. With L^{12} as a commutator and $\mathsf{Q} = 1 - \mathsf{P}$, we find

$$\mathsf{PL}^{12}\mathsf{Q}\mathscr{X} = \mathsf{PL}^{12}\mathscr{X} - \mathsf{PL}^{12}\mathsf{P}\mathscr{X} = \mathsf{P}\frac{1}{\hbar}[\mathscr{H}^{12}, \mathscr{X}] - \mathsf{P}\frac{1}{\hbar}[\langle\mathscr{H}^{12}\rangle_2, \mathscr{X}]$$

$$= \mathsf{P}\frac{1}{\hbar}[\mathscr{H}^{12} - \langle\mathscr{H}^{12}\rangle_2, \mathscr{X}] \ ;$$

$$\mathsf{QL}^{12}\mathsf{P}\mathscr{X} = \mathsf{L}^{12}\mathsf{P}\mathscr{X} - \mathsf{PL}^{12}\mathsf{P}\mathscr{X} = \frac{1}{\hbar}[\mathscr{H}^{12}, \mathsf{P}\mathscr{X}] - \frac{1}{\hbar}[\langle\mathscr{H}^{12}\rangle_2, \mathsf{P}\mathscr{X}]$$

$$= \frac{1}{\hbar}[\mathscr{H}^{12} - \langle\mathscr{H}^{12}\rangle_2, \mathsf{P}\mathscr{X}] \ .$$

***17.3**(p. 294). The time derivative of $\rho_{\mathrm{rel}}(t)$ (17.2.59), together with the von Neumann equation and $\mathsf{P} + \mathsf{Q} = 1$, leads to

$$\frac{d}{dt}\rho_{\mathrm{rel}}(t) = \frac{d}{dt}\mathsf{P}(t)\rho(t) = \mathsf{P}(t)\dot{\rho}(t) + \dot{\mathsf{P}}(t)\rho(t) = -\mathrm{i}\{\mathsf{P}(t)\mathsf{L}(t) + \mathrm{i}\dot{\mathsf{P}}(t)\}\rho(t)$$

$$= -\mathrm{i}\{\mathsf{P}(t)\mathsf{L}(t) + \mathrm{i}\dot{\mathsf{P}}(t)\}\mathsf{P}(t)\rho(t) - \mathrm{i}\{\mathsf{P}(t)\mathsf{L}(t) + \mathrm{i}\dot{\mathsf{P}}(t)\}\mathsf{Q}(t)\rho(t) \ .$$

The first term already contains $\rho_{\mathrm{rel}}(t) = \mathsf{P}(t)\rho(t)$; the second term, $\mathsf{Q}(t)\rho(t) = \mathsf{Q}(t)\mathsf{U}(t, t_0)\rho(t_0)$, can be expressed in terms of $\rho_{\mathrm{rel}}(t')$ using the identity (3.4.4b). Choosing

$$\mathsf{L}_1(t) = \mathsf{Q}(t)\mathsf{L}(t) - \mathrm{i}\dot{\mathsf{P}}(t) \ , \qquad \mathsf{L}_2(t) = \mathsf{L} - \mathsf{L}_1 = \mathsf{P}(t)\mathsf{L}(t) + \mathrm{i}\dot{\mathsf{P}}(t)$$

and using the symbol $\mathsf{V}(t, t')$ instead of $\mathsf{U}_1(t, t')$, and the fact that

$$\mathsf{L}_2(t')\mathsf{U}(t', t_0)\rho(t_0) = \{\mathsf{P}(t')\mathsf{L}(t') + \mathrm{i}\dot{\mathsf{P}}(t')\}\rho(t')$$

$$= \mathrm{i}\frac{d}{dt'}\mathsf{P}(t')\rho(t') \quad \text{for} \quad \mathsf{Q}(t)\rho(t)$$

we find

$$\mathsf{Q}(t)\rho(t) = \mathsf{Q}(t)\left\{\mathsf{V}(t, t_0)\rho(t_0) + \int_{t_0}^{t} \mathsf{V}(t, t')\frac{\partial}{\partial t'}\{\mathsf{P}(t')\rho(t')\}\,dt'\right\} .$$

Since, according to Exercise 3.5, $\mathsf{V}(t, t')$ obeys the differential equation

$$\frac{\partial}{\partial t'}\mathsf{V}(t, t') = \mathsf{V}(t, t')\mathrm{i}\mathsf{L}_1(t') = \mathsf{V}(t, t')\mathrm{i}\{\mathsf{Q}(t')\mathsf{L}(t') - \mathrm{i}\dot{\mathsf{P}}(t')\} ,$$

after partial integration over t' and taking into account $\mathsf{V}(t, t) = 1$, (17.2.62), and $\mathsf{Q}(t)\,\mathsf{P}(t) = 0$, we obtain

$$\mathsf{Q}(t)\rho(t) = -\mathrm{i}\int_{t_0}^{t} \mathsf{Q}(t)\mathsf{V}(t, t')\{\mathsf{Q}(t')\mathsf{L}(t') - \mathrm{i}\dot{\mathsf{P}}(t')\}\,\mathsf{P}(t')\rho_{\mathrm{rel}}(t')dt' .$$

If we insert this expression for $\mathsf{Q}(t)\rho(t)$ into the time derivative $d\rho_{\mathrm{rel}}/dt$, which was already calculated, we arrive at (17.2.63) for $\rho_{\mathrm{rel}}(t)$, which we were seeking. Here, we use the fact that we can write $\mathsf{P}(t)\dot{\mathsf{P}}(t)\mathsf{Q}(t)$ instead of $\dot{\mathsf{P}}(t)\mathsf{Q}(t)$ due to the relations $\mathsf{P}(t)^2 = \mathsf{P}(t)$ and $\mathsf{P}(t)\dot{\mathsf{P}}(t) + \dot{\mathsf{P}}(t)\mathsf{P}(t) = \dot{\mathsf{P}}(t)$.

As in the case of a constant operator P [cf. (17.2.26)], the time-evolution operator $\mathsf{Q}(t)\mathsf{V}(t, t')\mathsf{Q}(t')$ in the integral kernel can be represented in different forms due to $\mathsf{Q}(t)^2 = \mathsf{Q}(t)$. Employing the differential equation in t' and $\mathsf{P}\dot{\mathsf{P}} + \dot{\mathsf{P}}\mathsf{P} = \dot{\mathsf{P}}$ and the initial conditions for $t = t'$, it can be shown that

$$\mathsf{Q}(t)\mathsf{V}(t, t')\mathsf{Q}(t') = \mathsf{V}(t, t')\mathsf{Q}(t') = \mathsf{W}(t, t')\mathsf{Q}(t') ,$$

where $\mathsf{W}(t, t')$ is defined like V, but with $\mathsf{L}_1(t)\mathsf{Q}(t)$, by

$$\frac{\partial}{\partial t}\mathsf{W}(t, t') = -\mathrm{i}\{\mathsf{Q}(t)\mathsf{L}(t)\mathsf{Q}(t) - \mathrm{i}\dot{\mathsf{P}}(t)\mathsf{Q}(t)\}\mathsf{W}(t, t') .$$

This result follows from the differential equations

$$\frac{\partial}{\partial t'}\{\mathsf{V}(t, t')\mathsf{Q}(t')\} = \mathsf{V}(t, t')\mathrm{i}\{\mathsf{Q}(t')\mathsf{L}(t')\mathsf{Q}(t') - \mathrm{i}\dot{\mathsf{P}}(t')\mathsf{Q}(t') - \mathrm{i}\dot{\mathsf{Q}}(t')\}$$

$$= \mathsf{V}(t, t')\mathsf{Q}(t')\mathrm{i}\{\mathsf{L}(t')\mathsf{Q}(t') + \mathrm{i}\dot{\mathsf{P}}(t')\mathsf{P}(t')\}$$

and

$$\frac{\partial}{\partial t'}\{\mathsf{W}(t, t')\mathsf{Q}(t')\} = \mathsf{W}(t, t')\mathrm{i}\{\mathsf{Q}(t')\mathsf{L}(t')\mathsf{Q}(t') - \mathrm{i}\dot{\mathsf{P}}(t')\mathsf{Q}(t') - \mathrm{i}\dot{\mathsf{Q}}(t')\}$$

$$= \mathsf{W}(t, t')\mathsf{Q}(t')\mathrm{i}\{\mathsf{L}(t')\mathsf{Q}(t') + \mathrm{i}\dot{\mathsf{P}}(t')\mathsf{P}(t')\}$$

with

$$\mathsf{V}(t,t')\mathsf{Q}(t')|_{t=t'} = \mathsf{W}(t,t')\mathsf{Q}(t')|_{t=t'} = \mathsf{Q}(t)$$

or

$$\mathsf{Q}(t)\mathsf{V}(t,t')\mathsf{Q}(t')|_{t=t'} = \mathsf{Q}(t) \ .$$

18.1(p. 306). To calculate (18.2.21), we begin with the definition (18.2.19):

$$\mathsf{P}[\mathscr{R}(t)]\mathscr{Y} = \left\{\mathscr{R}(t) - \sum_{\nu=1}^{n} \frac{\partial\mathscr{R}(t)}{\partial\langle\mathscr{G}_\nu\rangle(t)}\langle\mathscr{G}_\nu\rangle(t)\right\}\mathrm{Tr}\,\mathscr{Y} + \sum_{\nu=1}^{N} \frac{\partial\mathscr{R}}{\partial\langle\mathscr{G}_\nu\rangle(t)}\mathrm{Tr}(\mathscr{G}_\nu\mathscr{Y}) \ ,$$

and set $\mathscr{Y}$ equal to

$$\mathscr{Y} = \mathsf{P}[\mathscr{R}(t')]\mathscr{X} = \left\{\mathscr{R}(t') - \sum_{\mu=1}^{N} \frac{\partial\mathscr{R}(t')}{\partial\langle\mathscr{G}_\mu\rangle(t')}\langle\mathscr{G}_\mu\rangle(t')\right\}\mathrm{Tr}\,\mathscr{X} + \sum_{\mu=1}^{N} \frac{\partial\mathscr{R}(t')}{\partial\langle\mathscr{G}_\mu\rangle(t')}\mathrm{Tr}(\mathscr{G}_\mu\mathscr{X}) \ .$$

Exchanging the order of the traces and the partial derivatives and considering that $\mathrm{Tr}\,\mathscr{R} = 1$ and $\mathrm{Tr}(\mathscr{G}_\mu\mathscr{R}(t')) = \langle\mathscr{G}_\mu\rangle(t')$, we find

$$\mathrm{Tr}\frac{\partial\mathscr{R}(t')}{\partial\langle\mathscr{G}_\mu\rangle(t')} = \frac{\partial}{\partial\langle\mathscr{G}_\mu\rangle(t')}\mathrm{Tr}\,\mathscr{R} = 0 \qquad \text{and}$$

$$\mathrm{Tr}\,\mathscr{G}_\nu\frac{\partial\mathscr{R}(t')}{\partial\langle\mathscr{G}_\mu\rangle(t')} = \frac{\partial}{\partial\langle\mathscr{G}_\mu\rangle(t')}\mathrm{Tr}(\mathscr{G}_\nu\mathscr{R}(t')) = \delta_{\nu\mu} \ ;$$

with these relations and the definition of $\mathscr{Y}$, we obtain:

$$\mathrm{Tr}\,\mathscr{Y} = \{\mathrm{Tr}\mathscr{R}(t)\}\,\mathrm{Tr}\,\mathscr{X} + 0 = \mathrm{Tr}\,\mathscr{X} \ ;$$

$$\mathrm{Tr}(\mathscr{G}_\nu\mathscr{Y}) = \left\{\langle\mathscr{G}_\nu\rangle(t') - \sum_{\mu=1}^{n} \delta_{\nu\mu}\langle\mathscr{G}_\mu\rangle(t')\right\}\mathrm{Tr}\,\mathscr{X} + \sum_{\mu=1}^{N} \delta_{\nu\mu}\mathrm{Tr}(\mathscr{G}_\mu\mathscr{X}) = \mathrm{Tr}(\mathscr{G}_\nu\mathscr{X}) \ ,$$

and therefore

$$\mathsf{P}[\mathscr{R}(t)]\mathscr{Y} = \mathsf{P}[\mathscr{R}(t)]\mathsf{P}[\mathscr{R}(t')]\mathscr{X} = \mathsf{P}[\mathscr{R}(t)]\mathscr{X} \ .$$

18.2(p. 306). Specializing (17.3.28) to $\rho_{\mathrm{rel}}(t) = \mathscr{R}(t)$ yields the difference of the statistical operator and the generalized canonical density operator $\mathscr{R}_{\{\mathscr{G}\}}(t)$:

$$\rho(t) = \mathscr{R}_{\{\mathscr{G}\}}(t) - \mathrm{i}\int_{t_0}^{t} \mathsf{T}(t,t')\mathsf{Q}[\mathscr{R}(t')]\,\mathsf{L}\mathscr{R}_{\{\mathscr{G}\}}(t')dt' \ .$$

Taking the trace with the observable $\mathscr{F}$ leads to

$$\mathrm{Tr}\{\mathscr{F}\rho(t)\} = \mathrm{Tr}\{\mathscr{F}\mathscr{R}_{\{\mathscr{G}\}}(t)\} - \mathrm{i}\int_{t_0}^{t} \mathrm{Tr}\{\mathscr{F}\mathsf{T}(t,t')\mathsf{Q}[\mathscr{R}(t')]\mathsf{L}\mathscr{R}_{\{\mathscr{G}\}}(t')\}\,dt' \ .$$

The question as to whether $\mathscr{R}_{\{\mathscr{G}\}}(t)$ is sufficient for the calculation of $\mathscr{F}$ depends on the magnitude of the integral term.

***18.3**(p. 306). On expanding the right-hand side of (18.2.23) in terms of the deviations from equilibrium $\langle\mathscr{G}_\nu(t)\rangle - \langle\mathscr{G}_\nu\rangle_\beta$, we obtain the first-order terms by taking into account $\mathsf{L}\mathsf{R} = \mathsf{L}(\mathscr{R}_\beta + \mathscr{R}^1 + \ldots) = \mathsf{L}\mathscr{R}^1 + \ldots$ to first order, and all the projectors which occur to zeroth order (i.e. in equilibrium):

$$\mathsf{P}[\mathscr{R}] \to \mathsf{P}[\mathscr{R}_\beta] \ .$$

Here, the projector $\mathsf{P}[\mathscr{R}_\beta]$ is given according to (18.2.19) by

$$\mathsf{P}[\mathscr{R}_\beta]\mathscr{Y} = \mathscr{R}_\beta \mathrm{Tr}\,\mathscr{Y} + \sum_\nu \frac{\partial\mathscr{R}}{\partial\langle\mathscr{G}_\nu\rangle}\bigg|_\beta \mathrm{Tr}\{(\mathscr{G}_\nu - \langle\mathscr{G}_\nu\rangle_\beta)\mathscr{Y}\} \ .$$

Since $\mathsf{P}[\mathscr{R}_\beta]$ is constant in time, it follows for the evolution operator T in equilibrium

$$\mathsf{T}(t,t') = \mathrm{e}^{-\mathrm{i}\mathsf{Q}[\mathscr{R}_\beta]\mathsf{L}(t-t')}$$

and thus for the linear approximation of (18.2.23) with $t_0 = 0$,

$$\begin{aligned}\frac{d}{dt}\langle\mathscr{G}_\nu\rangle(t) = &-\mathrm{i}\sum_\lambda \mathrm{Tr}\left(\mathscr{G}_\nu\mathsf{L}\frac{\partial\mathscr{R}}{\partial\langle\mathscr{G}_\lambda\rangle}\bigg|_\beta\right)(\langle\mathscr{G}_\lambda\rangle(t) - \langle\mathscr{G}_\lambda\rangle_\beta)\\ &-\int_0^t\sum_\lambda \mathrm{Tr}\left(\mathscr{G}_\nu\mathsf{L}\mathrm{e}^{-\mathrm{i}\mathsf{Q}[\mathscr{R}_\beta]\mathsf{L}(t-t')}\mathsf{Q}[\mathscr{R}_\beta]\mathsf{L}\frac{\partial\mathscr{R}}{\partial\langle\mathscr{G}_\lambda\rangle}\bigg|_\beta\right)\\ &\times(\langle\mathscr{G}_\lambda\rangle(t') - \langle\mathscr{G}_\lambda\rangle_\beta)\,dt \ .\end{aligned}$$

The result therefore has the form

$$\begin{aligned}\frac{d}{dt}\langle\mathscr{G}_\nu\rangle(t) = &\sum_\lambda(\langle\mathscr{G}_\lambda\rangle(t) - \langle\mathscr{G}_\lambda\rangle_\beta)\mathrm{i}\hat{\Omega}_{\lambda\nu}\\ &-\int_0^t\sum_\lambda(\langle\mathscr{G}_\lambda\rangle(t') - \langle\mathscr{G}_\lambda\rangle_\beta)\hat{\gamma}_{\lambda\nu}(t-t')\,dt' \ ,\end{aligned}$$

with the notation

$$\hat{\Omega}_{\mu\nu} = -\mathrm{Tr}\left(\mathscr{G}_\nu\mathsf{L}\frac{\partial\mathscr{R}}{\partial\langle\mathscr{G}_\mu\rangle}\bigg|_\beta\right) \quad \text{and}$$

$$\hat{\gamma}_{\mu\nu}(t) = \mathrm{Tr}\left(\mathscr{G}_\nu\mathsf{L}\mathrm{e}^{-\mathrm{i}\mathsf{Q}[\mathscr{R}_\beta]\mathsf{L}t}\mathsf{Q}[\mathscr{R}_\beta]\mathsf{L}\frac{\partial\mathscr{R}}{\partial\langle\mathscr{G}_\mu\rangle}\bigg|_\beta\right) \ .$$

The equations obtained are to be compared with the Mori equations (11.4.30); the structures of the two systems of equations are the same. It remains to show that the frequency and memory matrices are equal.

We begin with the derivatives

$$\left.\frac{\partial \mathcal{R}}{\partial \langle \mathcal{G}_\mu \rangle}\right|_\beta = \sum_\nu \left.\frac{\partial \mathcal{R}}{\partial \lambda_\nu}\right|_\beta \left.\frac{\partial \lambda_\nu}{\partial \langle \mathcal{G}_\mu \rangle}\right|_\beta .$$

Since, on the one hand, the following relation holds:

$$\sum_\nu \left.\frac{\partial \langle \mathcal{G}_\varrho \rangle}{\partial \lambda_\nu}\right|_\beta \left.\frac{\partial \lambda_\nu}{\partial \langle \mathcal{G}_\mu \rangle}\right|_\beta = \delta_{\varrho\mu} ,$$

and, on the other, we can write $(\partial \langle \mathcal{G}_\varrho \rangle / \partial \lambda_\nu)|_\beta$ with

$$\left.\frac{\partial \mathcal{R}}{\partial \lambda_\nu}\right|_\beta = -\frac{1}{\beta}\int_0^\beta \mathrm{e}^{-\alpha\mathcal{H}}(\mathcal{G}_\nu - \langle \mathcal{G}_\nu \rangle_\beta)\mathrm{e}^{\alpha\mathcal{H}}\,\mathcal{R}_\beta d\alpha$$

as a Mori scalar product,

$$\left.\frac{\partial \langle \mathcal{G}_\varrho \rangle}{\partial \lambda_\nu}\right|_\beta = -(\delta\mathcal{G}_\varrho | \delta\mathcal{G}_\nu) ,$$

it follows that $(\partial \lambda_\nu / \partial \langle \mathcal{G}_\mu \rangle)|_\beta$ represents the inverse matrix to $-(\delta\mathcal{G}_\varrho|\delta\mathcal{G}_\nu)$, which was denoted in the Mori theory by $g_{\nu\mu}$,

$$\left.\frac{\partial \lambda_\nu}{\partial \langle \mathcal{G}_\mu \rangle}\right|_\beta = -g_{\nu\mu} = -g_{\mu\nu} .$$

We have thus obtained the relation

$$\left.\frac{\partial \mathcal{R}}{\partial \langle \mathcal{G}_\mu \rangle}\right|_\beta = -\sum_\varrho \left.\frac{\partial \mathcal{R}}{\partial \lambda_\varrho}\right|_\beta g_{\mu\varrho} .$$

Here, we have again used the abbreviation $\delta\mathcal{G}_\nu = \mathcal{G}_\nu - \langle \mathcal{G}_\nu \rangle_\beta$.

Next, we consider the frequency matrix. After inserting $\partial\mathcal{R}/\partial\langle \mathcal{G}_\mu \rangle_\beta$, we find using (3.1.15) and the definition of the Mori scalar product the following relation:

$$\hat{\Omega}_{\mu\nu} = \sum_\varrho \mathrm{Tr}\left(\mathcal{G}_\nu \mathsf{L} \left.\frac{\partial \mathcal{R}}{\partial \lambda_\varrho}\right|_\beta\right) g_{\mu\varrho} = \sum_\varrho - g_{\mu\varrho}\mathrm{Tr}\left(\left.\frac{\partial \mathcal{R}}{\partial \lambda_\varrho}\right|_\beta \mathsf{L}\mathcal{G}_\nu\right) = \sum_\varrho g_{\mu\varrho}(\delta\mathcal{G}_\varrho | \mathsf{L}\delta\mathcal{G}_\nu) .$$

The frequency matrix $\hat{\Omega}_{\mu\nu}$ is therefore identical with that defined in (11.2.15) in the Mori theory,

$$\hat{\Omega}_{\mu\nu} = \Omega_{\mu\nu} .$$

To rearrange the memory matrix $\hat{\gamma}_{\nu\mu}(t)$, we express the effect of $\mathsf{P}\{\mathcal{R}_\beta\}$ in terms of Mori scalar products. For arbitrary $\mathcal{X}$ and $\mathcal{Y}$ we have

$$\mathrm{Tr}(\mathscr{Y}\mathsf{P}[\mathscr{R}_\beta]\mathscr{X}) = \mathrm{Tr}(\mathscr{Y}\mathscr{R}_\beta)\mathrm{Tr}(\mathscr{X}) - \sum_{\varrho\nu}\mathrm{Tr}\left(\mathscr{Y}\frac{\partial\mathscr{R}}{\partial\lambda_\varrho}\bigg|_\beta g_{\nu\varrho}\mathrm{Tr}(\delta\mathscr{G}_\nu\mathscr{X})\right)$$

$$= \mathrm{Tr}(\mathscr{X}\mathit{1})(\mathit{1}|\mathscr{Y}) + \sum_{\nu\varrho}\mathrm{Tr}(\mathscr{X}\delta\mathscr{G}_\nu)g_{\nu\varrho}(\delta\mathscr{G}_\varrho|\mathscr{Y})$$

$$= \mathrm{Tr}(\mathscr{X}\mathsf{P}_\mathrm{M}\mathscr{Y}) \ .$$

Here, P_M is the Mori projector introduced previously,

$$\mathsf{P}_\mathrm{M} = |\mathit{1})(\mathit{1}| + \sum_{\nu\varrho}|\delta\mathscr{G}_\nu)g_{\nu\varrho}(\delta\mathscr{G}_\varrho| \ .$$

For $\mathsf{LQ}[\mathscr{R}_\beta]$ it thus follows with (3.1.15) that

$$\mathrm{Tr}(\mathscr{Y}\mathsf{LQ}[\mathscr{R}_\beta]\mathscr{X}) = -\mathrm{Tr}(\mathscr{X}\mathsf{Q}_\mathrm{M}\mathsf{L}\mathscr{Y}) \ .$$

Expanding

$$\mathsf{L}\mathrm{e}^{-\mathrm{i}\mathsf{Q}[\mathscr{R}_\beta]\mathsf{L}t}\mathsf{Q}[\mathscr{R}_\beta] = \mathsf{LQ}[\mathscr{R}_\beta]\mathrm{e}^{-\mathrm{i}\mathsf{LQ}[\mathscr{R}_\beta]t}$$

in powers of $\mathsf{LQ}[\mathscr{R}_\beta]$, we obtain an expression for the memory function $\hat{\gamma}_{\mu\nu}(t)$:

$$\hat{\gamma}_{\mu\nu}(t) = -\sum_\varrho g_{\mu\varrho}\mathrm{Tr}\left(\frac{\partial\mathscr{R}}{\partial\lambda_\varrho}\bigg|_\beta \mathsf{L}\mathrm{e}^{\mathrm{i}\mathsf{Q}_\mathrm{M}\mathsf{L}t}\mathsf{Q}_\mathrm{M}\mathsf{L}\mathscr{G}_\nu\right) = \sum_\varrho g_{\mu\varrho}(\delta\mathscr{G}_\varrho|\mathsf{L}\mathrm{e}^{\mathrm{i}\mathsf{Q}_\mathrm{M}\mathsf{L}t}\mathsf{Q}_\mathrm{M}\mathsf{L}\delta\mathscr{G}_\nu) \ .$$

Comparing with (11.2.23), we see that $\hat{\gamma}_{\mu\nu}(t)$ is identical to the Mori memory matrix $\gamma_{\mu\nu}(t)$,

$$\hat{\gamma}_{\mu\nu}(t) = \gamma_{\mu\nu}(t) \ .$$

Finally, we note that these results, obtained here using the linearized Robertson equation, can also be derived with the Nakajima–Zwanzig equation, starting with a set of observables $\mathscr{F}_\nu \to \mathit{1}, \cdot\mathscr{G}_\varrho$ and a basis set $\rho_\nu \to \mathscr{R}_\beta, \partial\mathscr{R}/\partial\lambda_\varrho|_\beta$. The projector which is then obtained from (17.2.16) via $v_{\nu\mu}$ is just $\mathsf{P}[\mathscr{R}_\beta]$ (for the relationship to the Mori theory, see also Appendix A).

18.4(p. 306). We have

$$\mathsf{L}\mathscr{R} = \frac{1}{\hbar}[\mathscr{H},\mathscr{R}] = -\frac{1}{\hbar}\int_0^1 \frac{d}{dx}(\mathscr{R}^x\mathscr{H}\mathscr{R}^{1-x})\,dx$$

so that with $\mathscr{R} = \mathrm{e}^{-\mathscr{A}}$, after carrying out the differentiation with respect to x, the result

$$\mathsf{L}\mathscr{R} = -\int_0^1 \mathscr{R}^x(\mathsf{L}\mathscr{A})\mathscr{R}^{1-x}\,dx$$

$$= -\sum_\mu\int_0^1 \mathscr{R}^x(\mathsf{L}\mathscr{G}_\mu)\mathscr{R}^{1-x}\,dx\,\lambda_\mu$$

is obtained.

***18.5**(p. 306). Setting $\mathsf{L}(t)=\mathsf{L}^0(t)+\mathsf{L}^1$ in the Robertson equation (18.2.22), and using the result of Exercise 18.4, we can write the terms $\mathsf{L}^0(t)\mathscr{R}(t)$ in the form

$$\mathsf{L}^0(t)\mathscr{R}(t) = -\sum_{\mu}\int_0^1 \mathscr{R}^x(t)(\mathsf{L}^0(t)\mathscr{G}_\mu)\mathscr{R}^{1-x}(t)\,dx\,\lambda_\mu(t) ,$$

or, with the given assumption that L^0 applied to $\mathscr{G}_\mu$ does not lead out of the space $\mathfrak{L}_{\mathscr{G}}$,

$$\mathsf{L}^0(t)\mathscr{R}(t) = -\sum_{\mu\nu}\int_0^1 \mathscr{R}^x(t)\mathscr{G}_\nu\mathscr{R}^{1-x}(t)\,dx\,\Omega^0_{\nu\mu}(t)\lambda_\mu(t) .$$

Employing (18.1.3) and the identity (3.4.10), we can also write this in the form:

$$\mathsf{L}^0(t)\mathscr{R}(t) = \sum_{\mu\nu}\left(\frac{\partial\mathscr{R}(t)}{\partial\lambda_\nu} - \langle\mathscr{G}_\nu\rangle(t)\mathscr{R}(t)\right)\Omega^0_{\nu\mu}(t)\lambda_\mu(t) .$$

The application of the Kawasaki–Gunton operator (18.2.19) to the expression in parentheses converts the latter into itself. Here, we have used the fact that $\mathrm{Tr}\,\mathscr{R} = \mathit{1}$ and $\mathrm{Tr}(\mathscr{G}_\nu\mathscr{R}) = \langle\mathscr{G}_\nu\rangle$, and have applied the chain rule for differentiation. It therefore follows that

$$\mathsf{P}[\mathscr{R}(t)]\,\mathsf{L}^0(t)\mathscr{R}(t) = \mathsf{L}^0(t)\mathscr{R}(t) ,$$

and from this, that

$$\mathsf{Q}[\mathscr{R}(t)]\,\mathsf{L}^0(t)\mathscr{R}(t) = \mathcal{O} .$$

We can thus replace the term with $\mathsf{L}(t')\mathscr{R}(t')$ in (18.2.22) by $\mathsf{L}^1\mathscr{R}(t')$. However, in the integral, the term

$$\mathsf{P}[\mathscr{R}(t)]\,\mathsf{L}^0(t)\mathsf{T}(t,t')\mathsf{Q}[\mathscr{R}(t')]\mathscr{X}$$
$$= -\sum_{\mu\nu}\mathrm{Tr}(\mathscr{G}_\nu\mathsf{T}(t,t')\mathsf{Q}[\mathscr{R}(t')]\mathscr{X})\Omega^0_{\nu\mu}(t)\frac{\partial\mathscr{R}(t)}{\partial\langle\mathscr{G}_\mu\rangle(t)} ,$$

also vanishes, since from formal integration of (17.3.25), it follows that

$$\mathsf{T}(t,t')\mathsf{Q}[\mathscr{R}(t')] = \mathsf{Q}[\mathscr{R}(t')] - \mathrm{i}\int_{t'}^{t}\mathsf{Q}[\mathscr{R}(t'')]\,\mathsf{L}(t'')\mathsf{T}(t'',t')\,dt''\,\mathsf{Q}[\mathscr{R}(t')]$$

and, using (17.3.12),

$$\mathrm{Tr}(\mathscr{G}_\nu\mathsf{Q}[\mathscr{R}(t)]\mathscr{Y})$$

is zero.

With the given assumption $\mathsf{L}^0(t)\mathscr{G}_\mu = \sum_\nu\mathscr{G}_\nu\Omega^0_{\nu\mu}(t)$, we can therefore write the Robertson equation in the form:

$$\frac{d}{dt}\mathcal{R}(t) = -\mathrm{i}\mathsf{L}^0(t)\mathcal{R}(t) - \mathrm{i}\mathsf{P}[\mathcal{R}(t)]\mathsf{L}^1\mathcal{R}(t)$$

$$-\int_{t_0}^{t} \mathsf{P}[\mathcal{R}(t)]\mathsf{L}^1\mathsf{T}(t,t')\mathsf{Q}[\mathcal{R}(t')]\mathsf{L}^1\mathcal{R}(t')dt' \ .$$

We see that, similarly to the Mori theory, L^1 enters the integral only in second order. The second term, $\mathrm{i}\mathsf{P}[\mathcal{R}_\beta]\,\mathsf{L}^1\mathcal{R}(t)$, can frequently be interpreted as a renormalization of the L^0 dynamics by L^1. If L^1 is a small perturbation on $\mathsf{L}^0(t)$, where $\mathsf{L}^0(t)$ itself may contain a strong external field, then in the manner of an iteration process we can replace $\mathcal{R}(t')$ within the integral by the unperturbed dynamics $\mathsf{U}^0(t',t)\mathcal{R}(t)$. We thus have a starting point for the approximate solution of nonlinear equations for expectation values $\langle\mathcal{G}_\nu\rangle(t)$ in specific physical problems.

***18.6**(p. 306). On neglecting the integral term in the Robertson equation and taking $\mathcal{G}_\nu \to a_1^\dagger a_2$, we find the equations of motion

$$\frac{d}{dt}\langle a_1^\dagger a_2\rangle(t) = -\mathrm{i}\,\mathrm{Tr}\{a_1^\dagger a_2 \mathsf{L}\mathcal{R}(t)\} = \frac{\mathrm{i}}{\hbar}\mathrm{Tr}\{[\mathcal{H}, a_1^\dagger a_2]\mathcal{R}(t)\} \ .$$

Here, $\mathcal{R}(t)$ has the form

$$\mathcal{R}(t) = \frac{1}{Z}\exp\left[-\sum_{34}\lambda_{34}(t)a_4^\dagger a_3\right] ,$$

and the Lagrange multipliers are determined as functions of the expectation values $\langle a_1^\dagger a_2\rangle(t)$ by

$$\langle a_1^\dagger a_2\rangle(t) = \mathrm{Tr}\{a_1^\dagger a_2\mathcal{R}(t)\} \ .$$

The right-hand side of the above equation of motion can be explicitly evaluated as a function of the expectation values. With

$$\mathcal{H} = \sum_3 \varepsilon_3 a_3^\dagger a_3 + \tfrac{1}{2}\sum_{33'34'} V_{34,3'4'}a_3^\dagger a_4^\dagger a_{4'}a_{3'}$$

and the commutation relations for fermion operators, one obtains for the commutator $[\mathcal{H}, a_1^\dagger a_2]$

$$[\mathcal{H}, a_1^\dagger a_2] = [\mathcal{H}, a_1^\dagger]a_2 + a_1^\dagger[\mathcal{H}, a_2]$$

$$= (\varepsilon_1 - \varepsilon_2)a_1^\dagger a_2 + \sum_{344'} V_{34,14'}a_3^\dagger a_4^\dagger a_{4'}a_2$$

$$- \sum_{3'44'} V_{24,3'4'}a_1^\dagger a_4^\dagger a_{4'}a_{3'} \ .$$

By diagonalizing the Hermitian matrix $\lambda_{34}(t)$, we can prove (analogously to equilibrium statistics) that

$$\mathrm{Tr}\{a_1^\dagger a_2 \mathscr{R}(t)\} = (1 + \mathrm{e}^{\lambda(t)})_{12}^{-1} \qquad \text{and}$$

$$\mathrm{Tr}\{a_1^\dagger a_2^\dagger a_{2'} a_{1'} \mathscr{R}(t)\} = \langle a_1^\dagger a_{1'}\rangle(t)\langle a_2^\dagger a_{2'}\rangle(t) - \langle a_1^\dagger a_{2'}\rangle(t)\langle a_2^\dagger a_{1'}\rangle(t) \;.$$

Using this factorization property and the symmetry $V_{34,3'4'} = V_{43,4'3'}$, and renaming, we thus find the closed nonlinear system of equations,

$$\begin{aligned}\frac{d}{dt}\langle a_1^\dagger a_2\rangle(t) = \frac{\mathrm{i}}{\hbar}\mathrm{Tr}\{[\mathscr{H}, a_1^\dagger a_2]\mathscr{R}(t)\} &= \frac{\mathrm{i}}{\hbar}(\varepsilon_1 - \varepsilon_2)\langle a_1^\dagger a_2\rangle(t)\\ &+ \frac{\mathrm{i}}{\hbar}\sum_{344'}(V_{34,14'} - V_{43,14'})\langle a_4^\dagger a_{4'}\rangle(t)\langle a_3^\dagger a_2\rangle(t)\\ &- \frac{\mathrm{i}}{\hbar}\sum_{3'44'}(V_{24,3'4'} - V_{42,3'4'})\langle a_4^\dagger a_{4'}\rangle(t)\langle a_1^\dagger a_{3'}\rangle(t) \;.\end{aligned}$$

These equations may be interpreted as equations of motion under an effective single-particle operator. Thus, for

$$\tilde{\mathscr{H}}(t) = \sum_3 \varepsilon_3 a_3^\dagger a_3 + \sum_{33'} V_{33'}(t) a_3^\dagger a_{3'} \;,$$

the equations

$$\begin{aligned}\frac{d}{dt}\langle a_1^\dagger a_2\rangle(t) = \frac{\mathrm{i}}{\hbar}\langle[\tilde{\mathscr{H}}, a_1^\dagger a_2]\rangle &= \frac{\mathrm{i}}{\hbar}(\varepsilon_1 - \varepsilon_2)\langle a_1^\dagger a_2\rangle(t)\\ &+ \frac{\mathrm{i}}{\hbar}\left\{\sum_3 V_{31}\langle a_3^\dagger a_2\rangle(t) - \sum_{3'} V_{23'}\langle a_1^\dagger a_{3'}\rangle(t)\right\}\end{aligned}$$

are exactly valid. By comparison with the nonlinear system obtained, one sees that this can be generated by the effective Hamiltonian

$$\mathscr{H}_{\mathrm{eff}}(t) = \sum_3 \varepsilon_3 a_3^\dagger a_3 + \sum_{33'}\left\{\sum_{44'}(V_{34,3'4'} - V_{43,3'4'})\langle a_4^\dagger a_4\rangle(t)\right\} a_3^\dagger a_{3'}$$

when the single-particle potential is self-consistently determined (time-dependent Hartree–Fock equations).

Bibliography

Supplementary Textbooks*

1) Quantum Theory

Dirac, P. A. M.: *Principles of Quantum Mechanics* (Oxford U. Press, Oxford 1982)

Fick, E.: *Einführung in die Grundlagen der Quantentheorie* (AULA, Wiesbaden 1988; Akad. Verlagsgesellschaft, Leipzig 1981)

Ludwig, G.: *Foundations of Quantum Mechanics, Vols. I and II*, Texts and Monographs in Physics (Springer, Berlin, Heidelberg, New York 1983 and 1985)

Messiah, A.: *Quantum Mechanics I and II* (Wiley, New York 1958)

Roman, P.: *Advanced Quantum Theory* (Addison-Wesley, Reading, MA 1965)

Shankar, R.: *Principles of Quantum Mechanics* (Plenum, New York 1980)

2) General Statistical Physics

Balescu, R.: *Equilibrium and Nonequilibrium Statistical Mechanics* (Wiley, New York, London 1975)

Becker, R.: *Theorie der Wärme*, Heidelberger Taschenbücher, Band 10, 3. Aufl. (Springer, Berlin, Heidelberg, New York 1985)

Jancel, R.: *Foundations of Classical and Quantum Statistical Mechanics* (Pergamon, Oxford, London 1969)

Landau, L. D., L. P. Pitaevskii: *Statistical Physics* (Pergamon, Oxford 1980)

Reichl, L. E.: *A Modern Course in Statistical Physics* (U. Texas Press, Austin 1980)

3) Information Theory and Statistical Physics

Denbigh, K. G., J. S. Denbigh: *Entropy in Relation to Incomplete Knowledge* (Cambridge University Press, Cambridge 1985)

Hobson, A.: *Concepts in Statistical Mechanics* (Gordon and Breach, New York 1971)

---

* This list makes no claim to completeness.

Jaglom, A. M., I. M. Jaglom: *Wahrscheinlichkeit und Information* (VEB Deutscher Verlag der Wissenschaften, Berlin 1965)

Katz, A.: *Principles of Statistical Mechanics* (W. H. Freeman, San Francisco 1967)

Levine, R. D., M. Tribus (eds.): *The Maximum Entropy Formalism* (MIT-Press, Cambridge, MA 1978)

4) Equilibrium Statistics

Brenig, W.: *Statistical Theory of Heat* (Springer, Berlin, Heidelberg, New York 1989)

Holman, J. P.: *Thermodynamics*, (McGraw-Hill, New York 1980)

Morrill, B.: *Introduction to Equilibrium Thermodynamics* (Pergamon, New York, London 1972)

Stumpf, H., A. Rieckers: *Thermodynamik I* (Vieweg, Braunschweig 1976)

Toda, M., R. Kubo, N. Saitô.: *Statistical Physics I. Equilibrium Statistical Mechanics*, Springer Series in Solid-State Sciences, Vol. 30 (Springer, Berlin, Heidelberg, New York, Tokyo 1983)

Weidlich, W.: *Thermodynamik und Statistische Mechanik* (Akad. Verlagsgesellschaft, Wiesbaden 1976)

5) Linear Response and Relaxation Theory

Forster, D.: *Hydrodynamic Fluctuations, Broken Symmetry, and Correlation Functions* (Benjamin, Reading, MA 1975)

Lovesey, S. W.: *Condensed Matter Physics. Dynamic Correlations* (Benjamin, Reading, MA 1980)

Martin, P. C.: *Measurements and Correlation Functions* (Gordon and Breach, New York, 1968)

Review Articles

Berne, B. J., D. Forster: Topics in Time-Dependent Statistical Mechanics. Ann. Rev. of Phys. Chem. **22**, 563–596 (1971)

Kubo, R.: Response, Relaxation, and Fluctuation. Lecture Notes in Physics **31**, 75–122 (1974)

Zwanzig, R.: Linear Response Theory, in *Molecular Fluids*, Les Houches Lectures, 1973 (London, 1976, pp. 1–36)

6) Thermodynamics of Irreversible Processes

de Groot, S. R., P. Mazur: *Nonequilibrium Thermodynamics* (North-Holland, Amsterdam 1969)

Haase, R.: *Thermodynamics of Irreversible Processes* (Addison-Wesley, Reading, MA 1969)

7) Nonequilibrium Statistics, General

Grabert, H.: *Projection Operator Techniques in Nonequilibrium Statistical Mechanics* (Springer, Berlin 1982)

Kreuzer, H. J.: *Nonequilibrium Thermodynamics and Its Statistical Foundations* (Oxford U. Press, Oxford 1981)

Kubo, R., M. Toda, N. Hashitsume: *Statistical Physics II. Nonequilibrium Statistical Mechanics*, Springer Series in Solid-State Sciences, Vol. 31 (Springer, Berlin, Heidelberg, New York, Tokyo 1985)

Rieckers, A., H. Stumpf: *Thermodynamik II* (Vieweg, Braunschweig 1977)

Risken, H.: *The Fokker–Planck Equation. Methods of Solution and Applications*, Springer Series in Synergetics, Vol. 18 (Springer, Berlin, Heidelberg, New York 1989)

Röpke, G.: *Statistische Mechanik für das Nichtgleichgewicht* (Physik-Verlag, Weinheim 1987)

Zubarev, D. N.: *Nonequilibrium Statistical Thermodynamics* (Plenum, New York, London 1974)

Subject Index

Springer Series in Solid-State Sciences

Editors: M. Cardona P. Fulde K. von Klitzing H.-J. Queisser

1 **Principles of Magnetic Resonance** 3rd Edition By C. P. Slichter
2 **Introduction to Solid-State Theory** 2nd Printing By O. Madelung
3 **Dynamical Scattering of X-Rays in Crystals** By Z. G. Pinsker
4 **Inelastic Electron Tunneling Spectroscopy** Editor: T. Wolfram
5 **Fundamentals of Crystal Growth I** Macroscopic Equilibrium and Transport Concepts. 2nd Printing By F. Rosenberger
6 **Magnetic Flux Structures in Superconductors** By R. P. Huebener
7 **Green's Functions in Quantum Physics** 2nd Edition By E. N. Economou
8 **Solitons and Condensed Matter Physics** 2nd Printing Editors: A. R. Bishop and T. Schneider
9 **Photoferroelectrics** By V. M. Fridkin
10 **Phonon Dispersion Relations in Insulators** By H. Bilz and W. Kress
11 **Electron Transport in Compound Semiconductors** By B. R. Nag
12 **The Physics of Elementary Excitations** By S. Nakajima, Y. Toyozawa, and R. Abe
13 **The Physics of Selenium and Tellurium** Editors: E. Gerlach and P. Grosse
14 **Magnetic Bubble Technology** 2nd Edition By A. H. Eschenfelder
15 **Modern Crystallography I** Symmetry of Crystals. Methods of Structural Crystallography By B. K. Vainshtein
16 **Organic Molecular Crystals** Their Electronic States. By E. A. Silinsh
17 **The Theory of Magnetism I** Statics and Dynamics. 2nd Printing By D. C. Mattis
18 **Relaxation of Elementary Excitations** Editors: R. Kubo and E. Hanamura
19 **Solitons.** Mathematical Methods for Physicists. 2nd Printing By G. Eilenberger
20 **Theory of Nonlinear Lattices** 2nd Edition By M. Toda
21 **Modern Crystallography II** Structure of Crystals By B. K. Vainshtein, V. M. Fridkin, and V. L. Indenbom
22 **Point Defects in Semiconductors I** Theoretical Aspects By M. Lannoo and J. Bourgoin
23 **Physics in One Dimension** Editors: J. Bernasconi, T. Schneider
24 **Physics in High Magnetic Fields** Editors: S. Chikazumi and N. Miura
25 **Fundamental Physics of Amorphous Semiconductors** Editor: F. Yonezawa
26 **Elastic Media with Microstructure I** One-Dimensional Models. By I. A. Kunin
27 **Superconductivity of Transition Metals** Their Alloys and Compounds By S. V. Vonsovsky, Yu. A. Izyumov, and E. Z. Kurmaev
28 **The Structure and Properties of Matter** Editor: T. Matsubara
29 **Electron Correlation and Magnetism in Narrow-Band Systems** Editor: T. Moriya
30 **Statistical Physics I** Equilibrium Statistical Mechanics By M. Toda, R. Kubo, N. Saito
31 **Statistical Physics II** Nonequilibrium Statistical Mechanics By R. Kubo, M. Toda, N. Hashitsume
32 **Quantum Theory of Magnetism** 2nd Edition By R. M. White
33 **Mixed Crystals** By A. I. Kitaigorodsky
34 **Phonons: Theory and Experiments I** Lattice Dynamics and Models of Interatomic Forces. By P. Brüesch
35 **Point Defects in Semiconductors II** Experimental Aspects By J. Bourgoin and M. Lannoo
36 **Modern Crystallography III** Crystal Growth 2nd Edition By A. A. Chernov
37 **Modern Crystallography IV** Physical Properties of Crystals Editor: L. A. Shuvalov
38 **Physics of Intercalation Compounds** Editors: L. Pietronero and E. Tosatti
39 **Anderson Localization** Editors: Y. Nagaoka and H. Fukuyama
40 **Semiconductor Physics** An Introduction 4th Edition By K. Seeger
41 **The LMTO Method** Muffin-Tin Orbitals and Electronic Structure By H. L. Skriver
42 **Crystal Optics with Spatial Dispersion, and Excitons** 2nd Edition By V. M. Agranovich and V. L. Ginzburg
43 **Structure Analysis of Point Defects in Solids by Multiple Magnetic Resonance Spectroscopy** By J.-M. Spaeth, J. R. Niklas, and R. H. Bartram
44 **Elastic Media with Microstructure II** Three-Dimensional Models By I. A. Kunin
45 **Electronic Properties of Doped Semiconductors** By B. I. Shklovskii and A. L. Efros
46 **Topological Disorder in Condensed Matter** Editors: F. Yonezawa and T. Ninomiya
47 **Statics and Dynamics of Nonlinear Systems** Editors: G. Benedek, H. Bilz, and R. Zeyher
48 **Magnetic Phase Transitions** Editors: M. Ausloos and R. J. Elliott
49 **Organic Molecular Aggregates** Electronic Excitation and Interaction Processes Editors: P. Reineker, H. Haken, and H. C. Wolf

GPSR Compliance

The European Union's (EU) General Product Safety Regulation (GPSR) is a set of rules that requires consumer products to be safe and our obligations to ensure this.

If you have any concerns about our products, you can contact us on ProductSafety@springernature.com

In case Publisher is established outside the EU, the EU authorized representative is:

Springer Nature Customer Service Center GmbH
Europaplatz 3
69115 Heidelberg, Germany

Batch number: 10682323

Printed by Printforce, the Netherlands